Numerical Methods with MATLAB®

A Resource for Scientists and Engineers

G. J. BORSE

Lehigh University

 PWS Publishing Company

I(T)P **AN INTERNATIONAL THOMSON PUBLISHING COMPANY**

Boston • Albany • Bonn • Cincinnati • Detroit • London • Madrid • Melbourne • Mexico City
New York • Paris • San Francisco • Singapore • Tokyo • Toronto • Washington

PWS PUBLISHING COMPANY
20 Park Plaza, Boston, MA 02116-4324

Copyright © 1997 by PWS Publishing Company, a division of International Thomson Publishing Inc.

MATLAB and PC MATLAB are trademarks of The Mathworks, Inc. The MathWorks, Inc. is the developer of MATLAB, the high-performance computational software introduced in this book. For further information on MATLAB and other MathWorks products — including SIMULINK^{TM} and MATLAB Application Toolboxes for math and analysis, control system design, system identification, and other disciplines — contact The MathWorks at 24 Prim Park Way, Natick, MA 01760 (phone: 508-647-7000; fax: 508-647-7001; email: info@mathworks.com). You can also sign up to receive the MathWorks quarterly newsletter and register for the user group. Macintosh is a trademark of Apple Computer, Inc. MS-DOS is a trademark of Microsoft Corporation.

I(T)P ^{TM}

International Thomson Publishing
The trademark ITP is used under license.

For more information, contact:

PWS Publishing Company
20 Park Plaza
Boston, MA 02116

International Thomson Publishing Europe
Berkshire House I68–I73
High Holborn
London WC1V 7AA
England

Thomas Nelson Australia
102 Dodds Street
South Melbourne, 3205
Victoria, Australia

Nelson Canada
1120 Birchmount Road
Scarborough, Ontario
Canada M1K 5G4

International Thomson Editores
Campos Eliseos 385, Piso 7
Col. Polanco
11560 Mexico C.F., Mexico

International Thomson Publishing GmbH
Königswinterer Strasse 418
53227 Bonn, Germany

International Thomson Publishing Asia
221 Henderson Road
#05–10 Henderson Building
Singapore 0315

International Thomson Publishing Japan
Hirakawacho Kyyowa Building, 31
2-2-1 Hirakawacho
Chiyoda-ku, Tokyo 102
Japan

Sponsoring Editor: *Bill Barter*
Marketing Manager: *Nathan Wilbur*
Manufacturing Buyer: *Andrew Christensen*
Designer/Production Editor: *Pamela Rockwell*
Text Printer: *Courier/Westford*
Cover Printer: *Phoenix Color*

Printed and bound in the United States of America.

97 98 99 00 — 10 9 8 7 6 5 4 3 2

Library of Congress Cataloging-in-Publication Data

Borse, G. J. (Garold J.)
   Numerical methods with MATLAB : a resource for scientists and engineers / G.J. Borse.
     p. cm.
   Includes bibliographical references (p. –) and index.
   ISBN 0-534-93822-1
   1. Numerical analysis—Data processing. 2. MATLAB.
I. Title.
   QA297.B647 1997
   519.4'0285'53042—dc21   96-39300
                                     CIP

Contents

Part I   Linear Analysis and Matrices

---

*Sections and chapters marked with (§) may be omitted without loss of continuity.

Part III   Functions of Two or More Variables

Part IV   (§) Data Analysis and Modeling

Appendix   Programming in MATLAB

Preface

This book details the main topics of numerical methods, such as linear analysis, analysis of functions of one or several variables, data analysis and modeling, and the solution of ordinary differential equations. However, the arrangement of topics and the focus of the treatment have been specifically designed to incorporate the implementation of these techniques by means of the numerical software tool called MATLAB. Readers familiar with the numerical solution of problems using the standard programming languages such as FORTRAN, Pascal, and C++ will be pleased to find that the majority of such problems can be solved in a fraction of the programming time using MATLAB.

MATLAB seeks to address three particular concerns familiar to programmers experienced in more traditional higher-level languages:

1. Most computer programs are heavily dependent upon a variety of "looping" procedures to execute summations or to implement iterative techniques. Often, a single program will contain essentially identical looping structures tens of times, greatly complicating the writing of the code.
2. The majority of problems repeatedly call upon a relatively small set of "subfunctions," usually the same set for a variety of problems. Thus, the programmer must have routines available, for example, to evaluate Bessel functions, to integrate or find the roots of a function, or to sort a list.
3. Often the best guide to a problem's solution is provided by the graphical display of data or the behavior of a function over a limited range. Each of the higher-level languages mentioned is woefully lacking in user-friendly, easy-to-use graphics capabilities.

The last item, graphical display capability, is MATLAB's most spectacular feature. Elementary two- or three-dimensional graphs are readily constructed by even the newest users. And the potential for ever more elaborate displays exists as the user becomes more experienced.

The second strength of MATLAB is its encyclopedic collection of subprograms, called M-files, for the solution of nearly any numerical problem imaginable. The number of such subprograms is very large and is continually growing with each edition of MATLAB. Each subprogram has a short help file detailing its use that may be accessed by entering **help** followed by the name of the subprogram. Making use of these subprograms to handle the often time-consuming, but perhaps straightforward, aspects of a problem dramatically reduces the effort associated with most numerical tasks, leaving the user free for the more challenging and creative aspects of problem solving.

However, it is MATLAB's approach to the first listed item that truly distinguishes it from similar software packages. FORTRAN and subsequent higher-level computer languages associate a computer address (and therefore a stored number) with a letter or a name, and elementary computer arithmetic operations with symbols like *, /, +, -. Thus, numbers and operations can be combined into a computational algebra. MATLAB takes this association a step further. Whereas in FORTRAN each name represents a single numerical value, in MATLAB each name represents a complete *matrix* of numbers, and the elementary arithmetic operations are likewise generalized to apply to the arithmetic of matrices. If $\vec{x}, \vec{y}$ are each column vector matrices of the same length, then a frequently encountered operation like

$$s = \sum_{i=1}^{n} x_i y_i$$

would be programmed in FORTRAN or similar languages as a loop that successively adds each addition term to the accumulating sum. But, in MATLAB, the same value is obtained by simply multiplying the two column vector matrices in the form of a matrix product, $s = \vec{x}^T \cdot \vec{y}$, where $\vec{x}^T$ is the transpose of the column matrix $\vec{x}$. This "vectorization" of the elementary operations of arithmetic has the dual advantage of producing faster execution of operations as well as a program structure that is substantially more transparent and easier to follow and to write.

Of course, because matrix operations are fundamental to MATLAB, any book on numerical methods using MATLAB must begin with a detailed discussion of the elements of matrix algebra; i.e., linear analysis. To take full advantage of the power of MATLAB, it is essential that the reader be comfortable with the arithmetic of matrices, particularly the replacement of a variety of sequential operations like summations by single line products of matrices. For this reason, Part I of this text spends considerable time illustrating numerous mathematical topics by means of matrices.

Because different readers will have a varying degree of familiarity with MATLAB, this text has been designed to be useful to users at both extremes. Those experienced in MATLAB should be able to follow the development of each topic without interruption. Novice MATLAB users are urged to read the parallel development of MATLAB features in the Appendix. This lengthy Appendix follows the sequence of topics in the main body of the text and contains numerous MATLAB programming problems to test your proficiency. An indication of the relevant Appendix sections is indicated at the beginning of each chapter in which new MATLAB features are introduced.

Part II contains a collection of the most common numerical procedures applied to functions of a single variable; namely, series expansions, roots of functions, extrema, and numerical integration. An important MATLAB feature introduced at this point is automatic two-dimensional graphing of functions and data by means of the function `plot`.

The procedures of Part II, when generalized to functions of more than one variable, are substantially more complex and more subtle than their one-dimensional cousins and are, therefore, treated separately in Part III. Once again, graphical displays of the functions of interest are found to be particularly useful in understanding the numerical techniques employed. In this case, the graphs are three dimensional and make use of the variety of MATLAB capabilities in three-dimensional perspective plots as well as contour graphs. Part III also includes a description of Monte Carlo techniques for evaluating multidimensional integrals.

Part IV is devoted to the important topic of data modeling and analysis. The statistical analysis of data is described and facilitated using several MATLAB toolbox functions. A variety of techniques, both linear and nonlinear, are outlined for the modeling or curve fitting of data, including a short introduction to the vast library of MATLAB routines for spline fits and Fourier analysis.

Part V is devoted to the numerical solution of ordinary differential equations. Numerous stepping procedures are described for solving initial value equations culminating in the illustration of the excellent MATLAB programs for this purpose. Two additional features characterize this part. First, because initial value differential equations are so frequently encountered in numerical work, and because realistic equations are often fraught with program-killing pathologies, an alternative algorithm is developed for their solution which may prove useful in rare instances when MATLAB toolbox codes fail. Second, the extremely important topic of numerical solutions to boundary value differential equations is absent from the MATLAB library. As a result, in Part V considerable time is spent developing a variety of techniques for solving these boundary value equations. This section concludes with an introduction to the numerical solution of partial differential equations. Elementary methods for solving each of the three types of PDEs (hyperbolic, parabolic, and elliptic) are described and illustrated. Of particular note is the introduction of MATLAB sparse matrix techniques for solving the resulting matrix equations involving very large matrices.

GJB

ACKNOWLEDGMENTS

The following reviewers contributed to the development of this text:

Professor Robert N. Eli
West Virginia University

Professor Stephen G. Nash
George Mason Univeristy

Professor Robert L. Rankin
Arizona State University

Professor John R. White
University of Massachusetts, Lowell

Notes to the Instructor

The objective of this text is to introduce engineering and science students from a variety of disciplines and backgrounds to the vast array of problems that are amenable to numerical solution. It is expected that the students have taken the normal calculus sequence, including an introduction to ordinary differential equations. It is not assumed that the students have a background in linear algebra, beyond what is normally encountered in an algebra class. Some topics, such as the Taylor series, integration, partial derivatives, and elementary techniques for solving ordinary differential equations, are included as concise reviews. The emphasis is on application rather than theory, which, although kept to a minimum and presented in a mostly heuristic and intuitive manner, is sufficient for the student to understand the workings, efficiency, and failings of each technique. The computer language MATLAB is used throughout the text. A successful incorporation of MATLAB into a numerical methods text requires (1) that the order of topics be adjusted to coincide with the growing sophistication the student is simultaneously acquiring in MATLAB, and (2) that there be ample demonstration of MATLAB's potential for obtaining relatively easy solutions to complex problems from traditionally more advanced areas.

The most important requirement for effective use of MATLAB is ease and proficiency in elementary matrix manipulation. Thus, in Part I, the text begins with a review of linear algebra and matrices. The emphasis is on the manipulation of matrices: multiplication, addition, transpose, etc. This topic may be familiar to some students and can be condensed down to a review of matrix algebra. There is an extensive gallery of problems that should be tried to test the students' proficiency before continuing.

A unique feature of the text is that it is suitable both for students with some previous experience in MATLAB and for students who are totally new to the subject. Students new to MATLAB should be following the parallel development of MATLAB programming in the Appendix. This extensive Appendix has been designed to introduce MATLAB topics in the same sequence in which they are required in the numerical methods sections. Although MATLAB is designed to be self-taught, it is usually helpful to provide a laboratory-style tutorial session for the first few classes while the students are becoming comfortable with MATLAB.

The last two chapters in Part I outline, first, the solution of the equation $\mathbf{A}\vec{x} = \vec{b}$, and, second, the numerical solution of matrix eigenvalue equations. The basic LU algorithm is used for the former and is quite close to the actual procedures used by MATLAB. However, the MATLAB treatment of eigenvalue equations is via singular value decomposition techniques, a topic that is likely beyond the scope of most beginning numerical methods classes. For that reason, a simpler procedure, based upon Jacobi rotations, is presented to provide the student with one example of a technique that is easy to understand. This is followed by a cursory explanation of the ideas of singular value decomposition and an illustration of the solution methods of MATLAB.

The material of Part I is essential to most of the remaining text and, with the exception of sections marked by (§), should be mastered before continuing on to Part II. It is recommended that only one or two of the sections of Chapter 1 illustrating matrix multiplication be covered in a one-semester course.

The material of Part II contains the normal sequence of topics related to the numerical analysis of a function of a single variable: function evaluation, roots, extrema, and integration. In Part II, students are expected to be able to code M-file functions and to apply the various ideas and codes contained in the text to these functions. Also, the two-dimensional plotting capabilities of MATLAB are highlighted in this part.

Parts I and II are preliminary to the remainder of the text. In a typical one-semester course at the junior/senior level, roughly half of a semester is required to adequately cover the material of these two parts. Clearly, the remaining material in the text is far too extensive to fit comfortably in the remaining half-semester. The course instructor will then have to select from the remaining topics in order to suit the needs of his/her students. A possible selection of topics for the remainder of the course is suggested as follows. (The sections marked by (§) may be skipped without loss of continuity.)

In Part III, the techniques for finding roots and extrema and for numerical integration are extended to functions of more than one variable. This topic, normally not a part of an introductory numerical methods course, is included in this text for a variety of reasons. Foremost among them is the pedagogical value associated with the three-dimensional plotting capabilities of MATLAB. The ease with which complicated multivariable functions can be evaluated and displayed, makes it possible to clearly present problems that have been previously difficult to comprehend. In order to feature these capabilities while still devoting sufficient time to additional topics, it is possible to skip the material of Chapters 10 (extrema) and 12 (Monte Carlo integration) without a loss of continuity.

Part IV, Data Analysis and Modeling, is clearly a topic of varying importance to different engineering and scientific disciplines. The entire five chapters could be deferred to a later course, or a more likely selection would be to cover linear least squares fitting (Chapter 14) and perhaps one of the remaining four chapters. It should be noted that very interesting problems that relate to a particular discipline can be constructed by the instructor to illustrate the ideas of modeling and data analysis described in this part.

Part V, Differential Equations, is a key element of any course in numerical methods, and certainly the chapter on initial value equations should be included. The principles behind the MATLAB codes for solving initial value problems; i.e., the basic Runge-Kutta algorithm plus adaptive step sizing, are clearly outlined and illustrated. Extending the ideas to include the methods of Bulirsch-Stoer makes use of material contained in earlier sections on interpolation and can be omitted in an introductory survey course. Chapter 20, Boundary Value Problems, is an important topic not usually covered in a first course in numerical methods. However, the treatment is designed to be suitable as an introduction to the subject, and should prove useful when the student later encounters all the theoretical subtleties associated with these types of equations. Chapter 21 outlines a few special techniques that can be applied to linear equations only, techniques often superior to the more general methods of Chapter 20, that the reader should consider as alternatives for linear equations. Finally, Chapter 22 utilizes the basic matrix features of MATLAB, particularly the ease of implementing sparse matrix techniques, to illustrate the numerical solution of the three basic types of partial differential equations. The treatment in Chapter 22 is specifically intended as an introduction to the subject for students at the junior/senior level and does not presuppose any prior exposure to PDEs.

Many of the sections that would be omitted in a first survey course will later prove useful to the student as his/her sophistication grows with increased experience, and the text should continue to serve as a valuable reference. Even the most advanced MATLAB routines are designed to be useable by novices in a "black-box" manner. The suggested sequence of topics appropriate to a one-semester, junior/senior level course is indicated in the diagrams on the next two pages.

In any course whose goal is the attainment of skills, in this case, computational skills, substantial practice is essential. There are ample problems at the end of each chapter which should be attempted to reinforce the reading matter. It is intended that most of these problems be executed in MATLAB code. To encourage this, many of the answers to the problems are given in terms of MATLAB script files and graphics that are found on the accompanying program disk. A self-running, interactive tutorial is also available on the disk to aid in the introduction of elementary MATLAB skills.

Some universities offer a full-year course in numerical methods, and clearly there is sufficient material contained in this text to adequately serve such a course. For such a course, Parts I, II, and III would be appropriate for the first semester. The second semester would continue with Parts IV and V, followed by either a third of a semester devoted to one of the specialized toolboxes available in MATLAB or by an introduction to partial differential equations.

FUNDAMENTALS: PARTS I AND II
(approximately one-half semester)

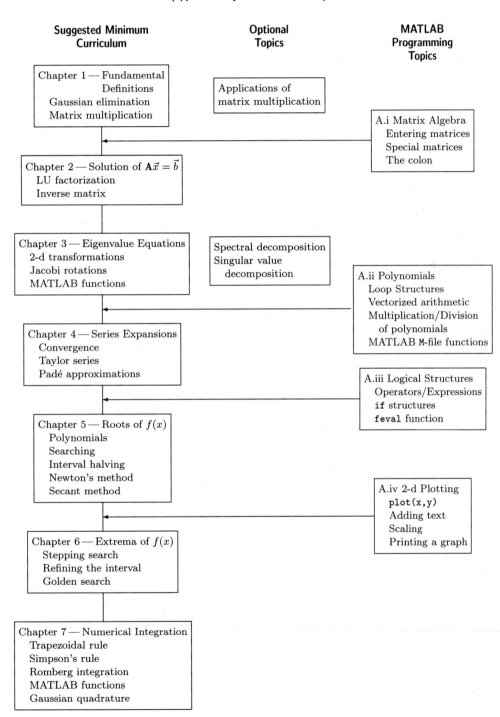

Suggested Minimum Curriculum

Chapter 1 — Fundamental Definitions
Gaussian elimination
Matrix multiplication

Chapter 2 — Solution of $\mathbf{A}\vec{x} = \vec{b}$
LU factorization
Inverse matrix

Chapter 3 — Eigenvalue Equations
2-d transformations
Jacobi rotations
MATLAB functions

Chapter 4 — Series Expansions
Convergence
Taylor series
Padé approximations

Chapter 5 — Roots of $f(x)$
Polynomials
Searching
Interval halving
Newton's method
Secant method

Chapter 6 — Extrema of $f(x)$
Stepping search
Refining the interval
Golden search

Chapter 7 — Numerical Integration
Trapezoidal rule
Simpson's rule
Romberg integration
MATLAB functions
Gaussian quadrature

Optional Topics

Applications of matrix multiplication

Spectral decomposition
Singular value decomposition

MATLAB Programming Topics

A.i Matrix Algebra
Entering matrices
Special matrices
The colon

A.ii Polynomials
Loop Structures
Vectorized arithmetic
Multiplication/Division of polynomials
MATLAB M-file functions

A.iii Logical Structures
Operators/Expressions
if structures
feval function

A.iv 2-d Plotting
plot(x,y)
Adding text
Scaling
Printing a graph

SELECTED ADVANCED TOPICS: Parts III, IV, AND V
(approximately one-half semester)

Suggested Curriculum	Optional Topics	MATLAB Programming Topics

Chapter 8 — Derivatives of Functions
of Several Variables
Partial derivatives
The gradient
2-d Taylor series

A.v Constructing and Plotting
Multivariable Functions
Constructing a grid
3-d perspective graphs
Contour graphs
Non-Cartesian coordinates
3-d trajectories

Chapter 9 — Roots of $f(x, y)$
Zero contours
Simultaneous nonlinear
equations

Chapter 10 — Extrema of $f(x, y)$
Simplex methods
Directional methods
Quasi-Newton methods

Chapter 11 — Multiple Integrals
MATLAB code for
iterated integrals

Chapter 12 — Monte Carlo Integrals
Using random sampling
Complex boundaries

A.vi Input/Output in MATLAB
fopen statement
fscanf statement
fprintf statement

Chapter 14 — Linear Least Squares
Fitting straight lines
Polynomials
General least squares

Chapter 13 — Statistical Description
of Data
Moments and modes
Distribution functions

Chapter 15 — Nonlinear Least
Squares

Chapter 16 — Interpolation and
Spline Fits

Ch. 17 — Fourier Analysis
of Data

Chapter 18 — Differential Equations
First-order equations
Second-order equations
as coupled first order

Chapter 19 — Initial Value Equations
Elementary stepping
procedures
Improvements
Runge-Kutta methods

Chapter 20 — Boundary Value
Equations
Shooting
Relaxation
Finite element

Chapter 21 — Superposition Method

Chapter 22 — Introduction to PDEs
Stepping methods
Boundary value methods

PART I / Linear Analysis and Matrices

1 Fundamental Definitions
2 Solution of the Matrix Equation $A\vec{x} = \vec{b}$
3 Matrix Eigenvalue Equations

PREVIEW

The effective and efficient use of the powerful numerical analysis package called MATLAB relies on your ability to construct, manipulate, and solve matrix equations. MATLAB will assume, unless it is told otherwise, that every variable name it encounters refers to a matrix of values. Therefore, contrary to the normal order of topics in a text on numerical methods, we must begin with a review of the elementary aspects of matrix algebra, as well as touch upon a few more advanced ideas.

In Part I you will see how matrices of various shapes are defined, how the ordinary operations of arithmetic are generalized to matrix arithmetic, and how sets of equations written in matrix form can be solved. Part I describes and illustrates the standard techniques for solving matrix equations. However, because the principal aim is to put you at ease with the formulation of the problem and the procedures for its solution so that you can use the powerful programs available in the MATLAB toolbox, a number of matrix problems will be solved using MATLAB routines.

Part I also provides an elementary introduction to a different class of matrix equations, eigenvalue equations. This type of equation is one of the most important in all of science and engineering. The emphasis will be on a geometrical interpretation of the eigenvalue equation, a cursory understanding of its numerical solution, and a detailed examination of the MATLAB techniques available for solving it.

INTRODUCTION

A first step in analyzing a problem in science and engineering is finding an adequate description of the problem in terms of mathematical equations governing the pertinent variables that characterize the phenomena. These equations describe the effect that changes in one variable have on all the remaining variables. A clever engineer or scientist will draw on his or her background in numerous fields such as fluid mechanics, properties of materials, electromagnetism, thermodynamics, and others to postulate relationships in the form of equations involving the parameters. These

1

relationships can be empirical relations that summarize the results of experiments, or they can be drawn from first principles by relating the problem to idealized conditions. In any event, getting the description of a problem into the form of a set of equations is the most difficult part of any problem. The art is to formulate the simplest mathematical description that still retains all of the significant features of the phenomena in question. The actual solution of these equations is then, very likely, straightforward. There have been significant advances in mathematical analysis in the last century that make possible closed-form or exact solutions to many problems. And for those problems for which an analytic or exact solution is either not possible or is unwieldy, modern computers will often provide a numerical solution. Usually problems of this type are solved by using a higher-level computer language such as FORTRAN, C, Pascal, Ada, or others, and, frequently, the numerical results are displayed using additional graphics packages to plot the results. In this text we will use a later generation of computer software called MATLAB. MATLAB is a computational language like FORTRAN but it also includes user-friendly graphics capabilities and thousands of library functions that can be of use in the solution of most numerical problems. The ability to make use of these functions to quickly move through the rudimentary to the heart of a problem, and then to easily display the results in a variety of graphical forms, makes MATLAB a valuable tool for the numerical practitioner. The purpose of this text is to outline the common mathematical techniques used by scientists and engineers and to illustrate their implementation via MATLAB. But first let's review some basic terminology.

1 / *Fundamental Definitions**

*This chapter makes use of MATLAB programming techniques described in Appendix Section A.i.

PREVIEW Modern analysis or modern numerical analysis begins by categorizing equations as belonging to different types and then devises methods of solution for generic problems of each type. Thus, we must first review some of the labels used to characterize different forms of equations. Next, we summarize the common techniques for solving a set of *linear, coupled, algebraic* equations. We will see that these techniques can lead to a number of outcomes:

1. A single, *unique* solution for all variables
2. An *infinite* number of valid solutions
3. A determination that *no* solution whatsoever exists
4. An evaluation of the equations as *ill conditioned*, i.e., too sensitive to small variations in the coefficients to provide a reliable solution.

We will then attempt to understand the characteristic features of the equations that will give rise to each possibility.

At this point, to simplify the manipulation of several simultaneous equations, we introduce the notation of *matrices* and define and illustrate matrix multiplication. To facilitate later applications that rely heavily on matrix manipulation, this chapter ends with several examples of applications of matrix multiplication using MATLAB.

Classification of Equations

An equation is simply an expression of equality between constants, variables, functions of variables, or sums, products, derivatives, or integrals of variables. The various classifications of equations are:

Linear

An equation is linear in x if x only appears as x raised to the first power. That is,

$$y = ax + b$$

Algebraic

An algebraic equation in x contains x raised to any finite, nonnegative integer power n. That is, a polynomial in x is an algebraic expression in x. An equation is *homogeneous* if the coefficient of x^0 is zero. Thus $f(x) = 3x^3 + x^2 - 2x = 0$ is an *algebraic (degree 3) homogeneous* equation in x. Also, n distinct equations in m independent variables are called *coupled*. Thus, the following set of three equations in three unknowns is called *coupled linear homogeneous*.

$$3x - 2y - 8z = 0$$
$$-x + 5y - 2z = 0$$
$$7x + 7y + 7z = 0$$

Transcendental

A transcendental equation in x is one in which the dependence on x cannot be represented by a finite number of algebraic operations on x. For example, $\sin(x)$ is a transcendental function because its simplest representation in terms of algebraic operations on x is that of an infinite series. The following equations are a *coupled transcendental* set:

$$\sin x \ \cos y - xy = 0$$
$$\tan\left(\frac{x}{y}\right) - \frac{x}{y} = 0$$

In addition, equations may contain derivatives of the variables, in which case they are called *differential* equations. (We postpone discussion of differential equations until Chapter 18.) All these classes can be further broadened by generalizing the variables to be complex, or by including inequalities, as well as equalities, as part of our sets of equations. To proceed to a solution, we must begin by limiting our class of equations. We could start, as is often done, by solving algebraic and transcendental equations in a single real variable and move on to solving two or three simultaneous such equations in a like number of variables. However, for reasons that will become clear shortly, we choose instead to start from a different direction by limiting the equation types to the simplest possible, i.e., *linear*, but permitting any number of simultaneous variables. Thus, we will first develop techniques to solve n simultaneous linear equations in m different variables.

Linear Simultaneous Equations

In this section, we will review the elementary techniques for solving coupled linear equations as a prelude for recasting the equations into matrix form later in this chapter.

We first consider an elementary problem of two equations in two unknowns. Two equations that are not contradictory[1] and are not simply multiples of one another are called *consistent* and *independent*. Such equations have a unique, well-defined solution.

Consistent, independent equations

$$3x + 2y = 4$$
$$2x + \ y = 1$$

You can readily solve these equations (say, by solving the second for y ($y = 1 - 2x$) and inserting this expression for y into the first) to obtain $(x, y) = (-2, 5)$ as the only possible values of x and y that satisfy both equations. Figure 1.1 shows the solution of two consistent equations.

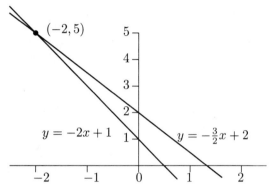

Figure 1.1 *Two consistent equations have a single point of intersection.*

An alternative, but equivalent, procedure is to perform a sequence of algebraic operations on the equations to yield a new set that is more readily solved. For example, by multiplying the first equation by $-\frac{2}{3}$ and then adding it to the second equation, we obtain

$$3x + \ 2y = \ \ 4$$
$$-\tfrac{1}{3}y = -\tfrac{5}{3}$$

The solution for y is then 5 and the value for x is obtained by substituting this value into the first equation.

However, a slight modification of these equations, for example,

Inconsistent equations

$$2x + y = 2$$
$$2x + y = 1$$

---

[1]For example, the equations $x + y = 1$ and $x + y = 2$ are obviously contradictory because the same quantity, $x + y$, cannot equal both 1 and 2 simultaneously. Such equations are called *inconsistent*. Equations that are not inconsistent are called *consistent*.

yields a set that is clearly contradictory. These two equations are graphed in Figure 1.2, and it is evident that there are *no* values of x and y that can satisfy both equations simultaneously. The equations are said to be inconsistent. (That is, $2x + y$ cannot equal both 2 and 1 simultaneously.) These equations then have a solution set that is empty. Performing the same algebraic operations on these equations would yield

$$
\begin{aligned}
2x + \ y &= \ \ 2 \\
0y &= -1
\end{aligned}
$$

which clearly demonstrates that there is no value of y that can be a solution.

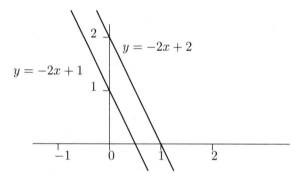

Figure 1.2   *Two inconsistent equations have no points of intersection.*

Another slight variation in the equations could be

Dependent
equations
$$
\begin{aligned}
4x + 2y &= 4 \\
2x + \ y &= 2
\end{aligned}
$$

In this instance, the algebraic solution of the equations yields

$$
\begin{aligned}
4x + 2y &= 4 \\
0y &= 0
\end{aligned}
$$

Thus, any value of y would solve the equations, and the corresponding value of x would be $x = 1 - \frac{1}{2}y$. The solution set for these equations would be infinite, as shown in Figure 1.3. Of course, in this instance, the first equation is simply twice the second equation, and the two equations are not independent. (Two equations are said to be linearly dependent if one is a multiple of the other.)

Finally, consider the effect of uncertainties in any of the coefficients or in the elements of the terms, b_i. For example, if the coefficient of x in the first of the dependent equations above is replaced by $4 + \epsilon$, a unique solution $(x, y) = (0, 2)$ is easily obtained for any value of $\epsilon \neq 0$, no matter how small. Such uncertainties can be the result of experimental error or may have been introduced by computer round-off errors in the computation of the coefficients. In any event, the

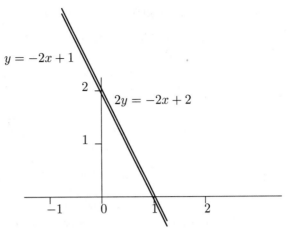

$y = -2x + 1$

$2y = -2x + 2$

Figure 1.3 *Two dependent equations correspond to the same line and thus intersect at an infinite number of points.*

numbers used in a computer calculation will have limited precision.[2] This fact will have profound implications with respect to the accuracy of solutions and will significantly affect the methods used to obtain a solution.

A set of two equations in two unknowns is termed *ill conditioned* if a "small" change in any of the coefficients will change the solution set from unique to either infinite or empty. Of course, the definition of small will have to be more precise. This will be addressed later in the context of n equations in n unknowns. Thus, if ϵ is a very small number, the equations

Ill-conditioned equations

$$(4 + \epsilon)x + 2y = 4$$
$$2x + y = 2$$

are precariously close to being inconsistent. Following the same steps, we obtain

$$(4 + \epsilon)x + 2y = 4$$
$$\epsilon y = -4 + \epsilon$$

The solution for y is then $y = 1 - 4/\epsilon$, and small changes in ϵ will result in large changes in y.

These properties concerning the solution of two equations in two unknowns can be readily generalized to n equations in n unknowns by a procedure known as *Gaussian elimination*.

---

[2]Purely integer values used in a computer calculation are not subject to round-off errors. Such numbers are written without a decimal point and are called *fixed-point* numbers. Fixed-point numbers are used primarily as counters. Numbers written with a decimal point, with or without an exponent, as 3.14159 or $1.0e - 6$, are labeled *floating-point* numbers and are generally imprecise because the computer will only retain a finite number of digits. In MATLAB, all numbers are assumed to be floating-point numbers.

THE CASE OF n EQUATIONS IN n UNKNOWNS

The set of n variables is written as $x_1, x_2, \ldots, x_n$. The coefficients of the x's in the first equation are labeled as $a_{11}, a_{12}, \ldots, a_{1n}$, and the coefficients in the second equation are $a_{21}, a_{22}, \ldots, a_{2n}$, and so on. Assuming we have n equations in n unknowns, the generalized set of equations takes the form:

$$a_{11}x_1 + a_{12}x_2 + \cdots + a_{1n}x_n = b_1$$
$$a_{21}x_1 + a_{22}x_2 + \cdots + a_{2n}x_n = b_2$$
$$\vdots \qquad \vdots \qquad \ddots \qquad \vdots \qquad \vdots$$
$$a_{n1}x_1 + a_{n2}x_2 + \cdots + a_{nn}x_n = b_n$$

where b_i is the constant on the right-hand side of the ith equation.

The solution of these equations (if one exists) will specify values for each of the variables x; that is

$$x_1 = c_1$$
$$x_2 = c_2$$
$$\vdots \qquad \vdots$$
$$x_n = c_n$$

Techniques to solve these equations are generalizations of the methods used for two equations in two unknowns and make use of *elementary row operations*.

ELEMENTARY ROW OPERATIONS

Clearly, the solution set of a set of equations will not be changed if the same quantity is added to both sides of any particular equation. And, because both sides of an equation represent the same numerical value, we can easily conclude that any of the following operations will not change the values of the solution to the original set of equations:

1. Interchange of any two equations
2. Addition of a multiple of an equation to another equation
3. Multiplication of any equation by a nonzero constant.

Gaussian Elimination

A procedure employing *elementary row operations*, whereby the original set of equations is systematically transformed into an equivalent set that is more readily solved, is known as *Gaussian elimination*. The method is quite easy to understand and almost as easy to program for a computer, provided problems such as division by zero are not encountered. We will see how to handle potential problems in the next section; here we'll simply outline the basic method.

FIRST PASS

In the first pass, the top row is called the *pivot row*, and the leftmost coefficient, a_{11}, is called the *pivot element*. In this pass, *all* the equations below the first are replaced by adding to each a multiple of the pivot equation with the factor chosen such that the resulting coefficients of x_1 are *zero*. Thus, if we replace the kth equation by [Equation k] $- \left(\dfrac{a_{k1}}{a_{11}} \right) \times$ [Equation 1], the coefficient of x_i is replaced by

$$a_{ki} \Longrightarrow a_{ki} - a_{1i} \frac{a_{k1}}{a_{11}}$$

and the right-hand-side coefficients are likewise replaced by

$$b_k \Longrightarrow b_k - b_1 \frac{a_{k1}}{a_{11}}$$

then the coefficient of x_1 in each equation but the first becomes zero. (Set $i = 1$ in this replacement.) Denoting the new set of coefficients as a', this system can be written as:

$$
\begin{aligned}
a_{11}x_1 + a_{12}x_2 + \cdots + a_{1n}x_n &= b_1 \\
0 \quad + a'_{22}x_2 + \cdots + a'_{2n}x_n &= b'_2 \\
\vdots \qquad \vdots \qquad \ddots \qquad \vdots \qquad \vdots \\
0 \quad + a'_{n2}x_2 + \cdots + a'_{nn}x_n &= b'_n
\end{aligned}
$$

SECOND PASS

Equation 2 is now the pivot row, and the pivot element is a'_{22}. Replace *all* equations *below* the second by adding to each a multiple of the pivot equation (now Equation 2) with the factor chosen such that the resulting coefficients of x_2 are *zero*. After completing this step, and again denoting the resulting set of coefficients as a'', the equations now resemble

$$
\begin{aligned}
a_{11}x_1 + a_{12}x_2 + a_{13}x_3 + \cdots + a_{1n}x_n &= b_1 \\
0 \quad + a'_{22}x_2 + a'_{23}x_3 + \cdots + a'_{2n}x_n &= b'_2 \\
0 \quad + \quad 0 \quad + a''_{33}x_3 + \cdots + a''_{3n}x_n &= b''_3 \\
\vdots \qquad \vdots \qquad \vdots \qquad \ddots \qquad \vdots \qquad \vdots \\
0 \quad + \quad 0 \quad + a''_{n3}x_3 + \cdots + a''_{nn}x_n &= b''_n
\end{aligned}
$$

Clearly, after completing n passes, the original set of equations will be transformed into an equivalent set of equations of the form

$$
\begin{aligned}
\alpha_{11}x_1 + \alpha_{12}x_2 + \alpha_{13}x_3 + \cdots + \alpha_{1n}x_n &= c_1 \\
0 \quad + \alpha_{22}x_2 + \alpha_{23}x_3 + \cdots + \alpha_{2n}x_n &= c_2 \\
0 \quad + \quad 0 \quad + \alpha_{33}x_3 + \cdots + \alpha_{3n}x_n &= c_3 \\
\vdots \qquad \vdots \qquad \vdots \qquad \ddots \qquad \vdots \qquad \vdots \\
0 \quad + \quad 0 \quad + \quad 0 \quad + \cdots + \alpha_{nn}x_n &= c_n
\end{aligned}
$$

It is then an easy matter to solve these equations by back substitutions: that is, by inserting the value obtained for x_n from the last equation into the $(n-1)$st equation to then obtain x_{n-1}. These two values are then used to obtain x_{n-2} from the $(n-2)$nd equation, etc.

The Gaussian elimination procedure is very straightforward and, though tedious, is easy to carry out, provided that problems are not encountered, such as one of the pivots being zero.

ILLUSTRATIVE PROBLEM 1.1

Solution of Three Equations in Three Unknowns by the Gaussian Elimination Method

As an example of this procedure, consider the following three equations in three unknowns:

$$
\begin{aligned}
2x + y + z &= 2 \\
x \qquad\quad + 2z &= 3 \\
x + 2y + 4z &= 4
\end{aligned}
$$

The solution of these equations by the Gaussian elimination method proceeds as follows.

FIRST PASS

Eliminate first-column numbers in Equations 2 and 3.

$$
\left.\begin{aligned}
2x + y + z &= 2 \\
x \qquad + 2z &= 3 \\
x + 2y + 4z &= 4
\end{aligned}\right\}
\begin{aligned}
&\text{Replace by} \\
\Longrightarrow\ &[\text{row}_2] - \tfrac{1}{2}[\text{row}_1] \\
&[\text{row}_3] - \tfrac{1}{2}[\text{row}_1]
\end{aligned}
\left\{\begin{aligned}
2x + y + z &= 2 \\
-\tfrac{1}{2}y + \tfrac{3}{2}z &= 2 \\
+\tfrac{3}{2}y + \tfrac{7}{2}z &= \tfrac{7}{2}
\end{aligned}\right.
$$

SECOND PASS

Eliminate second-column numbers in row 3 by replacing this equation by subtracting 3 times row 2, the pivot row.

$$
\Longrightarrow
\left\{\begin{aligned}
2x + y + z &= 2 \\
-\tfrac{1}{2}y + \tfrac{3}{2}z &= 2 \\
+ 8z &= 9
\end{aligned}\right.
$$

Thus, the solution for z is $z = \tfrac{9}{8}$. Inserting this value in the previous equation, we obtain $y = -4 + \tfrac{27}{8} = -\tfrac{5}{8}$. Finally, inserting these two values into the remaining equation, the solution for x is $x = \tfrac{3}{4}$. The solution set for (x, y, z) can be verified by plugging these values into the original equations. ∎

As with long division, it is instructive to carry out this rather tedious procedure for a few examples to ensure that you understand the method; but once you are comfortable with the method, you will prefer an alternative to actually performing all the arithmetic. In long division, the alternative is a pocket calculator; for the solution of linear simultaneous equations, the alternative will be found in the discussion of matrices that follows, along with the computer software package MATLAB to carry out the arithmetic. But first we should continue the study of Gaussian elimination to anticipate some problems that may develop when executing the method.

Potential Problems in Gaussian Elimination

If we begin with n equations in n unknowns, the only catastrophic problem that can occur during Gaussian elimination is that one of the pivot elements, a'_{kk}, turns out to be zero. Because further steps in the reduction of the set of equations rely on dividing by this quantity, the procedure fails. The equations may still have a unique solution, but the method to the solution will have to be altered.

If the current pivot element is zero, there are three possible outcomes:

1. A **unique** solution exists.

In pass k, the pivot row is row k, and the pivot element is a'_{kk}. The reduction steps in this pass require adding row k, multiplied by a'_{mk}/a'_{kk}, to each row m for $m > k$. Clearly, if $a'_{kk} = 0$, the procedure fails. However, as the sequence of the equations is of no significance, we could replace the offending equation by any equation below it in the set. It must be an equation below because the equations above contain nonzero elements to the left of column k. For example, if the coefficient of y in the second of the equations above were $\frac{1}{2}$ instead of zero, you would find that the second pivot element would be identically zero. To proceed with the calculation, simply interchange Equations 2 and 3 and continue from there. You should obtain the solutions $(x, y, z) = (\frac{8}{9}, -\frac{10}{9}, \frac{12}{9})$ without difficulty. Again, these are verified by inserting them back into the original set of equations.

Pivoting Strategy   All the examples in this chapter involve a small number of simultaneous equations, usually three, in a like number of unknowns. Problems of this type can easily be solved using *exact* arithmetic and the question of zero pivots is that either a pivot is precisely zero, or it isn't. For a larger set of equations, a computer solution is most often sought. Because computer arithmetic with floating-point numbers is not exact, potential problems associated with round-off errors must be addressed. In the present context this would mean that if the current pivot were *approximately* zero, the method might be expected to fail, or at least to be susceptible to errors. Thus, in computer implementations of the Gaussian elimination method, the switching of rows must be performed whenever a pivot is dangerously small. (The meaning of small of course must be specified.) In fact, it is found that to reduce the overall accumulated round-off error in the solution, it is best to switch rows in *every* pass, replacing the current pivot row with the row below that contains the maximum magnitude coefficient in the current pivot column. This step slows the method only slightly, and can improve the accuracy dramatically. The reader should be aware that there are numerous variations on this strategy. Some computer codes will first scale each row by the maximum element in that row to avoid problems caused by a single or a few very large elements in a row.

2. The equations are **not independent**.

If a zero pivot is unavoidable, and if the equations are not contradictory (that is, they are *consistent*), an *infinite* number of solutions are obtained. For example, replace the second equation by $x - y - 3z = 1$ so that the set now becomes

$$2x + y + z = 2$$
$$x - y - 3z = 1$$
$$x + 2y + 4z = 1$$

Notice that simply adding Equations 2 and 3 yields Equation 1, indicating that the equations are not independent.

LINEARLY INDEPENDENT EQUATIONS

A set of n equations in n unknowns is termed *linearly dependent* if any one of the equations can be written as a linear, homogeneous combination of the remaining equations. Equations that are not linearly dependent are termed *linearly independent*.

Carrying out the Gaussian elimination method then yields, after the second pass,

$$2x + y + z = 2$$
$$-\tfrac{3}{2}y + -\tfrac{7}{2}z = 0$$
$$0z = 0$$

The last equation suggests that any value of z will be acceptable. The solution set is then $(x, y, z) = (1 + \tfrac{2}{3}c, -\tfrac{7}{3}c, c)$ for *any* value of c. (This can again be verified by substitution.) The conclusion is that the equations are consistent but not independent.

3. The equations are **inconsistent**.

Altering any of the numbers on the right-hand side of the equations will yield an inconsistent set of equations. For example, replacing the right-hand-side 2 by a 1 in Equation 1 gives us

$$2x + y + z = 1$$
$$x - y - 3z = 1$$
$$x + 2y + 4z = 1$$

After completing two passes the equations become

$$2x + y + z = 2$$
$$-\tfrac{3}{2}y - \tfrac{7}{2}z = \tfrac{1}{2}$$
$$0z = 1$$

This time there is no solution possible for z and the set of equations must be contradictory. (The contradiction is that Equations 2 plus 3 state that $2x + y + z = 2$, which contradicts Equation 1.)

A set of consistent equations in which none of the equations can be expressed as a linear combination of the remaining equations is called *linearly independent*. A fundamental theorem of algebra is that any linearly independent set of n equations in n unknowns always has a unique solution set. However, as we shall see, the approximate nature of computer arithmetic will render this statement much less precise. In particular, if one of the pivots in the Gaussian elimination method is not precisely zero but is very small compared with other coefficients in that equation, dividing by this pivot may introduce considerable round-off error in the numerical solution and may invalidate the solution entirely. We shall deal with this problem shortly.

Just as with the solution to two equations in two unknowns, the solution set for a set of n equations in n unknowns is either unique, infinite, or empty. We could further generalize the discussion to consider m equations in n unknowns with $m > n$. Clearly then, methods developed to solve multiple equations must be able to distinguish each of these three situations before proceeding to a solution. To this end, we next introduce a notation that will enable us to manipulate sets of equations in a manner less cumbersome than continually writing out the complete set of equations. This is the notation of matrices.

Matrix Notation and Multiplication

About two centuries ago, England began to lose her clear leadership position in science to Germany and France. One possible reason for this can be traced to the fact that England, out of pride in her native son, continued to use the awkward notation for calculus invented by Newton, whereas on the Continent the notation of Leibnitz, the notation in use today, was becoming universally used. Whether or not this fact had such a profound impact on the progress of science and technology can be debated; however, it does remind us that quite often improvements in seemingly secondary items, such as notation, can have very significant consequences. Examples such as analytic geometry, vector notation, and Boolean algebra that we now take for granted enable us to more easily concentrate on the central problem and not be distracted by less important details—to distinguish the forest from the trees.

Perhaps the notational improvement that has had the greatest consequence is related to simplifications in the writing and manipulation of many simultaneous linear equations, the notation of *matrices*. The notation of matrices makes it possible for ordinary people to juggle hundreds of equations simultaneously. In addition, matrix notation has direct application in the modern implementation of *vectorized* computer algorithms, which is at the heart of MATLAB and very likely at the core of future advancements in computational mathematics.

This section defines matrices and illustrates their manipulation, particularly their multiplication. Because this topic is so important to an efficient use of MATLAB, numerous applications of matrix multiplication are illustrated.

MULTIPLICATION OF MATRICES

The notation of matrices facilitates the easy manipulation of multiple equations in a manner less cumbersome than continually writing out the complete set of equations.

Consider again the two equations in two unknowns that appear on page 5. We can group the set of coefficients, the two unknowns, and the two inhomogeneous terms on the right-hand side of these equations into boxes as

$$\begin{pmatrix} 3 & 2 \\ 2 & 1 \end{pmatrix} \begin{pmatrix} x \\ y \end{pmatrix} = \begin{pmatrix} 4 \\ 1 \end{pmatrix}$$

This structure will represent the original set of equations if we define the multiplication of the rectangular arrays in the following way:

$$\begin{pmatrix} 3 & 2 \\ 2 & 1 \end{pmatrix} \begin{pmatrix} x \\ y \end{pmatrix} = \left(\boxed{\begin{matrix} 3 & 2 \end{matrix}} \right) \left(\boxed{\begin{matrix} x \\ y \end{matrix}} \right) = \left(\boxed{\begin{matrix} 3x + 2y \end{matrix}} \\ 2x + y \right) = \left(\boxed{4} \\ 1 \right)$$

That is, the first equation results from the product of the elements of the first row of the array of coefficients and the first (and only) column of the array of variables. Similarly, the second equation results from the second row of coefficients times the column of variables. The array of coefficients is called a *matrix*. In this case the coefficient matrix has the same number of rows and columns and is termed a *square* matrix. The array of variables is a single column matrix, usually called a column vector.

If we call the matrix of coefficients collectively **A**, we can identify the elements of **A** as a_{ij}, where the subscripts refer to the row-column location of the element in the matrix. Thus, the identification would be

$$\begin{pmatrix} a_{11} & a_{12} \\ a_{21} & a_{22} \end{pmatrix} = \begin{pmatrix} 3 & 2 \\ 2 & 1 \end{pmatrix}$$

Further, we could generalize the column vector of variables as

$$\vec{x} = \begin{pmatrix} x \\ y \end{pmatrix} = \begin{pmatrix} x_1 \\ x_2 \end{pmatrix}$$

and the column vector of right-hand-side numbers as

$$\vec{b} = \begin{pmatrix} b_1 \\ b_2 \end{pmatrix}$$

Thus the entire set of equations could be succinctly written as

$$\mathbf{A}\vec{x} = \vec{b}$$

where the multiplication implied in this equation[3] was defined earlier and can

---

[3]Of course, an expression of equality between matrices, $\mathbf{A} = \mathbf{B}$, requires that both matrices be of the same size and that $a_{ij} = b_{ij}$ for all i and j.

be further expressed as

$$\sum_{j=1}^{2} a_{ij}x_j = a_{i1}x_1 + a_{i2}x_2 = b_i$$

This last expression then represents the ith equation of the set.

This notation can now be extended to matrices of arbitrary size, to describe simultaneous equations involving a large number of variables. Further, the multiplication defined in $\mathbf{A}\vec{x}$ can be generalized to other forms of matrices. For example, $\mathbf{AB}$, where $\mathbf{A}$ and $\mathbf{B}$ are rectangular arrays of numbers or variables, is defined as

$$\mathbf{C} = \mathbf{AB}$$

where the element c_{ij} of $\mathbf{C}$ is obtained from the ith row of $\mathbf{A}$ and the jth column of $\mathbf{B}$ by multiplying one by the other, element by element, and summing the products, i.e.,

$$c_{ij} = \sum_k a_{ik}b_{kj} \tag{1.1}$$

Thus, for this operation to be defined, the width (i.e., the number of columns) of the first array in the product, $\mathbf{A}$, must match the height (i.e., the number of rows) of the second array, $\mathbf{B}$.

For example, you should use MATLAB to verify that the elements of the product of the following two 3×3 matrices are as indicated:

$$\begin{pmatrix} 1 & 2 & 3 \\ 2 & 2 & 2 \\ 3 & 2 & 1 \end{pmatrix} \begin{pmatrix} 1 & 0 & 1 \\ 0 & 3 & 0 \\ 2 & 1 & 2 \end{pmatrix} = \begin{pmatrix} 7 & 9 & 7 \\ 6 & 8 & 6 \\ 5 & 7 & 5 \end{pmatrix}$$

or in MATLAB

```
A = [1 2 3;2 2 2;3 2 1];
B = [1 0 1;0 3 0;2 1 2];
C = A*B                    ;

C =
        7      9      7
        6      8      6
        5      7      5
```

DEFINITIONS RELATING TO SQUARE MATRICES

Diagonal The elements a_{ii} of a square matrix are those positioned along the main diagonal of the matrix. A square matrix that has nonzero elements only along the main diagonal is called a *diagonal* matrix. Other diagonals are defined by the elements $a_{i,i+1}, a_{i,i+2}, a_{i,i-1}, a_{i,i-2}$, etc., and consist of lines of elements running parallel to the main diagonal. A matrix in which the nonzero elements only appear on the main diagonal, plus a few parallel diagonals, is said to be *banded*.

A banded, tridiagonal matrix

$$
\begin{pmatrix}
a_1 & b_1 & 0 & \cdots & 0 & 0 \\
c_1 & a_2 & b_2 & \cdots & 0 & 0 \\
0 & c_2 & a_3 & \cdots & 0 & 0 \\
\vdots & \vdots & \vdots & \ddots & \vdots & \vdots \\
0 & 0 & 0 & \cdots & a_{n-1} & b_{n-1} \\
0 & 0 & 0 & \cdots & c_{n-1} & a_n
\end{pmatrix}
$$

Triangular   A square matrix with all zeros below the main diagonal is said to be upper triangular. Lower-triangular matrices are defined similarly.

An upper-triangular matrix

$$
\begin{pmatrix}
a_{11} & a_{12} & a_{13} & \cdots & a_{1,n-1} & a_{1n} \\
0 & a_{22} & a_{23} & \cdots & a_{2,n-1} & a_{2n} \\
0 & 0 & a_{33} & \cdots & a_{3,n-1} & a_{3n} \\
\vdots & \vdots & \vdots & \ddots & \vdots & \vdots \\
0 & 0 & 0 & \cdots & a_{n-1,n-1} & a_{n-1,n} \\
0 & 0 & 0 & \cdots & 0 & a_{nn}
\end{pmatrix}
$$

Symmetric   A square matrix that remains the same if the rows and columns are interchanged is called *symmetric*. This is equivalent to the condition $a_{ij} = a_{ji}$ for all i, j. The first of the two matrices on page 14 is symmetric. An *antisymmetric* or *skew-symmetric* matrix is one satisfying the condition $a_{ij} = -a_{ji}$. Thus, every diagonal matrix is symmetric, and every antisymmetric matrix must have zeros on the main diagonal.

Identity   A diagonal $n \times n$ matrix with only 1s along the diagonal is called an *identity* matrix of rank n. This matrix is designated as $\mathbf{I}_n$ and has the property that

$$
\mathbf{AI}_n = \mathbf{I}_n \mathbf{A} = \mathbf{A}
$$

where $\mathbf{A}$ is any $n \times n$ matrix.

Null   An $n \times n$ matrix of all zeros is called a *null* matrix of size n and is designated as $\mathbf{O}_n$ or simply $\mathbf{O}$. It has the property $\mathbf{OA} = \mathbf{AO} = \mathbf{O}$. But beware, $\mathbf{AB} = \mathbf{O}$ does not imply that either $\mathbf{A}$ or $\mathbf{B}$ is a null matrix. For example,

$$
\begin{pmatrix} 1 & 1 \\ 1 & 1 \end{pmatrix} \begin{pmatrix} 1 & 1 \\ -1 & -1 \end{pmatrix} = \begin{pmatrix} 0 & 0 \\ 0 & 0 \end{pmatrix}
$$

Commutative   Two square matrices $\mathbf{A}$ and $\mathbf{B}$ are said to *commute*, i.e., be *commutative*, if $\mathbf{AB} = \mathbf{BA}$. In general, square matrices do not commute. For example, neither the 2×2 nor the 3×3 matrices given earlier commute. However, you should demonstrate that diagonal matrices of the same size always commute.

MULTIPLICATION OF RECTANGULAR MATRICES

The definition of matrix multiplication involving square matrices can be generalized to nonsquare, i.e., rectangular, matrices, provided the number of columns of the first matrix matches the number of rows of the second. Thus, the product of a 2×3 matrix times a 3×2 matrix can be defined, but that of a 2×3 matrix times a 2×3 cannot. For example,

$$\begin{pmatrix} 1 & 0 & 2 \\ 0 & 3 & 1 \end{pmatrix} \begin{pmatrix} 1 & 0 \\ 0 & 3 \\ 2 & 1 \end{pmatrix} = \begin{pmatrix} 5 & 2 \\ 2 & 10 \end{pmatrix}$$

and

$$\begin{pmatrix} 1 & 0 \\ 0 & 3 \\ 2 & 1 \end{pmatrix} \begin{pmatrix} 1 & 0 & 2 \\ 0 & 3 & 1 \end{pmatrix} = \begin{pmatrix} 1 & 0 & 2 \\ 0 & 9 & 3 \\ 2 & 3 & 5 \end{pmatrix}$$

Obviously, nonsquare matrices can never commute. Nor can they be symmetric, triangular, or diagonal. Also, an identity matrix cannot be defined for nonsquare matrices.

The transpose of a matrix $\mathbf{A}$ is denoted as $\mathbf{A}^T$ and is obtained from $\mathbf{A}$ by interchanging the rows and columns of $\mathbf{A}$, i.e., if $[\mathbf{A}]_{ij} = a_{ij}$, then $[\mathbf{A}]_{ij}^T = a_{ji}$. For a square matrix, this is the same as reflecting the elements across the main diagonal. In the preceding example, if $\mathbf{A}$ is the 2×3 matrix, then the two products represent $\mathbf{A}\mathbf{A}^T$ and $\mathbf{A}^T\mathbf{A}$, respectively.

The transpose operation has some interesting properties. Obviously $[\mathbf{A}^T]^T = \mathbf{A}$ and $(\mathbf{A} + \mathbf{B})^T = \mathbf{A}^T + \mathbf{B}^T$. Not so obvious is the identity $[\mathbf{A}\mathbf{B}]^T = \mathbf{B}^T\mathbf{A}^T$. That is, the transpose of the product of two matrices is the same as the product of the transposed matrices in reverse order. This is easily proved by writing the product of two matrices, $\mathbf{A}$ and $\mathbf{B}$, in terms of the elements of each matrix as indicated in Equation 1.1.

That is, if

$$[\mathbf{C}]_{ij} = \sum_{k=1}^{n} [\mathbf{A}]_{ik} [\mathbf{B}]_{kj}$$

then

$$[\mathbf{C}^T]_{ij} = [\mathbf{C}]_{ji} = \sum_{k=1}^{n} [\mathbf{A}]_{jk} [\mathbf{B}]_{ki}$$

$$= \sum_{k=1}^{n} [\mathbf{A}^T]_{kj} [\mathbf{B}^T]_{ik}$$

$$= \sum_{k=1}^{n} [\mathbf{B}^T]_{ik} [\mathbf{A}^T]_{kj}$$

$$= \left(\mathbf{B}^T \mathbf{A}^T \right)_{ij}$$

Notice that the proof is valid for either square or rectangular matrices. For example, using the preceding matrices,

$$
\begin{pmatrix} 5 & 2 \\ 2 & 10 \end{pmatrix}^T = \left(\begin{pmatrix} 1 & 0 & 2 \\ 0 & 3 & 1 \end{pmatrix} \begin{pmatrix} 1 & 0 \\ 0 & 3 \\ 2 & 1 \end{pmatrix} \right)^T
$$

$$
= \begin{pmatrix} 1 & 0 \\ 0 & 3 \\ 2 & 1 \end{pmatrix}^T \begin{pmatrix} 1 & 0 & 2 \\ 0 & 3 & 1 \end{pmatrix}^T
$$

$$
= \begin{pmatrix} 1 & 0 & 2 \\ 0 & 3 & 1 \end{pmatrix} \begin{pmatrix} 1 & 0 \\ 0 & 3 \\ 2 & 1 \end{pmatrix}
$$

This computation illustrates the fact that the product $\mathbf{AA}^T$ always results in a symmetric matrix (see Problem 1.10). The transpose of a matrix is sometimes designated with a *prime* instead of the superscript T. It is especially important that we be comfortable with both types of notation, because MATLAB uses A' for the transpose.

Arithmetic of Matrices

Matrices and matrix multiplication have been defined to simplify the awkward notation associated with writing out numerous simultaneous equations. We have seen that the operation of matrix multiplication has a few odd features that are not encountered in the arithmetic of ordinary numbers:

AB is defined only if the width of **A** is equal to the height of **B**.

AB does not, in general, equal **BA**.

AB = **0** does not imply that either **A** or **B** is a null matrix.

The other basic arithmetic operation, addition, when applied to matrices, is, however, quite well behaved and introduces no surprises.

MATRIX ADDITION

1. If **A** and **B** are two matrices of the same size (both $m \times n$), then the sum, **A** + **B**, is defined by

$$
[\mathbf{A} + \mathbf{B}]_{ij} = [\mathbf{A}]_{ij} + [\mathbf{B}]_{ij}
$$

It then follows from this definition that the multiplication of a matrix by a scalar, such as 3**A**, is obtained by multiplying each element of **A** by the scalar 3.

$$
(c\mathbf{A})_{ij} = ca_{ij}
$$

2. Subtraction of matrices is defined as $\mathbf{A} - \mathbf{B} = \mathbf{A} + (-1)\mathbf{B}$. With these definitions many of the ordinary properties of arithmetic of ordinary numbers can be shown to be valid for matrices as well.

Commutative	$\mathbf{A} + \mathbf{B} = \quad \mathbf{B} + \mathbf{A}$
Associative	$\mathbf{A} + (\mathbf{B} + \mathbf{C}) = (\mathbf{A} + \mathbf{B}) + \mathbf{C}$
Additive inverse	$\mathbf{A} + (-\mathbf{A}) = \quad \mathbf{0}$
Additive identity	$\mathbf{A} + \mathbf{0} = \quad \mathbf{A}$

3. The addition of a *scalar* to a matrix can cause some initial confusion, but usually the intended operation is clear from the context. The addition of a scalar to a matrix, such as $\mathbf{A} + c$, will add the value c to each of the elements of $\mathbf{A}$. More formally, the intended meaning for an operation like $\mathbf{A} + c$ is a matrix addition of the form $\mathbf{A} + \mathbf{C}$, where $\mathbf{C}$ is the same size as $\mathbf{A}$ with each of its elements equal to c.

Multiplication of Row and Column Vector Matrices

The most important example of nonsquare matrix multiplication involves products of row and column vectors. For example, if

$$\vec{x} = \begin{pmatrix} 4 \\ 0 \\ 3 \end{pmatrix}, \quad \vec{x}^T = \begin{pmatrix} 4 & 0 & 3 \end{pmatrix}$$

then

$$\vec{x}^T \vec{x} = \begin{pmatrix} 4 & 0 & 3 \end{pmatrix} \begin{pmatrix} 4 \\ 0 \\ 3 \end{pmatrix} = 25$$

and

$$\vec{x}\vec{x}^T = \begin{pmatrix} 4 \\ 0 \\ 3 \end{pmatrix} \begin{pmatrix} 4 & 0 & 3 \end{pmatrix} = \begin{pmatrix} 16 & 0 & 12 \\ 0 & 0 & 0 \\ 12 & 0 & 9 \end{pmatrix}$$

Notice that the product of an n-component row-vector matrix times an n-component column-vector matrix is a single number, whereas the product in the reverse order generates an $n \times n$ matrix.

THE DOT PRODUCT OF TWO VECTORS

The product of a row vector times a column vector plays a particularly important role in linear analysis and is given a special name: the *dot product*.

If $\vec{x}$ and $\vec{y}$ are both column vectors containing n real elements, then the dot product of the two vectors is defined as

$$\vec{x} \cdot \vec{y} = \vec{x}^T \vec{y} = \sum_{i=1}^{n} x_i y_i = \vec{y}^T \vec{x}$$

As with ordinary three-dimensional vectors, the dot product is used to define both the length of a vector and the angle between vectors.

If $\vec{a}$ and $\vec{b}$ are three-dimensional vectors in a Cartesian coordinate system, i.e.,

$$\vec{a} = \begin{pmatrix} a_x \\ a_y \\ a_z \end{pmatrix}, \quad \vec{b} = \begin{pmatrix} b_x \\ b_y \\ b_z \end{pmatrix}$$

then $\vec{a} \cdot \vec{b} = a_x b_x + a_y b_y + a_z b_z = \vec{a}^T \vec{b}$.

The Norm of a Vector   One use of the dot product is in the definition of the *norm* or "length" of a vector. The length of a three-dimensional vector, $\vec{a}$, is given in terms of its components by the Pythagorean theorem as

$$|\vec{a}| = \sqrt{a_x^2 + a_y^2 + a_z^2}$$
$$= \sqrt{\vec{a} \cdot \vec{a}}$$

This definition is generalized in an obvious manner to define the norm of an n-component vector.

The Angle Between Vectors   If we choose the direction of the x axis to be along the direction of the three-dimensional vector $\vec{a}$, then in terms of components we have $\vec{a} = (a,\, 0,\, 0)^T$, where a is then the length of the vector. The dot product of an arbitrary second vector $\vec{b}$ is then $\vec{a} \cdot \vec{b} = ab_x$. But b_x is simply the projection of vector $\vec{b}$ along the direction of the x axis; i.e., along the direction of $\vec{a}$. If we define the angle between vectors $\vec{a}$ and $\vec{b}$ to be θ_{ab}, the dot product of the two vectors can be expressed as

$$\vec{a} \cdot \vec{b} \equiv |\vec{a}||\vec{b}| \cos \theta_{ab}$$

None of the quantities on the right-hand side of this equation depends upon our choice of coordinate system, thus the expression must remain valid in all coordinate systems. This geometrical expression for the dot product is illustrated in Figure 1.4.

Generalizing these definitions to n-component column vectors, we define the length of a vector as $|x| = \sqrt{\vec{x} \cdot \vec{x}}$, and the angle between vectors $\vec{x}$ and $\vec{y}$ as $\theta_{xy} = \cos^{-1}\left(\frac{\vec{x} \cdot \vec{y}}{|\vec{x}||\vec{y}|}\right)$. If $\vec{x} \cdot \vec{y} = 0$, the vectors are said to be *orthogonal*, whereas if $\vec{x} \cdot \vec{y} = \pm|\vec{x}||\vec{y}|$, the vectors are *parallel (+)* or *antiparallel (−)*. For example,

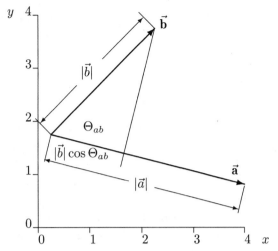

Figure 1.4 *A geometrical picture of the dot product of two vectors.*

the vectors

$$\vec{x} = \begin{pmatrix} 1 \\ 0 \\ 1 \\ 0 \end{pmatrix} \qquad \vec{y} = \begin{pmatrix} 0 \\ 1 \\ 0 \\ 1 \end{pmatrix}$$

both have a length of $\sqrt{2}$ and are orthogonal.

UNIT VECTORS

Vectors that have length 1 are called *unit* vectors and are usually written with a "hat" notation. Thus

$$\hat{e}_1 = \frac{1}{\sqrt{2}} \begin{pmatrix} 1 \\ 0 \\ 1 \\ 0 \end{pmatrix} \qquad \hat{e}_2 = \frac{1}{\sqrt{2}} \begin{pmatrix} 0 \\ 1 \\ 0 \\ 1 \end{pmatrix}$$

are *orthogonal unit* vectors. (Orthogonal unit vectors are called *orthonormal*.)

PROJECTION OPERATORS

Linear algebra describes the effect of *operations*, represented by matrices, on *vectors*, represented by column matrices, and is particularly well suited to geometric interpretations. One type of operation that is especially amenable to such visualization is that of projecting a vector onto another vector or onto a plane or another subspace. These projection operations also provide an excellent illustration of the utility of matrix and vector multiplication.

The matrix, $\mathbf{P}_1$, is constructed from the unit vector, $\hat{e}_1$, in the following way:

$$\mathbf{P}_1 = \hat{e}_1 \cdot \hat{e}_1^T = \frac{1}{2} \begin{pmatrix} 1 \\ 0 \\ 1 \\ 0 \end{pmatrix} \cdot \begin{pmatrix} 1 & 0 & 1 & 0 \end{pmatrix}$$

$$= \frac{1}{2} \begin{pmatrix} 1 & 0 & 1 & 0 \\ 0 & 0 & 0 & 0 \\ 1 & 0 & 1 & 0 \\ 0 & 0 & 0 & 0 \end{pmatrix}$$

The matrix has the interesting property[4] that $\mathbf{P}_1^2 = \mathbf{P}_1$. This matrix times an arbitrary 4×1 column vector will produce the projection of that vector along the direction of $\hat{e}_1$. That is, $\mathbf{P}_1 \cdot \vec{y}$ is parallel to $\hat{e}_1$. (Verify this.)

The vector $\hat{e}_2$ is also of unit length and is orthogonal to $\hat{e}_1$. Thus, the projection represented by $\mathbf{P}_1 \cdot \hat{e}_2$ must yield zero. Further, the projection operator represented by $\mathbf{P}_{12} = \mathbf{P}_1 + \mathbf{P}_2$, where $\mathbf{P}_2 = \hat{e}_2 \cdot \hat{e}_2^T$, would represent a projection onto the plane defined by the two vectors $\hat{e}_1$ and $\hat{e}_2$.

Finally, two additional vectors,

$$\hat{e}_3 = \frac{1}{\sqrt{2}} \begin{pmatrix} 1 \\ 0 \\ -1 \\ 0 \end{pmatrix} \qquad \hat{e}_4 = \frac{1}{\sqrt{2}} \begin{pmatrix} 0 \\ 1 \\ 0 \\ -1 \end{pmatrix}$$

are easily verified to have unit length and to be orthogonal to each other and to $\hat{e}_1$ and $\hat{e}_2$. The projection operator defined by $\mathbf{P} = \mathbf{P}_1 + \mathbf{P}_2 + \mathbf{P}_3 + \mathbf{P}_4$ is then

$$\mathbf{P} = \sum_{i=1}^{4} \hat{e}_i \cdot \hat{e}_i^T$$

$$= \frac{1}{2} \begin{pmatrix} 1 & 0 & 1 & 0 \\ 0 & 0 & 0 & 0 \\ 1 & 0 & 1 & 0 \\ 0 & 0 & 0 & 0 \end{pmatrix} + \frac{1}{2} \begin{pmatrix} 0 & 0 & 0 & 0 \\ 0 & 1 & 0 & 1 \\ 0 & 0 & 0 & 0 \\ 0 & 1 & 0 & 1 \end{pmatrix}$$

$$+ \frac{1}{2} \begin{pmatrix} 1 & 0 & -1 & 0 \\ 0 & 0 & 0 & 0 \\ -1 & 0 & 1 & 0 \\ 0 & 0 & 0 & 0 \end{pmatrix} + \frac{1}{2} \begin{pmatrix} 0 & 0 & 0 & 0 \\ 0 & 1 & 0 & -1 \\ 0 & 0 & 0 & 0 \\ 0 & -1 & 0 & 1 \end{pmatrix}$$

$$= \begin{pmatrix} 1 & 0 & 0 & 0 \\ 0 & 1 & 0 & 0 \\ 0 & 0 & 1 & 0 \\ 0 & 0 & 0 & 1 \end{pmatrix} = \mathbf{I}_4$$

---

[4] A matrix $\mathbf{P}$ satisfying the property $\mathbf{P}^2 = \mathbf{P}$ is called idempotent. Every such matrix can be interpreted as a projection matrix.

This result is a statement that the set of four vectors, $\hat{e}_1, \hat{e}_2, \hat{e}_3$, and $\hat{e}_4$, is *complete*, meaning that any 4×1 column vector can be written in terms of these unit vectors. These four vectors are also said to "span" this four-dimensional space.[5] For example, the vector $\vec{y} = [3 \quad 2 \quad 2 \quad 1]^T$ would be expanded in terms of the $\hat{e}$'s as

$$\vec{y} = \left(\sum_{i=1}^{4} \mathbf{P}_i\right)\vec{y} \quad (= \mathbf{I}_4\vec{y})$$

$$= \left(\sum_{i=1}^{4} \hat{e}_i\hat{e}_i^T\right)\vec{y}$$

$$= \sum_{i=1}^{4} (\hat{e}_i^T \cdot \vec{y})\hat{e}_i$$

The four dot products, $(\hat{e}_i^T \cdot \vec{y})$, are easily evaluated using the explicit expressions given for $\vec{y}$ and the unit vectors $\hat{e}_i$. The result is

$$\vec{y} = \frac{1}{\sqrt{2}}(5\hat{e}_1 + 3\hat{e}_2 + \hat{e}_3 + \hat{e}_4)$$

Use MATLAB to construct the vectors $\hat{e}_i, i = 1, \ldots, 4$, and the vector $\vec{y}$. Next, compute the projection operators P_i and verify the preceding results.

(§) Applications of Matrix Multiplication*

PREVIEW The remainder of this chapter illustrates the utility of matrix multiplication and reinforces the reader's skills in matrix manipulation by exploring several common applications based primarily on matrix multiplication.

The examples make use of MATLAB utilities for constructing and multiplying matrices. The reader should be familiar with the contents of Section A.i of the Appendix dealing with elementary MATLAB matrix manipulation techniques.

The first topic, Markov matrices, describes the transitions between members of a fixed population and is important in several areas of engineering and science. The

---

[5]More formally, if $\hat{e}_i, i = 1, \ldots, n$, are all n-component, mutually orthogonal unit vectors, then *any* n-component vector, i.e., any vector in this n-dimensional space, can be written in terms of the $\hat{e}_i$, which are said to *span* the n-dimensional space. The completeness condition on the unit vectors is then

$$\sum_{i=1}^{n} \hat{e}_i \cdot \hat{e}_i^T = \mathbf{I}_n$$

Interestingly, this definition can be extended to spaces of infinite dimension.
*Sections labeled with (§) may be omitted without loss of continuity. These sections are either more advanced or more exploratory in nature.

topics that follow deal with vectors in an n-dimensional space and are central to many topics in applied mathematics. The applications in this chapter are progressively more difficult and should develop a facility in visualizing the products of matrices and vectors. Once again, we emphasize that the reader should be comfortable with matrix and vector manipulation to make full use of MATLAB.

(§) Markov Matrices

Consider a fixed population of N individuals, which could be persons, atoms, machine parts, etc. Assume that there are n states available to each individual. These could be jobs for persons, energy levels for atoms, or stages of assembly for machine parts. As time progresses, each of the individuals has the possibility of switching from its current state to one of the other available states (or not). The *probability* of a transition in one time step from state a to state b, $(\text{Prob}(b \leftarrow a))$, is called p_{ba}. Defining a probability of 1 to indicate absolute certainty of occurrence, and a probability of 0 to indicate impossibility, the sum of the probabilities of all possible transitions that begin in state a and end up in any other state (including state a itself) must be 1.

$$\sum_{b=1}^{n} p_{ba} \equiv 1$$

The collection of all possible transitions between the n states is called a *Markov* matrix. The fundamental requirements on a Markov matrix are then:

- *All* the elements of $\mathbf{P}$ are nonnegative; i.e., $p_{ij} \geq 0$ for all i, j.
- The sum of each of the columns is 1; i.e., $\sum_{j=1}^{n} p_{ji} = 1$ for every value of i.

For example, a college fraternity has 50 members currently at various stages in their college careers (freshman, sophomore, junior, senior). There are also the potential stages each will be in as of next year (freshman, sophomore, junior, senior, graduated, or left without degree). Suppose the probabilities of transitions are contained in the matrix $\mathbf{P}$ given by

			From			
To	Fr.	So.	Jr.	Sr.	grad	lwd
Fr.	0.1	0	0	0	0	0
So.	0.6	0.1	0	0	0	0
Jr.	0	0.8	0.1	0	0	0
Sr.	0	0	0.8	0.1	0	0
grad	0	0	0	0.7	1	0
lwd	0.3	0.1	0.1	0.2	0	1

That is, the probability that a freshman will leave without a degree (lwd) is 30%, and the probability that a senior will graduate (grad) is 70%.

If the initial population at each level is x0 = [4 18 12 16 0 0]', the same states will have a population of $x_1 = \mathbf{P}x_0$ after 1 year, and $x_2 = \mathbf{P}x_1$ after 2 years. After a certain number of years, all the current population will either have left or have graduated, and applying the transition matrix again will not change this population. This situation is called the *equilibrium* state.

In this example, there were no possibilities of "retreating" from, say, junior to sophomore, and so everyone had to eventually end up in one of two states: graduated or left without degree. Also, once in these final states, the transition to any other state was not possible; these final states are thus called *absorbing* states.

More commonly, the transition matrix will have a possibility of transition in both directions, and the final equilibrium state will be either a mixture of several states or a distribution among the absorbing states, if any.

Consider next the problem of a *one-dimensional random walk*. Initially, a particle can be in any of, say, five positions along a string. If the particle is currently at position i, then in one transition, the particle can either stay put (with probability p_{ii}), jump to the next position ($p_{i+1,i}$), or fall back to the previous position ($p_{i-1,i}$). The probabilities of staying at the end positions are p_{11} and p_{55}. If these are equal to 1, they are absorbing positions, and the particle will be locked into that end. If they are 0, they are called *reflecting*, as it is impossible for a particle landing on one of the ends to remain there. With $p_{11} = p_{55} = 0$, the remaining probabilities are given by the matrix

$$\mathbf{P} = \begin{pmatrix} 0 & 0.4 & 0 & 0 & 0 \\ 1 & 0.2 & 0.4 & 0 & 0 \\ 0 & 0.4 & 0.2 & 0.4 & 0 \\ 0 & 0 & 0.4 & 0.2 & 1 \\ 0 & 0 & 0 & 0.4 & 0 \end{pmatrix}$$

Use MATLAB to construct the matrix and repeatedly apply this matrix to the initial population x = [10 10 10 10 10]' to try to find the final equilibrium state. (*Answer*: [5.263 13.158 13.158 13.158 5.263])

Notice that the equilibrium state vector is a solution of the equation

$$\mathbf{P}\vec{x}_{\text{eq}} = \vec{x}_{\text{eq}}$$

which is called an *eigenvalue* equation; it is described in more detail in Chapter 3.

(§) Rotation of Coordinate Systems

In three dimensions, a vector $\vec{a}$ is often written in terms of unit vectors $\hat{i}, \hat{j}$, and $\hat{k}$ as $\vec{a} = a_x\hat{i} + a_y\hat{j}a + a_z\hat{k}$, where

$$\hat{i} \cdot \hat{i} = \hat{j} \cdot \hat{j} = \hat{k} \cdot \hat{k} = 1$$
$$\hat{i} \cdot \hat{j} = \hat{i} \cdot \hat{k} = \hat{j} \cdot \hat{k} = 0$$

See Figure 1.5 (p. 26). Thus, $a_x = \vec{a} \cdot \hat{i}$, etc., is the projection of $\vec{a}$ along the direction of the unit vector $\hat{i}$. These equations have a simple representation in

Figure 1.5   *The unit vectors $\hat{\imath}, \hat{\jmath}, \hat{k}$ span three-dimensional space.*

terms of column vectors. Defining

$$\hat{\imath} = \begin{pmatrix} 1 \\ 0 \\ 0 \end{pmatrix}; \qquad \hat{\jmath} = \begin{pmatrix} 0 \\ 1 \\ 0 \end{pmatrix}; \qquad \hat{k} = \begin{pmatrix} 0 \\ 0 \\ 1 \end{pmatrix}$$

we easily see that

$$\hat{\imath} \cdot \hat{\imath} = \hat{\imath}^T \hat{\imath} = \begin{pmatrix} 1 & 0 & 0 \end{pmatrix} \begin{pmatrix} 1 \\ 0 \\ 0 \end{pmatrix} = 1$$

$$\hat{\imath} \cdot \hat{\jmath} = \hat{\imath}^T \hat{\jmath} = \begin{pmatrix} 1 & 0 & 0 \end{pmatrix} \begin{pmatrix} 0 \\ 1 \\ 0 \end{pmatrix} = 0$$

so that

$$\vec{a} = a_x \begin{pmatrix} 1 \\ 0 \\ 0 \end{pmatrix} + a_y \begin{pmatrix} 0 \\ 1 \\ 0 \end{pmatrix} + a_z \begin{pmatrix} 0 \\ 0 \\ 1 \end{pmatrix}$$

Next, consider the representation of the same vector in a different coordinate system, (x', y', z'), which is obtained by a rotation of the original coordinates and represented by unit vectors $\hat{e}_1, \hat{e}_2$, and $\hat{e}_3$. Clearly, the same vector can be represented as $\vec{a} = \begin{pmatrix} a_{x'} \\ a_{y'} \\ a_{z'} \end{pmatrix}$, with $a_{x'} = \vec{a} \cdot \hat{e}_1$, etc. The question is: How are the elements $(a_{x'}, a_{y'}, a_{z'})$ related to the original set of components (a_x, a_y, a_z)?

To simplify the notation, we will designate the unit vectors $\hat{\imath}, \hat{\jmath}$, and $\hat{k}$ in the original coordinate system as $\hat{\imath}_1, \hat{\imath}_2$, and $\hat{\imath}_3$. Because these vectors are a complete set (meaning that any three-dimensional vector can be written in terms of these vectors) we can form the projection operator identity

$$\mathbf{I}_3 = \sum_{m=1}^{3} \hat{\imath}_m \hat{\imath}_m^T$$

and write

$$\hat{e}_j = \left(\sum_{m=1}^{3} \hat{\imath}_m \hat{\imath}_m^T \right) \hat{e}_j$$

$$= \sum_{m=1}^{3} \hat{\imath}_m (\hat{\imath}_m^T \hat{e}_j) \equiv \sum_{m=1}^{3} U_{jm}^T \hat{\imath}_m$$

where $U_{jm}^T \equiv \hat{\imath}_m^T \hat{e}_j$ is the matrix of dot products between all combinations of the unit vectors and defines the relationship between the two sets of vectors. This matrix can be symbolically written as

$$\mathbf{U}^T = \left(\begin{array}{c} \overline{\hat{\imath}_1} \\ \overline{\hat{\imath}_2} \\ \overline{\hat{\imath}_3} \end{array} \right) \cdot \left(\left(\begin{array}{c} \\ e_1 \\ \\ \end{array} \right) \left(\begin{array}{c} \\ e_2 \\ \\ \end{array} \right) \left(\begin{array}{c} \\ e_3 \\ \\ \end{array} \right) \right)$$

Or, noting that the transpose of a product of matrices is the product of the individual transposes in reverse order, $(\mathbf{AB})^T = \mathbf{B}^T \mathbf{A}^T$, the symbolic expression for the matrix $\mathbf{U}$ is

$$\mathbf{U} = \left(\begin{array}{c} \hat{e}_1^T \\ \hat{e}_2^T \\ \hat{e}_3^T \end{array} \right) (\hat{\imath}_1 \ \hat{\imath}_2 \ \hat{\imath}_3)$$

If we denote the matrix of unit column vectors $\vec{e}_i$ as simply $\mathbf{E}$, the matrix of corresponding unit *row* vectors is then $\mathbf{E}^T$. The product $\mathbf{E}^T \mathbf{E}$ is clearly equal to $\mathbf{I}_3$ because its nine elements are just the nine combinations of dot products of the unit vectors $\hat{e}_i$. Interestingly, the product in reverse order, $\mathbf{EE}^T$, is also equal to the identity. More explicitly, the product $\mathbf{EE}^T$ can be evaluated as

$$\mathbf{EE}^T = (\hat{e}_1 \ \hat{e}_2 \ \hat{e}_3) \left(\begin{array}{c} \hat{e}_1^T \\ \hat{e}_2^T \\ \hat{e}_3^T \end{array} \right)$$

$$= \mathbf{P}_1 + \mathbf{P}_2 + \mathbf{P}_3$$

$$= \mathbf{I}_3$$

The sum of the projections, $\mathbf{P}_k = \hat{e}_k \hat{e}_k^T$, must be 1 if the set of vectors $\hat{e}_k$ is a complete set.

This result implies that $\mathbf{U}^T$ is the inverse matrix of $\mathbf{U}$, or

$$\mathbf{UU}^T = \mathbf{E}^T \mathbf{I}_3 \mathbf{I}_3^T \mathbf{E}$$

$$= \mathbf{E}^T \mathbf{E}$$

$$= \mathbf{I}_3$$

Thus, $\mathbf{U}^{-1} = \mathbf{U}^T$. A real matrix whose transpose is equal to its own inverse is called *orthogonal*. Finally, the expressions for an arbitrary vector, $\vec{a}$, in the two coordinate systems are

$$\vec{a} = \mathbf{I}_3 \begin{pmatrix} a_x \\ a_y \\ a_z \end{pmatrix} ; \quad \vec{a} = \mathbf{E} \begin{pmatrix} a_{x'} \\ a_{y'} \\ a_{z'} \end{pmatrix}$$

Using the fact that $\mathbf{E}^T\mathbf{E} = \mathbf{I}_3$, the relation between the components of an arbitrary vector in the two coordinate systems is given by

$$\begin{pmatrix} a_{x'} \\ a_{y'} \\ a_{z'} \end{pmatrix} = \mathbf{E}^T\mathbf{I}_3 \begin{pmatrix} a_x \\ a_y \\ a_z \end{pmatrix} = \mathbf{U} \begin{pmatrix} a_x \\ a_y \\ a_z \end{pmatrix}$$

REFLECTIONS

An orthogonal transformation represented by the matrix $\mathbf{U}$, where $\mathbf{U}^T\mathbf{U} = \mathbf{U}\mathbf{U}^T = \mathbf{I}$, maps vectors $\vec{x}$ into vectors $\vec{x}'$, such that the length of the vector is preserved. That is, $|\vec{x}'| = \sqrt{\vec{x}^T\mathbf{U}^T\mathbf{U}\vec{x}} = |\vec{x}|$. Clearly, any rotation of a vector satisfies this condition, so all rotations can be represented by orthogonal transformations.

However, there are other transformations on a vector that will preserve its length. One is to *reflect* one or more of the components of the vector.[6] That is, $x_i \Rightarrow -x_i$. The coordinate systems (x, y, z) and $(-x, y, z)$ cannot be related by a rotation; therefore, it is apparent that not all orthogonal transformations are represented by rotations. For example, a rotation in the (x, y) plane of the vector $\begin{pmatrix} x \\ y \end{pmatrix}$ by an angle θ is represented as

$$\begin{pmatrix} x' \\ y' \end{pmatrix} = \begin{pmatrix} \cos\theta & \sin\theta \\ -\sin\theta & \cos\theta \end{pmatrix} \begin{pmatrix} x \\ y \end{pmatrix}$$

whereas a rotation plus a reflection is

$$\begin{pmatrix} x' \\ y' \end{pmatrix} = \begin{pmatrix} \cos\theta & \sin\theta \\ \sin\theta & -\cos\theta \end{pmatrix} \begin{pmatrix} x \\ y \end{pmatrix}$$
$$= \begin{pmatrix} 1 & 0 \\ 0 & -1 \end{pmatrix} \begin{pmatrix} \cos\theta & \sin\theta \\ -\sin\theta & \cos\theta \end{pmatrix} \begin{pmatrix} x \\ y \end{pmatrix}$$

This latter transformation represents a rotation followed by a reflection wherein the y' coordinate is reflected. We return to this topic in Chapter 3.

---

[6]A third transformation that preserves the length of a vector is to simply *translate* the vector in space corresponding to a shift of the coordinate axes. However, essentially all the vectors we consider will be characterized by only a *length* and a *direction*, not a *position*, and will therefore be invariant to such a shift.

(§) Gram-Schmidt Orthogonalization

Starting with a given set of n vectors, $\vec{x}_i, i = 1, \ldots, n$, the *space* of these vectors is defined to include *all* vectors that can be expressed as linear combinations of the elements of the set. In general, the vectors $\vec{x}_i$ will not be orthogonal and may not even be linearly independent,[7] so it is useful to instead construct a set of $m \leq n$ mutually orthogonal unit vectors, $\hat{e}_i$, that completely span the same space. These vectors are then said to be a *basis* for this vector space.

The process known as *Gram-Schmidt orthogonalization* is a systematic procedure whereby the orthogonal unit vectors are constructed, one at a time, in the following way:

1. Start by selecting one of the vectors, say, $\vec{x}_1$. Normalize this vector by defining

$$\hat{e}_1 = \vec{x}_1 / \sqrt{\vec{x}_1 \cdot \vec{x}_1}$$

2. Select a second vector, $\vec{x}_2$, and subtract from this vector the part that is parallel to $\hat{e}_1$, i.e.,

$$\vec{e}_2 = (\mathbf{I}_n - \mathbf{P}_1)\vec{x}_2$$

where $\mathbf{I}_n$ is the $n \times n$ identity matrix and $\mathbf{P}_1 = \hat{e}_1 \hat{e}_1^T$ is the projection operator associated with the first unit vector. Thus, $\mathbf{P}_1 \vec{x}_2$ is the component of $\vec{x}_2$ that is parallel to $\hat{e}_1$. The vector $\vec{e}_2$ is then normalized; $\hat{e}_2 = \vec{e}_2 / \sqrt{\vec{e}_2 \cdot \vec{e}_2}$.

3. For the third vector, $\vec{x}_3$, subtract out the components parallel to the unit vectors already obtained,

$$\vec{e}_3 = (\mathbf{I}_n - \mathbf{P}_1 - \mathbf{P}_2)\vec{x}_3$$

and once again normalize: $\hat{e}_3 = \vec{e}_3 / \sqrt{\vec{e}_3 \cdot \vec{e}_3}$. Notice that if $\vec{x}_3$ is simply a linear combination of $\vec{x}_1$ and $\vec{x}_2$, i.e., $\vec{x}_3$ lies in the plane defined by $\vec{x}_1$ and $\vec{x}_2$, then this projection operation will yield a null vector for $\vec{e}_3$, and this vector will not be included in the set of orthonormal basis vectors.

4. Continue the procedure until the entire set of vectors, $\vec{x}_i$, has been exhausted. The number of unit vectors obtained by implementing this process is equal to the *dimensionality* of the space containing the original vectors. Note that if the original vectors are n-component column vectors (or row vectors), the maximum dimensionality of the space of any combination of such vectors is clearly n.

---

[7] A set of n vectors is said to be *linearly dependent* if there exists a set of n scalars, c_i, not all of which are zero, such that

$$c_1 \vec{x}_1 + c_2 \vec{x}_2 + \cdots + c_n \vec{x}_n = \mathbf{0}$$

This is equivalent to saying that at least one of the vectors can be written as a linear combination of the remaining vectors. A set of vectors that is *not* linearly dependent is termed *linearly independent*.

The example that follows is an illustration of the use of MATLAB to find an orthonormal basis set of vectors formed from a nonorthogonal collection of starting vectors.

EXAMPLE 1.1   Let

**MATLAB Exam-
ple of Orthogonal-
ization of a Set of
Vectors**

$$\vec{x}_1 = \begin{pmatrix} 1 \\ 1 \\ 1 \end{pmatrix}, \quad \vec{x}_2 = \begin{pmatrix} 1 \\ 1 \\ 0 \end{pmatrix}, \quad \vec{x}_3 = \begin{pmatrix} 0 \\ 1 \\ 1 \end{pmatrix}$$

The MATLAB code to form an orthonormal basis from these vectors would be

```
MATLAB Script
x1 = [1 1 1]';            %
e1 = x1/sqrt(x1'*x1);     % Normalize the first vector to
                          % unit length.
x2 = [1 1 0]';            %
e2 = (eye(3) - e1*e1')*x2; % This projects out only the part
                          % of x2 that is orthogonal to e1.
                          % Note, eye(3) is the 3x3 identity
                          % matrix.
e2 = e2/sqrt(e2'*e2);     % e2 is then normalized.
                          %
x3 = [ 0 1 1]';           % Finally, construct e3.
e3 = (eye(3) - e1*e1' - e2*e2')*x3;
e3 = e3/sqrt(e3'*e3);     %
E  = [e1 e2 e3];          % E is a matrix whose columns are
                          % orthonormal basis vectors.
```

The resulting matrix of unit vectors is then computed as

$$\mathbf{E} = \begin{pmatrix} \frac{1}{\sqrt{3}} & \frac{1}{\sqrt{6}} & -\frac{1}{\sqrt{2}} \\ \frac{1}{\sqrt{3}} & \frac{1}{\sqrt{6}} & \frac{1}{\sqrt{2}} \\ \frac{1}{\sqrt{3}} & -\frac{2}{\sqrt{6}} & 0 \end{pmatrix}$$

It is a simple matter to check that, to the limits of machine accuracy, this matrix is indeed orthogonal:

$$\text{Check} = \mathbf{E}' * \mathbf{E}$$

$$\text{Check} =$$

```
1.0000   0.0000   0.0000
0.0000   1.0000   0.0000
0.0000   0.0000   1.0000
```

The expansion of an arbitrary vector, $\vec{x}_k$, in terms of this basis set can be written as

$$\vec{x}_k = \sum_{j=1}^{4} a_j^{(k)} \hat{e}_j$$

Multiplying both sides of this equation by $\hat{e}_i^T$ and using the orthonormal properties of the vectors, $\hat{e}_i$, the expansion coefficients are determined to be

$$a_i^{(k)} = \hat{e}_i^T \vec{x}_k$$

Clearly then, the complete set of expansion coefficients for the expansion of the set of vectors $\vec{x}_1, \vec{x}_2$,and $\vec{x}_3$ in terms of the basis set is contained in the matrix

$$\mathbf{A} = \mathbf{E}^T\mathbf{X} = \begin{pmatrix} \frac{3}{\sqrt{3}} & \frac{2}{\sqrt{3}} & \frac{2}{\sqrt{3}} \\ 0 & \frac{2}{\sqrt{6}} & -\frac{1}{\sqrt{6}} \\ 0 & 0 & \frac{1}{\sqrt{2}} \end{pmatrix}$$

Notice that using an orthonormal basis set that is constructed by the Gram-Schmidt method will ensure that the matrix of expansion coefficients is always upper triangular. Thus, the expansion of $\vec{x}_2$ in terms of the $\hat{e}_i$ is

$$\vec{x}_2 = \frac{2}{\sqrt{3}}\hat{e}_1 + \frac{2}{\sqrt{6}}\hat{e}_2 = \begin{pmatrix} 1 \\ 1 \\ 0 \end{pmatrix}$$ ●

WHAT IF? If the starting set of vectors is not linearly independent, the number of basis vectors will be less than the number of original vectors, but the procedure to obtain an expansion of the vectors in terms of the reduced size basis set is very nearly the same. Duplicate the preceding calculation, replacing $\vec{x}_3$ by

$$\vec{x}_3 = \begin{pmatrix} 0 \\ 0 \\ 1 \end{pmatrix}$$ ●

Although the Gram-Schmidt procedure is straightforward, it quickly grows in complexity as the total number of vectors, $\vec{x}_i$, increases. For these situations, you may prefer to use the MATLAB function orth(X), which will compute and return a set of unit vectors[8] that *span* the space of the $\vec{x}_i$'s. Thus, if $\mathbf{X}$ is a matrix whose columns contain the column vectors $\vec{x}_i, i = 1, \ldots, p$ (note that $\mathbf{X}$ need not be square),

$$X = \begin{bmatrix} x1 & x2 & \ldots & xp \end{bmatrix};$$

then the MATLAB statement

$$U = \text{orth}(X);$$

will return a matrix, $\mathbf{U}$, whose columns represent orthogonal unit vectors that span the space of the $\vec{x}_i$'s and that satisfies the equation $\mathbf{U}^T\mathbf{U} = \mathbf{I}_n$, where $n \le p$ is the dimensionality of the space of the vectors, $\vec{x}_i$.

---

[8]Although the MATLAB function orth(X) returns an orthonormal basis set formed from the vectors $\vec{x}_i$, it computes these vectors using a procedure based on *singular value decomposition* (see Chapter 3, page 115) and, as a result, will most often return vectors different from those obtained through a Gram-Schmidt procedure. Both methods return valid basis sets that are related by a rotation; i.e., by an orthogonal transformation matrix.

CROSS CHECK   The matrix of expansion coefficients of the vectors $\vec{x}_i$ in terms of the unit vectors is obtained from the matrix $\mathbf{A} = \mathbf{U}^T\mathbf{X}$. Thus, the expansion of the kth vector in terms of the basis vectors contained in $\mathbf{U}$ is written as

$$\text{xk} = \text{U} * \text{A}(:, \text{k});$$

The product U*A should then duplicate the original matrix X.   ●

For example, starting with the following four three-dimensional vectors:

$$\vec{x}_1 = \begin{pmatrix} 1 \\ 1 \\ 1 \end{pmatrix} \quad \vec{x}_2 = \begin{pmatrix} 1 \\ 0 \\ 1 \end{pmatrix} \quad \vec{x}_3 = \begin{pmatrix} 0 \\ 0 \\ 1 \end{pmatrix} \quad \vec{x}_4 = \begin{pmatrix} 0 \\ 1 \\ 1 \end{pmatrix}$$

the matrix $\mathbf{X}$ is first constructed as X = [x1 x2 x3 x4]. The function U = orth(X) then yields only three unit vectors,

$$\mathbf{U} = \begin{pmatrix} 0.4544 & -0.7071 & -0.5418 \\ 0.4544 & 0.7071 & -0.5418 \\ 0.7662 & 0.0000 & 0.6426 \end{pmatrix} = \left(\begin{pmatrix} \hat{e}_1 \end{pmatrix} \begin{pmatrix} \hat{e}_2 \end{pmatrix} \begin{pmatrix} \hat{e}_3 \end{pmatrix} \right)$$

indicating that the set $\vec{x}_i$ is not independent.

What are the expansion coefficients for the expansion of $\vec{x}_2$ in terms of the $\hat{e}_i$'s? Answer: a2 = U'*x2.

(§) Reciprocal Basis

We will turn next to the general problem of obtaining the multiplicative inverse of an arbitrary matrix in the context of orthogonal unit vectors.

Consider, once again, a set of three-dimensional vectors that are *not* orthogonal and are *not* of unit length, but that are linearly independent, such as

$$\text{x1} = [1 \ 1 \ 1]';$$
$$\text{x2} = [1 \ 1 \ 0]';$$
$$\text{x3} = [0 \ 1 \ 1]';$$
$$\text{X} = [\text{x1} \ \text{x2} \ \text{x3}];$$

The matrix of the nine combinations of dot products is then

$$\text{N} = \text{X}' * \text{X}$$
$$= \begin{pmatrix} 3 & 2 & 2 \\ 2 & 2 & 1 \\ 2 & 1 & 2 \end{pmatrix}$$

Thus, the set is clearly not a set of orthogonal unit vectors.

Is it possible to find a set of *reciprocal* vectors, $\vec{y}$, that are orthogonal to the $\vec{x}$'s? That is, the set of vectors $\vec{y}$ should satisfy the equations

$$\vec{y}_1 \cdot \vec{x}_1 = 1 \quad \vec{y}_1 \cdot \vec{x}_2 = 0 \quad \vec{y}_1 \cdot \vec{x}_3 = 0$$
$$\vec{y}_2 \cdot \vec{x}_1 = 0 \quad \vec{y}_2 \cdot \vec{x}_2 = 1 \quad \vec{y}_2 \cdot \vec{x}_3 = 0$$
$$\vec{y}_3 \cdot \vec{x}_1 = 0 \quad \vec{y}_3 \cdot \vec{x}_2 = 0 \quad \vec{y}_3 \cdot \vec{x}_3 = 1$$

or

$$\vec{y}_i \cdot \vec{x}_j = \vec{y}_i^T \vec{x}_j = \delta_{ij}$$

where δ_{ij} is a *Kronecker* delta function and is defined to be equal to $+1$ if $i = j$, and 0 if $i \neq j$.

This relationship between the two sets of vectors can be written symbolically as

$$\begin{pmatrix} \overline{\overline{y_1}} \\ \overline{\overline{y_2}} \\ y_3 \end{pmatrix} \cdot \left(\begin{pmatrix} x_1 \end{pmatrix} \begin{pmatrix} x_2 \end{pmatrix} \begin{pmatrix} x_3 \end{pmatrix} \right) = \begin{pmatrix} 1 & 0 & 0 \\ 0 & 1 & 0 \\ 0 & 0 & 1 \end{pmatrix}$$

That is, if $\mathbf{X}$ is the matrix of the *column* vectors $\vec{x}$, the matrix, $\mathbf{Y}^T$, of the *row* vectors, $\vec{y}$, is none other than the *inverse* of this matrix.

We begin with a set of n nonorthogonal vectors, $\vec{x}_i, i = 1, \ldots, n$, which we assume are nonetheless linearly independent. The construction, then, of an orthonormal basis set by the Gram-Schmidt orthogonalization procedure will yield n basis vectors, $\hat{e}_i$. The matrix containing these vectors as columns will be designated as $\mathbf{E}$ and will be of size $n \times n$. Using the Gram-Schmidt method guarantees that the matrix of expansion coefficients, $\mathbf{A} = \mathbf{E}^T \mathbf{X}$, is upper triangular. The matrix $\mathbf{X}$ may then be written as

$$\mathbf{X} = \mathbf{EA} = \left(\begin{pmatrix} \hat{e}_1 \end{pmatrix} \begin{pmatrix} \hat{e}_2 \end{pmatrix} \cdots \begin{pmatrix} \hat{e}_n \end{pmatrix} \right) \begin{pmatrix} a_{11} & a_{12} & \cdots & a_{1,n} \\ 0 & a_{22} & \cdots & a_{2,n} \\ \vdots & \vdots & \ddots & \vdots \\ 0 & 0 & \cdots & a_{n,n} \end{pmatrix}$$

However, the vectors, $\vec{y}_i, i = 1, \ldots, n$, may also be expanded in terms of the unit vectors, $\hat{e}_i, i = 1, \ldots, n$, with expansion coefficients contained in the matrix $\mathbf{B}$; that is

$$\mathbf{Y} = \mathbf{EB}$$

Finally, requiring that the set of vectors $\vec{y}_i$ be orthogonal to the set $\vec{x}_j$, we obtain the relation

$$\mathbf{Y}^T \mathbf{X} = \mathbf{B}^T \mathbf{E}^T \mathbf{EA}$$
$$= \mathbf{B}^T \mathbf{A}$$
$$= \begin{pmatrix} b_{11} & b_{21} & \cdots & b_{n1} \\ b_{12} & b_{22} & \cdots & b_{n2} \\ \vdots & \vdots & \ddots & \vdots \\ b_{1n} & b_{2n} & \cdots & b_{nn} \end{pmatrix} \begin{pmatrix} a_{11} & a_{12} & \cdots & a_{1,n} \\ 0 & a_{22} & \cdots & a_{2,n} \\ \vdots & \vdots & \ddots & \vdots \\ 0 & 0 & \cdots & a_{n,n} \end{pmatrix} = \mathbf{I}_n$$

This last expression may then be used to solve for the expansion coefficients of the vectors $\vec{y}_i$ in the following way:

$$b_{11}a_{11} \qquad\qquad = 1 \Rightarrow b_{11} = \frac{1}{a_{11}}$$

$$a_{12}/a_{11} + b_{21}a_{22} = 0 \Rightarrow b_{21} = -\frac{a_{12}}{a_{11}a_{22}}$$

$$\vdots \qquad\qquad\qquad \vdots$$

etc.

If the number of vectors, n, is large, this procedure will be quite inefficient and will be supplanted in later chapters by more efficient MATLAB techniques for obtaining the inverse of a matrix.

EXAMPLE 1.2

Example Calculation of a Reciprocal or Dual Basis

We begin with the nonorthogonal set of vectors used earlier and construct an orthonormal basis set using the Gram-Schmidt orthogonalization procedure

```
x1= [1 1 1]';      x2= [1 1 0]';      x3= [0 1 1]';
X = [x1 x2 x3];
e1 = x1/norm(x1);
e2 = (eye(3) - e1*e1')*x2;           e2 = e2/norm(e2);
e3 = (eye(3) - e1*e1' - e2*e2')*x3; e3 = e3/norm(e3);
E = [e1 e2 e3];
A = E'*X;
```

The orthogonal unit vectors computed by this means are, as before,

$$\hat{e}_1 = \frac{1}{\sqrt{3}}\begin{pmatrix} 1 \\ 1 \\ 1 \end{pmatrix} \qquad \hat{e}_2 = \frac{1}{\sqrt{6}}\begin{pmatrix} 1 \\ 1 \\ -2 \end{pmatrix} \qquad \hat{e}_3 = \frac{1}{\sqrt{2}}\begin{pmatrix} -1 \\ 1 \\ 0 \end{pmatrix}$$

and

$$\mathbf{A} = \mathbf{E}^T \cdot X$$

$$= \frac{1}{\sqrt{6}}\begin{pmatrix} 3\sqrt{2} & 2\sqrt{2} & 2\sqrt{2} \\ 0 & 2 & -1 \\ 0 & 0 & \sqrt{3} \end{pmatrix}.$$

which, you will notice, is upper triangular.

Thus, the set of vectors, $\vec{x}_i$, written as columns of the matrix $\mathbf{X}$ can be written in terms of the unit vectors, $\hat{e}_i$, written as columns of the matrix $\mathbf{E}$, or

$$\mathbf{X} = \mathbf{EA}$$

The expansion of the complementary vectors $\vec{y}_i$ that are to be orthogonal to the $\vec{x}$'s is then given by $\mathbf{Y} = \mathbf{EB}$, where the matrix $\mathbf{B}$ must satisfy the equation

$$\frac{1}{\sqrt{6}} \begin{pmatrix} b_{11} & b_{21} & b_{31} \\ b_{12} & b_{22} & b_{32} \\ b_{13} & b_{23} & b_{33} \end{pmatrix} \begin{pmatrix} 3\sqrt{2} & 2\sqrt{2} & 2\sqrt{2} \\ 0 & 2 & -1 \\ 0 & 0 & \sqrt{3} \end{pmatrix} = \mathbf{I}_3$$

which can be solved algebraically for the elements of the matrix $\mathbf{B}$, yielding

$$\mathbf{B} = \begin{pmatrix} \frac{1}{\sqrt{3}} & 0 & 0 \\ -\sqrt{\frac{2}{3}} & \sqrt{\frac{3}{2}} & 0 \\ -\sqrt{2} & \frac{1}{\sqrt{2}} & \sqrt{2} \end{pmatrix}$$

Finally, the matrix of complementary vectors, $\vec{y}_i$, is obtained from

$$\mathbf{Y} = \mathbf{EB}$$

$$= \begin{pmatrix} \frac{1}{\sqrt{3}} & \frac{1}{\sqrt{6}} & -\frac{1}{\sqrt{2}} \\ \frac{1}{\sqrt{3}} & \frac{1}{\sqrt{6}} & \frac{1}{\sqrt{2}} \\ \frac{1}{\sqrt{3}} & -\frac{2}{\sqrt{6}} & 0 \end{pmatrix} \begin{pmatrix} \frac{1}{\sqrt{3}} & 0 & 0 \\ -\sqrt{\frac{2}{3}} & \sqrt{\frac{3}{2}} & 0 \\ -\sqrt{2} & \frac{1}{\sqrt{2}} & \sqrt{2} \end{pmatrix}$$

$$= \begin{pmatrix} 1 & 0 & -1 \\ -1 & 1 & 1 \\ 1 & -1 & 0 \end{pmatrix}$$

It is then a trivial matter to verify that $\mathbf{Y}^T\mathbf{X} = \mathbf{I}_3$. ●

Problems

REINFORCEMENT EXERCISES

P1.1 The Gauss Elimination Method of Solving Simultaneous Equations. Solve the following sets of equations using the methods of this section:

a. $\quad x + 2y = 1$
$\quad\quad 2x \quad\;\; = 1$

b. $\quad 2x_1 - \;\; x_2 \quad\quad\;\; = 1$
$\quad\quad -x_1 + 2x_2 - \;\; x_3 = 1$
$\quad\quad\quad\;\; - \;\; x_2 + 2x_3 = 1$

c. $\quad 2x_1 - \;\; x_2 \quad\quad\quad\quad\;\; = 1$
$\quad\quad -x_1 + 2x_2 - \;\; x_3 \quad\quad\;\; = 1$
$\quad\quad\quad\;\; - \;\; x_2 + 2x_3 - \;\; x_4 = 1$
$\quad\quad\quad\quad\quad\quad\; - \;\; x_3 + 2x_4 = 1$

d. $\quad 2x_1 + \;\; x_2 + \;\; x_3 = 1$
$\quad\quad -x_1 + 2x_2 + \;\; x_3 = 1$
$\quad\quad -x_1 - \;\; x_2 + 2x_3 = 1$

e.
$$x_2 + x_3 = 0$$
$$-x_1 \quad\quad + x_3 = 1$$
$$-x_1 - x_2 \quad\quad = 1$$

f.
$$2x_1 + x_2 + 2x_3 = 1$$
$$2x_1 + x_2 + 4x_3 = 2$$
$$x_1 - x_2 \quad\quad = 3$$

P1.2 Dependent Equations.   Attempt to solve the following four equations in three unknowns by the Gaussian elimination method:

$$2x_1 + \quad x_2 + \quad x_3 = 3$$
$$x_1 \quad\quad + \quad x_3 = 1$$
$$-x_1 + 2x_2 + 2x_3 = 2$$
$$x_2 - \quad x_3 = 1$$

Show that a unique solution is possible. You should then conclude that one of the equations is unnecessary and is simply a linear combination of the other equations. Thus, show that the last equation is a linear combination of two other equations.

P1.3 Solutions in Terms of Parameters.   Use the Gaussian elimination method to solve the following three equations in five unknowns:

$$x_1 + 2x_2 + \quad x_3 + x_4 + \quad x_5 = 4$$
$$2x_1 + \quad x_2 + 2x_3 + x_4 \quad\quad = 1$$
$$x_1 + 2x_2 + \quad x_3 \quad\quad + 2x_5 = 0$$

Show that the solution can be expressed in terms of two free parameters, α and β, as

$$x_1 = \alpha - 2 \quad\quad x_3 = -\alpha \quad\quad x_5 = -\beta$$
$$x_2 = \beta + 1 \quad\quad x_4 = 4 - \beta$$

P1.4 Inconsistent Equations.   In the following set of three equations in three unknowns, c is a fixed constant:

$$2x_1 + \quad x_2 + 2x_3 = c$$
$$x_1 + 2x_2 + \quad x_3 = 1$$
$$2x_1 + \quad x_2 + 2x_3 = 0$$

a. The first equation would be inconsistent with the third if $c \neq 0$, and we expect the Gaussian elimination method to confirm this. Carry out the method and show that a solution is possible *only* if $c = 0$.

b. With $c = 0$, show that there are an infinite number of solutions possible and that they can be expressed in terms of the parameter γ as

$$x_1 = \gamma - \frac{1}{3}$$

$$x_2 = \frac{2}{3}$$

$$x_3 = -\gamma$$

P1.5 Matrix Multiplication. True or False?

a. The product of two symmetric matrices is a symmetric matrix.

b. The null matrix, $\mathbf{O}_n$, commutes with all square $n \times n$ matrices.

c. Two diagonal matrices always commute.

d. An upper-triangular matrix times an upper-triangular matrix is an upper-triangular matrix.

e. The product of an upper-triangular matrix and a lower-triangular matrix is always a symmetric matrix.

f. $(\mathbf{A} - \mathbf{B})(\mathbf{A} + \mathbf{B}) = \mathbf{A}^2 - \mathbf{B}^2$.

P1.6 Multiplication of Rectangular Matrices. Consider all possible combinations of products of three matrices constructed from the following matrices and their transposes:

$$
\mathbf{A} = \begin{pmatrix} \square & \square \\ \square & \square \\ \square & \square \end{pmatrix}, \quad
\mathbf{B} = \begin{pmatrix} \square & \square & \square & \square \\ \square & \square & \square & \square \end{pmatrix}, \quad
\mathbf{C} = \begin{pmatrix} \square & \square \\ \square & \square \\ \square & \square \\ \square & \square \end{pmatrix}
$$

Demonstrate that there are only one 2×2, one 3×3, and one 4×4 (and their transposes) that can be constructed from three elements of this set.

P1.7 Anticommuting Matrices. If $\mathbf{AB} = -\mathbf{BA}$, the matrices $\mathbf{A}$ and $\mathbf{B}$ are said to anticommute.

a. Find a real, nondiagonal matrix that anticommutes with

$$
\mathbf{A} = \begin{pmatrix} 0 & 1 \\ 1 & 0 \end{pmatrix}
$$

b. Show that the set of matrices (called the *Pauli matrices*)

$$
\sigma_1 = \begin{pmatrix} 0 & 1 \\ 1 & 0 \end{pmatrix}, \quad
\sigma_2 = \begin{pmatrix} 0 & -i \\ i & 0 \end{pmatrix}, \quad
\sigma_3 = \begin{pmatrix} 1 & 0 \\ 0 & -1 \end{pmatrix}
$$

satisfy the following relations:

$$
\sigma_i^2 = \mathbf{I}_2; \quad \sigma_1 \sigma_2 = -\sigma_2 \sigma_1
$$

plus all cyclic permutations.

P1.8 Diagonal Matrices and Matrix Diagonals.

a. Find a matrix $\mathbf{B}$ that satisfies the equation $\mathbf{BA} = \mathbf{AB} = \mathbf{I}_3$, where

$$
\mathbf{A} = \begin{pmatrix} a_{11} & 0 & 0 \\ 0 & a_{22} & 0 \\ 0 & 0 & a_{33} \end{pmatrix}
$$

Assume that the elements a_{ii} are nonzero. The matrix $\mathbf{B}$ is then the multiplicative inverse of $\mathbf{A}$.

b. The *trace* of a matrix is defined as the sum of the diagonal elements of the matrix $[\mathrm{Tr}(\mathbf{A}) = \sum a_{ii}]$. Show that for arbitrary square matrices of the same size, $\mathrm{Tr}(\mathbf{AB}) = \mathrm{Tr}(\mathbf{BA})$, and that $\mathrm{Tr}(\mathbf{ABC}) = \mathrm{Tr}(\mathbf{CAB}) = \mathrm{Tr}(\mathbf{BCA})$.

c. If $\mathbf{A}$ and $\mathbf{B}$ are square matrices of the same size, prove that if $\mathbf{A}$ is symmetric and $\mathbf{B}$ is antisymmetric, then $\mathrm{Tr}(\mathbf{AB}) = 0$.

P1.9 Nilpotent Matrices. If $\mathbf{A}^p = \mathbf{O}$ for some positive integer power p, the matrix $\mathbf{A}$ is said to be *nilpotent*. Show that the matrix

$$\mathbf{A} = \begin{pmatrix} 0 & 1 & 1 & 1 \\ 0 & 0 & 1 & 1 \\ 0 & 0 & 0 & 1 \\ 0 & 0 & 0 & 0 \end{pmatrix}$$

is nilpotent with $p = 4$. Use this fact to explicitly evaluate $e^{\mathbf{A}}$ from the series expansion for e^x; i.e., $e^x \equiv 1 + x + x^2/2! + \cdots + x^n/n! + \cdots$.

P1.10 Properties of the Transpose. Prove each of the following:

a. The transpose operation is *linear*; i.e.,

$$(\mathbf{A} + \mathbf{B})^T = \mathbf{A}^T + \mathbf{B}^T$$

b. $(c\mathbf{A})^T = c\mathbf{A}^T$, where c is a scalar.

c. The products of a matrix and its transpose, in either order, always result in a square, symmetric matrix.

d. Show that the combinations $\mathbf{A} + \mathbf{A}^T$ and $\mathbf{A} - \mathbf{A}^T$ are, respectively, symmetric and antisymmetric for any square matrix $\mathbf{A}$. Use this to demonstrate that any matrix can be written as the sum of a symmetric plus an antisymmetric matrix. Give an example.

P1.11 Markov Matrices.

a. It was stated that for each Markov matrix there is a corresponding *equilibrium state* that is reached by applying the Markov transition matrix over and over again. That is,

$$\vec{x}_{\mathrm{eq}} \equiv \lim_{n \to \infty} \mathbf{M}^n \vec{x}_0$$

where $\vec{x}_0$ is an arbitrary starting vector. Show that the vector $\vec{x}_{\mathrm{eq}}$ satisfies the equation $\mathbf{M}\vec{x}_{\mathrm{eq}} = \vec{x}_{\mathrm{eq}}$.

b. Suppose that there are two distinct Markov matrices, $\mathbf{M}_1$ and $\mathbf{M}_2$, summarizing the transition probabilities between the same set of states. Prove that the product $\mathbf{M}_1\mathbf{M}_2$ is also a Markov matrix. Illustrate that this is true using the following Markov matrices:

$$\mathbf{M}_1 = \begin{pmatrix} 0.1 & 0.7 & 0.3 \\ 0.3 & 0.2 & 0.3 \\ 0.6 & 0.1 & 0.4 \end{pmatrix} \qquad \mathbf{M}_2 = \begin{pmatrix} 0.8 & 0.1 & 0 \\ 0.1 & 0.8 & 0 \\ 0.1 & 0.1 & 1 \end{pmatrix}$$

P1.12 Markov Matrices and Equilibrium University Populations.

    a. A university has three undergraduate colleges, (a) Arts and Sciences, (b) Business, and (c) Engineering. Each year there are changes in the student populations according to the following patterns (in percentages):

From

To	(a)	(b)	(c)
(a)	70	8	20
(b)	18	75	15
(c)	12	17	65

    Assuming initial populations of 5000, 3000, 2000 in the three colleges, determine the equilibrium populations.

    b. In a more realistic model, each year some students leave and new students enter the university, so that the population of individuals is no longer static. However, assuming the total net population of the university remains constant, this possibility can be accommodated by adding a fourth class, Outside, to our transition matrix. Assume that each year, 2500 leave the university to be replaced by new students. Also, take the percentage leaving to be the same for each college, but the fractions admitted each year are in proportions, 55%, 25%, and 20%, respectively, for colleges (a), (b), and (c). Again determine the equilibrium populations.

P1.13 Operations with Vectors. Use the following vectors:

$$\vec{a} = \begin{pmatrix} 1 \\ 2 \\ 3 \end{pmatrix} \qquad \vec{b} = \begin{pmatrix} 4 \\ -3 \\ 2 \end{pmatrix} \qquad \vec{c} = \begin{pmatrix} 0 \\ -1 \\ 0 \end{pmatrix}$$

    a. Evaluate $\vec{a}^T \vec{b}$, $\vec{a}^T \vec{c}$, $\vec{b}^T \vec{c}$, and the norms of each vector.

    b. Determine the angles (in degrees) between any two vectors.

P1.14 Rotations in the Plane. In Cartesian coordinates, a vector is expressed in terms of the $x, y,$ and z axes by means of the unit vectors, $\hat{i}, \hat{j},$ and $\hat{k}$ along these axes.

    a. Show that in a coordinate system (x', y', z), obtained from (x, y, z) by a rotation about z by an angle θ, the components of a vector $\vec{a}$ in the two coordinate systems are related by $a_i' = \sum_{j=1}^{3} U_{ij} a_j$, where $(a_1, a_2, a_3) \equiv (a_x, a_y, a_z)$, $(a_1', a_2', a_3') \equiv (a_{x'}, a_{y'}, a_{z'})$ and

$$\mathbf{U} = \begin{pmatrix} \cos\theta & \sin\theta & 0 \\ -\sin\theta & \cos\theta & 0 \\ 0 & 0 & 1 \end{pmatrix}$$

b. What is the matrix, **V**, that will rotate the vector back to its original position?

c. Let $\theta = 2\pi/n$ and use MATLAB to show that n successive rotations bring the vector back to the original position. (Try $n = 4, 8, 16$.)

P1.15 Invariance of the Dot Product.   The vectors $\vec{a}'$ and $\vec{b}'$ are related to the vectors $\vec{a}$ and $\vec{b}$ by a rotation. Show that the value of the dot product remains unchanged by a rotation; i.e., $\vec{a}'^T \vec{b}' = \vec{a}^T \vec{b}$.

P1.16 Gram-Schmidt Orthogonalization.   Apply the Gram-Schmidt orthogonalization method to find a set of orthogonal unit vectors starting with the four vectors $x_1 = [1\ 0\ 0\ 0]^T$, $x_2 = [1\ 1\ 0\ 0]^T$, $x_3 = [1\ 1\ 1\ 0]^T$, $x_4 = [1\ 1\ 1\ 1]^T$. Can you generalize this result to n similar vectors? Does the MATLAB function `ortho([x1 x2 x3 x4])` give the same result?

P1.17 Properties of Projection Operators.

a. Show that if $\mathbf{P}_1$ and $\mathbf{P}_2$ are projection operators, then $\mathbf{P}_1 + \mathbf{P}_2$ is a projection operator only if $(\mathbf{P}_1\mathbf{P}_2 + \mathbf{P}_2\mathbf{P}_1) = \mathbf{O}$.

b. Show that if $\mathbf{P}$ is a projection operator, then $\mathbf{I} - \mathbf{P}$ is also a projection operator.

c. Prove that the trace of a projection matrix must be zero. [*Hint*: If the projection operator is written as $\mathbf{P} = \hat{a}\hat{a}^T$, where $\hat{a}$ is some unit vector, then in a basis specified by the orthonormal unit vectors, $\hat{e}_i$, $i = 1, \ldots, n$, the trace would be written $\text{Tr}(\mathbf{P}) = \sum_{i=1}^{n} \hat{e}_i^T \mathbf{P} \hat{e}_i$.]

P1.18 Projection Operators.   Given the four column vectors $\vec{x}_i, i = 1, \ldots, 4$,

$$\vec{x}_1 = \begin{pmatrix} 1 \\ 1 \\ 1 \\ 1 \end{pmatrix}, \quad \vec{x}_2 = \begin{pmatrix} 1 \\ -1 \\ 0 \\ 0 \end{pmatrix}, \quad \vec{x}_3 = \begin{pmatrix} 0 \\ 0 \\ 1 \\ -1 \end{pmatrix}, \quad \text{and} \quad \vec{x}_4 = \begin{pmatrix} 1 \\ 1 \\ -1 \\ -1 \end{pmatrix}$$

a. Determine the normalization of each vector and thereby construct four unit vectors, $\hat{e}_i$, $i = 1, \ldots, 4$. Using the unit vectors, construct four projection matrices, $\mathbf{P}_i$, and show that $\mathbf{P}_i^2 = \mathbf{P}_i$.

b. Show that the set of four unit vectors is *orthonormal* and *complete*.

c. Expand the vector $\vec{y}^T = [1\ 2\ 3\ 4]$ in terms of the $\hat{e}_i$'s.

P1.19 Perpendicular Components of a Vector.   Start with a projection operator, $\mathbf{P} = \hat{e}\hat{e}^T$, where $\hat{e}^T = (1, -1, 2)/\sqrt{6}$, and a vector, $\vec{a}$, given by

$$\vec{a} = \begin{pmatrix} 4 \\ -2 \\ 1 \end{pmatrix}$$

a. Show that $\mathbf{P}\vec{a}$ is a vector proportional to $\hat{e}$ and find the proportionality constant.

b. Find the component of $\vec{a}$ that is orthogonal to $\hat{e}$.

P1.20 Complementary Vectors. Find a set of four vectors, $\vec{y}_i$, $i = 1, \ldots, 4$ that is complementary to the following set of vectors:

$$\vec{x}_1 = \begin{pmatrix} 1 \\ 1 \\ 1 \\ 1 \end{pmatrix}, \quad \vec{x}_2 = \begin{pmatrix} 1 \\ -2 \\ 2 \\ -1 \end{pmatrix}, \quad \vec{x}_3 = \begin{pmatrix} 2 \\ 1 \\ -1 \\ -2 \end{pmatrix}, \quad \vec{x}_4 = \begin{pmatrix} 1 \\ -2 \\ -1 \\ 2 \end{pmatrix}$$

P1.21 Transformation of a Nilpotent Matrix. If $\mathbf{A}$ is a nilpotent matrix; i.e., $\mathbf{A}^P = \mathbf{O}$ for some positive integer p, prove that $\mathbf{A}' = \mathbf{U}^T \mathbf{A} \mathbf{U}$, where $\mathbf{U}$ is an orthogonal matrix, is also nilpotent for the same value of p.

EXPLORATION PROBLEMS

P1.22 Functions of a Matrix. The series expansion for the exponential function

$$e^x = \sum_{n=0}^{\infty} \frac{x^n}{n!}$$

remains valid even if x is replaced by a square matrix, $\mathbf{A}$. For example, show that if $\mathbf{P} = \begin{pmatrix} 0 & 1 \\ 1 & 0 \end{pmatrix}$, then

$$e^{i\theta \mathbf{P}} = \mathbf{I}_2 \cos\theta + i\mathbf{P} \sin\theta$$

You will need the series expansions for $\sin x$ and $\cos x$,

$$\sin x = \sum_{n=0}^{\infty} (-1)^n \frac{x^{2n+1}}{(2n+1)!}$$

$$\cos x = \sum_{n=0}^{\infty} (-1)^n \frac{x^{2n}}{(2n)!}$$

and the fact that $i^{2n} = (-1)^n$.

P1.23 The Norm of a Vector. The length of a vector, defined in terms of the dot product, is also called the *norm*, $|\vec{x}|$, of the vector $\vec{x}$. The mathematical requirements for the norm of a vector $\vec{x}$ are:

- $|c\vec{x}| = |c||\vec{x}|$, where c is a scalar and $|c|$ is its magnitude.
- $|\vec{x}| > 0$ for all $\vec{x} \neq 0$ and $|\vec{x}| = 0$ if $\vec{x} = 0$.
- $|\vec{x} + \vec{y}| \leq |\vec{x}| + |\vec{y}|$.

    a. Show that the definition used for the norm, $|\vec{x}| = \sqrt{\vec{x} \cdot \vec{x}}$, satisfies these conditions.

    b. However, other possible definitions for norms will also satisfy these conditions. Show that each of the following definitions also satisfies the mathematical definition of a norm:

$$\text{1-norm: } |\vec{x}| \equiv \sum_{i=1}^{n} |x_i|$$

$$\text{Infinity norm: } |\vec{x}| \equiv \max |x_i| = \text{absolute maximum element of } \vec{x}$$

The norm of a *matrix* can also be defined and is used to get an idea of the magnitude of the matrix. We return to this in Chapter 2.

P1.24 Kirchhoff's Laws for DC Circuits.   If a current i flows through a resistor, the voltage drop across the resistor in the direction of the current is $V = iR$. (When traversing the resistor in a direction opposite to that of the current, the voltage will then *increase* by an amount equal to iR, corresponding to a *negative* voltage drop.) This leads to the first Kirchhoff law:

1. The sum of voltage drops around any closed circuit loop must equal zero.

Next, a battery whose voltage is $\mathcal{E}$ in a circuit will give a *boost* to the voltage if we step across the battery from its *minus* to its *plus* side. Thus, in the equation for the circuit shown in Figure 1.6, $V_0 - iR = 0$.

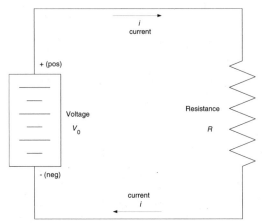

Figure 1.6   *Around every closed loop:*

$$\underset{\substack{\text{Voltage} \\ \text{boosts}}}{\sum \mathcal{E}} - \underset{\substack{\text{Voltage} \\ \text{drops}}}{\sum iR} = 0.$$

The next law is a statement of conservation of charge, namely, the current flow into a junction must equal the current flow out of that junction.

2. At a junction of three or more wires, $\sum i_{\text{in}} = \sum i_{\text{out}}$.

The prescription is then:

a. Assign a current (including direction) along each current element in the circuit. Usually, the unknowns are then the n current values i_k, $k = 1, \ldots, n$.

b. At each junction; i.e., a point where three or more currents meet, use the second rule to obtain a relation between the unknown currents. Of course, not all junctions will yield independent equations. For example, in the circuit elements shown in Figure 1.7, the equation for junction a is $i_1 = i_2 + i_3$, and at junction b, it is $i_2 + i_3 = i_1$.

c. After obtaining m independent junction equations, choose $n - m$ independent closed loops within the circuit and equate the sum of the current drops and rises around each loop to zero.

You now have n equations in the n unknown i's. These are then solved using the Gaussian elimination method. Apply this procedure to obtain the currents in the circuits of Figures 1.7 and 1.8.

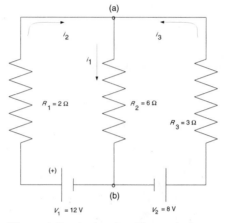

Figure 1.7 *A circuit with two independent loops.*

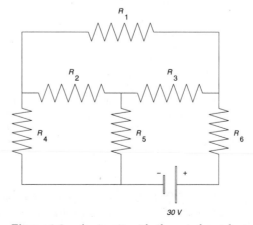

Figure 1.8 *A circuit with three independent loops. Use* $R_1 = 12\,\Omega$, $R_2 = 8\,\Omega$, $R_3 = R_4 = R_5 = 4\,\Omega$.

***P1.25* Statics Problems.** The requirement that a rigid object remain stationary can be expressed in terms of three equations:

  a. $\sum F_x = 0$. The sum of the x *components* of all outside forces acting on the object must be zero.

  b. $\sum F_y = 0$. The sum of the y *components* of all outside forces acting on the object must be zero.

  c. $\sum \tau_0 = 0$. The sum of all *torques* on the object caused by the outside forces must be zero.[9]

In Figures 1.9 and 1.10, use the rules for static equilibrium to obtain a sufficient number of independent equations to define a solution. Solve these equations using the Gaussian elimination method.

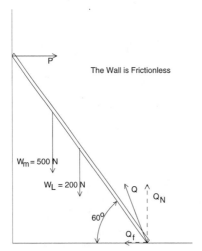

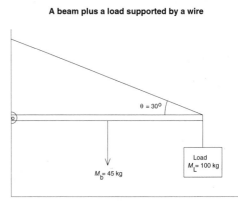

Figure 1.9 *Man on a ladder. With the man positioned 60% of the way up the ladder and with $Q_f = \frac{1}{5}Q_N$, solve for the value of the force, P, of the wall on the ladder.*

Figure 1.10 *Cable, beam, and a load. The beam can rotate freely about the hinge on the left. The force of the cable on the beam is along the cable. Determine this force.*

***P1.26* Indeterminate Statics Problem.** In Figure 1.11, if the wall is *not* frictionless, the direction of the force of the wall on the ladder is no longer perpendicular to the wall.

---

[9]A torque, τ, can be computed about any convenient point, P_0, according to the following prescription: $\tau = |\vec{F}|d_\perp$, where $|\vec{F}|$ is the magnitude of the applied force, and $d_\perp$ is the *perpendicular distance* from the *line of action* of the vector $\vec{F}$ and the point P_0. If the torque tends to twist the object in a *counterclockwise* direction, it is *positive*; a clockwise torque is *negative*.

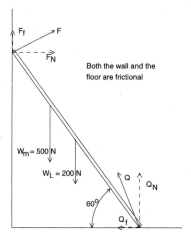

Figure 1.11 *Man on a ladder, frictional wall and floor.*

The man, as in the previous problem, is 60% of the way up the ladder, and once again there are three equations, but in this case there are *four* unknowns. Define the ratio $\mu = \mathbf{F}_y^{[\text{wall}]}/\mathbf{F}_x^{[\text{wall}]}$, and solve the problem as a function of μ.

P1.27 Markov Processes with Rewards. A company has three levels of employees, president, office manager, and file clerk, and the employees are continually being promoted or demoted from one level to another as a result of their performances. The probability that an employee currently at level j will be at level i by next year is p_{ij}. Suppose the matrix of possible transitions is given by

$$\mathbf{P} = \begin{pmatrix} \frac{2}{3} & \frac{1}{6} & \frac{1}{6} \\ \frac{1}{6} & \frac{2}{3} & \frac{1}{6} \\ \frac{1}{6} & \frac{1}{6} & \frac{2}{3} \end{pmatrix}$$

Next, suppose that there are financial consequences to each of these transitions. That is, a promotion from level 1 to level 2 might result in a bonus of 5 (hundred dollars), whereas a demotion from level 3 to level 2 could result in a pay cut of 8 (hundred dollars). Thus, a matrix of *rewards* is associated with the transitions. This matrix could be given by

$$\mathbf{R} = \begin{pmatrix} 1 & -4 & -12 \\ 5 & 1 & -3 \\ 15 & 8 & 1 \end{pmatrix}$$

with the units in hundreds of dollars. If an individual starts at level i, the average financial reward after one year is $g_i = \sum_{j=1}^{3} p_{ij} r_{ij}$; that is, the financial reward associated with a transition from level i to level j times the probability of such a transition. Note that g is a vector and it can also be written as $\vec{g} = \text{diag}(\mathbf{PR}^T)$.

Let $x_i(1) = g_i$ be the net financial gain after one year for an employee starting at level i. Then the gain after 2 years, $x_i(2)$, would be

$$x_i(2) = \sum_{j=1}^{3} p_{ij} x_j(1) + x_i(1)$$

or in matrix notation

$$\vec{x}(2) = \mathbf{P}\vec{x}(1) + \vec{x}(1)$$

which can be iterated for successive years to yield

$$\vec{x}(n) = \mathbf{P}\vec{x}(n-1) + \vec{x}(n-1)$$
$$= (\mathbf{P}^n + \mathbf{P}^{n-1} + \cdots + \mathbf{P} + 1)\vec{g}$$

Construct matrices $\mathbf{P}$ and $\mathbf{R}$ in MATLAB and

    a. Compute $\vec{g}, \vec{x}(5)$, and $\vec{x}(10)$, thereby determining whether it is better to start out as file clerk or as president if you intend to quit in 5 or 10 years.

    b. Compute $\mathbf{P}^n$ for $n = 10, 20$ and determine the limiting distribution of employees at each level.

P1.28 Parametric Functions. A curve in three dimensions is usually given *parametrically* as $(x, y, z) = (x(t), y(t), z(t))$, implying that each value of t uniquely determines values for (x, y, z), thereby tracing out a curve for a range of t values. For example, if the relations between (x, y, z) and t are all linear, the resulting curve is a straight line in three-dimensional space. If this straight line is specified by the relations

$$x(t) = 4t - 1$$
$$y(t) = -t + 1$$
$$z(t) = 2 - 2t$$

the vector $\vec{r}$ from the origin to the point (x, y, z) could be written as

$$\vec{r} = \vec{d} + \vec{v}\, t = \begin{pmatrix} -1 \\ 1 \\ 2 \end{pmatrix} + \begin{pmatrix} 4 \\ -1 \\ -2 \end{pmatrix} t$$

where $\vec{d}$ represents a constant *displacement* of the line from the origin, and $\vec{v}$ is a vector along the direction of the line. Normalizing the latter, we can say that the line is parallel to the vector

$$\hat{e} = \frac{1}{\sqrt{21}} \begin{pmatrix} 4 \\ -1 \\ -2 \end{pmatrix}$$

a. Determine the *length* of the component of $\vec{a} = \begin{pmatrix} 3 \\ -2 \\ 2 \end{pmatrix}$ that is parallel to the given line.

b. Determine the *minimum* distance from the line to the origin.

P1.29 Reciprocal Lattice Vectors. A crystalline material is characterized by translational periodicity in three separate directions, which is described by means of three *primitive* lattice vectors, $\vec{a}_1, \vec{a}_2$, and $\vec{a}_3$, whose lengths and directions correspond to the smallest repeat distances in the crystal. In general, the three vectors are not of the same length and are not orthogonal. In defining functions over the space of the crystal, it is convenient to express quantities in terms of a *reciprocal* basis, $\vec{b}_1, \vec{b}_2$, and $\vec{b}_3$, defined to be orthogonal to the set $\vec{a}_i$; i.e., $\vec{b}_i \cdot \vec{a}_j = \delta_{ij}$. The fundamental requirement on all functions defined over the space of the crystal is

$$V(\vec{r}) = V(\vec{r} + \vec{\delta})$$

where $\vec{\delta}$ is a displacement in the crystal by an arbitrary (integer) number of primitive repeat distances $\vec{a}_i$; i.e.,

$$\vec{\delta} = n_1\vec{a}_1 + n_2\vec{a}_2 + n_3\vec{a}_3$$
$$= \mathbf{A}\vec{n}$$

the elements of the vector $\vec{n}$ are integers, and the matrix $\mathbf{A}$ contains the vectors $\vec{a}_i$ as columns. A corresponding *reciprocal lattice vector* is defined as

$$\vec{k} = m_1\vec{b}_1 + m_2\vec{b}_2 + m_3\vec{b}_3$$
$$= \mathbf{B}\vec{m}$$

where $\vec{m}$ is also a vector with integer components.

The function $V(\vec{r})$ can be expanded as the following three-dimensional Fourier series

$$V(\vec{r}) = \sum_{\vec{n}} V_{\vec{n}} \exp(2\pi i\vec{k} \cdot \vec{r})$$

where the summation is over all integer values of n_1, n_2, and n_3, and the symbol $V_{\vec{n}}$ stands for V_{n_1,n_2,n_3}. Show that this function satisfies the basic requirement of periodicity.

Summary

CHARACTERISTICS OF EQUATIONS IN SEVERAL VARIABLES

Linear

An equation is *linear* in x if it contains x to only the first power.

Homogeneous

An equation is *homogeneous* in the variables if each term contains one or more of the variables to nonzero powers. Equations with constant terms are then *inhomogeneous*.

Coupled

Two or more equations are *coupled* if each cannot be solved separately from the remaining equations.

Consistent   Two or more equations are consistent if they have a nonempty solution set.

Dependent   Two equations are *dependent* if one can be expressed as a nonzero multiple of the other. Equations that are not dependent are *independent*.

Transcendental   A transcendental equation, if expressed only in terms of elementary arithmetic operations, requires an *infinite* series of terms.

SPECIAL MATRIX CHARACTERISTICS

Square Matrix   The number of rows equals the number of columns.

Column Vector   A matrix is a column vector if it has only one column.

Row Vector   A matrix is a row vector if it has only one row.

Diagonal Matrix   The elements a_{ij} of the square matrix $\mathbf{A}$ are zero if $i \neq j$.

Banded Matrix   The elements a_{ij} of the square matrix $\mathbf{A}$ are zero except for the principal diagonal, a_{ii}, and a few nearby parallel diagonals.

Triangular Matrix   A matrix is upper triangular if $a_{ij} = 0$ for all $i > j$, and lower triangular if $a_{ij} = 0$ for all $i < j$.

Symmetric Matrix   A square matrix is symmetric if $a_{ij} = a_{ji}$ for all i, j.

Identity Matrix   A square, $n \times n$ matrix with 1s on the principal diagonal and 0s elsewhere is called the nth-degree identity, $\mathbf{I}_n$.

Commuting Matrices   Two matrices, $\mathbf{A}$ and $\mathbf{B}$, commute if $\mathbf{AB} = \mathbf{BA}$.

Matrix Transpose   The transpose of $\mathbf{A}$, designated by $\mathbf{A}^T$ or $\mathbf{A}'$, corresponds to interchanging the rows and columns of $\mathbf{A}$, i.e., $[\mathbf{A}^T]_{ij} = [\mathbf{A}]_{ji}$.

MATRIX ALGEBRA

Equality   For the statement $\mathbf{A} = \mathbf{B}$ to be valid, both matrices $\mathbf{A}$ and $\mathbf{B}$ must be of the same size and all their corresponding elements must be equal; i.e., $a_{ij} = b_{ij}$ for all i, j.

Multiplication   The multiplication of two matrices, $\mathbf{AB}$, can be defined, provided the number of *columns* of the *first*, $\mathbf{A}$, is the same as the number of *rows* of the *second*, $\mathbf{B}$. The product is then a matrix, $\mathbf{C}$, whose elements c_{ij} are obtained by multiplying, element-by-element, the ith row of $\mathbf{A}$ by the jth column of $\mathbf{B}$ and summing: $c_{ij} = \sum_{k=1}^{n} a_{ik} b_{kj}$.

Dot Product

If a vector, $\vec{x}$, is represented by a single column matrix, then the *dot product* of $\vec{x}$ with a similar vector of the same length, $\vec{y}$, is represented variously as $\vec{x} \cdot \vec{y}$ or $(\vec{x}, \vec{y})$ or $\langle \vec{x} \mid \vec{y} \rangle$, and is obtained by the matrix product, $\vec{x}^T \vec{y}$, which yields a single number. Two vectors that have a zero dot product are termed *orthogonal*.

Norm of a Vector

The norm, or *length*, of a vector is designated as $|\vec{x}|$ and is computed as $\sqrt{\vec{x}^T \vec{x}}$. The norm of a vector must always be nonnegative.

Unit Vectors

A vector of length 1 is a unit vector and is often designated using a hat notation: $\hat{e}$. Orthogonal unit vectors are called *orthonormal* vectors.

Projection Operators

A square, nonnull matrix, $\mathbf{P}$, that satisfies the condition $\mathbf{P}^2 = \mathbf{P}$ is called a projection matrix or projection operator. Projection matrices may be constructed from unit vectors by the operation $\mathbf{P} = \hat{e}\hat{e}^T$.

Completeness

If the set of projection operators, $\mathbf{P}_i$, constructed from a set of unit vectors, $\hat{e}_i$, $i = 1, \ldots, n$, satisfies the condition $\sum_{i=1}^{n} \mathbf{P}_i = \mathbf{I}_n$, the unit vectors are said to form a complete set, meaning that *any* vector of length n can be expanded in terms of the $\hat{e}_i$'s.

SOLUTIONS OF n-COUPLED, INHOMOGENEOUS, LINEAR EQUATIONS

Elementary Row Operations

If the n equations are written as successive rows

$$
\begin{aligned}
a_{11}x_1 + a_{12}x_2 + \cdots + a_{1n}x_n &= b_1 \\
a_{21}x_1 + a_{22}x_2 + \cdots + a_{2n}x_n &= b_1 \\
&\vdots \\
a_{n1}x_1 + a_{n2}x_2 + \cdots + a_{nn}x_n &= b_1
\end{aligned}
$$

where the n independent variables are x_i, $i = 1, \ldots, n$, and a_{ij}, $i, j = 1, \ldots, n$, is the coefficient of x_j in the ith equation, the solution is obtained by performing *elementary row operations* on this set until the following form is obtained:

$$
\begin{aligned}
\alpha_{11}x_1 + \alpha_{12}x_2 + \alpha_{13}x_3 + \cdots + \alpha_{1n}x_n &= c_1 \\
0 + \alpha_{22}x_2 + \alpha_{23}x_3 + \cdots + \alpha_{2n}x_n &= c_2 \\
0 + 0 + \alpha_{33}x_3 + \cdots + \alpha_{3n}x_n &= c_3 \\
\vdots \qquad \vdots \qquad \vdots \qquad \ddots \qquad \vdots \qquad &\vdots \\
0 + 0 + 0 + \cdots + \alpha_{nn}x_n &= c_1
\end{aligned}
$$

This set of equivalent equations may then be solved by back substitution.

Elementary row operations are represented by any of the following:

1. Interchanging any two equations.
2. Adding a multiple of an equation to another equation.
3. Multiplying any equation by a nonzero constant.

Gaussian
Elimination
Method

The Gaussian elimination method systematically performs a set of elementary row operations on the set of n equations in n variables to reduce them to the preceding form. These operations, starting with the topmost equation, consist of a sequence of $n-1$ passes. In pass k, the pivot row is row k, the pivot element is a_{kk}, and elementary row operations are performed on each row $m > k$ so that all the elements in column k below the pivot row are zero. This results in an upper-triangular set of equations that can be solved by back substitution.

MARKOV MATRICES

Definition

The elements of an $n \times n$ Markov matrix $\mathbf{M}$ are designated as p_{ab}, which represents the probability of a transition of a single individual from state a to state b.

Requirements

The elements of a Markov matrix must satisfy the conditions (1) $p_{ji} \geq 0$ for all $i, j = 1, \ldots, n$, and (2) $\sum_{j=1}^{n} p_{ji} = 1$ for all $i = 1, \ldots, n$.

Properties

The equilibrium state, $\vec{x}_{\mathrm{eq}}$, satisfies the equation $\lim_{n \to \infty} [\mathbf{M}^T]^n \vec{x}_{\mathrm{eq}} = \vec{x}_{\mathrm{eq}}$ and can be found by starting with any nonzero initial population vector and repeatedly multiplying by $\mathbf{M}$.

ROTATIONS

Basis Set of Unit
Vectors

A coordinate system (in n dimensions) can be described in terms of n orthogonal unit vectors, $\hat{e}_i$, $i = 1, \ldots, n$. Any vector in that coordinate system may then be written in terms of those unit vectors as $\vec{a} = \sum_{i=1}^{n} (\vec{a} \cdot \hat{e}_i) \hat{e}_i$, or in three dimensions, $\vec{a} = a_x \hat{\imath} + a_y \hat{\jmath} + a_z \hat{k}$, where $\hat{\imath}, \hat{\jmath}$, and $\hat{k}$ are the unit vectors in the x, y, and z directions, and $a_x = \vec{a} \cdot \hat{\imath}$, etc.

Rotation of
Coordinate Axes

One set of orthogonal unit vectors, $\hat{e}_i$, $i = 1, \ldots, n$, can be related to a different set, $\hat{e}_i'$, $i = 1, \ldots, n$, by a *rotation* of the coordinate axes. If $\mathbf{E}$ and $\mathbf{E}'$ are the matrices consisting of the column vectors of the two sets of orthogonal unit vectors, respectively, then the rotation matrix, $\mathbf{U} = \mathbf{E}'\mathbf{E}$, relates the two coordinate axes by $\mathbf{E}' = \mathbf{E}\mathbf{U}^T$. The rotation matrix satisfies the orthogonality condition $\mathbf{U}^T \mathbf{U} = \mathbf{U}\mathbf{U}^T = \mathbf{I}_n$.

Rotation of
Vectors

If a vector, $\vec{a}$, is expressed in terms of components, a_i, relative to one coordinate system, the components of the vector in a rotated coordinate system are given by $\mathbf{U}\vec{a}$.

The Group
Property for
Rotations

The geometric property that any two successive rotations can be represented by a single rotation is equivalent to the following condition: If $\mathbf{U}_3 = \mathbf{U}_2 \mathbf{U}_1$, then $\mathbf{U}_3^T \mathbf{U}_3 = \mathbf{U}_3 \mathbf{U}_3^T = \mathbf{I}_n$.

ORTHOGONAL BASES

Gram-Schmidt
Orthogonalization

A procedure that starts with a set of vectors $\vec{x}_i$, $i = 1, \ldots, p$ and constructs, one at a time, a set of n $(n \leq p)$ orthogonal unit vectors, $\hat{e}_i$, $i = 1, \ldots, n$. These vectors then span the space of the $\vec{x}_i$'s. The expansion coefficients, $a_i^{(k)}$, in the equation

$$\vec{x}_k = \sum_{i=1}^{n} a_i^{(k)} \hat{e}_i$$

are then contained in the matrix $\mathbf{A} = \mathbf{E}^T \mathbf{X}$, where $\mathbf{E}$, $\mathbf{X}$ are matrices whose columns are the vectors $\vec{x}_i$ and $\hat{e}_j$, respectively. The matrix $\mathbf{A}$ is upper triangular.

Orthogonal Basis
Computed with
`orth(X)`

The MATLAB function `orth(X)` will return an equivalent set of basis vectors.

MATLAB Functions Used

`orth(X)`

If $\mathbf{X}$ is an $m \times n$ matrix, then `orth(X)` will return an $m \times p$ matrix with $p \leq n$ with columns representing p orthogonal unit column vectors that span the same space as the columns of $\mathbf{X}$.

2 / *Solution of the Matrix Equation* $\mathbf{A}\vec{x} = \vec{b}$

PREVIEW   In Chapter 1, we saw that the solution of two simultaneous linear equations could be viewed geometrically as the intersection point of two straight lines. There were then four classes of solutions possible: (1) The two lines intersect at a single unique point, (2) the lines are parallel and never intersect, (3) the lines are the same and intersect at an infinite number of points, or (4) the two equations are ill conditioned, meaning that the two lines are nearly parallel and that a small variation in the coefficients may cause a large change in the intersection point or may cause the lines to no longer intersect at all. The situation of n linear equations in n unknowns, written as a matrix equation $\mathbf{A}\vec{x} = \vec{b}$, yields four similar classes of solutions. A critical element in any solution of the equations is to establish criteria that will predict whether or not a given set of equations has a solution, and, if so, whether or not the solution is unique. These criteria are most often expressed in terms of the *rank* of a matrix, which we will describe in the context of solving the equations by a Gaussian elimination procedure.

There are then several alternative methods of solving the set of linear simultaneous equations. The first, and most direct method, is to recast the Gaussian elimination procedure to apply to matrix equations. A variation is to use Gaussian elimination to factor the coefficient matrix into a product of a lower-triangular and an upper-triangular matrix. This method, called LU factorization, is useful when there are several matrix equations to be solved, all with the same coefficient matrix, but with differing right-hand-side vectors. It also leads to a third approach, which is to find the multiplicative inverse of the coefficient matrix, so that the solution can be written simply as $\vec{x} = \mathbf{A}^{-1}\vec{b}$. A final method, using determinants, is familiar from elementary courses in linear algebra and is called *Cramer's rule*.

Because each of these methods represents solutions to the same problem, there is considerable overlap in their underlying ideas. The methods based upon Gaussian elimination are all highly efficient and ideally suited to computer solutions and will be described in detail in this chapter. In addition, there are MATLAB toolbox functions implementing each of these methods. Methods requiring the evaluation of determinants are only of theoretical interest, are prohibitively inefficient when the number of equations exceeds three or four, and as a result are only mentioned in passing.

Finally, there is a substantially different avenue of attack to the problem, based on iterative techniques, whereby a solution to the matrix equation is first "guessed" and the matrix equation is then repeatedly used to improve the guess. This method, called the Gauss-Seidel method, is particularly well suited to matrix equations when the matrix is both extremely large and sparse; i.e., it contains a large fraction of zero elements. The Gauss-Seidel method is also briefly outlined in this chapter.

Existence and Uniqueness of Solutions to A$\vec{x} = \vec{b}$

The solution of n linear equations in n unknowns, written in matrix form as $\mathbf{A}\vec{x} = \vec{b}$, is a central problem in linear analysis. One of the reasons for its importance is that it is possible to establish criteria both for the existence of a solution to the problem and for its uniqueness. This is usually not possible for nonlinear problems, and even if it is, the criteria are often extremely complicated. For this reason, the first line of attack on any problem is frequently to express the problem in terms of linear equations, at least as a first approximation. Procedures for linear problems are quite tractable and, more importantly, come with a prescription for determining beforehand whether or not to proceed; i.e., whether or not a solution exists. To establish existence/uniqueness conditions we begin by formulating the Gaussian elimination method for n equations in terms of matrices.

GAUSSIAN ELIMINATION IN MATRIX FORM

When the system of equations was formally solved in Chapter 1, we began with a set of n equations in m unknowns,

$$
\begin{aligned}
a_{11}x_1 + a_{12}x_2 + a_{13}x_3 + \cdots + a_{1m}x_m &= b_1 \\
a_{21}x_1 + a_{22}x_2 + a_{23}x_3 + \cdots + a_{2m}x_m &= b_2 \\
a_{31}x_1 + a_{32}x_2 + a_{33}x_3 + \cdots + a_{3m}x_m &= b_3 \\
\vdots \qquad \vdots \qquad \vdots \qquad \ddots \qquad \vdots &= \vdots \\
a_{n1}x_1 + a_{n2}x_2 + a_{n3}x_3 + \cdots + a_{nm}x_m &= b_n
\end{aligned}
$$

which we now write in matrix form as

$$
\begin{pmatrix}
a_{11} & a_{12} & a_{13} & \cdots & a_{1m} \\
a_{21} & a_{22} & a_{23} & \cdots & a_{2m} \\
a_{31} & a_{32} & a_{33} & \cdots & a_{3m} \\
\vdots & \vdots & \vdots & \ddots & \vdots \\
a_{n1} & a_{n2} & a_{n3} & \cdots & a_{nm}
\end{pmatrix}
\begin{pmatrix}
x_1 \\
x_2 \\
x_3 \\
\vdots \\
x_m
\end{pmatrix}
=
\begin{pmatrix}
b_1 \\
b_2 \\
b_3 \\
\vdots \\
b_n
\end{pmatrix}
$$

The Gaussian elimination method consisted of a set of algebraic operations applied to the equations that left the solution set (if any existed) unaltered. These operations were known as *elementary row operations* and included the following:

1. Multiplying a particular equation by a nonzero constant,
2. Replacing a particular equation by that equation plus a multiple of another equation (to achieve zeros in a particular column), and

3. Interchanging the position of two equations (e.g., switching rows if the current pivot were zero or very small).

No doubt, you have noticed that elementary row operations—and therefore the Gaussian elimination method itself—do not affect the position of the variables x_i. Thus, as a shorthand notation we could eliminate the x_i's entirely from the preceding equations and simply write down the coefficient matrix augmented by the right-hand-side vector b_i.

$$\left(\begin{array}{ccccc|c} a_{11} & a_{12} & a_{13} & \cdots & a_{1m} & b_1 \\ a_{21} & a_{22} & a_{23} & \cdots & a_{2m} & b_2 \\ a_{31} & a_{32} & a_{33} & \cdots & a_{3m} & b_3 \\ \vdots & \vdots & \vdots & \ddots & \vdots & \vdots \\ a_{n1} & a_{n2} & a_{n3} & \cdots & a_{nm} & b_n \end{array} \right)$$

The Gaussian elimination procedure is then applied directly to this arrangement of numbers.

Before continuing, however, we will trade a small decrease in efficiency for an increase in the clarity of the discussion. We do this by *normalizing* each row of the upper-triangular matrix obtained in the Gaussian elimination method by dividing all the elements of a given row by its corresponding pivot element, a'_{kk}. This adds nothing to the calculation, but results in a matrix with 1s along the principal diagonal, making the following description somewhat easier to follow.

First, consider a possible outcome for the situation where $n > m$; i.e., there are more equations than unknowns. The final step might then yield

$$\left(\begin{array}{cccccc|c} 1 & \square & \square & \cdots & \square & \square & c_1 \\ 0 & 1 & \square & \cdots & \square & \square & c_2 \\ 0 & 0 & 1 & \cdots & \square & \square & c_3 \\ \cdots & \cdots & \cdots & \ddots & \cdots & \cdots & \ddots \\ 0 & 0 & 0 & \cdots & 0 & 1 & c_m \\ 0 & 0 & 0 & \cdots & 0 & 0 & c_{m+1} \\ \cdots & \cdots & \cdots & \ddots & \cdots & \cdots & \ddots \\ 0 & 0 & 0 & \cdots & 0 & 0 & c_{n-1} \\ 0 & 0 & 0 & \cdots & 0 & 0 & c_n \end{array} \right) \qquad n > m$$

where the symbol $\square$ represents an entry that is, in general, not zero. Notice that the last few lines now represent equations of the form $0x_i = c_i$ for $i = m+1, \ldots, n$, suggesting that, unless the remaining c_i's are zero, there is *no solution* for these x's. Thus, if there are more equations than unknowns, and if the equations are independent, the problem has no solution.

If there are more unknowns than equations, $n \le m$, and, say $n = m - 1$, the final step might be

$$
\left(
\begin{array}{ccccccc|c}
1 & \square & \square & \cdots & \square & \square & a'_{1m} & c_1 \\
0 & 1 & \square & \cdots & \square & \square & a'_{2m} & c_2 \\
0 & 0 & 1 & \cdots & \square & \square & a'_{3m} & c_3 \\
\cdots & \cdots & \cdots & \ddots & \cdots & \cdots & \cdots & \ddots \\
0 & 0 & 0 & \cdots & 1 & \square & a'_{m-2,m} & c_{m-1} \\
0 & 0 & 0 & \cdots & 0 & 1 & a'_{m-1,m} & c_m
\end{array}
\right) \quad n = m - 1
$$

In this case there is no way to continue the process to obtain a unique solution. The last equation, $x_{m-1} + a'_{m-1,m} x_m = c_m$, then has an infinite number of solutions. Letting $x_m = \gamma$, each of the remaining x_i's can then be obtained by back substitution and will in turn depend upon γ. That is, a set of $m - 1$ independent equations in m unknowns will have a *family* of solutions that depend upon a single parameter, γ.

Even if the number of equations equals the number of unknowns, we know that two or more equations could be (1) *dependent* (i.e., simply combinations of other equations), in which case there are effectively more unknowns than independent equations, and the last step of the Gaussian elimination method would yield a result similar to the second example, or (2) *inconsistent* (e.g., $x_1 + x_2 = 1$ and $x_1 + x_2 = 2$), and the last step of the method then resembles the first example. These ideas are the basis for a general prescription for the *uniqueness/existence* classifications of the solutions of n in m unknowns. But first, we will describe a minor variation to the basic Gaussian elimination method, called the Gauss-Jordan method.

GAUSS-JORDAN ELIMINATION

The basic Gaussian elimination procedure to reduce a matrix equation to upper-triangular form, followed by back substitution to obtain the final answer, is certainly the most popular solution method for the equation $\mathbf{A}\vec{x} = \vec{b}$. It is highly efficient, stable, and easy to understand and to code as a computer algorithm.[1] There is, however, a popular, somewhat less efficient, variation to the basic Gaussian elimination method that does not require back substitution.

---

[1]The Gaussian elimination method is not the only efficient method for solving the linear equation, $\mathbf{A}\vec{x} = \vec{b}$. The method used by MATLAB is based on a technique called LU decomposition, which in turn uses a variation of Gaussian elimination to write an arbitrary square matrix as the product of a lower-triangular matrix, ($\mathbf{L}$), and an upper-triangular matrix, ($\mathbf{U}$). The inverses of these matrices are then easily computed by back substitution and the inverse of the original matrix is then expressed as $\mathbf{A}^{-1} = \mathbf{U}^{-1}\mathbf{L}^{-1}$. The LU decomposition method is about as efficient as Gaussian elimination but has the advantage that the inverse matrix is computed independently of a particular right-hand-side vector $\vec{b}$. The details of this method will be discussed beginning on page 60.

If, instead of "zeroing" just the elements in the *pivot* column below the current pivot row, we zero all the elements in this column, both *above and below*, the method is called *Gauss-Jordan elimination*. Thus, starting with a matrix equation of the form

$$
\begin{pmatrix}
a_{11} & a_{12} & a_{13} & \cdots & a_{1n} \\
a_{21} & a_{22} & a_{23} & \cdots & a_{2n} \\
a_{31} & a_{32} & a_{33} & \cdots & a_{3n} \\
\vdots & \vdots & \vdots & \ddots & \vdots \\
a_{n1} & a_{n2} & a_{n3} & \cdots & a_{nn}
\end{pmatrix}
\begin{pmatrix}
x_1 \\ x_2 \\ x_3 \\ \vdots \\ x_n
\end{pmatrix}
=
\begin{pmatrix}
b_1 \\ b_2 \\ b_3 \\ \vdots \\ b_n
\end{pmatrix}
\tag{2.1}
$$

if the equations have a unique solution, the final step of the Gauss-Jordan method will yield an equation such as

$$
\begin{pmatrix}
1 & 0 & 0 & \cdots & 0 \\
0 & 1 & 0 & \cdots & 0 \\
0 & 0 & 1 & \cdots & 0 \\
\vdots & \vdots & \vdots & \ddots & \vdots \\
0 & 0 & 0 & \cdots & 1
\end{pmatrix}
\begin{pmatrix}
x_1 \\ x_2 \\ x_3 \\ \vdots \\ x_n
\end{pmatrix}
=
\begin{pmatrix}
c_1 \\ c_2 \\ c_3 \\ \vdots \\ c_n
\end{pmatrix}
$$

and the solution for the x_i's is simply $x_i = c_i$.

Furthermore, in those situations described previously where a unique solution is not possible, this method will yield a result similar to the diagrams on page 55, except that entries designated by □ will be replaced by zeros.

Thus, the final step of the Gauss-Jordan method will either yield the unique solution, if one exists, *or* can be used to define a general prescription for the classification of the solutions of n linear equations in m unknowns. The structure of the matrix in the final step of the Gauss-Jordan elimination procedure is termed the *reduced row echelon form* of the matrix.

REDUCED ROW ECHELON FORM

The form of this matrix is called **R***educed* **R***ow* **E***chelon* **F***orm* (hereafter abbreviated as **RREF**). A matrix is in RREF if it has the following properties:

1. Reading terms from left to right, the first nonzero entry in each row is a 1. This is called the leading 1 of that row.
2. Each leading 1 is the *only* nonzero entry in its column.
3. The column positions of successive leading 1s are to the *right* of those above them.
4. Any rows of *all* zeros are positioned at the bottom of the matrix.

Examples of Matrices in RREF

a. $\begin{pmatrix} 1 & 0 & 0 \\ 0 & 1 & 0 \\ 0 & 0 & 1 \end{pmatrix}$ b. $\begin{pmatrix} 0 & 0 \\ 0 & 0 \\ 0 & 0 \end{pmatrix} = \mathbf{0}_{3 \times 2}$

c. $\begin{pmatrix} 1 & 0 & 0 & 3 \\ 0 & 1 & 0 & 5 \\ 0 & 0 & 1 & 7 \end{pmatrix}$ d. $\begin{pmatrix} 1 & 0 \\ 0 & 1 \\ 0 & 0 \\ 0 & 0 \end{pmatrix}$

We state, without proof, that each and every matrix has a unique RREF and that this RREF can be obtained by elementary row operations as outlined in the Gaussian elimination procedures.

RANK OF A MATRIX

The rank of a matrix **A**, written as rank(**A**), is equal to the number of nonzero rows of the RREF obtained from **A**.

Using this definition, we can now state requirements concerning the uniqueness and existence of solutions to the set of n equations $A\vec{x} = \vec{b}$ in m unknowns.

First, let $[\mathbf{A}|b]$ be a matrix consisting of the matrix, **A**, augmented by the column vector, $\vec{b}$, added as a column on the right of **A**. Next, by Gaussian elimination, obtain the RREF of the augmented matrix, thereby determining the *rank* of $[\mathbf{A}|b]$ (as well as the *rank* of **A**). Thus, if **A** is an $n \times m$ matrix, and $[\mathbf{A}|b]$ is the $n \times (m+1)$ augmented matrix, then:

1. When rank(**A**) = rank(**A**|b) = m, a *unique* solution exists for the m unknowns x_i, $i = 1, \ldots, m$.
2. When rank(**A**) = rank(**A**|b) = $r < m$, a unique solution *is not possible.* However, the solution set can be expressed in terms of $(m - r)$ free parameters.
3. When rank(**A**) $\neq$ rank(**A**|b), *no solution exists.*

First, a few examples. (To save on needless algebra, we will begin with matrices nearly in RREF.)[2]

1. $\begin{pmatrix} 1 & 0 & 0 \\ 0 & 1 & 2 \\ 0 & 1 & 1 \end{pmatrix} \begin{pmatrix} x \\ y \\ z \end{pmatrix} = \begin{pmatrix} 1 \\ 2 \\ 3 \end{pmatrix}$, so that $n = m = 3$.

Computing the RREF, we obtain

$$[\mathbf{A}]_{\text{RREF}} = \begin{pmatrix} 1 & 0 & 0 \\ 0 & 1 & 0 \\ 0 & 0 & 1 \end{pmatrix} \qquad [\mathbf{A}|b]_{\text{RREF}} = \left(\begin{array}{ccc|c} 1 & 0 & 0 & 1 \\ 0 & 1 & 0 & 4 \\ 0 & 0 & 1 & -1 \end{array} \right)$$

Thus, because rank(**A**) = rank(**A**|b) = 3, the solution is $(x, y, z) = (1, 4, -1)$, and this solution is unique.

---

[2]MATLAB has a toolbox function that will compute the RREF of a matrix **A**. Simply type `rref(A)`.

2. $\begin{pmatrix} 1 & 0 & 0 \\ 0 & 1 & 2 \\ 0 & 0 & 0 \end{pmatrix} \begin{pmatrix} x \\ y \\ z \end{pmatrix} = \begin{pmatrix} 1 \\ 2 \\ 3 \end{pmatrix}$

In this case, rank($\mathbf{A}$) = 2, and

$$[\mathbf{A}|b]_{\mathrm{RREF}} = \left(\begin{array}{ccc|c} 1 & 0 & 0 & 0 \\ 0 & 1 & 2 & 0 \\ 0 & 0 & 0 & 1 \end{array} \right)$$

Because rank($\mathbf{A}|b$) = 3 > rank($\mathbf{A}$), no solution exists.

3. $\begin{pmatrix} 1 & 0 & 0 \\ 0 & 1 & 2 \\ 0 & 0 & 0 \end{pmatrix} \begin{pmatrix} x \\ y \\ z \end{pmatrix} = \begin{pmatrix} 1 \\ 2 \\ 0 \end{pmatrix}$

This time rank($\mathbf{A}$) = rank($\mathbf{A}|b$) = 2 < n = 3, and any value of z will solve the equations.

The proof of the preceding uniqueness theorem is simply a matter of generalizing the examples to matrices of arbitrary size and shape and should present no problem to the reader.

EXAMPLE 2.1

A MATLAB Example of rank(A)

First, construct an arbitrary matrix; e.g., an 8×8. To reduce the labor associated with entering 64 elements, you may wish to take advantage of several of MATLAB's built-in matrices, like `magic(8)`, `hilb(8)`[3] or `rand(8)`, which generates an 8×8 matrix of random numbers in the range $0 \to 1$. Next construct an arbitrary right-hand-side vector, $\vec{b}$.

We can determine the rank of $\mathbf{A}$ and $[\mathbf{A}|b]$ from their RREF computed by MATLAB by using a few simple statements. First, the function `sum(B)` will return a row vector whose elements contain the sum of the columns of the matrix $\mathbf{B}$. Here we wish to determine how many *rows* of a matrix are all zeros. To accomplish this we use the MATLAB logical comparison operator ~=, (meaning *not equal*).[4] Thus,

$$\texttt{sum(B')} \; \texttt{\~{}= 0}$$

will return a row vector of 1s and 0s corresponding to whether or not the sum of the elements of a column is equal to zero. Thus, the rank[5] of a matrix could

---

[3]The matrix returned by the MATLAB statement `hilb(n)` has elements of the form $h_{ij} = 1/(i + j - 1)$, for $i, j = 1, \ldots, n$. The *exact* inverse, for $n \leq 15$, is returned by the function `invhilb(n)`.

[4]MATLAB logical operators are described in detail in Appendix Section A.iii.

[5]The MATLAB function `rank(A)` will directly return the rank of a matrix. However, this function computes the rank in a somewhat different manner using a procedure known as *singular value decomposition(SVD)* and is outside the scope of this discussion. We will, however, return to this type of rank calculation when SVD is briefly described in Chapter 3.

be computed as

$$\texttt{rank = sum(sum(rref(A)) } \tilde{} \texttt{= 0)}$$

Of course, this assumes that the elements of `rref(A)` are specified exactly, without rounding or truncation errors. It would be more prudent to amend this statement to read

$$\texttt{rank = sum(sum(abs(rref(A)') > 10*eps))}$$

Using this procedure to determine the ranks of $\mathbf{A}$ and $[\mathbf{A}|b]$, we can ascertain whether or not a unique solution of the equation $\mathbf{A}\vec{x} = \vec{b}$ is possible. Thus,

```
A = hilb(8);
b = [1 2 1 2 1 2 1 2]';
rankA = sum(sum(abs(rref(A)')) > 10*eps);
rankAb = sum(sum(abs(rref([A b])')) > 10*eps)
disp([rankA rankAb])
```

              8   8

Because the two ranks are equal to $n = 8$, the equation has a unique solution, which is obtained using `x = A\b`.

WHAT IF?   Can you think of a way to alter the equations so that two of them will be inconsistent? Two equations are inconsistent if one is a multiple of the other, but the right-hand-side values are not related by the same factor. For example, inserting the following two lines

$$\texttt{A}(3,:) = 2 * \texttt{A}(5,:);$$
$$\texttt{b}(3) \quad = 4 * \texttt{b}(5);$$

in the preceding code yields

$$\texttt{disp([rankA rankAb])}$$

              7   8

indicating that there is no solution to the equation. The statement `x = A\b` now yields $\vec{x} = \infty$.

Finally, alter the right-side vector so that two of the equations are dependent, say, $\texttt{b}(3) = 2 * \texttt{b}(5)$. The ranks of both $\mathbf{A}$ and $(\mathbf{A}|b)$ are now equal to 7, which is less than $n = 8$, indicating that an infinite number of solutions are possible. That is, the solution may be expressed in terms of a parameter. To solve this equation, obtain the RREF of $(\mathbf{A}|b)$ using the MATLAB function `rref`. Except for the last row and last two columns, this matrix will equal the identity matrix $\mathbf{I}_7$. Letting the last variable, x_8, be represented by a parameter γ, the solution for the remaining x_i's should then be apparent. ●

LU Factorization of a Matrix

A problem frequently encountered when solving the equation $\mathbf{A}\vec{x} = \vec{b}$ is the need to obtain solutions for a variety of right-hand-side vectors, $\vec{b}$, while the coefficient matrix, $\mathbf{A}$, remains unchanged. Using the Gaussian elimination method developed to this point, we would have to separately solve $\mathbf{A}\vec{x} = \vec{b}$ for each vector, $\vec{b}$, while employing essentially the same arithmetic operations in each calculation up to the point of back substitution. A process known as LU factorization addresses this problem by concentrating on the coefficient matrix only.

Assume that it is possible to *factor* an arbitrary, square matrix, $\mathbf{A}$, into a product of two matrices, $\mathbf{L}$, $\mathbf{U}$, where $\mathbf{L}$ is lower triangular, and $\mathbf{U}$ is upper triangular. Using a 4×4 for demonstration purposes,

$$\begin{pmatrix} \alpha_{11} & 0 & 0 & 0 \\ \alpha_{12} & \alpha_{22} & 0 & 0 \\ \alpha_{13} & \alpha_{23} & \alpha_{33} & 0 \\ \alpha_{14} & \alpha_{24} & \alpha_{34} & \alpha_{44} \end{pmatrix} \begin{pmatrix} \beta_{11} & \beta_{12} & \beta_{13} & \beta_{14} \\ 0 & \beta_{22} & \beta_{23} & \beta_{24} \\ 0 & 0 & \beta_{33} & \beta_{34} \\ 0 & 0 & 0 & \beta_{44} \end{pmatrix} \equiv \begin{pmatrix} a_{11} & a_{12} & a_{13} & a_{14} \\ a_{21} & a_{22} & a_{23} & a_{24} \\ a_{31} & a_{32} & a_{33} & a_{34} \\ a_{41} & a_{42} & a_{43} & a_{44} \end{pmatrix}$$

The solution of $\mathbf{A}\vec{x} = \vec{b}$ can then be segmented as follows:

$$\mathbf{A}\vec{x} = \mathbf{L}\,\mathbf{U}\vec{x}$$
$$= \mathbf{L}\vec{y} = \vec{b}$$

where we have defined the vector $\vec{y} \equiv \mathbf{U}\vec{x}$.

Because $\mathbf{L}$ is lower triangular, the solution of $\mathbf{L}\vec{y} = \vec{b}$ is easily obtained by *forward* substitution; that is,

$$y_1 = b_1/\alpha_{11}$$
$$y_2 = (b_2 - \alpha_{21}y_1)/\alpha_{22}$$
$$\vdots \qquad \vdots$$

Because $\mathbf{U}$ is upper triangular, once the elements of $\vec{y}$ are obtained, the equation $\mathbf{U}\vec{x} = \vec{y}$ can next be solved by *back* substitution. Thus, once the LU factorization of the matrix $\mathbf{A}$ is obtained, the equation $\mathbf{A}\vec{x} = \vec{b}$ can easily be solved for a variety of vectors $\vec{b}$.

To accomplish the factorization, the matrices $\mathbf{L}$ and $\mathbf{U}$ are multiplied and equated to the corresponding elements a_{ij}. Before doing that, we first note that an $n \times n$ coefficient matrix $\mathbf{A}$ contains n^2 independent elements, and both matrices $\mathbf{L}$ and $\mathbf{U}$ contain $\frac{1}{2}n(n+1)$ undetermined elements, or $n^2 + n$ elements in all. We are thus free to arbitrarily specify n of the elements of $\mathbf{L}$ or of $\mathbf{U}$. The standard choice is to specify all the diagonal elements of $\mathbf{L}$ to be 1s. The equation then reads

$$\begin{pmatrix} 1 & 0 & 0 & 0 \\ \alpha_{12} & 1 & 0 & 0 \\ \alpha_{13} & \alpha_{23} & 1 & 0 \\ \alpha_{14} & \alpha_{24} & \alpha_{34} & 1 \end{pmatrix} \begin{pmatrix} \beta_{11} & \beta_{12} & \beta_{13} & \beta_{14} \\ 0 & \beta_{22} & \beta_{23} & \beta_{24} \\ 0 & 0 & \beta_{33} & \beta_{34} \\ 0 & 0 & 0 & \beta_{44} \end{pmatrix} = \begin{pmatrix} a_{11} & a_{12} & a_{13} & a_{14} \\ a_{21} & a_{22} & a_{23} & a_{24} \\ a_{31} & a_{32} & a_{33} & a_{34} \\ a_{41} & a_{42} & a_{43} & a_{44} \end{pmatrix} \qquad (2.2)$$

Obtaining the relations for the α's and the β's is now straightforward, but not very illuminating. The result, however, is surprisingly simple. The upper-triangular matrix, $\mathbf{U}$, is exactly the same as the upper-triangular matrix obtained from a standard Gaussian elimination procedure. Thus, it is an easy matter to first compute $\mathbf{U}$. Having obtained the elements β_{ij} by multiplying the matrices in Equation 2.2, it is not difficult to show that the equations for the α's in terms of the known β's and a_{ij} are given by the following equation:

$$\alpha_{mk} = \left(a_{mk} - \sum_{i=1}^{k-1} \alpha_{mi}\beta_{ik} \right) / \beta_{kk} \qquad \text{for } k < m \qquad (2.3)$$

When using this equation, because α's appear on the right as well, it is essential that the calculation start with $k = 1$ and proceed in succession to $k = m - 1$ for each value of $m \geq 2$.

PIVOTING STRATEGY AND LU FACTORIZATION

Because LU factorization is based upon standard Gaussian elimination to obtain the matrix $\mathbf{U}$, the same pivoting strategy is used. That is, when computing $\mathbf{U}$, in each pass the current pivot row is interchanged with the row below that contains the largest magnitude element in the pivot column. Switching of rows has no effect on the ordering of the elements of the vector $\vec{x}$, so there was no reason to keep track of the row switching in an ordinary Gaussian elimination solution of $\mathbf{A}\vec{x} = \vec{b}$. However, row switching will alter the equations for the elements α_{mk} in Equation 2.3, thus requiring us to keep track of the row switching that takes place in obtaining the matrix $\mathbf{U}$. This is usually done by means of a *permutation* matrix, $\mathbf{P}$. For example, if the coefficient matrix, $\mathbf{A}$, is a 4×4, and in the second step of the Gaussian elimination rows 2 and 4 are interchanged, this is equivalent to multiplying the matrix $\mathbf{A}$ by the permutation matrix

$$\mathbf{P} = \begin{pmatrix} 1 & 0 & 0 & 0 \\ 0 & 0 & 0 & 1 \\ 0 & 0 & 1 & 0 \\ 0 & 1 & 0 & 0 \end{pmatrix}$$

If then, in the next pass, rows 3 and 4 are interchanged, this permutation is recorded by interchanging rows 3 and 4 of the permutation matrix as well:

$$\mathbf{P} = \begin{pmatrix} 1 & 0 & 0 & 0 \\ 0 & 0 & 0 & 1 \\ 0 & 1 & 0 & 0 \\ 0 & 0 & 1 & 0 \end{pmatrix}$$

The combined set of permutations performed on the matrix $\mathbf{A}$ is then represented by the matrix $\mathbf{P}$. In effect, the equation that has been solved is $\mathbf{P}\mathbf{A}\vec{x} = \mathbf{P}\vec{b}$, where

$$\mathbf{L}\,\mathbf{U} = \mathbf{P}\,\mathbf{A} \qquad (2.4)$$

EXAMPLE 2.2

LU Factorization of a 4 × 4 Matrix

The Gaussian elimination method applied to the matrix

$$A = \begin{pmatrix} 1 & 0 & 0 & 1 \\ 2 & 2 & 2 & 2 \\ 1 & 2 & 1 & 2 \\ 2 & 0 & 0 & 1 \end{pmatrix}$$

proceeds as follows:

1 0 0 1 2 2 2 2 1 2 1 2 2 0 0 1	Interchange rows 1 and 2 and record in permutation matrix **P**

2 2 2 2 1 0 0 1 1 2 1 2 2 0 0 1	$P = \begin{pmatrix} 0 & 1 & 0 & 0 \\ 1 & 0 & 0 & 0 \\ 0 & 0 & 1 & 0 \\ 0 & 0 & 0 & 1 \end{pmatrix}$

2 2 2 2 0 −1 −1 0 0 1 0 1 0 −2 −2 −1	Zero elements of col. 1 by adding multiples of row 1, then interchange rows 2 and 4

2 2 2 2 0 −2 −2 −1 0 1 0 1 0 −1 −1 0	$P = \begin{pmatrix} 0 & 1 & 0 & 0 \\ 0 & 0 & 0 & 1 \\ 0 & 0 & 1 & 0 \\ 1 & 0 & 0 & 0 \end{pmatrix}$

2 2 2 2 0 −2 −2 −1 0 0 −1 $\frac{1}{2}$ 0 0 0 $\frac{1}{2}$	Zero the elements of col. 2 This is the matrix **U**

The elements of the matrix **L** are next specified by Equation 2.4 as

$$\begin{pmatrix} 1 & 0 & 0 & 0 \\ \alpha_{21} & 1 & 0 & 0 \\ \alpha_{31} & \alpha_{32} & 1 & 0 \\ \alpha_{41} & \alpha_{42} & \alpha_{43} & 1 \end{pmatrix} \begin{pmatrix} 2 & 2 & 2 & 2 \\ 0 & -2 & -2 & -1 \\ 0 & 0 & -1 & \frac{1}{2} \\ 0 & 0 & 0 & \frac{1}{2} \end{pmatrix} = \begin{pmatrix} 0 & 1 & 0 & 0 \\ 0 & 0 & 0 & 1 \\ 0 & 0 & 1 & 0 \\ 1 & 0 & 0 & 0 \end{pmatrix} \begin{pmatrix} 1 & 0 & 0 & 1 \\ 2 & 2 & 2 & 2 \\ 1 & 2 & 1 & 2 \\ 2 & 0 & 0 & 1 \end{pmatrix}$$

or, using Equation 2.3, as

$$\alpha_{21} = (PA)_{21}/U_{11} \qquad\qquad = 1$$

$$\alpha_{31} = (PA)_{31}/U_{11} \qquad\qquad = \tfrac{1}{2}$$

$$\alpha_{32} = [(PA)_{32} - \alpha_{31}U_{12}]/U_{22} \qquad = -\tfrac{1}{2}$$

$$\alpha_{41} = (\mathbf{PA})_{41}/\mathbf{U}_{11} \qquad\qquad = \tfrac{1}{2}$$

$$\alpha_{42} = \left[(\mathbf{PA})_{42} - \alpha_{41}\mathbf{U}_{12}\right]/\mathbf{U}_{22} \qquad = +\tfrac{1}{2}$$

$$\alpha_{43} = \left[(\mathbf{PA})_{43} - (\alpha_{41}\mathbf{U}_{13} + \alpha_{42}\mathbf{U}_{23})\right]/\mathbf{U}_{33} = \quad 0$$

so that

$$\mathbf{L} = \begin{pmatrix} 1 & 0 & 0 & 0 \\ 1 & 1 & 0 & 0 \\ \tfrac{1}{2} & -\tfrac{1}{2} & 1 & 0 \\ 1 & \tfrac{1}{2} & 0 & 1 \end{pmatrix}$$

It is easily verified that $\mathbf{LU} = \mathbf{PA}$.

Finally, to solve the equation $\mathbf{A}\vec{x} = \vec{b}$ for a particular choice of right-hand-side vector, $\vec{b}$, we solve instead the equation $\mathbf{PA}\vec{x} = \mathbf{P}\vec{b}$ by first obtaining the solutions to $\mathbf{L}\vec{y} = \mathbf{P}\vec{b}$, followed by a solution for $\vec{x}$ using $\mathbf{U}\vec{x} = \vec{y}$.

Thus, using $\vec{b} = (1\ 2\ 0\ 1)'$, the first step is to solve

$$\begin{pmatrix} 1 & 0 & 0 & 0 \\ 1 & 1 & 0 & 0 \\ \tfrac{1}{2} & -\tfrac{1}{2} & 1 & 0 \\ 1 & \tfrac{1}{2} & 0 & 1 \end{pmatrix} \begin{pmatrix} y_1 \\ y_2 \\ y_3 \\ y_4 \end{pmatrix} = \begin{pmatrix} 2 \\ 1 \\ 0 \\ 1 \end{pmatrix}$$

The results are easily obtained by forward substitution to be

$$\vec{y} = \begin{pmatrix} 2 \\ -1 \\ -\tfrac{3}{2} \\ \tfrac{1}{2} \end{pmatrix}$$

The solution for the elements of $\vec{x}$ are then obtained by solving

$$\begin{pmatrix} 2 & 2 & 2 & 2 \\ 0 & -2 & -2 & -1 \\ 0 & 0 & -1 & \tfrac{1}{2} \\ 0 & 0 & 0 & \tfrac{1}{2} \end{pmatrix} \begin{pmatrix} x_1 \\ x_2 \\ x_3 \\ x_4 \end{pmatrix} = \begin{pmatrix} 2 \\ -1 \\ -\tfrac{3}{2} \\ \tfrac{1}{2} \end{pmatrix}$$

by back substitution, to obtain

$$\vec{x} = \begin{pmatrix} 0 \\ -2 \\ 2 \\ 1 \end{pmatrix}$$

$\bullet$

THE TOOLBOX FUNCTION lu

As you might expect, there is a MATLAB toolbox function that will perform all of the arithmetic involved in the LU factorization of a square matrix. The use is simply

$$[\mathtt{L}, \mathtt{U}, \mathtt{P}] = \mathtt{lu}(\mathtt{A})$$

where **L** and **U** are the triangular matrices, and **P** is the permutation matrix, corresponding to the LU factorization of **A**.

You should verify the results of the previous example using this function.

The Inverse Matrix

If the coefficient matrix in the set of equations, $\mathbf{A}\vec{x} = \vec{b}$, is a square $n \times n$ matrix and has rank n, then a unique solution exists. In this case it is also possible to obtain a second matrix, labeled $\mathbf{A}^{-1}$, that satisfies $\mathbf{A} \cdot \mathbf{A}^{-1} = \mathbf{A}^{-1} \cdot \mathbf{A} = \mathbf{I}_n$. The matrix $\mathbf{A}^{-1}$ is called the inverse of **A**. If the rank of **A** is less than n, the matrix is said to be *rank deficient* and no inverse exists for such a matrix.

The inverse matrix, $\mathbf{A}^{-1}$, can be obtained by a rather simple extension of the Gaussian elimination method combined with LU factorization of the matrix **A**. If we label the elements of $\mathbf{A}^{-1}$ as c_{ij}, then the c_{ij}'s must be a solution of the equation $\mathbf{A} \cdot \mathbf{A}^{-1} = \mathbf{I}_n$, or

$$\begin{pmatrix} a_{11} & a_{12} & \cdots & a_{1n} \\ a_{21} & a_{22} & \cdots & a_{2n} \\ \vdots & \vdots & \ddots & \vdots \\ a_{n1} & a_{n2} & \cdots & a_{nn} \end{pmatrix} \begin{pmatrix} c_{11} & c_{12} & \cdots & c_{1n} \\ c_{21} & c_{22} & \cdots & c_{2n} \\ \vdots & \vdots & \ddots & \vdots \\ c_{n1} & c_{n2} & \cdots & c_{nn} \end{pmatrix} = \begin{pmatrix} 1 & 0 & \cdots & 0 \\ 0 & 1 & \cdots & 0 \\ \vdots & \vdots & \ddots & \vdots \\ 0 & 0 & \cdots & 1 \end{pmatrix} \quad (2.5)$$

This represents n^2 equations that can be solved for the n^2 unknown elements of $\mathbf{A}^{-1}$. However, a more efficient procedure is to first obtain the LU factorization of **A**,

$$\mathbf{LU} = \mathbf{PA}$$

or, equivalently,

$$\mathbf{A} = \mathbf{P}^{-1}\mathbf{LU}$$

(It is an easy matter to prove that the inverse of any permutation matrix is equal to its transpose; i.e., $\mathbf{P}^{-1} = \mathbf{P}^T$, so that this step does not involve any additional arithmetic.)

The inverse of **A** may then be expressed as

$$\mathbf{A}^{-1} = \mathbf{U}^{-1}\mathbf{L}^{-1}\mathbf{P} \quad (2.6)$$

which may be verified by multiplying the two expressions. Thus, a procedure to obtain the inverses of the two matrices, **L** and **U**, will enable us to obtain the inverse of the matrix **A**. Although replacing the problem of finding $\mathbf{A}^{-1}$ by that of finding two inverse matrices may appear to be a step backwards, the triangular nature of **L** and **U** will redeem this.

We first proceed to obtain the inverse of **U**. If we identify the elements of **U** as u_{ij} and the elements of $\mathbf{U}^{-1}$ as γ_{ij}, then an equation analogous to Equation 2.5

would read

$$\begin{pmatrix} u_{11} & u_{12} & \cdots & u_{1n} \\ 0 & u_{22} & \cdots & u_{2n} \\ \vdots & \vdots & \ddots & \vdots \\ 0 & 0 & \cdots & u_{nn} \end{pmatrix} \begin{pmatrix} \gamma_{11} & \gamma_{12} & \cdots & \gamma_{1n} \\ \gamma_{21} & \gamma_{22} & \cdots & \gamma_{2n} \\ \vdots & \vdots & \ddots & \vdots \\ \gamma_{n1} & \gamma_{n2} & \cdots & \gamma_{nn} \end{pmatrix} = \begin{pmatrix} 1 & 0 & \cdots & 0 \\ 0 & 1 & \cdots & 0 \\ \vdots & \vdots & \ddots & \vdots \\ 0 & 0 & \cdots & 1 \end{pmatrix} \qquad (2.7)$$

Writing this equation in a more suggestive manner as

$$\begin{pmatrix} u_{11} & u_{12} & \cdots & u_{1n} \\ 0 & 0 & \cdots & u_{2n} \\ \vdots & \vdots & \ddots & \vdots \\ 0 & 0 & \cdots & u_{nn} \end{pmatrix} \left(\begin{bmatrix} \gamma_{11} \\ \gamma_{21} \\ \vdots \\ \gamma_{n1} \end{bmatrix} \begin{bmatrix} \gamma_{12} \\ \gamma_{22} \\ \vdots \\ \gamma_{n2} \end{bmatrix} \begin{bmatrix} \cdot \\ \cdot \\ \vdots \\ \cdot \end{bmatrix} \begin{bmatrix} \gamma_{1n} \\ \gamma_{2n} \\ \vdots \\ \gamma_{nn} \end{bmatrix} \right) = \left(\begin{bmatrix} 1 \\ 0 \\ \vdots \\ 0 \end{bmatrix} \begin{bmatrix} 0 \\ 1 \\ \vdots \\ 0 \end{bmatrix} \begin{bmatrix} \cdot \\ \cdot \\ \vdots \\ \cdot \end{bmatrix} \begin{bmatrix} 0 \\ 0 \\ \vdots \\ 1 \end{bmatrix} \right)$$

leads to an interpretation of the single equation as n matrix equations of the form

$$\begin{pmatrix} u_{11} & u_{12} & \cdots & a_{1n} \\ 0 & u_{22} & \cdots & a_{2n} \\ \vdots & \vdots & \ddots & \vdots \\ 0 & 0 & \cdots & u_{nn} \end{pmatrix} \begin{pmatrix} \vec{\gamma}_1 \end{pmatrix} = \begin{pmatrix} 1 \\ 0 \\ \vdots \\ 0 \end{pmatrix}$$

$$\begin{pmatrix} u_{11} & u_{12} & \cdots & a_{1n} \\ 0 & u_{22} & \cdots & a_{2n} \\ \cdots & \cdots & \ddots & \cdots \\ 0 & 0 & \cdots & u_{nn} \end{pmatrix} \begin{pmatrix} \vec{\gamma}_2 \end{pmatrix} = \begin{pmatrix} 0 \\ 1 \\ \vdots \\ 0 \end{pmatrix}$$

that are to be solved for the elements of the n vectors $\vec{\gamma}_i$. However, the solution to an equation in which the coefficient matrix is triangular is obtained directly by back substitution (forward substitution, if the matrix is lower triangular). Thus, the inverse of $\mathbf{U}$ is obtained by simply executing n separate back substitution operations, one for each of the vectors $\vec{\gamma}_i$. Obviously, the inverse of $\mathbf{L}$ is obtained in a similar manner, and the inverse of $\mathbf{A}$ is given by Equation 2.6. (See Example 2.3.)

From this discussion we conclude that if $\mathbf{A}$ is an $n \times n$ matrix and rank$(\mathbf{A}) = n$, then the inverse of $\mathbf{A}$ is guaranteed to exist, whereas if rank$(\mathbf{A}) \neq n$ the method will fail, suggesting that an inverse to $\mathbf{A}$ cannot be found.

EXAMPLE 2.3

Calculation of the Inverse Matrix by Gaussian Elimination

To find the inverse of the 3×3 matrix

$$\mathbf{A} = \begin{pmatrix} 1 & 2 & -1 \\ 1 & 0 & 1 \\ -1 & 2 & 1 \end{pmatrix}$$

we first obtain the LU factorization of the matrix using the MATLAB function lu.

```
[L,U,P] = lu(A)
```

yields

$$\mathbf{L} = \begin{pmatrix} 1 & 0 & 0 \\ -1 & 1 & 0 \\ 1 & -\frac{1}{2} & 1 \end{pmatrix} \qquad \mathbf{U} = \begin{pmatrix} 1 & 2 & -1 \\ 0 & 4 & 0 \\ 0 & 0 & 2 \end{pmatrix} \qquad \mathbf{P} = \begin{pmatrix} 1 & 0 & 0 \\ 0 & 0 & 1 \\ 0 & 1 & 0 \end{pmatrix}$$

Next, the inverse of $\mathbf{U}$ is obtained by solving the three equations

$$\begin{pmatrix} 1 & 2 & -1 \\ 0 & 4 & 0 \\ 0 & 0 & 2 \end{pmatrix} \begin{pmatrix} \gamma_{11} \\ \gamma_{21} \\ \gamma_{31} \end{pmatrix} = \begin{pmatrix} 1 \\ 0 \\ 0 \end{pmatrix}$$

$$\begin{pmatrix} 1 & 2 & -1 \\ 0 & 4 & 0 \\ 0 & 0 & 2 \end{pmatrix} \begin{pmatrix} \gamma_{12} \\ \gamma_{22} \\ \gamma_{32} \end{pmatrix} = \begin{pmatrix} 0 \\ 1 \\ 0 \end{pmatrix}$$

$$\begin{pmatrix} 1 & 2 & -1 \\ 0 & 4 & 0 \\ 0 & 0 & 2 \end{pmatrix} \begin{pmatrix} \gamma_{31} \\ \gamma_{32} \\ \gamma_{33} \end{pmatrix} = \begin{pmatrix} 0 \\ 0 \\ 1 \end{pmatrix}$$

by back substitution. The result is

$$\mathbf{U}^{-1} = \begin{pmatrix} 1 & -\frac{1}{2} & \frac{1}{2} \\ 0 & \frac{1}{4} & 0 \\ 0 & 0 & \frac{1}{2} \end{pmatrix}$$

The inverse of $\mathbf{L}$, obtained similarly, is

$$\mathbf{L}^{-1} = \begin{pmatrix} 1 & 0 & 0 \\ 1 & 1 & 0 \\ -\frac{1}{2} & \frac{1}{2} & 1 \end{pmatrix}$$

Finally, the inverse of $\mathbf{A}$ is obtained using Equation 2.6 to be

$$\mathbf{A}^{-1} = \mathbf{U}^{-1}\mathbf{L}^{-1}\mathbf{P}$$

$$= \begin{pmatrix} 1 & -\frac{1}{2} & \frac{1}{2} \\ 0 & \frac{1}{4} & 0 \\ 0 & 0 & \frac{1}{2} \end{pmatrix} \begin{pmatrix} 1 & 0 & 0 \\ 1 & 1 & 0 \\ -\frac{1}{2} & \frac{1}{2} & 1 \end{pmatrix} \begin{pmatrix} 1 & 0 & 0 \\ 0 & 0 & 1 \\ 0 & 1 & 0 \end{pmatrix}$$

$$= \frac{1}{4} \begin{pmatrix} 1 & 2 & -1 \\ 1 & 0 & 1 \\ -1 & 2 & 1 \end{pmatrix}$$

which, coincidentally, is simply $\frac{1}{4}\mathbf{A}$. As a check on the calculation, you can verify that $\mathbf{A}\mathbf{A}^{-1} = \mathbf{A}^{-1}\mathbf{A} = \mathbf{I}_3$.   ●

MATLAB has a *built-in* function that will compute the inverse. The function `inv(A)` will return the inverse of the square matrix $\mathbf{A}$, if it exists. Actually, it is rarely necessary to compute the inverse matrix to solve a numerical problem. Even if you wish to solve the equation $\mathbf{A}\vec{x} = \vec{b}$ for a variety of right-hand-side vectors, $\vec{b}$, it is usually more efficient, and certainly more accurate, to solve each individually by ordinary Gaussian elimination or, equivalently, by using MATLAB's *backslash* division. That is, the equation should ordinarily be solved as x = A\b or, if solutions are required for several right-hand-side vectors, x = A\[b1 b2 ... bn]. To begin to understand the reasons for this we need to have a cursory understanding of the comparable efficiencies of the numerical methods described thus far.

OPERATION COUNTS

The basic arithmetic operations of a computer, addition/subtraction and multiplication/division, are called floating-point operations. The MATLAB function flops keeps a running total of the number of floating-point operations, and flops(0) resets the counter to zero. This function can be used to evaluate the relative computational efficiency of various numerical methods used to solve the same problem. In addition, because the progress of an algorithm is quite predictable, we can easily count the number of *flops* that will be required before the calculation is attempted.

In solving the $n \times n$ matrix equation $\mathbf{A}\vec{x} = \vec{b}$ by the Gaussian elimination method, the augmented matrix $(\mathbf{A}|\vec{b})$ is first reduced to upper-triangular form. Taking into account the increasing number of zeros as the algorithm progresses, the reduction of the matrix to zeros below the diagonal will require $(n-1)^2 + (n-2)^2 + \cdots + 2^2 + 1$ multiplications and a like number of subtractions.[6] Finally, back substitution requires $n(n-1)/2$ multiplications and a like number of additions. Combining all these operations, we conclude that the minimum number of flops required in the Gaussian elimination method is

$$\frac{1}{6}n(n-1)(4n+1) \approx \frac{2n^3}{3}$$

This is a minimum because the algorithm for Gaussian elimination will very likely be designed to switch rows if small pivots are encountered. Even assuming that the algorithm swaps rows in *every* step,[7] the maximum number of flops is still asymptotic to $2n^3/3$.

If addition/subtraction operations are ignored relative to multiplication/division, this expression is asymptotic to $n^3/3$ instead of $2n^3/3$.

A similar counting can be done for the procedure to compute the complete inverse matrix by the Gaussian elimination method and the LU factorization method. The result is that $\approx n^3$ total floating-point operations are required.

---

[6]This sum can be evaluated using the identity $\displaystyle\sum_{i=1}^{n} i^2 = \frac{n(n+1)(2n+1)}{6}$.

[7]When switching rows, the current pivot element is compared to the same column elements in the remaining rows in order to find the maximum, and then the two corresponding rows are swapped. Because computer comparisons are much faster than floating-point operations, these comparisons are ignored. Then, to accomplish a swap of two elements, three separate replacement statements are required, which are also faster than a floating-point operation and are, therefore, not included in the count of flops.

EXAMPLE 2.4

A MATLAB
Example of
flops

If we once again construct an arbitrary matrix **A** of size $n \times n$ and an arbitrary, n-component, right-hand-side vector, $\vec{b}$, the number of floating-point operations used to obtain a solution via x = A\b could be displayed by the following:

```
for i = 1:5
    n = 8*i^2;
    A = rand(n);     b = ones(n,1);
    limit = 2*n^3/3;
    flops(0); x = A\b; f = flops;
    disp([n f/limit])
end
```

```
      8.0000    2.2500
     32.0000    1.3103
     72.0000    1.1343
    128.0000    1.0743
    200.0000    1.0482
```

The ratio of the actual operations count to the theoretical limit, $2n^3/3$, approaches 1 for larger and larger matrices. Repeat the calculation, this time evaluating the complete inverse for each value of n using the function inv.   ●

HOMOGENEOUS EQUATIONS

Consider the homogeneous matrix equation $\mathbf{A}\vec{x} = \vec{0}$, where the right-hand side is a column vector of zeros. We have seen earlier that just because the product of two matrices is equal to a null matrix, we may not conclude that either of the matrices is also a null matrix. In this equation as well, we cannot immediately conclude that $\vec{x}$ must be zero.

However, if the rank of the $n \times n$ matrix **A** is equal to n, the existence of the inverse matrix $\mathbf{A}^{-1}$ is guaranteed. Then, multiplying the homogeneous equation by $\mathbf{A}^{-1}$, we conclude that the *only* solution to the homogeneous equation is the null vector, $\vec{x} = 0$. On the other hand, if rank(**A**) $< n$, nonzero solutions to the homogeneous equation will be possible that will depend on various free parameters. A very common situation is that where the rank of the matrix is $n - 1$. That is, the equation, when reduced to RREF, is of the form

$$\left(\begin{array}{cccccc|c} 1 & 0 & 0 & \cdots & 0 & \square & 0 \\ 0 & 1 & 0 & \cdots & 0 & \square & 0 \\ 0 & 0 & 1 & \cdots & 0 & \square & 0 \\ \cdots & \cdots & \cdots & \ddots & \cdots & \cdots & \vdots \\ 0 & 0 & 0 & 0 & 1 & \square & 0 \\ 0 & 0 & 0 & 0 & 0 & 0 & 0 \end{array} \right)$$

and thus the value for x_n is undetermined; i.e., it can be assigned to be equal to a parameter, γ, and the remaining components of $\vec{x}$ are determined in terms of

γ. In Chapter 3, we encounter this problem when solving for the eigenvectors of a so-called matrix eigenvalue equation. In that case, the undetermined parameter is specified by adding an additional requirement to the vector $\vec{x}$, namely, that it also be of unit length.

For example, the matrix in the equation

$$\begin{pmatrix} 1 & 1 & 1 \\ 0 & 1 & 0 \\ 1 & 0 & 1 \end{pmatrix} \begin{pmatrix} x \\ y \\ z \end{pmatrix} = \begin{pmatrix} 0 \\ 0 \\ 0 \end{pmatrix}$$

is easily shown to have rank 2 and thus has no inverse. The solution of the equation $\mathbf{A}\vec{x} = 0$ is then obtained to be $[-\gamma, \ 0, \ \gamma]^T$ for any value of γ. Requiring the vector to also be of unit length then yields a solution of $[-1, \ 0, \ 1]/\sqrt{2}$.

DETERMINANTS

The determinant of a square $n \times n$ matrix $\mathbf{A}$ is a single number (i.e., a scalar) that is uniquely defined in terms of the elements of the matrix. The determinant is written as either $\det(A)$ or $|\mathbf{A}|$, and it can be computed using the MATLAB function `det(A)`.

The value of the determinant of an $n \times n$ matrix is equal to the product of the diagonal elements of the upper-triangular matrix obtained by Gaussian elimination applied to the matrix; i.e., the product of the pivots. Clearly, if any of these pivot elements are zero, the rank of the matrix is less than n and the inverse, $\mathbf{A}^{-1}$, does not exist. That is, a matrix with zero determinant is termed *singular* and has no inverse.

Determinants are usually introduced in an elementary course in algebra and are very likely to be familiar to the reader. However, they can be extremely costly to evaluate for even moderately large matrices and should never be used to obtain a solution in such problems. Nonetheless, they do permit the solution to the linear matrix equation, $\mathbf{A}\vec{x} = \vec{b}$, to be formally expressed in a closed form known as Cramer's rule, and are therefore often helpful in theoretical analysis. They are commonly encountered in discussions relating to the solutions of matrix equations, so we provide here a short review of their properties. However, we are primarily concerned with computational algorithms in this text, and procedures based on the evaluation of determinants are never competitive with the methods already outlined. Therefore, we refer the reader to any elementary text on linear algebra for a detailed description of determinants. (See Strang, p. 163.)

The procedure for the evaluation of the determinant of a matrix is defined recursively; i.e., the determinant of an $n \times n$ matrix can be expressed in terms of determinants of $(n-1) \times (n-1)$ matrices. Or, starting from the other direction, the determinant of a 1×1 is simply a_{11} and that of a 2×2 is defined as

$$\begin{vmatrix} a_{11} & a_{12} \\ a_{21} & a_{22} \end{vmatrix} = a_{11}a_{22} - a_{12}a_{21} = \begin{pmatrix} a_{11} & a_{12} \\ a_{21} & a_{22} \end{pmatrix}$$
$$(-) \qquad\qquad (+)$$

A similar construction is used to define the determinant of a 3×3.

$$\begin{vmatrix} a_{11} & a_{12} & a_{13} \\ a_{21} & a_{22} & a_{23} \\ a_{31} & a_{32} & a_{33} \end{vmatrix} = [a_{11}a_{22}a_{33} + a_{12}a_{23}a_{31} + a_{13}a_{32}a_{21}]$$

$$- [a_{11}a_{23}a_{32} + a_{21}a_{12}a_{33} + a_{31}a_{22}a_{13}]$$

$$= \begin{pmatrix} a_{11} & a_{12} & a_{13} & a_{11} & a_{12} & a_{13} \\ a_{21} & a_{22} & a_{23} & a_{21} & a_{22} & a_{23} \\ a_{31} & a_{32} & a_{33} & a_{31} & a_{32} & a_{33} \end{pmatrix}$$

The general prescription for the determinant of an $n \times n$ matrix is

$$|\mathbf{A}| = \sum_{j=1}^{n} a_{ij} |\mathbf{A}_{ij}^{c}| (-1)^{i+j} \tag{2.8}$$

where $\mathbf{A}_{ij}^{c}$ is a complete $(n-1) \times (n-1)$ matrix (not the i-j element of a matrix) obtained from $\mathbf{A}$ by deleting row i and column j. The sum is evaluated *for any value of i.*

For example, the following 4×4 determinant is evaluated with $i = 1$:

$$\begin{vmatrix} 1 & 2 & 3 & 4 \\ 2 & 4 & 6 & 3 \\ 3 & 6 & 4 & 2 \\ 4 & 3 & 2 & 1 \end{vmatrix} = \mathbf{1} \begin{vmatrix} 4 & 6 & 3 \\ 6 & 4 & 2 \\ 3 & 2 & 1 \end{vmatrix} - \mathbf{2} \begin{vmatrix} 2 & 6 & 3 \\ 3 & 4 & 2 \\ 4 & 2 & 1 \end{vmatrix} + \mathbf{3} \begin{vmatrix} 2 & 4 & 3 \\ 3 & 6 & 2 \\ 4 & 3 & 1 \end{vmatrix} - \mathbf{4} \begin{vmatrix} 2 & 4 & 6 \\ 3 & 6 & 4 \\ 4 & 3 & 2 \end{vmatrix}$$

The four 3×3 determinants are easily computed and have values of 0, 0, -25, -50, respectively.[8] Thus the value of the determinant is 125. From these definitions of the determinant, we can easily conclude that

- The determinant of the *identity* matrix is 1.
- The determinant of a *diagonal* matrix is simply the product of the diagonal elements.
- The determinant of an upper-triangular or a lower-triangular matrix is simply the product of the diagonal elements.
- The determinant of a matrix in RREF is either 0 or 1.

Clearly, the condition for the existence of an inverse to an $n \times n$ matrix, namely rank$(\mathbf{A}) = n$, is equivalent to $|\mathbf{A}| \neq 0$.

---

[8]Notice that the evaluation of an $n \times n$ determinant requires the evaluation of n determinants of size $(n-1) \times (n-1)$, etc. The complete evaluation of such a determinant will clearly require at least $n!$ multiplications. For this reason, in realistic problems, where n is larger than 4 or 5, the numerical evaluation of determinants is to be strongly avoided.

Properties of Determinants   The following are some useful properties of determinants that are given without proof.

1. If all elements of any row are zero, then $|\mathbf{A}| = 0$.
2. If any two rows are interchanged, then $|\mathbf{A}|$ simply changes sign.
3. If any row is replaced by c times that entire row, then $|\mathbf{A}| \Rightarrow c|\mathbf{A}|$.
4. If to any row we add a multiple of another row, the determinant remains unchanged.
5. If any row is proportional to another row, then $|\mathbf{A}| = 0$.
6. $|\mathbf{A}| = |\mathbf{A}^T|$. (As a consequence, all the preceding statements apply to columns as well.)
7. $|\mathbf{AB}| = |\mathbf{A}||\mathbf{B}|$, but $|\mathbf{A} + \mathbf{B}| \neq |\mathbf{A}| + |\mathbf{B}|$.

The Accuracy of a Solution

Although the numerical solution of a problem may be your primary concern, usually much more important is the reliability of the result you obtain. Unless the calculation involves exclusively integers or ratios of integers, computed numbers will always contain inaccuracies due to rounding or truncation, in addition to the normal experimental errors associated with data. The finest distinction between numbers that can be made in MATLAB is characterized by the quantity eps. That is, eps is the smallest magnitude solution to the relation $1 + \epsilon \neq 1$. Thus, for any computed number, c, its fractional error, $|\delta c|/c$, can never be specified to an accuracy better than eps. In addition, the procedures used to obtain a solution to an equation will inevitably amplify this minimum error. Division by a very small number, or the subtraction of nearly equal numbers in the course of a calculation, can have a dramatically deleterious effect on the accuracy of a result. The *norm* and the *condition number* of a matrix, defined in the following sections, are useful tools for estimating the amplification of errors resulting from a solution of the equation $\mathbf{A}\vec{x} = \vec{b}$.

We begin by assuming that the "exact" solution of the equation $\mathbf{A}\vec{x} = \vec{b}$ exists and is designated as $\vec{x}_e$, that the *numerical* solution to the equation is $\vec{x}$, and the difference between the two is defined as

$$\delta\vec{x} = \vec{x} - \vec{x}_e$$

In addition, because $\mathbf{A}(\vec{x} - \vec{x}_e) = \mathbf{A}\vec{x} - \vec{b}$, we can define

$$\delta\vec{b} \equiv \mathbf{A}\vec{x} - \vec{b}$$

which is usually called the *residual*.

Although $\vec{x}_e$ and, therefore, the actual error, $\delta\vec{x}$, are *unknowable*, we have precise numerical values for the quantity $\delta\vec{b}$. The goal is to then obtain an estimate of the relation between $\delta\vec{x}$ and $\delta\vec{b}$.

A measure of the size (or length) of a vector is provided by the norm,

$$\text{norm}(\vec{x}) \equiv |\vec{x}| = \sqrt{\vec{x} \cdot \vec{x}}$$

which, when applied to the linear equation, $\mathbf{A}\vec{x} = \vec{b}$, states that

$$|\mathbf{A}\vec{x}| = |\vec{b}|$$

or

$$|\mathbf{A}\delta\vec{x}| = |\delta\vec{b}|$$

Clearly then, the remaining step is to relate the norm $|\mathbf{A}\delta\vec{x}|$ to $|\delta\vec{x}|$. To do this, we next need to extend the definition of the norm of a vector to the norm of a matrix.

NORM OF A MATRIX

Multiplying an n-dimensional vector $\vec{x}$ by an $n \times n$ matrix results in a second n-component vector. That is, $\vec{y} = \mathbf{A}\vec{x}$ is a mapping of vectors $\vec{x}$ into vectors $\vec{y}$. We have seen that if the matrix $\mathbf{A}$ is *orthogonal*, i.e., $\mathbf{A}^T\mathbf{A} = \mathbf{I}_n$, then $\mathbf{A}\vec{x}$ simply represents a rotation or a rotation plus a reflection of the original vector $\vec{x}$, and we expect that the norm, or length of the vector, will be unchanged by such a transformation.

If $\mathbf{A}$ is a more general matrix, we expect that the effect of $\mathbf{A}\vec{x}$ could be a vector that is both rotated *and* whose length is changed. The definition of the norm of a matrix should then reflect this feature of mapping vectors into vectors of varying length. Of course, the norm of a matrix $\mathbf{A}$ should also be independent of the vector it multiplies. Thus, we are led to consider $|\mathbf{A}\vec{x}|/|\vec{x}|$. However, this quantity is still dependent upon the vector $\vec{x}$. For example, there may be some nonzero vectors that satisfy $\mathbf{A}\vec{x} = 0$ whereas others do not. To define the norm of a matrix so that it is independent of the particular vector it multiplies, we arrive at the definition

$$\text{norm}(\mathbf{A}) \equiv ||\mathbf{A}|| = \left[\frac{|\mathbf{A}\vec{x}|}{|\vec{x}|}\right]_{\max} \tag{2.9}$$

where the maximum is over the set of all possible nonzero vectors $\vec{x}$. Thus, the *norm* of a matrix determines the maximum possible factor by which $\mathbf{A}$ will increase the length of the vectors it multiplies. We can also write this relation as

$$||\mathbf{A}||^2 = \left[\frac{\vec{x}^T\mathbf{A}^T\mathbf{A}\vec{x}}{\vec{x}^T\vec{x}}\right]_{\max}$$

In Chapter 3, we see that if a matrix is real and symmetric, its norm corresponds to the magnitude of the largest of its *eigenvalues*, and various algorithms will be described to compute these eigenvalues. For the moment, it suffices to know that in the preceding expression $\mathbf{A}^T\mathbf{A}$ is symmetric if $\mathbf{A}$ is real, and thus there is a well-established procedure for calculating the norm of $\mathbf{A}$.

From the definition of $||\mathbf{A}||$, it is apparent that

$$|\mathbf{A}\vec{x}| \leq ||\mathbf{A}|| \cdot |\vec{x}|$$

We next use this relation to obtain limits on the accuracy of the solution to $\mathbf{A}\vec{x} = \vec{b}$.

THE CONDITION NUMBER OF A MATRIX

The sensitivity of solutions to $\mathbf{A}\vec{x} = \vec{b}$ to small uncertainties or changes in $\vec{b}$ is characterized by a quantity called the *condition number* of the coefficient matrix $\mathbf{A}$. The condition number, K, of a matrix is defined in terms of its norm and the norm of $\mathbf{A}^{-1}$ as follows.

Assuming that the right-hand-side vector, $\vec{b}$, contains errors designated by $\delta\vec{b}$ and that the inverse matrix, $\mathbf{A}^{-1}$, exists, the following four relations must hold:

$$\mathbf{A}^{-1}\vec{b} = \vec{x}; \qquad \delta\vec{b} = \mathbf{A}\delta\vec{x}; \qquad \mathbf{A}\vec{x} = \vec{b}; \qquad \delta\vec{x} = \mathbf{A}^{-1}\delta\vec{b}$$

so that

$$||\mathbf{A}^{-1}||\,|\vec{b}| \geq |\vec{x}|; \qquad |\delta\vec{b}| \leq ||\mathbf{A}||\,|\delta\vec{x}|; \qquad ||\mathbf{A}||\,|\vec{x}| \geq |\vec{b}|; \qquad |\delta\vec{x}| \leq ||\mathbf{A}^{-1}||\,|\delta\vec{b}|$$

and

$$\frac{1}{||\mathbf{A}^{-1}||\,|\vec{b}|} \leq \frac{1}{|\vec{x}|}; \qquad \frac{1}{||\mathbf{A}||\,|\vec{x}|} \leq \frac{1}{|\vec{b}|}$$

Thus

$$\frac{1}{||\mathbf{A}^{-1}||}\left(\frac{|\delta\vec{b}|}{|\vec{b}|}\right) \leq ||\mathbf{A}||\left(\frac{|\delta\vec{x}|}{|\vec{x}|}\right); \qquad \frac{1}{||\mathbf{A}||}\left(\frac{|\delta\vec{x}|}{|\vec{x}|}\right) \leq ||\mathbf{A}^{-1}||\left(\frac{|\delta\vec{b}|}{|\vec{b}|}\right)$$

where, in the third line, we have used the fact that if a and b are positive, and $a \leq b$, then $\frac{1}{a} \geq \frac{1}{b}$. Similarly, in the fourth line, if a, b, c, and d are positive, and $a \geq b$ and $c \geq d$, then $ac \geq bd$.

If we next define the *condition* of a matrix $\mathbf{A}$ as

$$K = \mathrm{cond}(\mathbf{A}) \equiv ||\mathbf{A}^{-1}||\,||\mathbf{A}||$$

these relations may be combined as

$$\frac{1}{K}\left(\frac{|\delta\vec{b}|}{|\vec{b}|}\right) \leq \left(\frac{|\delta\vec{x}|}{|\vec{x}|}\right) \leq K\left(\frac{|\delta\vec{b}|}{|\vec{b}|}\right) \tag{2.10}$$

Further, in Chapter 3, we show that if the eigenvalues of a symmetric $n \times n$ matrix, $\mathbf{A}$, are the set of numbers λ_i, $i = 1, \ldots, n$, then the eigenvalues of $\mathbf{A}^{-1}$ are simply λ_i^{-1}, $i = 1, \ldots, n$. Then, because $||\mathbf{A}||$ is determined from the maximum magnitude eigenvalue of the symmetric matrix $\mathbf{A}^T\mathbf{A}$, it is not surprising that the norm $||\mathbf{A}^{-1}||$ is determined from the reciprocal of the smallest magnitude eigenvalue. Thus, the condition of a matrix is the ratio of the largest "stretch" divided by the smallest "shrink" that will be effected on an arbitrary vector. Note that K is always greater than or equal to 1.

Equation 2.10 then gives us bounds on the fractional error of a solution in terms of the known fractional error ($|\delta\vec{b}|/|\vec{b}|$). Because $\delta\vec{b}$ is the result of rounding and truncation errors, this quantity can, at best, be expected to be of the size of the machine accuracy `eps`, so the expected error in the solution will certainly be unacceptably large if $K \cdot eps \approx 1$. More realistically, there will often be measurement errors associated with some of the quantities in the matrix equation, in which case the expected error in the solution will be approximately equal to the experimental error times the condition number. By either standard, matrices with an unacceptably large condition are called *ill conditioned*. Note that if the matrix is singular, $K = \infty$.

EXAMPLE 2.5

Accuracy of a Solution

The Hilbert matrix of order n,

$$
\mathbf{H} = \begin{pmatrix}
1 & \frac{1}{2} & \frac{1}{3} & \cdots & \frac{1}{n} \\
\frac{1}{2} & \frac{1}{4} & \frac{1}{5} & \cdots & \frac{1}{n+1} \\
\vdots & \vdots & \vdots & \ddots & \vdots \\
\frac{1}{n} & \frac{1}{n+1} & \frac{1}{n+2} & \cdots & \frac{1}{2n}
\end{pmatrix}
$$

provides the classic example of an ill-conditioned matrix equation. Because MAT-LAB provides an "exact" inverse to a Hilbert matrix by means of the function `invhilb(n)`, we can construct a test of the preceding ideas. Using the following MATLAB code, we can compute the condition, K, of the matrix and, from it, the approximate error, $K\epsilon$, where $\epsilon = $ `eps` of the machine accuracy. Next, the fractional error in the residual, $|\delta\vec{b}|/|\vec{b}|$, will be used to compute the maximum error from Equation 2.10, and this may be compared with the actual error using the solution obtained from `invhilb`.

```
n = 6;
H = hilb(n);  HI = invhilb(n);
b = ones(n,1);  x = H\b;
db = norm(H*x-b);
nb = norm(b);  nx = norm(x);
K = cond(H);
dx = x-HI*b;
err = norm(dx)/nx;
disp([n  err K*eps K*db/nb])
```

The results of this code for $n = 6, \ldots, 12$, are collected in Table 2.1.

n	Actual error $|\delta\vec{x}|/|\vec{x}|$	Approx. error $K\epsilon$	Maximum error $K|\delta\vec{b}|/|\vec{b}|$
6	0.000000000003	0.0000000033	0.0000014
7	0.00000000067	0.000000106	0.000149
8	0.0000000515	0.00000339	0.0342
9	0.00000468	0.000110	6.31
10	0.0001874	0.00356	637.5
11	0.00456	0.116	1.87×10^5
12	0.0532	3.73	2.20×10^7
13	1.820	116.3	2.27×10^9

Table 2.1 *Error computations for Hilbert matrices*

From the numbers in the last two columns, we conclude that solutions to the equation $\mathbf{H}\vec{x} = \vec{b}$, where $\mathbf{H}$ is a Hilbert matrix, will be increasingly unreliable if the size of the Hilbert matrix is greater than $n = 10$. The reader should duplicate these results using less pathological matrices, perhaps using `magic(n)` or `rand(n)` to see that, under normal circumstances, the error in a solution can be quite small. ●

Iterative Solutions to Matrix Equations

There is a special class of linear matrix equations for which the methods described thus far are not well suited. If a matrix $\mathbf{A}$ is very large, not only may there be substantial problems associated with storing the matrix, but the sheer number of arithmetic operations may cause the accumulation of round-off errors to invalidate any result that is obtained. Additionally, in most problems of this type the coefficient matrix is not only very large but usually *sparse*; i.e., composed of a large fraction of zero elements. Finally, as we shall see in Chapter 20, the most common occurrence of very large, sparse matrices is in the solution of boundary value differential equations, and these matrices have the added feature of being *banded*. It is, therefore, advantageous to design procedures that specifically utilize these special features.

However, as we shall see, the algorithms that result recast the problem as an algebraic problem, not as a matrix problem and are, therefore, only of secondary interest in this chapter. For this reason our discussion here is somewhat cursory; more details are given in Chapter 20.

THE GAUSS-SEIDEL ITERATION METHOD

If $\mathbf{A}$ is an $n \times n$ matrix and $\vec{b}$ is an n-component column vector, then $\mathbf{A}\vec{x} = \vec{b}$ represents n equations in n unknowns. Next, the ith equation can be formally solved for x_i, $i = 1, \ldots, n$, to obtain

$$a_{11}x_1 = b_1 - \sum_{i \neq 1} a_{1i}x_i$$

$$a_{22}x_1 = b_2 - \sum_{i \neq 2} a_{2i}x_i$$

$$\vdots \qquad \qquad \vdots$$

$$a_{nn}x_1 = b_n - \sum_{i \neq n} a_{ni}x_i$$

Then, if *none* of the diagonal elements, a_{ii}, is zero, these equations can be solved for the unknowns x_i, $i = 1, \ldots, n$. Clearly, this is not yet a useful solution, because to compute the values for the x_i's on the left, the values for the same set of x_i's must be used in the terms on the right. However, it does suggest the next best thing, namely, an iterative solution. The procedure is then quite simple. A complete set of values for the solution vector, $\vec{x}$, is first *guessed*. These values are then successively used in the preceding sequence of equations. Notice that in the first equation a *new* value for x_1 is computed. This value will, in general, be equal to the exact solution for x_1 only if all the other x_i's that enter this equation are also correct. However, under some criteria to be described shortly, this new value for x_1 will be an improvement on the earlier guess. This value for x_1 then enters into the calculation of x_2 in the second equation, and so on. After proceeding through the entire set of n equations, the difference between the current expression for the solution vector and the earlier guess is computed. If, by some prescribed criteria, this difference is judged to be small, the solution is accepted; if not, the most recent vector is used as the guess, and the entire process is repeated.

Diagonal Dominance   If the Gauss-Seidel method is to succeed, the diagonal elements must each be nonzero. If the iterative procedure is to *converge*, even stronger conditions must be satisfied. It can be shown (see the discussion in Chapter 20) that if the matrix **A** is *diagonally dominant*, the iterative process of Gauss-Seidel is *guaranteed* to converge, starting with *any* initial guess. (As with any iterative process, the rate of convergence will strongly depend upon the quality of the initial guess—the closer the initial guess to the exact solution, the faster the process will converge.)

A matrix is diagonally dominant if it satisfies the following condition for each and every row:

$$|a_{ii}| > \sum_{k \neq i} |a_{ik}| \qquad \text{for each value of } i$$

This is a very stringent condition. Often matrices in which the diagonal element is simply the largest magnitude element in each row are also found to converge, but this is not guaranteed.

If the original matrix is not diagonally dominant, it is frequently possible to rearrange the rows and/or the columns to bring it into diagonally dominant form. In the general case, this process can be quite complicated.

EXAMPLE 2.6

Iterative Solution of a Matrix Equation

The solution of the 20×20 matrix equation

$$\begin{pmatrix} 4 & -2 & 1 & 0 & \cdots & 0 & 0 \\ -2 & 4 & -2 & 1 & \cdots & 0 & 0 \\ 1 & -2 & 4 & -2 & \cdots & 0 & 0 \\ \vdots & \vdots & \vdots & \vdots & \ddots & \vdots & \vdots \\ 0 & 0 & 0 & 0 & \cdots & -2 & 4 \end{pmatrix} \begin{pmatrix} x_1 \\ x_2 \\ x_3 \\ \vdots \\ x_{20} \end{pmatrix} = \begin{pmatrix} 1 \\ 2 \\ 3 \\ \vdots \\ 20 \end{pmatrix}$$

is an excellent candidate for an iterative solution. The matrix consists of a large number of zeros, so any procedure based upon Gaussian elimination will spend an inordinate amount of time multiplying by zeros. The matrix is also banded and, although it is not diagonally dominant in the sense defined previously, at least the largest elements in each row appear along the main diagonal.

The matrix equation can be recast as algebraic equations, with the following result:

$$x_1 = \frac{1}{4}(1 + 2x_2 - x_3)$$

$$x_2 = \frac{1}{4}[2 + 2(x_3 + x_1) - x_4]$$

$$x_i = \frac{1}{4}[i + 2(x_{i-1} + x_{i+1}) - (x_{i-2} + x_{i+2})] \quad \text{for } i = 3, \ldots, 18$$

$$x_{19} = \frac{1}{4}[19 + 2(x_{20} + x_{18}) - x_{17}]$$

$$x_{20} = \frac{1}{4}[20 + 2x_{19} - x_{18}]$$

The MATLAB code to solve this set of equations would then resemble

```
N = 50;           b = [1:20]';
x0 = ones(20,1);
for iter = 1:N
 x = x0;
 x(1) = .25*(1+2*x(2)-x(3));
 x(2) = .25*(2+2*(x(3)+x(1))-x(4));
 for k = 3:18
    x(k) = .25*(k+2*(x(k-1)+x(k+1))-(x(k-2)+x(k+2)));
 end
 x(19) = .25*(19+2*(x(20)+x(18))-x(17));
 x(20) = .25*(20+2*x(19)-x(18));
 diff = norm(x-x0)/norm(x0);
 disp([iter diff])
 x0 = x;
end
```

The computed fractional changes in the solution vector, from one iteration to the next, are listed as follows:

n	Fractional change in successive solutions
5	0.0181...
10	0.00147...
15	0.000167...
20	0.0000199...
25	0.00000229...
30	0.000000251...
35	0.0000000260...
40	0.00000000257...
45	0.00000000027...
50	0.00000000002...

For this particular set of equations, the procedure is seen to converge. Also, it should be noted that problems related to round-off error are usually less troublesome in iterative methods, as the error in one iteration is, to some extent, corrected in the next step. You may wish to try this problem after reducing the magnitude of the diagonal elements to see how the rate of convergence is slowed as the matrix deviates even further from strict diagonal dominance.   ●

MATLAB Functions Related to Matrix Inversion

There are several toolbox functions that will perform many of the operations discussed in this chapter. A few of the more common ones are listed in Table 2.2.

lu(A)	If **A** is a square, $n \times n$ matrix, the MATLAB statement [L,U,P] = lu(A) will perform the LU factorization of the matrix. The matrices returned are **L**, a lower-triangular matrix with 1s on the main diagonal; **U**, an upper-triangular matrix that is identical to the matrix obtained by Gaussian elimination applied to **A**; and **P**, a permutation matrix that records the sequence of row switching that was employed in the Gaussian elimination process. The matrices satisfy the equation **LU = PA**.
inv(A)	If **A** is a square matrix with a nonzero determinant, the inverse is returned. If **A** is singular or *near* singular, a warning is displayed.
rref(A)	The reduced row echelon form of the matrix **A** is returned. The matrix need not be square.
rank(A)	The rank of the matrix **A** is returned.
norm(A)	The maximum value of the ratio of the norms, $|\vec{y}|/|\vec{x}|$, where $\vec{y} = \mathbf{A}\vec{x}$ and $\vec{x}$ is any nonzero vector, is returned. It is designated as $\mathrm{norm}(A) = ||\mathbf{A}||$.
cond(A)	The condition of a matrix is defined as $K = ||\mathbf{A}|| \, ||\mathbf{A}^{-1}||$.
det(A)	The determinant of the square matrix **A** is computed.
\	("backslash"). If the matrix multiplication $\mathbf{AB} = \mathbf{C}$ represents a valid matrix product; i.e., the three matrices are properly sized, then the backslash division operator can be used to solve this equation as, for example, $\mathbf{B} = \mathbf{A} \backslash \mathbf{C}$. The backslash division is then read as **A** *divided into* **C**.

Table 2.2   *Common MATLAB functions related to matrices*

Problems

REINFORCEMENT EXERCISES

P2.1 **Reduced Row Echelon Form.** Reduce each of the following matrices to RREF by hand, and verify by using the MATLAB function rref().

a. $\begin{pmatrix} 2 & 1 \\ 3 & 2 \end{pmatrix}$

b. $\begin{pmatrix} -1 & 2 & 0 \\ 2 & -1 & 2 \\ 0 & 2 & -1 \end{pmatrix}$

c. $\begin{pmatrix} 3 & 3 & 3 \\ 3 & 3 & 3 \\ 3 & 3 & 3 \end{pmatrix}$

d. $\begin{pmatrix} 3 & -2 & 12 \\ -2 & 3 & -13 \\ -1 & -2 & 4 \end{pmatrix}$

e. $\begin{pmatrix} 0 & 0 & 1 \\ 0 & 1 & 0 \\ 1 & 0 & 0 \end{pmatrix}$

f. $\begin{pmatrix} 1 & -2 & -2 & 0 \\ 0 & 0 & -1 & 3 \\ 0 & 1 & -1 & 1 \\ 1 & 2 & 0 & 4 \end{pmatrix}$

P2.2 Gaussian Elimination. Use the method of Gaussian elimination to solve, by hand, the following matrix equations:

a. $\begin{pmatrix} 1 & 2 & 3 \\ 3 & 2 & 1 \\ 1 & 1 & 0 \end{pmatrix} \begin{pmatrix} x_1 \\ x_2 \\ x_3 \end{pmatrix} = \begin{pmatrix} 1 \\ 2 \\ 3 \end{pmatrix}$

b. $\begin{pmatrix} 2 & 4 & -1 & -2 \\ 0 & 1 & -1 & -1 \\ 1 & 1 & 1 & 0 \\ 2 & 1 & -1 & 1 \end{pmatrix} \begin{pmatrix} x_1 \\ x_2 \\ x_3 \\ x_4 \end{pmatrix} = \begin{pmatrix} 18 \\ 5 \\ 4 \\ 3 \end{pmatrix}$

P2.3 Solutions to $\mathbf{A}\vec{x} = \vec{b}$ if A is *Rank Deficient* and No Inverse Exists. The matrix

$$\mathbf{A} = \begin{pmatrix} 0 & 2 & 2 \\ -2 & 0 & 2 \\ -2 & -2 & 0 \end{pmatrix}$$

is *singular*; i.e., rank$(\mathbf{A}) = 2$.

a. Use Gaussian elimination to determine the value of α for which there is a solution of the equation $\mathbf{A}\vec{x} = \vec{b}$, where

$$\vec{b} = \begin{pmatrix} 2 \\ 2 \\ \alpha \end{pmatrix}$$

[That is, for what value of α does rank$(\mathbf{A})$ = rank$([\mathbf{A}|b])$? For this value of α is the solution *unique*?]

b. The following coefficient matrix is also rank deficient, and α in the right-hand-side vector is a parameter. Apply the Gaussian elimination method to determine the only value of α that will permit a solution.

$$\begin{pmatrix} 2 & 1 & 0 & 0 \\ 1 & 2 & 1 & -3 \\ 0 & 1 & 2 & -2 \\ 0 & 0 & 1 & 0 \end{pmatrix} \begin{pmatrix} x_1 \\ x_2 \\ x_3 \\ x_4 \end{pmatrix} = \begin{pmatrix} -1 \\ -1 \\ 1 \\ \alpha \end{pmatrix}$$

P2.4 Rank of a Matrix. To determine whether or not the solution to the matrix equation $\mathbf{A}\vec{x} = \vec{b}$ is unique, we write down the matrix $[\mathbf{A}|\vec{b}]$ in augmented form, reduce to RREF, determine rank($\mathbf{A}$) and rank($[\mathbf{A}|\vec{b}]$), and apply the rules on page 57. Use this technique to determine whether the following matrix equations will yield: a unique solution, no solution, or an infinite number of solutions. (Use the MATLAB function `rref()`.)

a. $$\begin{pmatrix} 1 & -2 & -2 \\ 0 & 0 & -1 \\ 1 & 2 & 0 \end{pmatrix} \begin{pmatrix} x_1 \\ x_2 \\ x_3 \end{pmatrix} = \begin{pmatrix} 1 \\ 2 \\ 3 \end{pmatrix}$$

b. $$\begin{pmatrix} 1 & 2 & 2 & 4 \\ 0 & 0 & -1 & -3 \\ 0 & 0 & 0 & 0 \end{pmatrix} \begin{pmatrix} x_1 \\ x_2 \\ x_3 \\ x_4 \end{pmatrix} = \begin{pmatrix} 0 \\ 0 \\ 0 \end{pmatrix}$$

c. $$\begin{pmatrix} 1 & -2 & -1 & 5 & 4 & 4 \\ 2 & -2 & 0 & -2 & -6 & 2 \\ 0 & 1 & 0 & 0 & 3 & 0 \\ 2 & -2 & 1 & 4 & -8 & 5 \end{pmatrix} \begin{pmatrix} x_1 \\ x_2 \\ x_3 \\ x_4 \\ x_5 \\ x_6 \end{pmatrix} = \begin{pmatrix} -6 \\ -2 \\ 0 \\ -3 \end{pmatrix}$$

d. $$\begin{pmatrix} 1 & 2 & 3 & 4 & 5 & 6 \\ 2 & 3 & 4 & 5 & 6 & 7 \\ 3 & 4 & 5 & 6 & 7 & 8 \\ 9 & 0 & 1 & 2 & 3 & 4 \end{pmatrix} \begin{pmatrix} x_1 \\ x_2 \\ x_3 \\ x_4 \\ x_5 \\ x_6 \end{pmatrix} = \begin{pmatrix} 1 \\ -1 \\ 2 \\ 3 \end{pmatrix}$$

P2.5 Inverse Matrices. By hand calculation, for each of the following matrices: (1) obtain the LU factorization and verify by using the MATLAB function `lu`, and (2) use the matrices $\mathbf{L}$ and $\mathbf{U}$ to obtain the inverse of each matrix. Verify by using the MATLAB function `inv`.

a. $$\begin{pmatrix} 2 & 1 \\ 3 & 2 \end{pmatrix}$$

b. $$\begin{pmatrix} -1 & 2 & 0 \\ 2 & -1 & 2 \\ 0 & 2 & -1 \end{pmatrix}$$

$$
\text{c.} \quad \begin{pmatrix} 2 & -1 & 0 & 0 \\ -1 & 2 & 0 & 0 \\ 0 & 0 & 2 & -1 \\ 0 & 0 & -1 & 2 \end{pmatrix}
$$

$$
\text{d.} \quad \begin{pmatrix} 3 & -2 & 1 \\ -2 & 3 & 2 \\ 1 & 2 & 3 \end{pmatrix}
$$

$$
\text{e.} \quad \begin{pmatrix} 0 & 0 & 0 & 0 & 2 \\ 0 & 2 & 0 & 0 & 0 \\ 0 & 0 & 0 & 2 & 0 \\ 2 & 0 & 0 & 0 & 0 \\ 0 & 0 & 2 & 0 & 0 \end{pmatrix} \quad \textit{Note}: \text{This is two times a permutation matrix.}
$$

P2.6 Inverse of a Product. Express the inverse of $\mathbf{U} = \mathbf{ABC}$ in terms of the individual inverses of $\mathbf{A}$, $\mathbf{B}$, and $\mathbf{C}$.

P2.7 Inverse of an Exponentiated Matrix. If $\mathbf{A}$ is a square matrix, the expression $e^{\mathbf{A}}$ is then itself a square matrix and is defined in terms of the ordinary series expansion for the exponential. The inverse of the matrix $e^{\mathbf{A}}$ is $e^{-\mathbf{A}}$. Demonstrate that this is the case for the matrix

$$
\mathbf{A} = \begin{pmatrix} 3 & -1 & -1 \\ -1 & 3 & 1 \\ -1 & 1 & 3 \end{pmatrix}
$$

by using the MATLAB function `expm()` to compute $e^{\mathbf{A}}$ and $e^{-\mathbf{A}}$.

P2.8 Symmetric Matrices. If $\mathbf{A}$ is symmetric and $\mathbf{A}^{-1}$ exists, demonstrate by example that $\mathbf{A}^{-1}$ is also symmetric.

P2.9 Diagonal Matrices. Explicitly prove the results of the previous two problems for the case of diagonal matrices.

P2.10 Triangular Matrices. Let $\mathbf{L}$ be a 5×5, lower-triangular matrix.
    a. If $\ell_{ii} \neq 0$ for $i = 1, \ldots, 5$, prove that the rank of $\mathbf{L}$ is 5.
    b. If $\ell_{33} = 0$, but the remaining diagonal elements are all nonzero, show explicitly that a solution of $\mathbf{L}\vec{x} = 0$ exists.

EXPLORATION PROBLEMS

P2.11 Matrix Condition and Error Bounds. Consider a 2×2 matrix, $\mathbf{A}(\epsilon)$, and its inverse, $\mathbf{A}^{-1}(\epsilon)$, shown here:

$$
\mathbf{A}(\epsilon) = \begin{pmatrix} 4 & 2 \\ 2 & 1+\epsilon \end{pmatrix} \quad \mathbf{A}^{-1}(\epsilon) = \frac{1}{4\epsilon} \begin{pmatrix} 1+\epsilon & -2 \\ -2 & 4 \end{pmatrix}
$$

    a. For values of $\epsilon = 2^{-i}$ for $i = 1, \ldots, 6$, use the vectors $\vec{x}_1 = \begin{pmatrix} 2 \\ 1 \end{pmatrix}$ and $\vec{x}_2 = \begin{pmatrix} 1 \\ -2 \end{pmatrix}$ to compute the norms of $\mathbf{A}$ and $\mathbf{A}^{-1}$, respectively. (See Equation 2.9.)

    b. For each value of ϵ, estimate the condition of matrix $\mathbf{A}$ and compare with the value returned by the MATLAB function `cond`.

    c. If the right-hand-side vector $\vec{b}$ is equal to `[1 1]'`, estimate the range of the fractional error in the solution of $A\vec{x} = \vec{b}$ for each value of ϵ.

P2.12 Complex Matrices. MATLAB is quite adept at performing arithmetic, even when the elements are complex numbers. For example, demonstrate that the Gaussian elimination procedure can be applied to equations in which the matrices are complex, by solving the equation

$$\begin{pmatrix} 1 & \iota & 2 \\ 2\iota & 1 & 0 \\ 1 & -\iota & 3 \end{pmatrix} \begin{pmatrix} x_1 \\ x_2 \\ x_3 \end{pmatrix} = \begin{pmatrix} 3 - 2\iota \\ 0 \\ 4 + 2\iota \end{pmatrix}$$

where $\iota = \sqrt{-1}$. Verify your result using MATLAB backslash division; i.e., `x = A\b`.

P2.13 Fitting a Line Through Two Points. The general equation for a straight line is $ax + by + c = 0$. Use this to show that the requirements that the line go through the points (x_1, y_1), (x_2, y_2), and that the point (x, y) be on the line, are equivalent to requiring that the matrix

$$\begin{pmatrix} x & y & 1 \\ x_1 & y_1 & 1 \\ x_1 & y_2 & 1 \end{pmatrix}$$

have rank < 3; i.e., be rank deficient.

P2.14 Common Factors in Polynomials.

    a. The two lines, $y = a_1 x + a_0$ and $y = b_1 x + b_0$ will cross the x axis at the same point, x_r (i.e., have a common root), only if there exists a nonzero solution of the equation

$$\begin{pmatrix} a_1 & a_0 \\ b_1 & b_0 \end{pmatrix} \begin{pmatrix} x_r \\ 1 \end{pmatrix} = 0$$

    The condition that two quadratic functions have a common root would be

$$a_2 x_r^2 + a_1 x_r + a_0 = 0$$
$$b_2 x_r^2 + b_1 x_r + b_0 = 0$$

    By multiplying each of these equations by x_r to obtain two additional independent equations, show that the condition for a common root is equivalent to requiring that a nonzero solution exists to the equation

$$
\begin{pmatrix}
a_2 & a_1 & a_0 & 0 \\
0 & a_2 & a_1 & a_0 \\
b_2 & b_1 & b_0 & 0 \\
0 & b_2 & b_1 & b_0
\end{pmatrix}
\begin{pmatrix}
x_r^3 \\
x_r^2 \\
x_r \\
1
\end{pmatrix}
= 0
$$

b. Use the preceding discussion to determine whether or not the two quadratics, $y = x^2 + 2x - 3$ and $y = x^2 + 4x + 3$, have a common factor.

c. Generalize this procedure to cubic functions, and thereby determine whether the following sets of cubics have a common factor:

(i) $y = 2x^3 + 5x^2 + 6x + 8$
$y = -x^3 - 7x^2 - 9x + 2$

(ii) $y = 2x^3 + 5x^2 + 6x + 8$
$y = 2x^3 + 3x^2 + 5x + 4$

P2.15 Fitting a Polynomial Through Fixed Points.

a. Show that the coefficients of a line, $y = a_1 x + a_0$, through two points, (x_1, y_1) and (x_2, y_2), can be determined by solving

$$
\begin{pmatrix}
x_1 & 1 \\
x_2 & 1
\end{pmatrix}
\begin{pmatrix}
a_1 \\
a_0
\end{pmatrix}
=
\begin{pmatrix}
y_1 \\
y_2
\end{pmatrix}
$$

for the values of a_1 and a_0.

b. Show that the coefficients of a *quadratic*, $y = a_2 x^2 + a_1 x + a_0$, through three distinct points are solutions of

$$
\begin{pmatrix}
x_1^2 & x_1 & 1 \\
x_2^2 & x_2 & 1 \\
x_3^2 & x_3 & 1
\end{pmatrix}
\begin{pmatrix}
a_2 \\
a_1 \\
a_0
\end{pmatrix}
=
\begin{pmatrix}
y_1 \\
y_2 \\
y_3
\end{pmatrix}
$$

c. Generalize this result and use the MATLAB function `inv()` to find the coefficients of the 5th-degree polynomial that goes through the following 6 points:

i	x_i	y_i
1	−2.5	−2205
2	−1.5	−75
3	−0.5	−9
4	0.5	9
5	1.5	75
6	2.5	2205

d. The fitting of an nth-degree polynomial through $n + 1$ points is then possible, provided a solution exists for the equation

$$
\begin{pmatrix}
x_1^{n-1} & x_1^{n-2} & \cdots & x_1 & 1 \\
x_2^{n-1} & x_2^{n-2} & \cdots & x_2 & 1 \\
\vdots & \vdots & \ddots & \vdots & \vdots \\
x_{n-1}^{n-1} & x_{n-1}^{n-2} & \cdots & x_{n-1} & 1 \\
x_n^{n-1} & x_n^{n-2} & \cdots & x_n & 1
\end{pmatrix}
\begin{pmatrix}
a_n \\
a_{n-1} \\
\vdots \\
a_1 \\
a_0
\end{pmatrix}
=
\begin{pmatrix}
y_1 \\
ay_2 \\
\vdots \\
y_{n-1} \\
y_n
\end{pmatrix}
$$

This matrix is called the *Vandermonde* matrix and is full rank, provided no two values of x_i are equal. The MATLAB function **vander(x)** will compute a Vandermonde matrix based on the values of x_i contained in the input vector, **x**. Using this function, repeat the solution to part b, and show that the coefficients can be written as

```
a = vander(-2.5:1:2.5)\[-2205 -75 -9 9 75 2205]'
```

P2.16 Formal Solution to a 2 × 2. Use Gaussian elimination to obtain the formal solution to

$$
\begin{pmatrix}
a_{11} & a_{12} \\
a_{21} & a_{22}
\end{pmatrix}
\begin{pmatrix}
x_1 \\
x_2
\end{pmatrix}
=
\begin{pmatrix}
b_1 \\
b_2
\end{pmatrix}
$$

P2.17 Norm of a Matrix. Start with the matrix $\mathbf{A} = \begin{pmatrix} 5 & 1 \\ 1 & 2 \end{pmatrix}$ and a column vector $\vec{x} = \begin{pmatrix} z \\ 1 \end{pmatrix}$ and determine the function $f(z) = \vec{x}^T \mathbf{A}^T \mathbf{A} \vec{x} / \vec{x}^T \vec{x}$. Find the value of z that corresponds to a maximum of $f(z)$. Use the maximum value of f to estimate a value for the norm of the matrix. Compare this result with the value returned by MATLAB's **norm**.

P2.18 Norms of Special Matrices.
    a. Prove that the norm of an orthogonal matrix is +1.
    b. Prove that the norm of a projection matrix is +1.

P2.19 Condition of a Matrix.
    a. The condition of a matrix has a minimum value of $K = 1$. Give an example of a matrix with condition $K = 1$.
    b. Prove that the condition of a projection matrix is undefined; i.e., $K = \infty$.

P2.20 Gauss-Jordan Elimination. Gauss-Jordan elimination is a variation on ordinary Gaussian elimination wherein elementary row operations are used to "zero" all the elements in the *pivot* column, both above and below the pivot element. Use this method to solve the matrix equations in Problem 2.2.

P2.21 Gauss-Seidel Iteration. Rewrite the MATLAB code in Example 2.6 to handle a matrix of size $n \times n$ that is *banded* with a vector $\vec{a}$ of length n along the principal diagonal, a vector $\vec{c}$ of length $(n-1)$ along the nearest adjacent diagonals, and a vector

$\vec{d}$ of length $(n-2)$ along the next pair of diagonals. (Look up the use of the MATLAB function `diag`.) The right-hand-side vector should be contained in $\vec{b}$. Test the code using the matrix in Example 2.6. After obtaining a solution vector, determine the fractional change in the residual; i.e., $|\mathbf{A}\vec{x} - \vec{b}|/|\vec{b}|$ after 50 iterations.

Summary

REDUCED ROW ECHELON FORM (RREF)

Definition

The "final step" in a Gauss-Jordan reduction of a matrix. Each row of a matrix in RREF will begin with either a 0 or a 1 (the *leading* 1 in that row). Leading 1s in successive rows are always to the right of those above. All rows of all 0s are at the bottom of the matrix.

Use

The RREF of a matrix is used to determine the existence and uniqueness of the solutions of the equations. This is usually expressed in terms of the *rank* of the matrix.

MATLAB Function

`rref(A)` will return the reduced row echelon form of a rectangular matrix $\mathbf{A}$. In addition, `[R,k] = rref(A)` will return the RREF of $\mathbf{A}$ in the matrix $\mathbf{R}$ and the indices of the first r independent columns contained in $\mathbf{A}$. That is, `length(k)` = r = rank of $\mathbf{A}$, and the r column vectors `x(k)` are the only uniquely determined vectors in the solution of $\mathbf{A}\vec{x} = \vec{b}$.

RANK

Definition

The rank of a matrix is equal to the number of nonzero rows in the reduced row echelon form of the matrix.

Use

When solving the n equations in m unknowns, written as a matrix equation $\mathbf{A}\vec{x} = \vec{b}$, the following conditions apply:

If	Then
`rank(A)` = `rank([A b])` = m	A *unique* solution exists.
`rank(A)` = `rank([A b])` r < m	An *infinite* number of valid solutions exist in $(m-r)$ parameters.
`rank(A)` $\neq$ `rank([A b])`	No solution is possible.

MATLAB Function

`rank(A)` returns the rank of the matrix $\mathbf{A}$.

LU FACTORIZATION

Definition

A square $n \times n$ matrix, $\mathbf{A}$, can be factored into the product of a lower-triangular matrix, $\mathbf{L}$, and an upper-triangular matrix, $\mathbf{U}$, where $\mathbf{U}$ is identical to the result obtained by applying Gaussian elimination to the matrix, and $\mathbf{L}$ is then computed by means of Equation 2.3. If row switching is employed using Gaussian elimination, the record of rows switched is stored in the permutation matrix $\mathbf{P}$, and the matrices then satisfy the equation $\mathbf{LU} = \mathbf{PA}$.

Use

LU factorization is useful when repeatedly solving $\mathbf{A}\vec{x} = \vec{b}$ if the coefficient matrix remains constant while a variety of right-hand-side vectors $\vec{b}$ are used. It is also the most efficient means of computing $\mathbf{A}^{-1}$. (See Inverse Matrix.)

MATLAB Function

The MATLAB statement $[\mathtt{L}, \mathtt{U}, \mathtt{P}] = \mathtt{lu}(\mathtt{A})$ will return the three matrices that are the elements of the LU factorization of the matrix $\mathbf{A}$.

INVERSE MATRIX

Definition

If a square $n \times n$ matrix $\mathbf{A}$ is of rank n, then a second matrix $\mathbf{A}^{-1}$ of similar size exists such that $\mathbf{A}^{-1}\mathbf{A} = \mathbf{A}\mathbf{A}^{-1} = \mathbf{I}_n$. This matrix may be computed from $\mathbf{A}$ by first obtaining the LU factorization of the matrix $\mathbf{A} = \mathbf{LU}$ and then computing the inverses of $\mathbf{L}$ and $\mathbf{U}$ by forward and back substitution, respectively. The inverse of $\mathbf{A}$ is then given by $\mathbf{A}^{-1} = \mathbf{U}^{-1}\mathbf{L}^{-1}\mathbf{P}$.

Use

The solution of $\mathbf{A}\vec{x} = \vec{b}$ is $\vec{x} = \mathbf{A}^{-1}\vec{b}$.

MATLAB Function

$\mathtt{inv}(\mathtt{A})$ will return the inverse of $\mathbf{A}$ if it exists. If the condition of the matrix is unacceptably large, a warning is displayed. If the inverse does not exist, a matrix of "infinities" is returned.

THE DETERMINANT OF A MATRIX

Definition

The determinant of a matrix $\mathbf{A}$, denoted by $|\mathbf{A}|$, is equal to the product of the successive *pivots* in a Gaussian elimination procedure to obtain the inverse. Thus, if $|\mathbf{A}| = 0$, no inverse exists.

Use

To establish the existence or nonexistence of an inverse matrix.

MATLAB Function

$\mathtt{det}(\mathtt{A})$ will return the value of the determinant of a square matrix $\mathbf{A}$.

NORM OF A MATRIX

Definition

The norm of a matrix $\mathbf{A}$, designated as $||\mathbf{A}||$, is defined to be the maximum value of the expression

$$\mathrm{norm}(\mathbf{A}) \equiv ||\mathbf{A}|| = \left[\frac{|\mathbf{A}\vec{x}|}{|\vec{x}|}\right]_{\max} = \left[\frac{\vec{x}^T \mathbf{A}^T \mathbf{A}\vec{x}}{\vec{x}^T \vec{x}}\right]^{1/2}_{\max}$$

where the maximum is over the space of all nonzero vectors $\vec{x}$.

Use	The norm of a matrix is an indication of the maximum amount by which the length of a vector will be changed after multiplication by $\mathbf{A}$.
MATLAB Function	The function $\mathtt{norm(A)}$ returns the norm of the matrix $\mathbf{A}$. The matrix need not be square.

CONDITION OF A MATRIX

Definition	The condition of a matrix $\mathbf{A}$, designated as cond(A), is defined to be the product of the norms of $\mathbf{A}$ and $\mathbf{A}^{-1}$.
Use	A matrix condition is useful in determining the limits on the error in the solution of $\mathbf{A}\vec{x} = \vec{b}$. If the residual is defined as $\delta\vec{b} \equiv \mathbf{A}\vec{x} - \vec{b}$, and the condition of $\mathbf{A}$ is K, then the fractional error in the solution satisfies the inequality

$$\frac{1}{K}\left(\frac{|\delta\vec{b}|}{|\vec{b}|}\right) \leq \left(\frac{|\delta\vec{x}|}{|\vec{x}|}\right) \leq K\left(\frac{|\delta\vec{b}|}{|\vec{b}|}\right)$$

MATLAB Function	The function $\mathtt{cond(A)}$ returns the condition of the matrix $\mathbf{A}$. The matrix need not be square.

GAUSS-SEIDEL ITERATIVE METHOD

Definition	Instead of solving the equation $\mathbf{A}\vec{x} = \vec{b}$ as a matrix equation, the equations are written out algebraically, with the ith equation written as $x_i = \cdots$. A guess is made for the complete solution vector, and these values are successively inserted into the right-hand sides of the equations, with the elements of $\vec{x}$ being continually updated. The vector $\vec{x}$ obtained after using the last equation is then used as the starting vector in the next iteration.
Use	Very large, sparse matrices are usually well suited for a Gauss-Seidel procedure.

MATLAB Functions Used

$\mathtt{rref(A)}$	Returns the RREF of matrix $\mathbf{A}$. Can also be used to determine the rank of a matrix. See preceding text.
$\mathtt{lu(A)}$	The statement $\mathtt{[L,U,P] = lu(A)}$ returns the three matrices that form the LU factorization of the matrix $\mathbf{A}$; i.e., $\mathbf{LU} = \mathbf{PA}$.
$\mathtt{det(A)}$	If $\mathbf{A}$ is a square matrix, $\mathtt{det(A)}$ returns the value of the determinant of $\mathbf{A}$.
$\mathtt{inv(A)}$	If $\mathbf{A}$ is a square $n \times n$ matrix, $\mathtt{inv(A)}$ returns a matrix of the same size corresponding to the inverse matrix to $\mathbf{A}$. If the matrix $\mathbf{A}$ is singular or near singular, a warning is displayed.
$\mathtt{rank(A)}$	Returns the number of linearly independent columns of the matrix $\mathbf{A}$. See also $\mathtt{rref}$.

`norm(A)`   Returns the norm of either a matrix or a vector.

`cond(A)`   Returns the condition of the matrix **A**.

`\`   If **A**, **B**, and **C** are three matrices with appropriate sizes and shapes such that the multiplication **C** = **AB** can be defined, then MATLAB matrix division is likewise defined for the following pairs and operators:

Mathematical definition	MATLAB Operation	Comment
$\mathbf{A} = \mathbf{CB}^{-1}$	`A = C/B`	If $\mathbf{B}^{-1}$ exists, then the result of C/B is the same as the matrix **A**. However, if $\mathbf{B}^{-1}$ does not exist, the matrix returned will be the same size as **A**, but generally not equal to **A**. Its elements correspond to the *least squares best fit* to an over- or underdetermined set of equations. Thus, if $\mathbf{A}\vec{b} = \vec{c}$, then $\vec{c}/\vec{b}$ is a matrix with only column 3 nonzero and determined by $$\begin{pmatrix} 0 & 0 & a_1 \\ 0 & 0 & a_2 \\ 0 & 0 & a_3 \end{pmatrix} \begin{pmatrix} x_1 \\ x_2 \\ x_3 \end{pmatrix} = \begin{pmatrix} b_1 \\ b_2 \\ b_3 \end{pmatrix}$$
$\mathbf{B} = \mathbf{A}^{-1}\mathbf{C}$	`B = A\C`	This is read as **A** *divided into* **C**. If $\mathbf{A}^{-1}$ exists, the result is equal to **B**. Otherwise, a least squares fit to an over- or underdetermined set of equations is returned in a matrix of the same size but generally not equal to **B**. Thus, if $\mathbf{A}\vec{b} = \vec{c}$, then $\mathbf{A} \setminus \vec{c}$ is the solution to the matrix equation. However, if $\vec{a} = \begin{pmatrix} 1 \\ 2 \end{pmatrix}$ and $\vec{b} = \begin{pmatrix} 3 \\ 2 \end{pmatrix}$, then $$\vec{a}^T \setminus (\vec{a}^T\vec{b}) = \vec{a}^T \setminus 7 = \begin{pmatrix} 0 \\ \frac{7}{2} \end{pmatrix}$$ and $$\vec{a} \setminus (\vec{a}\vec{b}^T) = \vec{a} \setminus \begin{pmatrix} 3 & 2 \\ 6 & 4 \end{pmatrix}$$ $$= [3 \ 2] = \vec{b}^T$$

`expm(A)`   If **A** is a square matrix, then `expm(A)` is a *matrix* exponentiation and is equivalent to

$$e^{\mathbf{A}} = \mathbf{I}_n + \mathbf{A} + \frac{\mathbf{A}^2}{2!} + \frac{\mathbf{A}^3}{3!} + \cdots$$

vander(x)   If $\vec{x}$ is a vector of length n, then **vander(x)** returns the matrix

$$\mathbf{V} = \begin{pmatrix} x_1^{n-1} & \cdots & x_1^2 & x_1 & 1 \\ x_2^{n-1} & \cdots & x_2^2 & x_2 & 1 \\ \vdots & \ddots & \vdots & \vdots & \vdots \\ x_n^{n-1} & \cdots & x_n^2 & x_n & 1 \end{pmatrix}$$

3 / *Matrix Eigenvalue Equations*

PREVIEW   In this chapter, we are concerned with solutions of the matrix equation $\mathbf{A}\vec{x} = \lambda\vec{x}$, for the solution vectors $\vec{x}$ *and* for the parameter λ. As we shall see, for a given $n \times n$ matrix $\mathbf{A}$, solutions are possible only if λ is equal to one of a set of n values that depend on $\mathbf{A}$. These are called the *eigenvalues* of the matrix. The corresponding solution vectors are then the *eigenvectors*, and the equation $\mathbf{A}\vec{x} = \lambda\vec{x}$ is called an *eigenvalue equation*.

Eigenvalue equations are important in almost every area of applied mathematics. They are used to describe many types of processes: vibrations and buckling of structures, chemical kinetics, atomic physics, Markov processes, and the convergence of iterative procedures, just to name a few. In Chapter 20, we will see that an important class of differential equations can be structured as eigenvalue equations. Even if a problem does not quite match the form of an eigenvalue equation, it is frequently useful to recast the problem in terms of the eigenvectors of a related eigenvalue problem. Along with methods to solve the linear *inhomogeneous* equation, $\mathbf{A}\vec{x} = \vec{b}$, procedures to solve the *homogeneous* eigenvalue equation, $\mathbf{A}\vec{x} = \lambda\vec{x}$, form the basis of all of linear analysis.

In studying the solution of linear simultaneous equations, we have seen that the problem can be formulated in terms of matrix equations that may then be interpreted as operations on vectors in an n-dimensional space. The special cases of two or three dimensions then permit a visualization of the matrix equations in terms of geometrical operations, which can aid significantly in understanding the underlying mathematics. Therefore, we once again rely on geometrical arguments to guide the discussion and begin with the simplest situation, that of 2×2 eigenvalue equations. The results for the 2×2 problem are readily generalized to n dimensions.

Geometrical Transformations on Vectors in Two Dimensions

An *orthogonal* matrix in n dimensions is a square $n \times n$ matrix $\mathbf{O}$ that satisfies the requirement $\mathbf{O}^T\mathbf{O} = \mathbf{O}\mathbf{O}^T = \mathbf{I}_n$. We have seen that the product of an orthogonal matrix, $\mathbf{O}$, times a column vector, $\vec{x}$, represents a rotation or a rotation plus a reflection of the vector in the n-dimensional space. That is, a vector $\hat{e}$ of unit length retains its length after multiplication by $\mathbf{O}$. (If $\hat{e}' = \mathbf{O}\hat{e}$, then $|\hat{e}|^2 = \hat{e}^T\mathbf{O}^T\mathbf{O}\hat{e} = \hat{e}^T\hat{e} = 1$.) Because the vector $\hat{e}$ is of unit length, it can be represented as a point on a sphere of radius 1, and $\hat{e}' = \mathbf{O}\hat{e}$ represents moving the point to another location on the same sphere.

The operations represented by orthogonal matrices are, of course, simply a subset of the most general linear operations on a vector represented by the matrix product $\mathbf{A}\vec{x}$. We could expect this product to change the length of the vector, in addition to rotating it. To guide our thinking, we once again turn to the easily handled case of 2×2 matrices.

2×2 ORTHOGONAL MATRICES

The most general real 2×2 matrices that satisfy the condition for orthogonal matrices, $\mathbf{O}^T\mathbf{O} = \mathbf{I}_2$, can be expressed as

$$\mathbf{O}_1 = \begin{pmatrix} \cos\theta & \sin\theta \\ -\sin\theta & \cos\theta \end{pmatrix} \qquad \mathbf{O}_2 = \begin{pmatrix} \cos\theta & \sin\theta \\ \sin\theta & -\cos\theta \end{pmatrix}$$

The first form, $\mathbf{O}_1$, represents a rotation of the coordinate system by the angle θ. That is, if $\vec{r} = x\hat{i} + y\hat{j} = x\begin{pmatrix} 1 \\ 0 \end{pmatrix} + y\begin{pmatrix} 0 \\ 1 \end{pmatrix}$, then

$$\vec{r}' = \mathbf{O}_1\vec{r} = \begin{pmatrix} x\cos\theta + y\sin\theta \\ y\cos\theta - x\sin\theta \end{pmatrix}$$

from which it follows that $|\vec{r}'^2| = |\vec{r}'^2|$. The relationship between the vectors $\vec{x}$ and $\vec{x}'$ is illustrated graphically in Figure 3.1.

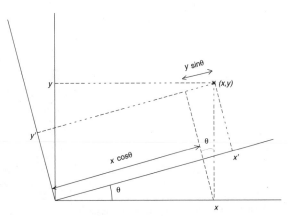

Figure 3.1 *The effects of a rotation of the coordinate system (x, y).*

The effect of the matrix $\mathbf{O}_2$ is to combine a rotation with a *reflection*. Thus, if $\theta \approx 0$, then $\cos\theta \approx 1$, $\sin\theta \approx 0$, $\mathbf{O}_2 \approx \begin{pmatrix} 1 & 0 \\ 0 & -1 \end{pmatrix}$, and $\mathbf{O}_2\vec{r}$ represents a reflection[1] of the vector $\vec{r}$ through the x axis; i.e., $\vec{r}' = \begin{pmatrix} x \\ -y \end{pmatrix}$.

SYMMETRIC 2 × 2 MATRICES

If multiplying an orthogonal matrix times a vector can be interpreted as a rotation or a rotation plus a reflection of the vector, is there a similar interpretation possible when multiplying a vector by an arbitrary *symmetric* matrix?

$$\vec{r}' = \mathbf{A}\vec{r}$$

$$\begin{pmatrix} x' \\ y' \end{pmatrix} = \begin{pmatrix} a & b \\ b & c \end{pmatrix} \begin{pmatrix} x \\ y \end{pmatrix}$$

$$= \begin{pmatrix} ax + by \\ cy + bx \end{pmatrix}$$

Because the matrix $\mathbf{A}$ is not, in general, orthogonal, we cannot expect that the operation $\mathbf{A}\vec{x}$ will preserve the length of the vector. Computing the ratio of the length of $\vec{x}'$ to $\vec{x}$ tells us how the vector has stretched or shortened. That is,

$$\frac{|\vec{r}'|^2}{|\vec{r}|^2} = \frac{(a^2 + b^2)x^2 + 2b(a + c)xy + (c^2 + b^2)y^2}{x^2 + y^2}$$

Consider a specific example:

$$\vec{r}' = \frac{1}{3}\begin{pmatrix} 9 & 2 \\ 2 & 6 \end{pmatrix} \begin{pmatrix} x \\ y \end{pmatrix} = \begin{pmatrix} x' \\ y' \end{pmatrix}$$

This represents a mapping of a point (x, y) into (x', y') and is illustrated for a selection of points on a circle in Figure 3.2.

This graph was produced by function `SPOKES(A,n)`, which is provided on the accompanying disk. You may wish to experiment with a variety of 2 × 2 matrices

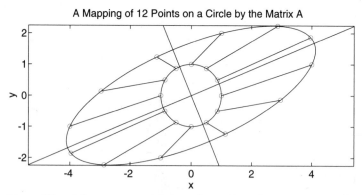

Figure 3.2   *A linear symmetric transformation of points on a circle.*

---

[1]Although both $\mathbf{O}_1$ and $\mathbf{O}_2$ have a norm of +1, $\det(\mathbf{O}_1) = 1$ but $\det(\mathbf{O}_2) = -1$. This distinction between orthogonal matrices that represent simple rotations and those representing rotations plus reflections remains true in multidimensions as well.

A to see how they map points on a circle. The quantity n is the number of equally spaced points on the circle that will be mapped. Note that if $\det(\mathbf{A}) < 0$, the mapping will include a reflection through one of the axes.

PRINCIPAL AXES OF A TRANSFORMATION

To further understand this rather complicated mapping, we might ask whether or not there exist any vectors that are simply lengthened or shortened but do not undergo any rotation. That is, we look for vectors that satisfy

Eigenvalue
Equation
$$\mathbf{A}\vec{x} = \lambda\vec{x}$$

This equation is known as an *eigenvalue* equation. The solution vector, $\vec{x}$, is called an *eigenvector*, and the scaling factor, λ, the *eigenvalue*. Vectors that satisfy this condition determine the *principal axes* of the transformation.

Clearly, the null vector satisfies every eigenvalue equation and, for this reason, is excluded from the desired solution set. In addition, the equation is homogeneous in $\vec{x}$; therefore, the length of the vector cannot be determined from the eigenvalue equation. We are thus free to require that all solutions be *normalized* to unit length. Writing out the equations in terms of the components of $\vec{x}$, we have, in our specific example:

$$\begin{pmatrix} 3 & \frac{2}{3} \\ \frac{2}{3} & 2 \end{pmatrix} \begin{pmatrix} x \\ y \end{pmatrix} = \lambda \begin{pmatrix} x \\ y \end{pmatrix}$$

Notice that for the n-dimensional case, this equation generalizes to n equations in $(n+1)$ unknowns (the n components of $\vec{x}$, plus λ). The normalization condition $(\sum_{i=1}^{n} x_i^2 = 1)$ supplies the final condition needed to specify a solution.

Solving for the Eigenvalues Rewriting the eigenvalue equation as

$$\begin{pmatrix} 3 - \lambda & \frac{2}{3} \\ \frac{2}{3} & 2 - \lambda \end{pmatrix} \begin{pmatrix} x \\ y \end{pmatrix} = 0$$

it is clear that the equation is of a form $(\mathbf{A} - \lambda\mathbf{I})\vec{x} = 0$; i.e., it is a homogeneous equation. Thus, if $\vec{x} \neq 0$, a solution is possible *only* if the rank of $(\mathbf{A} - \lambda\mathbf{I}) < 2$ or, equivalently, if $\det(\mathbf{A} - \lambda\mathbf{I}) = |\mathbf{A} - \lambda\mathbf{I}| = 0$. That is,

$$\begin{vmatrix} 3 - \lambda & \frac{2}{5} \\ \frac{2}{5} & 2 - \lambda \end{vmatrix} = 0$$

or

$$(3 - \lambda)(2 - \lambda) - \frac{4}{9} = \lambda^2 - 5\lambda + \frac{50}{9}$$

$$= \left(\lambda - \frac{5}{3}\right)\left(\lambda - \frac{10}{3}\right) = 0$$

so the equation can only have a solution if $\lambda = \frac{5}{3}$ or $\lambda = \frac{10}{3}$.

For the $n \times n$ case, the determinant, $|\mathbf{A} - \lambda \mathbf{I}_n|$, yields a polynomial of degree n in λ. The roots of the polynomial are then the eigenvalues. The equation for the roots,

Characteristic
Equation
$$|\mathbf{A} - \lambda \mathbf{I}_n| = 0$$

is called the *characteristic* equation. The roots of the characteristic equation for our 2×2 matrix are $\lambda_1 = \frac{5}{3}$, $\lambda_2 = \frac{10}{3}$.

Solving for the Eigenvectors for Each Eigenvalue We next take the solutions for λ_1, λ_2, one at a time, and solve for the corresponding eigenvectors.

$\lambda = \lambda_1 = \frac{5}{3}$

$$\begin{pmatrix} 3 - \frac{5}{3} & \frac{2}{3} \\ \frac{2}{3} & 2 - \frac{5}{3} \end{pmatrix} \begin{pmatrix} x \\ y \end{pmatrix} = \frac{1}{3} \begin{pmatrix} 4 & 2 \\ 2 & 1 \end{pmatrix} \begin{pmatrix} x \\ y \end{pmatrix} = 0$$

$$\Rightarrow y = -2x$$

so that $\vec{x}_1 = \begin{pmatrix} -x \\ 2x \end{pmatrix}$. And finally, normalizing this vector, we obtain

$$\hat{e}_1 = \frac{\vec{x}_1}{|\vec{x}_1|} = \frac{1}{\sqrt{5}} \begin{pmatrix} -1 \\ 2 \end{pmatrix}$$

The second eigenvalue, $\lambda = \frac{10}{3}$, yields $\hat{e}_2 = \frac{1}{\sqrt{5}} \begin{pmatrix} 2 \\ 1 \end{pmatrix}$. Notice that the requirement that the matrix be symmetric guarantees that the eigenvalues will be *real*.[2]

Furthermore, the two eigenvectors, $\hat{e}_1, \hat{e}_2$, that we have obtained are orthogonal and, therefore, *span* the two-dimensional space. (As we will see, this result is readily generalizable to the case of n dimensions.) Any two-dimensional vector can then be written in terms of $\hat{e}_1$ and $\hat{e}_2$ as

$$\begin{pmatrix} x \\ y \end{pmatrix} = a\hat{e}_1 + a\hat{e}_2$$

---

[2]The characteristic equation for a symmetric 2×2 matrix, $\begin{pmatrix} a & b \\ b & c \end{pmatrix}$, is

$$\lambda^2 - (a + c)\lambda + (ac - b^2) = 0$$

so that

$$\lambda = \frac{1}{2}\left[(a + c) \pm \sqrt{(a - c)^2 + 4b^2}\right]$$

which is real for any real values of a, b, and c.

Using the expressions for $\hat{x}_1$ and $\hat{x}_2$, you can solve for the expressions for a and b to obtain

$$\begin{pmatrix} x \\ y \end{pmatrix} = \frac{-x + 2y}{\sqrt{5}} \hat{x}_1 + \frac{2x + y}{\sqrt{5}} \hat{x}_2 \tag{3.1}$$

Thus, for those vectors parallel to the lines $y = \frac{1}{2}x$ and $y = -2x$, the effect of the transformation **A** is to simply scale the vector in a parallel transformation. These two directions are known as the *principal axes* of the transformation and, for our example, are included in Figure 3.2.

Transformation of the Matrix A to Principal Axes   Finally, consider what the matrix **A** would look like in a coordinate system rotated so that its axes are along the perpendicular eigenvectors; that is, a coordinate system oriented along the principal axes of the transformation.

We have seen in Chapter 1 that if one coordinate system is specified in terms of a set of orthogonal unit vectors, $\hat{e}_i$, $i = 1, \ldots, n$, and a second in terms of a different set, $\hat{e}_i'$, $i = 1, \ldots, n$, and the two sets of unit vectors are combined as columns of the matrices **E** and **E'**, respectively,[3] then the relationship between the two coordinate systems is simply an orthogonal transformation (rotation or rotation plus reflection) effected by the orthogonal matrix $\mathbf{U} = \mathbf{E}'^T \mathbf{E}'$. That is, the elements of $\vec{x}$ in the coordinate system $\hat{e}_i'$ are related to components in the $\hat{e}_i$ coordinate system by

Rotation of a Vector
$$\vec{x}' = \mathbf{U}^T \vec{x}; \quad \mathbf{U}^T = \mathbf{E}'^T \mathbf{E}$$

and the inverse relation is

$$\vec{x} = \mathbf{U}\vec{x}'; \quad \mathbf{U} = \mathbf{E}^T \mathbf{E}'$$

This equation may then be used to relate any matrix or vector in the original coordinate system to its transformed image. Thus, the equation

$$\vec{y} = \mathbf{A}\vec{x}$$

becomes, in the transformed system,

$$\mathbf{U}\vec{y}' = \mathbf{A}\mathbf{U}\vec{x}'$$

Multiplying by $\mathbf{U}^T$ from the left, we obtain

$$\vec{y}' = \mathbf{A}'\vec{x}'$$

where the matrix **A'** has been defined as

Orthogonal Transformation of a Matrix
$$\mathbf{A}' = \mathbf{U}^T \mathbf{A}\mathbf{U}$$

---

[3]In the present context, **E'** means **E**-*prime*, not **E**-transpose.

In summary, to transform a matrix equation given in one coordinate system to a second, related to the first by a rotation or a rotation plus a reflection, first do the following:

- Compute the *transformation* matrix, $\mathbf{U} = \mathbf{E}^T \mathbf{E}'$, where $\mathbf{E}, \mathbf{E}'$ are the matrices of the unit vectors written as columns in the original and rotated systems, respectively. The matrix $\mathbf{U}$ is an orthogonal matrix.
- Replace column vectors by $\vec{x}' = \mathbf{U}^T \vec{x}$ and matrices by $\mathbf{A}' = \mathbf{U}^T \mathbf{A} \mathbf{U}$.

We next return to our task of finding the image of the matrix in a coordinate system rotated to the principal axes of the 2×2 matrix $\mathbf{A}$ on page 93. The original coordinate system is specified by the ordinary unit vectors, $\hat{\imath} = \begin{pmatrix} 1 \\ 0 \end{pmatrix}$ and $\hat{\jmath} = \begin{pmatrix} 0 \\ 1 \end{pmatrix}$, so that

$$\mathbf{E} = \begin{pmatrix} 1 & 0 \\ 0 & 1 \end{pmatrix}$$

The principal axes of $\mathbf{A}$ are its eigenvectors, $\hat{e}_1$ and $\hat{e}_2$, so that

$$\mathbf{E}' = \frac{1}{\sqrt{5}} \begin{pmatrix} -1 & 2 \\ 2 & 1 \end{pmatrix}$$

$$\mathbf{U} = \mathbf{E}^T \mathbf{E}'$$

$$= \frac{1}{\sqrt{5}} \begin{pmatrix} -1 & 2 \\ 2 & 1 \end{pmatrix}$$

$$\mathbf{A}' = \mathbf{U}^T \mathbf{A} \mathbf{U}$$

$$= \frac{1}{3} \begin{pmatrix} 5 & 0 \\ 0 & 10 \end{pmatrix}$$

As expected, the image of $\mathbf{A}$ in the principal axes is a *diagonal* matrix; i.e., a matrix that simply "stretches" the coordinates independently. The magnitude of the stretch along a principal axis is just the corresponding eigenvalue.

SUMMARY OF PROPERTIES OF REAL, SYMMETRIC 2×2 EIGENVALUE EQUATIONS

The properties of real, symmetric 2×2 eigenvalue equations are listed as follows:

1. All of the eigenvalues of symmetric matrices are *real*.
2. The complete set of eigenvectors forms a complete set of basis vectors in the two-dimensional space. Eigenvectors corresponding to distinct eigenvalues are automatically orthogonal.
3. An orthogonal transformation to the basis set of eigenvectors yields an eigenvalue equation with a diagonal coefficient matrix. The diagonal elements of this matrix are the eigenvalues of the original matrix.

4. An orthogonal transformation of coordinates cannot alter the eigenvalues; therefore, $|\mathbf{A}| = \prod \lambda_i$. (See also Problem 3.1c.)

5. The *trace* of a matrix is the sum of the diagonal elements. The trace of the coefficient matrix in an eigenvalue equation is equal to the sum of the eigenvalues.

GENERALIZATION TO $N \times N$ EIGENVALUE EQUATIONS

The generalization of each of the statements describing the results of the 2×2 problem presents few difficulties. For example, the statement that the eigenvectors corresponding to different eigenvalues are orthogonal can be proved as follows.

Proof That the Eigenvectors of a Real-Symmetric Matrix Are Orthogonal

- Solve the eigenvalue equation $\mathbf{A}\vec{x} = \lambda\vec{x}$ for the eigenvalues, λ_i, and the normalized eigenvectors, $\hat{e}_i$; i.e., $\mathbf{A}\hat{e}_i = \lambda_i\hat{e}_i$.
- Write the equation for the ith eigenvector and, below it, the transpose of the equation for the jth eigenvector. That is,

$$\mathbf{A}\hat{e}_i = \lambda_i\hat{e}_i$$
$$\hat{e}_j^T \mathbf{A}^T = \lambda_j \hat{e}_j^T$$

- Next multiply the first equation, from the left, by $\hat{e}_j^T$ and the second, from the right, by $\hat{e}_i$, and subtract the two equations. The matrix is assumed to be symmetric, $\mathbf{A} = \mathbf{A}^T$; thus we obtain

$$\hat{e}_j^T \mathbf{A}\hat{e}_i = \lambda_i \hat{e}_j^T \hat{e}_i$$
$$\underline{-[\hat{e}_j^T \mathbf{A}^T \hat{e}_i = \lambda_j \hat{e}_j^T \hat{e}_i]}$$
$$0 = (\lambda_i - \lambda_j)\hat{e}_j^T \hat{e}_i$$

- Thus, if $\lambda_i \neq \lambda_j$, the vectors $\hat{e}_i$ and $\hat{e}_j$ must be orthogonal.[4]

This proof can be generalized to *Hermitian* matrices; i.e., matrices that satisfy $\overline{\mathbf{A}}^T = \mathbf{A}$, where $\bar{c}$ represents the *complex conjugate* of the quantity c. A symmetric-real matrix is simply a special case of a Hermitian matrix that happens to be real. Hermitian matrices have the following very important characteristic: Their eigenvectors form a complete, orthonormal set, and their eigenvalues are all real.

We will not prove the remaining claims, but refer the reader to any text on linear algebra.

---

[4]We state without proof that in the situation where λ is a repeated root of the characteristic equation; i.e., two or more eigenvalues have the same value, corresponding eigenvectors can be found that are orthogonal, provided the matrix $\mathbf{A}$ is real-symmetric or complex-Hermitian.

Before moving on to the computer algorithms used to solve larger eigenvalue equations, it is worthwhile to convince you that the methods and results associated with the 2×2 problem can be readily generalized to larger problems. To this end, we now include an extended example of a 4×4 problem.

SOLUTION OF THE GENERAL EIGENVALUE PROBLEM

The *algebraic* procedure for solving eigenvalue problems is to form the characteristic equation, $|\mathbf{A} - \lambda \mathbf{I}_n| = 0$, solve for the roots of this equation, and then, for each value of λ_i, construct a normalized eigenvector by solving the homogeneous equation, $(\mathbf{A} - \lambda \mathbf{I}_n)\hat{e}_i = 0$. This is the method most often encountered in texts on linear algebra, and it is suitable for small matrices, in particular, 3×3 or 4×4. Example 3.1 shows this type of calculation for a 4×4 matrix.

EXAMPLE 3.1

**Solution of a
4 × 4 Eigenvalue
Equation**

Consider the matrix

$$\mathbf{A} = \begin{pmatrix} 5 & 2 & 1 & 0 \\ 2 & 5 & 0 & 1 \\ 1 & 0 & 5 & 2 \\ 0 & 1 & 2 & 5 \end{pmatrix} \tag{3.2}$$

The characteristic equation is:

$$|\mathbf{A} - \lambda| = \begin{vmatrix} 5-\lambda & 2 & 1 & 0 \\ 2 & 5-\lambda & 0 & 1 \\ 1 & 0 & 5-\lambda & 2 \\ 0 & 1 & 2 & 5-\lambda \end{vmatrix}$$

$$= (5-\lambda) \begin{vmatrix} 5-\lambda & 0 & 1 \\ 0 & 5-\lambda & 2 \\ 1 & 2 & 5-\lambda \end{vmatrix}$$

$$-2 \begin{vmatrix} 2 & 0 & 1 \\ 1 & 5-\lambda & 2 \\ 0 & 2 & 5-\lambda \end{vmatrix} + \begin{vmatrix} 2 & 5-\lambda & 1 \\ 1 & 0 & 2 \\ 0 & 1 & 5-\lambda \end{vmatrix}$$

$$= (5-\lambda)^4 - 10(5-\lambda)^2 + 9 = 0$$

which can be factored to yield $(5 - \lambda) = \pm 1, \pm 3$. Thus, the roots of the characteristic equation are

$$\lambda_1 = 2 \quad \lambda_2 = 4$$
$$\lambda_3 = 6 \quad \lambda_4 = 8$$

To find the eigenvector corresponding to, say λ_2, we must next solve the equation

$$(\mathbf{A} - 4\mathbf{I}_n)\hat{e}_2 = 0$$

or

$$\begin{pmatrix} 5-4 & 2 & 1 & 0 \\ 2 & 5-4 & 0 & 1 \\ 1 & 0 & 5-4 & 2 \\ 0 & 1 & 2 & 5-4 \end{pmatrix} \begin{pmatrix} u \\ v \\ x \\ y \end{pmatrix} = 0 \quad \text{or} \quad \begin{aligned} \begin{pmatrix} 1 & 2 \\ 2 & 1 \end{pmatrix} \begin{pmatrix} u \\ v \end{pmatrix} &= -\begin{pmatrix} x \\ y \end{pmatrix} \\ \begin{pmatrix} 1 & 2 \\ 2 & 1 \end{pmatrix} \begin{pmatrix} x \\ y \end{pmatrix} &= -\begin{pmatrix} u \\ v \end{pmatrix} \end{aligned}$$

These two equations are easily decoupled to yield

$$\begin{pmatrix} 1 & 2 \\ 2 & 1 \end{pmatrix}^2 \begin{pmatrix} x \\ y \end{pmatrix} = \begin{pmatrix} x \\ y \end{pmatrix} \quad \text{or} \quad \begin{pmatrix} 1 & 1 \\ 1 & 1 \end{pmatrix} \begin{pmatrix} x \\ y \end{pmatrix} = 0$$

$$\begin{pmatrix} 1 & 2 \\ 2 & 1 \end{pmatrix}^2 \begin{pmatrix} u \\ v \end{pmatrix} = \begin{pmatrix} u \\ u \end{pmatrix} \quad \text{or} \quad \begin{pmatrix} 1 & 1 \\ 1 & 1 \end{pmatrix} \begin{pmatrix} u \\ v \end{pmatrix} = 0$$

Thus, $\hat{e}_2 = \frac{1}{2} \begin{pmatrix} 1 \\ -1 \\ 1 \\ -1 \end{pmatrix}$

The remaining eigenvectors are obtained in a similar fashion.

$$\lambda_1 = 2 \qquad \hat{e}_1 = \frac{1}{2} \begin{pmatrix} -1 \\ 1 \\ 1 \\ -1 \end{pmatrix}$$

$$\lambda_3 = 6 \qquad \hat{e}_3 = \frac{1}{2} \begin{pmatrix} 1 \\ 1 \\ -1 \\ -1 \end{pmatrix}$$

$$\lambda_4 = 8 \qquad \hat{e}_4 = \frac{1}{2} \begin{pmatrix} 1 \\ 1 \\ 1 \\ 1 \end{pmatrix} \qquad \bullet$$

Although the method of Example 3.1 is adequate for 3×3 and some 4×4 matrix equations, evaluating determinants to obtain the characteristic equation in larger problems is prohibitively inefficient. There are basically two alternative approaches: (1) Determine the characteristic equation for the eigenvalues by means other than using determinants, and then the roots of this equation correspond to the eigenvalues; (2) find a coordinate system in which the image of the original matrix is diagonal so that the elements along the diagonal are the eigenvalues of the matrix. We next briefly discuss each of these approaches.

In the remainder of this chapter, it is assumed that the matrix, **A**, is both real and symmetric. This guarantees that the eigenvalues are real and that the corresponding eigenvectors are orthogonal. Although the standard MATLAB functions are designed to handle the more general case, the details of the analysis are beyond the scope of this text. The reader is referred to Strang (1980).

The Characteristic Equation and the Cayley-Hamilton Theorem

As we have seen, if $\mathbf{A}$ is an $n \times n$ matrix, the equation $\mathbf{A}\vec{x} = \lambda\vec{x}$ has a solution only for particular values of λ that correspond to the solution of the characteristic equation given by $|\mathbf{A} - \lambda\mathbf{I}_n| = 0$. Evaluating this determinant in the normal way will yield an nth-degree polynomial in λ. Thus, the characteristic equation can formally be written as

$$\lambda^n + a_2\lambda^{n-1} + a_3\lambda^{n-1} + \cdots + a_n\lambda + a_{n+1} = 0 \tag{3.3}$$

Once this equation is obtained, the roots of the polynomial can be found and will correspond to the eigenvalues of the matrix $\mathbf{A}$. To determine the specific values for the n coefficients, a_i, $i = 2, \ldots, n+1$, an $n \times n$ determinant must be algebraically evaluated, and would appear to require a substantial amount of effort in all but the most trivial situations.

There is, however, a procedure whereby the algebra can be replaced by simpler matrix operations. If we replace λ by the complete matrix $\mathbf{A}$ in the characteristic equation, $|\mathbf{A} - \lambda\mathbf{I}_n| = 0$, we obtain the determinant of a null matrix, which is clearly equal to zero. This suggests that if we replace λ by $\mathbf{A}$ in Equation 3.3, we should obtain an expression that is equal to the null matrix. This is known as the *Cayley-Hamilton* theorem, which states that the matrix $\mathbf{A}$ must satisfy its own characteristic equation. We will leave the proof to the reader. (See Problem 3.6). Equation 3.3, with this replacement, becomes

$$\mathbf{A}^n + a_2\mathbf{A}^{n-1} + a_3\mathbf{A}^{n-1} + \cdots + a_n\mathbf{A} + a_{n+1}\mathbf{I}_n = \mathbf{O}_n \tag{3.4}$$

Next, start with an arbitrary n-component vector, $\vec{y}_1$, not equal to zero and compute the following:

$$\vec{y}_2 = \mathbf{A}\vec{y}_1$$
$$\vec{y}_3 = \mathbf{A}\vec{y}_2$$
$$\vdots = \vdots$$
$$\vec{y}_{n+1} = \mathbf{A}\vec{y}_n$$

Equation 3.4 may then be rewritten in terms of the vectors $\vec{y}_i$ as

$$\vec{y}_{n+1} + a_2\vec{y}_n + a_3\vec{y}_{n-1} + \cdots + a_n\vec{y}_2 + a_{n+1}\vec{y}_1 = \mathbf{O}_n \tag{3.5}$$

That is, the coefficients, a_i, are the expansion coefficients when $\vec{y}_{n+1}$ is expanded in terms of the remaining vectors:

$$-\vec{y}_{n+1} = \sum_{i=1}^{n} a_{n+2-i}\vec{y}_i \tag{3.6}$$

If the set of vectors, $\vec{y}_i$, $i = 1, \ldots, n$ were orthonormal, the expansion coefficients would be obtained by multiplying Equation 3.6 successively by each of the $\vec{y}_i^T$. Of course, this is, in general, not the case. However, from Chapter 2, we have seen that, if the set $[\vec{y}_i]$ is merely linearly independent, there exists a set of vectors $\vec{x}_i$, $i = 1, \ldots, n$, that satisfy the orthogonality conditions

$$\vec{x}_i \cdot \vec{y}_j = \vec{x}_i^T \vec{y}_j = \delta_{ij}$$

This is the set of reciprocal vectors.

To obtain the set $[\vec{x}_i]$, we recall that if $\mathbf{Y}$ is a matrix whose columns contain the vectors $\vec{y}_i$, and if $\mathbf{X}$ is a matrix whose columns contain the vectors $\vec{x}_i$, then $\mathbf{X}^T \mathbf{Y} = \mathbf{I}_n$. That is, we only have to compute the inverse of $\mathbf{Y}$ to obtain all the vectors $\vec{x}_i$. The coefficients are then obtained by multiplying Equation 3.6 by each of the vectors $\vec{x}_i^T$ to obtain

$$a_{n+2-i} = -\vec{x}_i^T \vec{y}_{n+1}$$

The following MATLAB code is used to obtain the coefficients of the characteristic polynomial of the matrix $\mathbf{A}$. The matrix chosen for this example is the same one used in Example 3.1.

```
A  = [5 2 1 0;2 5 0 1;1 0 5 2;0 1 2 5];
y1 = [1 2 3 4]';
y2 = A*y1;
y3 = A*y2;
y4 = A*y3;
y5 = A*y4;
Y  = [y1 y2 y3 y4];
X  = inv(Y);
a  = -X*y5;
a  = [1;flipud(a)];
disp(a')
      1    -20    140   -400    384
```

In the last line we have used the MATLAB function `flipud` to reorder the elements of `a` in reverse order and have inserted a 1 as the first coefficient of λ^n to correspond with the normal ordering of polynomial coefficients in MATLAB.

Also, MATLAB provides a toolbox function that will compute the coefficients of the characteristic equation corresponding to a square matrix $\mathbf{A}$. The name of the function is `poly` and the use is `a = poly(A)`. The coefficients are returned in `a`. You should use this function to verify the previous result.

Now that the specific coefficients of the characteristic equation are known, the remaining task is to obtain the roots of the polynomial. This is the subject of Chapter 5. However, to complete the example, we will mention at this point that the MATLAB function `roots(a)` will return all the roots of the polynomial whose coefficients are contained in the vector `a`. With `a` assigned the values determined (and in the same order), `roots(a)` returns the values 2, 4, 6, and 8. These are the same eigenvalues computed earlier for this matrix.

Reducing a Symmetric Matrix to Diagonal Form by Jacobi Rotations

Recalling the geometric discussion at the beginning of this chapter, in which the eigenvalue problem was represented as a rotation to a coordinate system based upon the set of eigenvectors, we next attempt to arrange a sequence of transformations that will bring a symmetric matrix into diagonal form.

A common and readily understood numerical method for solving matrix eigenvalue problems employs a succession of rotations, called *Jacobi rotations*, that attempt to "zero," one by one, the off-diagonal elements. Because we have already seen that a rotation of coordinates will not affect the eigenvalues or the eigenvectors, and because a succession of individual rotations is equivalent to a single combined rotation, we can be confident that the eigenvalues of the eventual diagonal matrix; i.e., the final diagonal elements themselves, will remain unchanged.

To illustrate the method, consider a 4×4 matrix eigenvalue problem like the preceding example. First concentrate on the 2×2 upper-left-hand-corner submatrix. Our first rotation will force just this submatrix into diagonal form. Recall that, for a 2×2 matrix, a rotation of a vector $\vec{x} = \begin{pmatrix} x \\ y \end{pmatrix}$ by an angle θ is effected by the matrix equation

$$\vec{x}' = \begin{pmatrix} x' \\ y' \end{pmatrix} = \begin{pmatrix} \cos\theta & \sin\theta \\ -\sin\theta & \cos\theta \end{pmatrix} \begin{pmatrix} x \\ y \end{pmatrix} = \mathbf{R}^T \vec{x}$$

Thus, the rotation of a 2×2 matrix would be effected by $\mathbf{A}' = \mathbf{R}^T \mathbf{A} \mathbf{R}$, or

$$\begin{pmatrix} a'_{11} & a'_{12} \\ a'_{21} & a'_{22} \end{pmatrix} = \begin{pmatrix} \cos\theta & \sin\theta \\ -\sin\theta & \cos\theta \end{pmatrix} \begin{pmatrix} a_{11} & a_{12} \\ a_{21} & a_{22} \end{pmatrix} \begin{pmatrix} \cos\theta & -\sin\theta \\ \sin\theta & \cos\theta \end{pmatrix}$$

and we wish to force $a'_{12} = a'_{21} = 0$. Solving this equation for a'_{21} and using the fact that $\mathbf{A}$ is symmetric, we obtain

$$a'_{21} = (a_{22} - a_{11})\sin\theta\cos\theta + a_{12}(\cos^2\theta - \sin^2\theta) = 0$$

or

$$\tan(2\theta) = -\frac{2a_{12}}{a_{22} - a_{11}}$$

$$\theta = -\frac{1}{2}\tan^{-1}\left(\frac{2a_{12}}{a_{22} - a_{11}}\right) \qquad \text{if } [a_{11} \neq a_{22}]$$

$$= \pm\frac{\pi}{4} \qquad \text{if } [a_{11} = a_{22}]$$

If we use the same 4×4 matrix as in Equation 3.2, this particular submatrix is

$\begin{pmatrix} 5 & 2 \\ 2 & 5 \end{pmatrix}$, $a_{11} = a_{22} = 5$, so that $\theta = \frac{\pi}{4}$, and the first rotation of the full matrix is obtained by the rotation matrix

$$\mathbf{R}_{12} = \begin{pmatrix} \frac{1}{\sqrt{2}} & -\frac{1}{\sqrt{2}} & 0 & 0 \\ \frac{1}{\sqrt{2}} & \frac{1}{\sqrt{2}} & 0 & 0 \\ 0 & 0 & 1 & 0 \\ 0 & 0 & 0 & 1 \end{pmatrix}$$

which yields

$$\mathbf{A}' = \mathbf{R}_{12}^T \mathbf{A} \mathbf{R}_{12} = \begin{pmatrix} 7 & 0 & 0.7071 & 0.7071 \\ \cdots & 3 & -0.7071 & 0.7071 \\ \cdots & \cdots & 5 & 2 \\ \cdots & \cdots & \cdots & 5 \end{pmatrix}$$

where, because the matrix is symmetric, we have included only the upper-triangular portion.

Next, we proceed to rotate in the 1–3 plane to eliminate the element a'_{13}. The result, after rotating with the matrix

$$\mathbf{R}_{13} = \begin{pmatrix} \cos\theta_{13} & 0 & -\sin\theta_{13} & 0 \\ 0 & 1 & 0 & 0 \\ \sin\theta_{13} & 0 & \cos\theta_{13} & 0 \\ 0 & 0 & 0 & 1 \end{pmatrix}$$

with $\theta_{13} = -\frac{1}{2}\tan^{-1}\left(2a_{13}/(a_{33} - a_{11})\right) = \frac{\pi}{4}$ then yields

$$\mathbf{A}'' = \mathbf{R}_{13}^T \mathbf{A}' \mathbf{R}_{13} = \begin{pmatrix} 7.2247 & -0.2142 & 0 & 1.2797 \\ \cdots & 3 & -0.6739 & 0.7071 \\ \cdots & \cdots & 4.7753 & 1.6919 \\ \cdots & \cdots & \cdots & 5 \end{pmatrix}$$

Notice that the elements previously *zeroed*, a_{12} and a_{21}, have been replaced with smaller, but nonzero, values. We hope that this can be corrected in a second pass.

Proceeding through all remaining combinations of 2×2 submatrices, we attempt to zero, in succession, a_{14}, a_{23}, a_{24}, and a_{34}, and obtain the result of the *first pass*:

$$\mathbf{A}'_1 = \begin{pmatrix} 7.8080 & 0.2809 & 0.3485 & 0.5492 \\ \cdots & 2.1528 & -0.4051 & -0.3986 \\ \cdots & \cdots & 3.9338 & 0 \\ \cdots & \cdots & \cdots & 6.1055 \end{pmatrix}$$

This sequence of rotations is called a set of *Jacobi* rotations.

The matrix is still not diagonal, but the off-diagonal elements are somewhat smaller. The process is repeated for two more passes as follows.

$$\mathbf{A}'_2 = \begin{pmatrix} 7.9962 & -0.1463 & 0.0207 & 0.0153 \\ \cdots & 2.0036 & 0.0047 & 0.0001 \\ \cdots & \cdots & 4.0001 & 0 \\ \cdots & \cdots & \cdots & 6.0001 \end{pmatrix}$$

$$\mathbf{A}'_3 = \begin{pmatrix} 8 & 0 & 0 & 0 \\ \cdots & 2 & 0 & 0 \\ \cdots & \cdots & 4 & 0 \\ \cdots & \cdots & \cdots & 6 \end{pmatrix}$$

We have kept track of all the rotations to this point, and the product of these rotations is

$$\mathbf{R} = \frac{1}{2} \begin{pmatrix} 1 & -1 & 1 & 1 \\ 1 & 1 & -1 & 1 \\ 1 & 1 & 1 & -1 \\ 1 & -1 & -1 & -1 \end{pmatrix}$$

Thus, the combination of all the rotations applied to the matrix can be summarized as

$$\mathbf{A}' = \begin{pmatrix} 8 & 0 & 0 & 0 \\ 0 & 2 & 0 & 0 \\ 0 & 0 & 4 & 0 \\ 0 & 0 & 0 & 6 \end{pmatrix} = \mathbf{R}^T \begin{pmatrix} 5 & 2 & 1 & 0 \\ 2 & 5 & 0 & 1 \\ 1 & 0 & 5 & 2 \\ 0 & 1 & 2 & 5 \end{pmatrix} \mathbf{R}$$

This example illustrates a systematic procedure, employing a sequence of rotations, that will rotate a symmetric $n \times n$ matrix to diagonal form. If the combined sequence of all these rotations is represented by the matrix $\mathbf{R}$, the diagonal image of the matrix $\mathbf{A}$ is obtained by

$$\mathbf{R}^T \mathbf{A} \mathbf{R} = \begin{pmatrix} \lambda_1 & 0 & 0 & \cdots & 0 \\ 0 & \lambda_2 & 0 & \cdots & 0 \\ 0 & 0 & \lambda_3 & \cdots & 0 \\ \vdots & \vdots & \vdots & \ddots & \vdots \\ 0 & 0 & 0 & \cdots & \lambda_n \end{pmatrix}$$

Because rotations will not alter the eigenvalues of the matrix, we can conclude that the elements along the diagonal of the resulting matrix are the eigenvalues of the matrix, and further, that the columns of the combined rotation matrix are the eigenvectors. The proof of this generalization of our example is known as Schur's theorem and is beyond the scope of this text. The proof that the method of Jacobi rotations will converge to yield a diagonal method is also based upon Schur's theorem. We are content here to provide the reader with a demonstration of a solution of an eigenvalue equation in the context of a sequence of rotations.

The method of Jacobi rotations applied to an arbitrary, symmetric, $n \times n$ matrix is certainly much more complicated than the Gaussian elimination method used to solve the equation $\mathbf{A}\vec{x} = \vec{b}$. However, like the Gaussian method, it can be written into a computer code and used to solve any eigenvalue equation, $\mathbf{A}\vec{x} = \lambda\vec{x}$, where $\mathbf{A}$ is a symmetric or Hermitian matrix.[5] Of course, a profession-

---

[5]Included on the applications disk is a demonstration MATLAB function, `JcbyDemo`, that will illustrate the sequence of Jacobi rotations used to bring an arbitrary real, symmetric matrix into diagonal form.

ally written eigenvalue code would select the sequence of Jacobi rotations so as to optimize efficiency and minimize round-off error.

OPERATIONS COUNT FOR THE JACOBI METHOD

The Jacobi method applied to a real, symmetric matrix must seek to zero $n(n-1)/2$ off-diagonal elements, with each rotation requiring about 12 floating-point operations. Additionally, it is found that to achieve convergence, roughly n iterations are required. Thus, we can expect that the complete diagonalization will entail approximately $6n^3$ floating-point operations.

COMMENTS ON MODERN EIGENVALUE/ EIGENVECTOR ALGORITHMS

The great advantage of the Jacobi method is that it is quite easy to understand and, therefore, is easily adaptable to particular problems. The principal objection to the method is that each element that is zeroed in one step usually reappears as a smaller, but nonzero, element in a later step. The Jacobi method dates from the middle of the nineteenth century.

Because of the importance of the eigenvalue equation in all areas of science and engineering, a great deal of effort has been devoted to the refinement of numerical techniques to find the eigenvalues and eigenvectors of both symmetric and nonsymmetric real and complex matrices. These procedures generally construct a series of transformations of the original matrix, each of which is designed to leave the eigenvalues of the matrix intact. (These transformation matrices are simply orthogonal matrices if the matrix $\mathbf{A}$ is real and symmetric.) The goal is to seek noninterfering transformations; i.e., ones that will leave elements, once zeroed, as zero. It has been found that if the matrix is real and symmetric, a series of transformations, called Householder transformations, can be constructed that will reduce the matrix to tridiagonal form in a finite number of steps. This much-reduced form of the matrix is then diagonalized iteratively by a method known as the $\mathbf{QR}$ method, which consists of first factoring the tridiagonal reduced version of $\mathbf{A}$ as $\mathbf{A} = \mathbf{QR}$, where $\mathbf{Q}$ is a *unitary* matrix satisfying the condition $\overline{\mathbf{Q}}^T = \mathbf{Q}^{-1}$, and $\mathbf{R}$ is upper triangular, and then computing a transformation of $\mathbf{A}$ given by $\mathbf{A}' = \mathbf{RQ} = \overline{\mathbf{Q}}^T \mathbf{AQ}$. The matrix $\mathbf{A}'$ is then factored as $\mathbf{Q}'\mathbf{R}'$ and the process repeated until the transformed matrix is diagonal.

If the matrix is *not* symmetric, a series of Householder transformations can be found that will, in a finite number of steps, bring the matrix into *almost* upper-triangular form; i.e, the transformed matrix will be upper triangular with only one nonzero diagonal just below the principal diagonal. This form is then iteratively diagonalized by the $\mathbf{QR}$ method.

These modern methods are very interesting to study, and the reader is urged to consult the more detailed discussions that can be found in Press, et al., Wong, or Stoer and Bulirsch, which are listed in the Bibliography. However, our primary concern here is to achieve a cursory understanding of precisely how eigenvalue equations are solved using MATLAB, and then to freely apply the MATLAB procedures to a variety of problems of an eigenvalue nature.

MATLAB Functions for Eigenvalue Problems

The *toolbox* function eig is based on the modern methods of obtaining solutions to eigenvalue equations. It uses techniques that are more efficient than Jacobi rotations and that can be applied to nonsymmetric or even complex matrices, as well as to the more common real, symmetric situation. The operation of the MATLAB function eig is as follows.

The MATLAB statement

$$[\mathtt{V}, \mathtt{D}] = \mathtt{eig}(\mathtt{A})$$

will return the eigenvectors of the $n \times n$ matrix **A** as the columns of the square matrix V. The second matrix returned, D, is also square and contains the corresponding eigenvalues along the main diagonal, with zeros elsewhere. If only the eigenvalues of the matrix are required, it is more efficient to use a shortened form of the statement,

$$\mathtt{d} = \mathtt{eig}(\mathtt{A})$$

in which the eigenvalues of the matrix are returned in the n-component column vector, d. The operations count using eig for a real symmetric matrix is $\approx 17n^3$ if only eigenvalues are computed, and $\approx 28n^3$ if both eigenvectors and eigenvalues are desired.

In the language of MATLAB, a check of the solution of the eigenvalue equation $\mathbf{A}\vec{x} = \lambda\vec{x}$, where **A** is an arbitrary, perhaps random, matrix, would read

```
      n  =    10;
      A  =    rand(n);
   [V,D]  =    eig(A);
test(1)  =    A*V(:,1) - D(1,1)*V(:,1);
test(2)  =    A*V(:,2) - D(2,2)*V(:,2);
test(3)  =    A*V(:,3) - D(3,3)*V(:,3);
      etc.,
   test =
           0 0 0 ... 0
```

or, more concisely,

```
[V,D]  =    eig(A);
 test =    (A - D)*V;
 test =
           0 0 0 ... 0
```

Remember, if **A** is real and symmetric, the computed eigenvalues D(k,k) are real, and the eigenvectors V(:,k) are orthogonal.

In addition, the function eig can handle more general matrices than simply those that are real-symmetric. Indeed, eig will find the eigenvalues and eigenvectors of a completely general square matrix **A**, including the case where **A** is complex. Of course, if **A** is not real and symmetric (or if complex, not Hermitian), the eigenvalues are not guaranteed to be real, and the eigenvectors are not guaranteed to be either orthogonal or complete.

Perspective on Eigenvalue Equations

To illustrate the importance of eigenvalue equations in applied analysis, we next consider the problem of combining ordinary differential equations and matrices. A great many problems can be expressed in terms of a known linear operation on a set of to-be-determined vectors, $\vec{x}_i$'s, in the form of an eigenvalue equation, $\mathbf{A}\vec{x} = \lambda\vec{x}$. If the matrix **A** is real and symmetric or Hermitian, the matrix itself defines a *complete* set of orthonormal basis vectors, its eigenvectors. The problem is almost always most transparent when expressed in the space of these basis vectors, and the solution is then expanded in this basis. As we will see in later chapters, this same principle applies to a variety of linear operators other than matrices, such as differential equations and integral equations. Under certain conditions, these operators also define a *complete* basis set, which then becomes the natural basis vectors for the solution of the problem. An important illustration of this is given in Example 3.2.

EXAMPLE 3.2

Normal Modes in a Mass-Spring System

Newton's equation of motion, $\vec{F} = m\vec{a}$, is a differential equation that determines the position of the mass, m, as a function of time, $\vec{x}(t)$, once the expression for the force $\vec{F}$ has been specified and the *initial conditions*, $\vec{x}|_{t=0}$, $\vec{v}|_{t=0}$, are given. For motion in one dimension, the equations become

$$F_x = ma_x = m\frac{d^2x}{dt^2} = m\ddot{x}(t)$$

where the notation $\ddot{x}(t)$ is a shorthand notation for d^2x/dt^2.

If the only force acting on the mass is a spring, the expression for this force is $F_x = -kx$, where k is the *spring constant*, and x is a measure of the displacement of the mass from the unstretched position of the spring. Thus, for one mass and one spring, the equation of motion is

$$\ddot{x}(t) = -\frac{k}{m}x(t) = -\omega^2 x(t) \tag{3.7}$$

with $\omega = \sqrt{k/m}$.

This equation can be solved by trying an exponential form[6] for $x(t)$; i.e., $x(t) = Ae^{i\alpha t}$. Inserting this into the equation, we obtain

$$-A\alpha^2 e^{i\alpha t} = -\omega^2 A e^{i\alpha t}$$
$$\alpha^2 = \omega^2$$

---

[6]We could just as well have chosen $Ae^{\alpha t}$, but because we expect motion to be oscillatory, and because $e^{i\alpha t}$ can be written as $e^{i\alpha t} = \cos(\alpha t) + i\sin(\alpha t)$, the choice of $i\alpha$ will somewhat shorten the algebra.

Thus, there are two independent solutions corresponding to $\alpha = \pm\omega$, or

$$x(t) = A_1 e^{i\omega t} + A_2 e^{-i\omega t} \tag{3.8}$$

If the *initial conditions* are given as

$$x(t = 0) = x_0$$
$$v(t = 0) = \left.\frac{dx}{dt}\right|_{t=0} = v_0$$

the two parameters, A_1 and A_2, can be written in terms of x_0 and v_0 to obtain

$$A_1 = \frac{1}{2}\left(x_0 + \frac{v_0}{i\omega}\right) \qquad A_2 = \frac{1}{2}\left(x_0 - \frac{v_0}{i\omega}\right)$$

so that

$$\begin{aligned}
x(t) &= A_1 e^{i\omega t} + A_2 e^{-i\omega t} \\
&= \frac{x_0}{2}\left(e^{i\omega t} + e^{i\omega t}\right) + i\frac{v_0}{2i\omega}\left(e^{i\omega t} - e^{i\omega t}\right) \\
&= x_0 \cos(\omega t) + \frac{v_0}{\omega}\sin(\omega t)
\end{aligned} \tag{3.9}$$

which clearly indicates how the motion depends on the initial displacement, x_0, and the initial velocity, v_0.

Next consider the *three*-mass system shown in Figure 3.3.

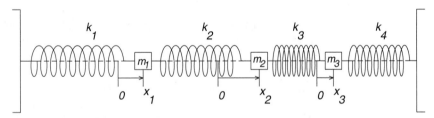

Figure 3.3 *A three-mass system connected by springs.*

If we assume that the masses are the same, $m_1 = m_2 = m_3 = m$, and that the displacements of each mass from its unstretched position are $x_1(t), x_2(t)$, and $x_3(t)$, respectively, the equations of motion for each of the three masses can be written as

$$\begin{aligned}
m\ddot{x}_1 &= -k_1 x_1 - k_2(x_1 - x_2) \\
m\ddot{x}_2 &= -k_2(x_2 - x_1) - k_3(x_2 - x_3) \\
m\ddot{x}_3 &= -k_3(x_3 - x_2) - k_4 x_3
\end{aligned}$$

These equations can be written in matrix form as

$$m \begin{pmatrix} \ddot{x}_1 \\ \ddot{x}_2 \\ \ddot{x}_3 \end{pmatrix} = - \begin{pmatrix} k_1 + k_2 & -k_2 & 0 \\ -k_2 & k_2 + k_3 & -k_3 \\ 0 & -k_3 & k_3 + k_4 \end{pmatrix} \begin{pmatrix} x_1 \\ x_2 \\ x_3 \end{pmatrix}$$

$$m \frac{d^2 \vec{x}}{dt^2} = -\mathbf{K}\vec{x}$$

Next, by assuming that the spring constants have the values $k_1 = k_4 = 10\text{N/m}$, and $k_2 = k_3 = 40\text{N/m}$, and that the masses are all equal to 1, the matrix $\mathbf{K}/\text{m}$ is then

$$\mathbf{K}/\text{m} = \begin{pmatrix} 50 & -40 & 0 \\ -40 & 80 & -40 \\ 0 & -40 & 50 \end{pmatrix} \tag{3.10}$$

Using the MATLAB function `eig` to determine the eigenvalues and eigenvectors of this matrix (i.e., `[V,D] = eig(K)`, and setting ω_i^2 equal to the diagonal elements of D) yields

$$\omega_1^2 = 6.477, \quad \hat{e}_1 = \begin{pmatrix} 0.5604 \\ 0.6098 \\ 0.5604 \end{pmatrix}$$

$$\omega_2^2 = 50.00, \quad \hat{e}_2 = \begin{pmatrix} 0.7071 \\ 0.0000 \\ -.7071 \end{pmatrix}$$

$$\omega_3^2 = 123.523, \quad \hat{e}_3 = \begin{pmatrix} 0.4319 \\ -.7926 \\ 0.4319 \end{pmatrix}$$

These eigenvectors define what we call the *normal coordinates* of the problem. To illustrate the utility of normal coordinates, we next expand the functions $\vec{x}(t)$ in terms of normal coordinates as

$$\vec{x}(t) = \sum_{i=1}^{3} X_i(t)\hat{e}_i \tag{3.11}$$

and rewrite the problem in terms of the $X_i(t)$'s.

$$\ddot{\vec{x}}(t) = -\frac{1}{m}\mathbf{K}\vec{x}$$

$$\sum_{i=1}^{3} \ddot{X}_i(t)\hat{e}_i = -\frac{1}{m}\sum_{i=1}^{3} X_i(t)\mathbf{K}\hat{e}_i$$

$$= -\sum_{i=1}^{3} X_i(t)\omega_i^2\hat{e}_i$$

and multiply from the left by $\hat{e}_k^T$, using $\hat{e}_k^T \hat{e}_i = \delta_{ik}$, to obtain

$$\ddot{X}_k(t) = -\omega_k^2 X_k(t) \quad \text{for } k = 1, 2, \text{ and } 3 \tag{3.12}$$

But this equation is of the same form as Equation 3.7, so that the solution for $X_k(t)$ is

$$X_k(t) = X_{0k} \cos(\omega_k t) + \frac{V_{0k}}{\omega_k} \sin(\omega_k t)$$

The solutions for the actual displacements of the masses is then obtained by using Equation 3.11.

For example, if each of the masses is started from rest ($v_{0i} = 0$, $i = 1, 2$, and 3), and if the initial displacements are $x_{01} = 0.1$, $x_{02} = 0.2$, and $x_{03} = 0.75$; i.e., $\vec{x}_0 = (0.1\ 0.2\ 0.75)^T$, we must first determine $\vec{X}_0$ from Equation 3.11 as $X_{0i} = \hat{e}_i^T \cdot \vec{x}_0$ (or in MATLAB, X0 = V'*x0). This yields

$$\vec{X}_0 = \begin{pmatrix} 0.5983 \\ -0.4596 \\ 0.2080 \end{pmatrix}$$

so that

$$\vec{X}(t) = \begin{pmatrix} 0.5983 \cos(\omega_1 t) \\ -0.4596 \cos(\omega_2 t) \\ 0.2080 \cos(\omega_3 t) \end{pmatrix}$$

Thus, the normal coordinate solutions are simply three independent cosine functions oscillating at the eigenfrequencies. These solutions are illustrated in Figure 3.4.

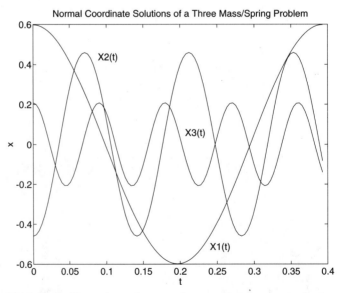

Figure 3.4 *Normal coordinate solutions of a three-mass/spring problem.*

The actual motions of the three masses are given by the quantities $x_i(t)$, which are then

$$\vec{x}(t) = \sum_{i=1}^{3} X_i(t)\hat{e}_i$$

$$= \begin{pmatrix} 0.335\cos(\omega_1 t) & -0.325\cos(\omega_2 t) & +0.090\cos(\omega_3 t) \\ 0.365\cos(\omega_1 t) & & -0.165\cos(\omega_3 t) \\ 0.335\cos(\omega_1 t) & +0.325\cos(\omega_2 t) & +0.090\cos(\omega_3 t) \end{pmatrix}$$

The motion of mass 1 is shown in Figure 3.5.

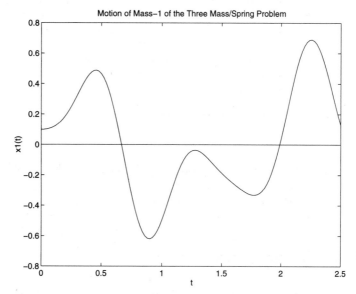

Figure 3.5   *The motion of mass 1 of a three-mass/spring system.*

Note that even though the computed motion of each of the masses is quite complicated, there are some conclusions we can draw from viewing the motion in terms of the normal coordinates, the eigenvectors of **K**:

- If $X_{01} = X_{03} = 0$, and $X_{02} = A$, the motion corresponds to masses 1 and 3 oscillating in opposite directions with equal amplitude at an angular frequency, ω_2, with mass 2 remaining stationary. This is one of the normal modes of oscillation of the system.
- If $X_{01} = X_{02} = 0$, and $X_{03} = A$, masses 1 and 3 oscillate in phase with angular frequency, ω_3, with equal amplitude, and mass 2 oscillates in an opposite direction with slightly less than twice the amplitude.
- The final normal mode, with $X_{02} = X_{03} = 0$, and $X_{01} = A$, has all three masses oscillating in phase with angular frequency, ω_1, and with similar amplitudes.

In short, the complicated motions of the three masses are readily understandable when expressed in terms of linear combinations of the normal coordinates.

The MATLAB statements required to obtain the solutions, $\vec{x}(t)$, for this problem are given in the following code:

```
MATLAB Script
x0 = [.1 .2 .75]';
k1 = 10;  k2 = 40;  k3 = 40;  k4 = 10;
K = [k1+k2 -k2 0;-k2 k2+k3 -k3;0 -k3 k3+k4];
[V,D] = eig(K);
X0 = V*x0
w = sqrt(diag(D))
                          %Make a table of x(i,t) values
T = 2*pi/min(w)           %for i = 1,2,3 and for 60 values
t = 0:T/60:T;             %of t from 0 to 2π/ω_min
X(1,:) = X0(1)*cos(w(1)*t);
X(2,:) = X0(2)*cos(w(2)*t);
X(3,:) = X0(3)*cos(w(3)*t);
x = V'*X;
```

●

WHAT IF?   The three-mass problem becomes a two-mass problem if two of the masses are rigidly connected. We can simulate this by making one of the spring constants extremely large. Duplicate the previous calculation with $k_3 = 10,000$.   ●

(§) Spectral Decomposition

Let $\hat{e}_k$ be the set of eigenvectors and λ_k the eigenvalues of the real-symmetric $n \times n$ matrix **A**. Then, as we have seen, this set is both complete and orthogonal, meaning that any vector can be expanded in this basis:

$$\vec{x} = \sum_{k=1}^{n} \hat{e}_k (\hat{e}_k^T \cdot \vec{x})$$

Next, applying the matrix **A** to this arbitrary vector,

$$\mathbf{A}\vec{x} = \sum_{k=1}^{n} \mathbf{A}\hat{e}_k (\hat{e}_k^T \cdot \vec{x})$$

$$= \sum_{k=1}^{n} \lambda_k \hat{e}_k (\hat{e}_k^T \cdot \vec{x}) \qquad \text{(using } \mathbf{A}\hat{e}_k = \lambda_k \hat{e}_k\text{)}$$

$$= \left[\sum_{k=1}^{n} \lambda_k \hat{e}_k \hat{e}_k^T \right] \vec{x}$$

we see that the term in the square brackets may be identified as the expansion of $\mathbf{A}$ in terms of the eigenvectors. Further, because $\hat{e}_k \hat{e}_k^T$ is the projection operator, $\mathbf{P}_k$, which projects a vector onto the direction of the unit vector, $\hat{e}_k$, the equation may also be written as

Spectral Decomposition

$$\mathbf{A} = \sum_{k=1}^{n} \lambda_k \hat{e}_k \hat{e}_k^T = \sum_{k=1}^{n} \lambda_k \mathbf{P}_k$$

This equation is known as the spectral decomposition of the matrix $\mathbf{A}$, and it is the genesis of numerous interesting and important matrix identities. A few of these are described in the following text. Using the defining property of a projection operator, namely $\mathbf{P}_k \mathbf{P}_{k'} = \mathbf{P}_k \delta_{kk'}$, along with the previous equation, $\mathbf{A}^m$ can be evaluated as follows:

$$\mathbf{A}^2 = \left(\sum_{k=1}^{n} \lambda_k \mathbf{P}_k \right) \left(\sum_{k'=1}^{n} \lambda_{k'} \mathbf{P}_{k'} \right)$$

$$= \sum_{k,k'=1}^{n} \lambda_k \lambda_{k'} \mathbf{P}_k \mathbf{P}_{k'}$$

$$= \sum_{k,k'=1}^{n} \lambda_k \lambda_{k'} \mathbf{P}_k \delta_{kk'}$$

$$= \sum_{k=1}^{n} \lambda_k^2 \mathbf{P}_k$$

And, by extension of this argument,

$$\mathbf{A}^m = \sum_{k=1}^{n} (\lambda_k)^m \mathbf{P}_k$$

Next, consider extending the definition of a function of a single variable, $f(x)$, to a function of a square, real-symmetric $n \times n$ matrix, $f(\mathbf{A})$. Now, every function is defined in terms of the elementary operations of arithmetic on the independent variable x. This is usually effected by obtaining a definition of the function in terms of a polynomial in x or a series expansion in x. We return to this concept in Chapter 4. For now, assume that the arbitrary function $f(x)$ can be so expanded as

$$f(x) = \sum_{i=1}^{\infty} a_i x^i$$

Then, extending this definition to $f(\mathbf{A})$ by using the spectral decomposition of $\mathbf{A}$, we have

$$f(\mathbf{A}) = \sum_{i=1}^{\infty} a_i \mathbf{A}^i$$

$$= \sum_{i=1}^{\infty} a_i \left(\sum_{k=1}^{n} \lambda_k^i \mathbf{P}_k \right)$$

$$= \sum_{k=1}^{n} \left(\sum_{i=1}^{\infty} a_i \lambda_k^i \right) \mathbf{P}_k$$

$$= \sum_{k=1}^{n} f(\lambda_k) \mathbf{P}_k$$

And so, we are led to useful identities like the following:

$$e^{\mathbf{A}} = \sum_{k=1}^{n} e^{\lambda_k} \mathbf{P}_k$$

$$\sin \mathbf{A} = \sum_{k=1}^{n} \sin(\lambda_k) \mathbf{P}_k$$

$$3^{\mathbf{A}} = \sum_{k=1}^{n} 3^{\lambda_k} \mathbf{P}_k$$

which, in turn, suggests how the MATLAB toolbox functions like `expm`, `logm`, `sqrtm`, and `scalar∧matrix` are constructed.

Finally, it is relatively easy to prove that the trace of any projection operator is $+1$ (see Problem 1.17c), and we can use this fact, along with the equation for spectral decomposition, to prove that

$$\det(\mathbf{A}) = |\mathbf{A}| = e^{\mathrm{Tr}(\ln \mathbf{A})} \tag{3.13}$$

We proceed as follows:

$$e^{\mathrm{Tr}(\ln \mathbf{A})} = e^{\mathrm{Tr}\left(\sum_k \ln \lambda_k \mathbf{P}_k \right)}$$

$$= e^{\left[\sum_k \ln \lambda_k \, \mathrm{Tr}(\mathbf{P}_k) \right]} \qquad [\textit{Note: } \ln(a) + \ln(b) = \ln(ab)]$$

$$= e^{\ln\left(\prod_k \lambda_k \right)} \qquad\quad [\textit{Note: } \prod_k \lambda_k = \lambda_1 \cdot \lambda_2 \cdots \lambda_n]$$

$$= \prod_k \lambda_k \qquad\qquad\quad [\textit{Note: } e^{\ln x} = x]$$

As the determinant of a matrix is indeed equal to the product of its eigenvalues, we have proved the relationship.

(§) Singular Value Decomposition

We have seen that the norm of a *square* $n \times n$ matrix $\mathbf{A}$ can be computed by evaluating the maximum of the expression

$$||\mathbf{A}||^2 \equiv \left[\frac{\vec{x}^T \mathbf{A}^T \mathbf{A} \vec{x}}{\vec{x}^T \vec{x}} \right]_{\max}$$

over the space of all nonzero vectors $\vec{x}$. Furthermore, if the inverse matrix $\mathbf{A}^{-1}$ exists, the norm of the vector $\vec{y} = \mathbf{A}\vec{x}$ is both nonzero and positive. The matrix $\mathbf{C} \equiv \mathbf{A}^T \mathbf{A}$ is then both symmetric and *positive definite*.[7] If $\mathbf{C}$ is then rotated to its principal axes; i.e., diagonalized, we obtain

$$\mathbf{C}' = \begin{pmatrix} \lambda_1 & 0 & 0 & \cdots & 0 \\ 0 & \lambda_2 & 0 & \cdots & 0 \\ 0 & 0 & \lambda_3 & \cdots & 0 \\ \vdots & \vdots & \vdots & \ddots & \vdots \\ 0 & 0 & 0 & \cdots & \lambda_n \end{pmatrix} = \mathbf{R}^T \mathbf{C} \mathbf{R}$$

where λ_i, $i = 1, \ldots, n$, are the positive real eigenvalues of $\mathbf{C} = \mathbf{A}^T \mathbf{A}$, and the rotation matrix $\mathbf{R}$ is an orthogonal matrix constructed by using the eigenvectors of $\mathbf{C}$ as successive columns. The norm of the matrix $\mathbf{A}$ is simply equal to the square root of the maximum eigenvalue. (See Problem 3.8.)

The inverse equation,

$$\mathbf{C} = \mathbf{R} \begin{pmatrix} \lambda_1 & 0 & 0 & \cdots & 0 \\ 0 & \lambda_2 & 0 & \cdots & 0 \\ 0 & 0 & \lambda_3 & \cdots & 0 \\ \vdots & \vdots & \vdots & \ddots & \vdots \\ 0 & 0 & 0 & \cdots & \lambda_n \end{pmatrix} \mathbf{R}^T$$

is a statement that any square $n \times n$ real-symmetric matrix $\mathbf{A}$ can be uniquely written in terms of its eigenvalues and eigenvectors.

By a similar argument, the matrix

$$\mathbf{C}^{-1} = \mathbf{R} \begin{pmatrix} \lambda_1^{-1} & 0 & 0 & \cdots & 0 \\ 0 & \lambda_2^{-1} & 0 & \cdots & 0 \\ 0 & 0 & \lambda_3^{-1} & \cdots & 0 \\ \vdots & \vdots & \vdots & \ddots & \vdots \\ 0 & 0 & 0 & \cdots & \lambda_n^{-1} \end{pmatrix} \mathbf{R}^T$$

would be obtained in the evaluation of the norm of $\mathbf{A}^{-1}$.

This clearly indicates that if any of the eigenvalues of $\mathbf{A}^T \mathbf{A}$ are zero, the matrix $\mathbf{A}$ is singular.

---

[7]A real and symmetric $n \times n$ matrix is called *positive definite* if $\vec{x}^T \mathbf{A} \vec{x} > 0$ for all nonzero vectors, $\vec{x}$.

SINGULAR VALUE DECOMPOSITION OF A NONSYMMETRIC SQUARE MATRIX

It is perhaps not surprising that the preceding result can be generalized to non-symmetric square matrices as well. That is, any square matrix, $\mathbf{C}$, can be decomposed into a product of three matrices,

$$
\mathbf{C} = \mathbf{U} \begin{pmatrix} \omega_1 & 0 & 0 & \cdots & 0 \\ 0 & \omega_2 & 0 & \cdots & 0 \\ 0 & 0 & \omega_3 & \cdots & 0 \\ \vdots & \vdots & \vdots & \ddots & \vdots \\ 0 & 0 & 0 & \cdots & \omega_n \end{pmatrix} \mathbf{V}^T
$$

where the matrices $\mathbf{U}$ and $\mathbf{V}$ are each orthogonal but, in general, different, and the diagonal matrix consists of strictly nonnegative values, ω_i, that are no longer equal to the eigenvalues of $\mathbf{C}$, and are now called the *singular values* of the matrix. In terms of these matrices, the inverse of $\mathbf{C}$ is then given by $\mathbf{C}^{-1} = \mathbf{V} \cdot [\omega_i^{-1}] \cdot \mathbf{U}^T$ with the diagonal matrix replaced by the reciprocals of the singular values. Once again, if any of the ω_i's are zero, the matrix $\mathbf{C}$ is singular.

The Situation for a Singular Matrix The situation where $\mathbf{C}^{-1}$ is singular is particularly interesting. If one or more of the singular values are zero, the equation $\mathbf{C}\vec{x} = \vec{b}$ has no unique solutions. But parameter-dependent *families* of solutions are still possible. To see this, consider the case where $\mathbf{C}$ is a 3×3 matrix whose singular value is $\omega_3 = 0$. That is,

$$
\mathbf{C} = \left(\begin{bmatrix} \vec{u}_1 \end{bmatrix} \begin{bmatrix} \vec{u}_2 \end{bmatrix} \begin{bmatrix} \vec{u}_3 \end{bmatrix} \right) \begin{pmatrix} \omega_1 & 0 & 0 \\ 0 & \omega_2 & 0 \\ 0 & 0 & 0 \end{pmatrix} \begin{pmatrix} \overline{\vec{v}_1^T} \\ \overline{\vec{v}_2^T} \\ \overline{\vec{v}_3^T} \end{pmatrix}
$$

where we have written the orthogonal matrices $\mathbf{U}$ and $\mathbf{V}$ in terms of the orthogonal unit vectors comprising their columns. The inverse of $\mathbf{C}$ would then *formally* be

$$
\mathbf{C}^{-1} = \left(\begin{bmatrix} \hat{v}_1 \end{bmatrix} \begin{bmatrix} \hat{v}_2 \end{bmatrix} \begin{bmatrix} \hat{v}_3 \end{bmatrix} \right) \begin{pmatrix} \omega_1^{-1} & 0 & 0 \\ 0 & \omega_2^{-1} & 0 \\ 0 & 0 & \infty \end{pmatrix} \begin{pmatrix} \overline{\hat{u}_1^T} \\ \overline{\hat{u}_2^T} \\ \overline{\hat{u}_3^T} \end{pmatrix}
$$

Next, consider the result of $\mathbf{C}^{-1}\vec{b}$, if $\vec{b}$ is a linear combination of the two vectors $\hat{u}_1$ and $\hat{u}_2$. For simplicity, we will take $\vec{b} = k\hat{u}_1$. Then

$$
\mathbf{C}^{-1}\vec{b} = \left(\begin{bmatrix} \hat{v}_1 \end{bmatrix} \begin{bmatrix} \hat{v}_2 \end{bmatrix} \begin{bmatrix} \hat{v}_3 \end{bmatrix} \right) \begin{pmatrix} \omega_1^{-1} & 0 & 0 \\ 0 & \omega_2^{-1} & 0 \\ 0 & 0 & \infty \end{pmatrix} \begin{pmatrix} k \\ 0 \\ 0 \end{pmatrix}
$$

$$
= \frac{k}{\omega_1}\hat{v}_1 + \frac{k}{\omega_2}\hat{v}_2 + \gamma\hat{v}_3 \tag{3.14}
$$

where $\gamma = 0 \cdot \infty$ is an *undefined*; i.e., free, constant.

Clearly, if the matrix $\mathbf{C}$ is singular, the only solutions to $\mathbf{C}\vec{x} = \vec{b}$ are if $\vec{b}$ is a linear combination of the basis vectors, $\hat{u}_i$, corresponding to the nonzero singular values, in which case the solution $\vec{x}$ is computed to be a linear combination of the corresponding basis vectors, $\hat{v}_i$, plus *any* combination of vectors proportional to the set, $\hat{v}_j$, corresponding to zero singular values. In the notation of Chapter 2, this corresponds to the case where $\text{rank}(C) = \text{rank}(C|b) < n$, where n is the size of the matrix.

Finally, what if $\vec{b}$ is a linear combination of the vectors $\hat{u}_j$ corresponding to the singular values that are zero? Unequivocally, there is then no solution to the matrix equation, $\mathbf{C}\vec{x} = \vec{b}$. It is, however, fair to ask: What is the vector $\vec{x}_{\text{best}}$ that comes closest to solving the equation; i.e., minimizes the norm of the difference $|\mathbf{C}\vec{x}_{\text{best}} - \vec{b}|$? In the preceding example, this would correspond to $\vec{b} = k\hat{u}_3$, and in Equation 3.14, we would no longer have $0 \cdot \infty$ but rather, $k \cdot \infty$, and thus, $\mathbf{C}^{-1}\vec{b}$ would not be possible. However, an infinitesimal variation in the matrix $\mathbf{C}$ that could lead to a solution would be to replace the zero singular values by infinitesimal elements, $\delta\omega$. In that case, the matrix of the reciprocal values in Equation 3.14 would be dominated by the reciprocals of these small elements,

$$\begin{pmatrix} \omega_1^{-1} & 0 & 0 \\ 0 & \omega_2^{-1} & 0 \\ 0 & 0 & 1/\delta\omega \end{pmatrix} \Rightarrow \begin{pmatrix} 0 & 0 & 0 \\ 0 & 0 & 0 \\ 0 & 0 & \gamma \end{pmatrix}$$

where γ is the reciprocal of the infinitesimal change to the matrix. The solution, $\vec{x}_{\text{best}}$, is then proportional to $\hat{v}_j$, the unit vector of $\mathbf{V}$ corresponding to the singular value. The proof is beyond the level of this text, but the generalized result is that the optimum solution, in a *least squares sense*, to the equation $\mathbf{C}\vec{x} = \vec{b}$ when *no* solution is possible; i.e., when $\text{rank}(C) \neq \text{rank}(C|b)$, is obtained by replacing the nonzero singular values by 0 and the zero singular values by 1 in Equation 3.14. As we see in Chapter 14, this is the preferred way to obtain a least squares fit to data.

SVD OF RECTANGULAR MATRICES

The truly surprising result, however, is that these results can even be generalized to arbitrary *rectangular* matrices. The proof of this is beyond the scope of this text,[8] but the result is as follows. If $\mathbf{M}$ is an $n \times m$ matrix with $n \geq m$, then $\mathbf{M}$ can be *decomposed* into the product of three matrices,

$$\mathbf{M} = \mathbf{U} \begin{pmatrix} \mu_1 & 0 & \cdots & 0 \\ 0 & \mu_2 & \cdots & 0 \\ 0 & 0 & \cdots & 0 \\ \vdots & \vdots & \ddots & \vdots \\ 0 & 0 & \cdots & \mu_m \\ 0 & 0 & \cdots & 0 \\ \vdots & \vdots & \ddots & \vdots \\ 0 & 0 & \cdots & 0 \end{pmatrix} \mathbf{V}^T \tag{3.15}$$

---

[8]See Stoer and Bulirsch, *Introduction to Numerical Analysis*, Springer-Verlag, New York, 1980, for a more complete discussion.

where $\mathbf{U}$ is a *square*-orthogonal $n \times n$ matrix, $\mathbf{V}$ is a *square*-orthogonal $m \times m$ matrix, and the matrix in the middle is $n \times m$ with the *singular* values of $\mathbf{M}$ along the principal diagonal. Most of the matrices encountered in linear analysis are square, so this may appear to be merely an interesting but somewhat exotic generalization. However, its true significance becomes apparent if we use this relation to define the "inverse" of a nonsquare matrix.

Generalized Inverse for Nonsquare Matrices   From the fact that the matrices $\mathbf{U}$ and $\mathbf{V}$ are orthogonal, we see that if none of the singular values is zero, the matrix defined as

$$\mathcal{M}^{-1} = \mathbf{V} \begin{pmatrix} 1/\mu_1 & 0 & \cdots & 0 & 0 & \cdots & 0 \\ 0 & 1/\mu_2 & \cdots & 0 & 0 & \cdots & 0 \\ \vdots & \vdots & \ddots & \vdots & 0 & \ddots & \vdots \\ 0 & 0 & \cdots & 1/\mu_m & 0 & \cdots & 0 \end{pmatrix} \mathbf{U}^T \qquad (3.16)$$

indeed generates the result

$$\mathcal{M}^{-1}\mathbf{M} = \mathbf{I}_m$$

The matrix $\mathcal{M}^{-1}$ is also called the *pseudoinverse* of $\mathbf{M}$. The pseudoinverse of a matrix $\mathbf{M}$ can be calculated directly by using the MATLAB function `pinv(M)`. The matrix, $\mathcal{M}^{-1}$, that is returned satisfies the equations

$$\mathcal{M}^{-1}\mathbf{M}\mathcal{M}^{-1} = \mathcal{M}^{-1}$$
$$\mathbf{M}\mathcal{M}^{-1}\mathbf{M} = \mathbf{M}$$

The matrix $\mathcal{M}^{-1}$ can then be applied to the matrix equation of the form $\mathbf{M}\vec{x}_{[m]} = \vec{b}_{[n]}$, where $\vec{x}_{[m]}, \vec{b}_{[n]}$ are column vectors with m and n elements, respectively, to *formally* obtain a "solution" $\vec{x}_{[m]} = \mathcal{M}^{-1}\vec{b}_{[n]}$. Because the matrix equation represents n equations in m unknowns with $n > m$, corresponding to the situation $\texttt{rank}(M) \neq \texttt{rank}(M|b)$, no solution is possible. This is analogous to the situation with singular square matrices, and the result obtained here by SVD is similar.

There is indeed no solution to the matrix equation $\mathbf{M}\vec{x}_{[m]} = \vec{b}_{[n]}$, but the result obtained by using the generalized inverse defined by singular value decomposition $\vec{x}_{[m]} = \mathcal{M}^{-1}\vec{b}_{[n]}$ corresponds to the set of x values that come closest to satisfying the n equations in a *least squares sense*. We return to this topic in Chapter 14, but, for the moment, consider the set of n equations in two unknowns a and b,

$$\begin{aligned} y_1 &= ax_1 + b \\ y_2 &= ax_2 + b \\ &\vdots \qquad \vdots \\ y_n &= ax_n + b \end{aligned}$$

written in matrix form

$$\begin{pmatrix} x_1 & 1 \\ x_2 & 1 \\ \vdots & \vdots \\ x_n & 1 \end{pmatrix} \begin{pmatrix} a \\ b \end{pmatrix} = \begin{pmatrix} y_1 \\ y_2 \\ \vdots \\ y_n \end{pmatrix}$$

$$\mathbf{M}_{[n \times 2]} \begin{pmatrix} a \\ b \end{pmatrix} = \vec{y}_{[n]}$$

According to the theory of singular value decomposition, the best-fit line to the *n data* points, (x_i, y_i), is obtained by solving the equation

$$\begin{pmatrix} a \\ b \end{pmatrix} = \mathcal{M}^{-1}_{[n \times 2]} \vec{y}_{[n]}$$

where $\mathcal{M}^{-1}_{[n \times 2]}$ is obtained from the SVD decomposition of the matrix $\mathbf{M}_{[n \times 2]}$. This is precisely the prescription used by MATLAB to implement backslash division. That is, a solution to a similar problem would be written in MATLAB as

```
x = [0.1 0.2 0.3 0.4 0.5 0.6 0.7 0.8]';
y = [1.0 1.5 2.2 2.8 3.0 3.4 3.9 4.4]';
M = [x ones(length(x),1)];
a = M\y

a =
    4.7143
    0.6536
```

That is, the best-fit line to these data is $y = 4.71435x + 0.65367$.

The situation with $n < m$ is similar. If there are fewer equations than there are unknowns, the solution to $\mathbf{M}\vec{x} = \vec{b}$ can, once again, be obtained by singular value decomposition. However, in this case, what is obtained is the *right* inverse;[9] i.e., $\mathbf{M}\mathcal{M}^{-1} = \mathbf{I}_n$.

Because there are more unknowns than equations, we know that there are an infinite number of possible solutions. The result that is returned by solving the equation using the SVD inverse corresponds to that solution vector with the smallest norm.

For example, the equation $\mathbf{M}\vec{a} = \vec{y}$, which was solved previously for the coefficients, $\vec{a}$, can be turned around to use the known values computed for $\vec{a}$ and the SVD expression for $\mathcal{M}^{-1}$ to now determine the vector $\vec{y}$.

---

[9]That is, when obtaining the generalized inverse of a rectangular matrix, the product $\mathbf{M}\mathcal{M}^{-1}$ or $\mathcal{M}^{-1}\mathbf{M}$ is constructed to equal the identity matrix $\mathbf{I}_p$ of rank p, where p is the *smaller* of the two dimensions of $\mathbf{M}$.

First, the generalized inverse to the matrix **M** is computed using the MAT-LAB function svd. The use is

$$[U, w, V] = \texttt{svd(M)}$$

The matrix of singular values, w, is then an 8×2 matrix. The generalized inverse matrix is then constructed as $\mathcal{M}^{-1} = \mathbf{V} \cdot (\omega_j^{-1}) \cdot \mathbf{U}^T$. That is,

```
[U,w,V] = svd(M);
winv = w';
winv(1,1)=1/w(1,1); winv(2,2)=1/w(2,2);
Minv = V*winv*U';
```

Next, in the equation $\mathcal{M}^{-1}\vec{y} = \vec{a}$, using the known values for the components of $\vec{a}$ computed previously, the solution for the elements of $\vec{y}$ would be obtained as

```
y = Minv\a;
disp(y')
-5.98784  0  0  0  0  0  0  6.64139
```

which is not the same as the original y values, indicating that $\mathcal{M}^{-1}\mathbf{M} \neq \mathbf{M}\mathcal{M}^{-1}$.

GENERALIZATION OF MATRIX DIVISION

Singular value decomposition provides a convenient method of finding a least squares best fit to data, but its significance to MATLAB is much more profound. Using SVD, *any* equation involving a valid product of matrices, $\mathbf{AB} = \mathbf{C}$; i.e., an equation with properly sized matrices, now has an equally valid companion equation involving *division*, $\mathbf{B} = \mathbf{A} \backslash \mathbf{C}$, where the backslash division is defined by the SVD prescription for the generalized inverse of **A**. The importance of this result cannot be exaggerated. With it, the arithmetic of matrices becomes complete. Any matrices that can be multiplied can also be divided. Needless to say, the meaning of the result depends on the meaning attributed to the generalized inverse.

MATLAB Toolbox Function for Singular Value Decomposition The MATLAB function svd(A) computes the singular value decomposition of the matrix **A**. The use is

$$[U, w, V] = \texttt{svd(A)}$$

where **A** is a real $n \times m$ matrix, **U** and **V** are square orthogonal matrices of size $n \times n$ and $m \times m$, respectively, and **w** is a matrix of the same size and shape as **A** with the *singular values* of **A** along the principal diagonal.

Problems

REINFORCEMENT EXERCISES

P3.1 Rotations.

a. The basic criterion for defining the rotation of a vector $\vec{x}$ into $\vec{x}'$ is that the *length* of the vector remain unchanged. That is, if $\vec{x}' = \mathbf{R}^T \vec{x}$, then $|\vec{x}| = |\vec{x}'| = |\vec{x}^T \mathbf{R} \mathbf{R}^T \vec{x}|$, so that $\mathbf{R}$ must be an orthogonal matrix.

   (i) Consider the eigenvalue equation for the matrix operator $\mathbf{R}$; i.e., $\mathbf{R}\hat{e}_k = \lambda_k \hat{e}_k$. Prove that $|\lambda_k|^2 = +1$; i.e., $\lambda = \pm 1$.

   (ii) Show that the determinant of an orthogonal matrix, $|\mathbf{R}|$, must have a value of ± 1.

b. In three dimensions the rotation of a vector $\vec{a} = (a_x, a_y, a_z)$ about the z axis by the angle θ is effected by the matrix equation

$$\vec{a}' = \mathbf{R}^T(\theta)\vec{a} = \begin{pmatrix} \cos\theta & \sin\theta & 0 \\ -\sin\theta & \cos\theta & 0 \\ 0 & 0 & 1 \end{pmatrix} \begin{pmatrix} a_x \\ a_y \\ a_z \end{pmatrix}$$

Use the trigonometric identities

$$\cos\theta_1 \cos\theta_2 - \sin\theta_1 \sin\theta_2 = \cos(\theta_1 + \theta_2)$$
$$\cos\theta_1 \sin\theta_2 + \sin\theta_1 \cos\theta_2 = \sin(\theta_1 + \theta_2)$$

to prove that $\mathbf{R}(\theta_2)\mathbf{R}(\theta_1) = \mathbf{R}(\theta_1 + \theta_1)$; that is, two successive rotations by θ_1 and θ_1 are equivalent to a single rotation by $\theta_1 + \theta_1$.

c. Not every orthogonal matrix is a rotation matrix. Show that the preceding orthogonal rotation matrix has determinant $+1$. Interpret the effect of the orthogonal matrix

$$\mathbf{P} = \begin{pmatrix} 1 & 0 & 0 \\ 0 & 1 & 0 \\ 0 & 0 & -1 \end{pmatrix}$$

which has determinant -1. By considering an arbitrary rotation by an *infinitesimal* angle, $\delta\theta$, explain why orthogonal matrices corresponding to rotations must have determinant $+1$. Upon a rotation, $\mathbf{R}$, of coordinates, the matrix $\mathbf{A}$ is transformed to $\mathbf{A}' = \mathbf{R}\mathbf{A}\mathbf{R}^T$. Show that the eigenvalues of $\mathbf{A}$ are also the eigenvalues of $\mathbf{A}'$.

P3.2 Eigenvalues and Eigenvectors. If $\mathbf{A}$ and $\mathbf{B}$ are two $n \times n$ real-symmetric matrices, and if:

a. The eigenvalues of $\mathbf{A}$ are equal to the eigenvalues of $\mathbf{B}$, does $\mathbf{A} = \mathbf{B}$?

b. The eigenvectors of $\mathbf{A}$ are equal to the eigenvectors of $\mathbf{B}$, does $\mathbf{A} = \mathbf{B}$?

c. Both the eigenvalues and the eigenvectors of $\mathbf{A}$ are equal to those of $\mathbf{B}$, does $\mathbf{A} = \mathbf{B}$?

P3.3 Calculation of Eigenvectors.   One of the eigenvalues of the matrix

$$\mathbf{A} = \begin{pmatrix} -1 & 5 & 2 \\ 5 & -1 & 2 \\ 2 & 2 & 2 \end{pmatrix}$$

is zero.

a. Determine the eigenvector, $\hat{e}_0$, associated with this eigenvalue.

b. Construct the projection operator, $\mathbf{P}_0 = \hat{e}_0 \hat{e}_0^T$, and show that $\mathbf{P}_0 \mathbf{A} = \mathbf{0}$. Will this always be the case for zero eigenvalues?

P3.4 Markov Matrices.   Each of the elements of an $n \times n$ Markov matrix must be nonnegative, and the sums of the elements of each column must be 1. (See Problems 1.11, 1.12, and 1.27 to review the properties of Markov matrices.)

a. Show that one of the eigenvalues of an arbitrary 2×2 Markov matrix must be $+1$.

b. It can be shown that an arbitrary Markov matrix of size $n \times n$ has one eigenvalue equal to $+1$, and the remaining eigenvalues are all positive and less than 1. Assuming that this is true, show that $\vec{x} = \lim_{p \to \infty} \mathbf{M}^p \vec{x}_0$, where $\vec{x}_0$ is an arbitrary nonnull vector, is the eigenvector of $\mathbf{M}$ with eigenvalue $+1$.

P3.5 Matrix Eigenvalue Problems.   By hand calculation, determine the eigenvalues and eigenvectors of the following matrices. Also, determine if the eigenvectors are orthogonal and if they form a complete set.

a. $\begin{pmatrix} 1 & 0 & 0 \\ 0 & 1 & 1 \\ 0 & 1 & 1 \end{pmatrix}$

b. $\begin{pmatrix} 0 & 0 & 1 \\ 0 & 0 & 1 \\ 1 & 1 & 1 \end{pmatrix}$

c. $\begin{pmatrix} 1 & 1 & 1 \\ 0 & 2 & 2 \\ 0 & 0 & 3 \end{pmatrix}$

d. $\begin{pmatrix} 0 & 1 & 0 \\ 1 & 0 & 1 \\ 0 & 1 & 0 \end{pmatrix}$

P3.6 The Cayley-Hamilton Theorem.   Because the eigenvectors of an arbitrary symmetric matrix, $\mathbf{A}$, form a complete set, an arbitrary column vector, $\vec{x}$, may be expanded in terms of these eigenvectors. Next, replace λ by $\mathbf{A}$ in the characteristic equation and apply the entire expression to the vector $\vec{x}$. Show that this expression must equal zero, thus proving the Cayley-Hamilton theorem.

P3.7 Spectral Decomposition.   Use the MATLAB functions `logm`, `trace`, and `det` to verify the identities on page 114 for each of the matrices of Problem 3.5.

P3.8 The Norm of a Matrix. Prove that the norm of a square matrix $\mathbf{A}$ is equal to the square root of the maximum eigenvalue of the matrix $\mathbf{C} = \mathbf{A}^T\mathbf{A}$.

P3.9 Reflections About a Line. See Figure 3.6 for the following problems:
   a. If $\vec{x}$ is an arbitrary vector, what are the components, $(x_{\parallel}, x_{\perp})$, parallel to, and perpendicular to, the unit vector $\hat{e}$?
   b. Determine the expression for the vector $\vec{x}_{\mathrm{ref}}$ that represents a reflection of $\vec{x}$ through a line parallel to the unit vector $\hat{e}$. Write this as an equation, $\vec{x}_{\mathrm{ref}} = \mathbf{R}\vec{x}$, and determine the expression for $\mathbf{R}$. Show that $\mathbf{R}$ is not a projection operator.

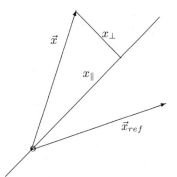

Figure 3.6 *A reflection of a vector about a line.*

   c. For vectors in a plane, the eigenvalues of $\mathbf{R}$ are ± 1. Show that the vectors, $x_{\parallel}$ and $x_{\perp}$, of part a are the eigenvectors of $\mathbf{R}$ with eigenvalues $+1$, -1, respectively.

P3.10 Singular Value Decomposition. For each of the following $n \times m$ rectangular matrices, use the MATLAB function `[U,w,V] = svd(M)` to determine the matrices $\mathbf{U}$, $\mathbf{w}$, and $\mathbf{V}$. Next, compute the generalized inverse by the prescription of Equation 3.16 and show that $\mathcal{M}^{-1}\mathbf{M} = \mathbf{I}_m$ if $n > m$, or $\mathbf{M}\mathcal{M}^{-1} = \mathbf{I}_n$ if $n < m$, and that $\mathcal{M}^{-1}\mathbf{M} \neq \mathbf{M}\mathcal{M}^{-1}$.

   a. $\begin{pmatrix} 1 & 2 \\ 2 & 3 \\ 3 & 4 \end{pmatrix}$

   b. $\begin{pmatrix} 1 & 1 & 1 \\ 0 & 2 & 0 \\ -1 & 0 & -1 \\ 2 & 0 & -2 \end{pmatrix}$

   c. $\begin{pmatrix} 1 & 1 & 1 & 1 \\ 0 & 2 & 0 & -2 \end{pmatrix}$

EXPLORATION PROBLEMS

P3.11 **Complex Matrices.** The most general 2×2 complex *Hermitian* matrix is of the form

$$\mathbf{M} = \begin{pmatrix} \alpha & \gamma + i\delta \\ \gamma - i\delta & \beta \end{pmatrix}$$

where $\alpha, \beta, \gamma,$ and δ are *real*. Show that:

    a. The eigenvalues, λ, of this matrix are real.

    b. Eigenvectors corresponding to different eigenvalues are orthogonal. [*Hint:* Construct a proof similar to that given for real matrices on page 97.]

    c. $\text{Tr}(\mathbf{M}) = \sum\limits_{i=1}^{2} \lambda_i$

P3.12 **Normal Modes in a Linear Chain.** Show that the differential equation defining the motion of the masses in the long chain shown in Figure 3.7 is given by

$$m \begin{pmatrix} \ddot{x}_1 \\ \ddot{x}_2 \\ \ddot{x}_3 \\ \vdots \\ \ddot{x}_n \end{pmatrix} = -k \begin{pmatrix} 2 & -1 & 0 & \cdots & 0 \\ -1 & 2 & -1 & \cdots & 0 \\ 0 & -1 & 2 & \cdots & 0 \\ \vdots & \vdots & \vdots & \ddots & \vdots \\ 0 & 0 & 0 & \cdots & 2 \end{pmatrix} \begin{pmatrix} x_1 \\ x_2 \\ x_3 \\ \vdots \\ x_n \end{pmatrix}$$

$$m \frac{d^2 \vec{x}}{dt^2} = -\mathbf{K}\vec{x}$$

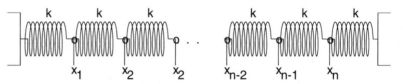

Figure 3.7 *A chain of n equal masses connected by equal springs.*

The motion of the masses in the chain can be described in terms of the n normal mode solutions corresponding to eigenvectors of the matrix $\mathbf{K}/m$. Many of these are, of course, extremely complicated. However, you should be able to guess the nature of at least two normal modes. Show that the vectors corresponding to your guesses are approximate eigenvectors of the matrix $\mathbf{K}/m$. One of these guesses will correspond to the mode with maximum frequency. What is that frequency? This analysis is of interest in the study of the fundamental excitations of a crystal lattice.

P3.13 **Two-Dimensional Normal Modes.** The mass in Figure 3.8 can move in the x and y directions and is held in place by the three springs. The unstretched length of each spring is $L_0 = 0.5$ m, and the springs are identical.

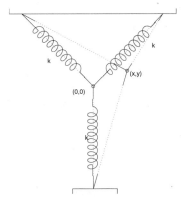

Figure 3.8 *A three-spring/one-mass configuration.*

We will consider only infinitesimal displacements from the origin of the mass, m. When the mass is at the origin, the directions of the three springs can be designated by unit vectors, $\hat{e}_1$, $\hat{e}_2$, and $\hat{e}_3$. Then, in terms of Cartesian unit vectors, we have $\hat{e}_1 = -\hat{i}$, $\hat{e}_2 = \frac{1}{\sqrt{2}}(\hat{i}+\hat{j})$, and $\hat{e}_3 = -\frac{1}{\sqrt{2}}(\hat{i}-\hat{j})$. The length of spring 1 after a displacement $\vec{dx}$ is equal to the length of the vector $\vec{L}_1 = -L_0\hat{e}_1 + \vec{dx}$, so that the force of spring 1 on the mass is

$$\vec{F}_1 = -k\vec{L}_1 \left(1 - \frac{L_0}{|\vec{L}_1|}\right)$$

Because $\vec{dx}$ is infinitesimal,

$$|\vec{L}_1|^2 \approx L_0^2 \left(1 - 2\frac{\hat{e}_1 \cdot \vec{dx}}{L_0}\right)$$

Using this expression in the equation for $\vec{F}_1$ and retaining only first-order terms in $|\vec{dx}|$, we obtain

$$\vec{F}_1 \approx -k(\hat{e}_1 \cdot \vec{dx})\hat{e}_1$$

and the equations for $\vec{F}_2$ and $\vec{F}_3$ are similar. Thus, the equation for the net force on the mass is approximately given by

$$\vec{F} = -k\sum_{i=1}^{3} \mathbf{P}_i \vec{dx}$$

where $\mathbf{P}_i$ is the projection matrix, $\mathbf{P}_i = \hat{e}_i\hat{e}_i^T$.

Resolve the spring equations into x and y components to obtain the two equations for $\ddot{x}$, $\ddot{y}$, written in matrix form and, assuming the following initial conditions, determine the motions of the mass, $x(t)$ and $y(t)$.

$$x_0 = x(0) = 0.1, \qquad v_{x0} = \left.\frac{dx}{dt}\right|_0 = 0$$

$$y_0 = y(0) = 0.2, \qquad v_{y0} = \left.\frac{dy}{dt}\right|_0 = 0$$

P3.14 Eigenvalues/Eigenvectors of Commuting Matrices. If $\mathbf{A}$ and $\mathbf{B}$ are real-symmetric matrices that also commute ($\mathbf{AB} = \mathbf{BA}$), show that if $\hat{e}_i$ is an eigenvector of $\mathbf{A}$; i.e., $\mathbf{A}\hat{e}_i = \lambda_i^{[A]}\hat{e}_i$, then $\mathbf{B}\hat{e}_i$ must be proportional to $\hat{e}_i$; that is, $\hat{e}_i$ must also be an eigenvector of $\mathbf{B}$. Thus, show that $\mathbf{A}$ and $\mathbf{B}$ have the *same* set of eigenvectors.

P3.15 Generalized Eigenvalue Equations. The matrix equation $\mathbf{A}\vec{x} = \lambda\mathbf{B}\vec{x}$, where both $\mathbf{A}$ and $\mathbf{B}$ are real-symmetric matrices, is known as a generalized eigenvalue problem. Although the general solution of this equation is beyond the scope of this text, it is noteworthy that the MATLAB toolbox function `eig` will also solve the generalized equation. The use for this case is

$$[\mathsf{V}, \mathsf{D}] = \mathtt{eig}(\mathsf{A}, \mathsf{B})$$

a. Using the two matrices

$$\mathbf{A} = \begin{pmatrix} 5 & 1 & 2 & 0 \\ 1 & 5 & 0 & 2 \\ 2 & 0 & 5 & 1 \\ 0 & 2 & 1 & 5 \end{pmatrix} \qquad \mathbf{B} = \begin{pmatrix} 7 & 0 & -1 & -2 \\ 0 & 7 & -2 & -1 \\ -1 & -2 & 7 & 0 \\ -2 & -1 & 0 & 7 \end{pmatrix}$$

and the MATLAB function `eig`, determine the eigenvalues and eigenvectors of the generalized eigenvalue equation.

b. If the two matrices $\mathbf{A}$ and $\mathbf{B}$ commute, use the results of Problem 3.14 to show that the eigenvalues of the generalized equation are $\lambda_i = \lambda_i^{[A]}/\lambda_i^{[B]}$, where $\lambda_i^{[A]}$ and $\lambda_i^{[B]}$ are the eigenvalues of $\mathbf{A}$ and $\mathbf{B}$, respectively. (What is the result if $\lambda_i^{[B]} = 0$?) Show that the two matrices $\mathbf{A}$ and $\mathbf{B}$ do indeed commute and verify this statement.

c. If there exists a matrix $\mathbf{B}^{\frac{1}{2}}$ that satisfies $\mathbf{B}^{\frac{1}{2}}\mathbf{B}^{\frac{1}{2}} = \mathbf{B}$, show that the generalized eigenvalue equation may be written as a *normal* eigenvalue equation $\mathbf{C}\vec{y} = \lambda\vec{y}$ with $\mathbf{C} = \mathbf{B}^{\frac{1}{2}}\mathbf{A}\mathbf{B}^{-\frac{1}{2}}$ and $\vec{y} = \mathbf{B}^{\frac{1}{2}}\vec{x}$. Use the MATLAB function `sqrtm()` to compute $\mathbf{B}^{\frac{1}{2}}$ and then the function `inv()` to compute $\mathbf{B}^{-\frac{1}{2}}$. Use these matrices and the function `eig` to solve the *normal* equation and verify that the eigenvalues and the eigenvectors are the same.

P3.16 Superconductivity. The matrix

$$\mathbf{H}_P = -G \begin{pmatrix} 1 & 1 & \cdots & 1 \\ 1 & 1 & \cdots & 1 \\ \vdots & \vdots & \ddots & \vdots \\ 1 & 1 & \cdots & 1 \end{pmatrix}$$

plays an important role in the theory of superconductivity, its eigenvalues corresponding to the energies of electron pair states. The constant G is the strength of the interaction. For such a matrix of arbitrary size, prove the following:

a. The sum of the eigenvalues is equal to $-nG$, where n is the size of the square matrix.

b. If $\mathbf{H}$ is a matrix of all 1s, what is $\mathbf{H}^2$? Use this result to prove that such a matrix is positive definite.

c. There is one eigenvector with eigenvalue $-nG$. Find this eigenvector.

d. The eigenvalues of all remaining eigenvectors must be 0.

e. This interaction then predicts that one combination of electron pair states, a *collective state*, is strongly affected by the interaction, whereas all others are unaffected.

P3.17 Inhomogeneous Eigenvalue Equation. The equation $\mathbf{A}\vec{x} = \lambda\vec{x} + \vec{c}$, where $\vec{c}$ is a constant vector and $\mathbf{A}$ is a real-symmetric matrix, can be solved by expanding both $\vec{x}$ and $\vec{c}$ in terms of the eigenvectors of $\mathbf{A}$:

$$\vec{x} = \sum_{i=1}^{n} \hat{e}_i [\hat{e}_i^T \vec{x}] \qquad \vec{c} = \sum_{i=1}^{n} \hat{e}_i [\hat{e}_i^T \vec{c}]$$

a. Show that if λ does not equal any of the eigenvalues of $\mathbf{A}$, then the solution of this equation can be written as

$$\vec{x} = \sum_{i=1}^{n} \hat{e}_i \left(\frac{\hat{e}_i^T \vec{c}}{\lambda_i - \lambda} \right)$$

b. If λ does equal one of the eigenvalues, say λ_6, and if $\vec{c} \cdot \hat{e}_6 = 0$, the solution is then

$$\vec{x} = \sum_{i \neq 6}^{n} \hat{e}_i \left(\frac{\hat{e}_i^T \vec{c}}{\lambda_i - \lambda} \right) + \alpha \hat{e}_6$$

where α is an undetermined parameter.

c. Solve the equation with

$$\mathbf{A} = \begin{pmatrix} 2 & -1 & -2 & 0 \\ -1 & 2 & 0 & -2 \\ -2 & 0 & 2 & -1 \\ 0 & -2 & -1 & 2 \end{pmatrix} \qquad \vec{c} = \begin{pmatrix} 2 \\ 1 \\ -1 \\ -2 \end{pmatrix}$$

first with $\lambda = 2$, and then with $\lambda = 3$.

Summary

MATRIX EIGENVALUE EQUATIONS: $\mathbf{A}\vec{X} = \vec{B}$

Definition
    If: $\mathbf{A}$ is a real-symmetric or complex-Hermitian $n \times n$ matrix,
    Then:

- There are n real eigenvalues of $\mathbf{A}$, λ_i, $i = 1, \ldots, n$, some of which may be the same.
- For each λ_i there is a corresponding eigenvector, $\hat{e}_i$, that satisfies the eigenvalue equation.
- In the set of n eigenvectors, $\hat{e}_i$, $i = 1, \ldots, n$, each eigenvector is of unit length and all are mutually orthogonal. This set of vectors completely *spans* the space of vectors, which can be multiplied by $\mathbf{A}$.

Characteristic Equation

The equation $|\mathbf{A} - \lambda \mathbf{I}_n| = 0$ is an nth-degree polynomial equation in λ. The roots of this equation are the eigenvalues of $\mathbf{A}$.

Solution for Eigenvectors

Once the eigenvalues λ_i, $i = 1, \ldots, n$, are determined, the eigenvectors are computed by taking each of the eigenvalues, one at a time, and solving the *homogeneous* vector equation, $(\mathbf{A} - \lambda_i \mathbf{I}_n)\vec{x} = 0$, for the vector $\vec{x}$. This vector, once normalized, is the eigenvector, $\hat{e}_i$, corresponding to the eigenvalue λ_i.

Principal Axes

The eigenvectors of a real, symmetric matrix $\mathbf{A}$ define a coordinate system with a set of axes called the principal axes. The image of $\mathbf{A}$ in this coordinate system is a diagonal matrix with the eigenvalues of $\mathbf{A}$ along the diagonal, and it is obtained by a rotation of coordinates defined by

$$\mathbf{A}' = \begin{pmatrix} \lambda_1 & 0 & 0 & \cdots & 0 \\ 0 & \lambda_2 & 0 & \cdots & 0 \\ 0 & 0 & \lambda_3 & \cdots & 0 \\ \vdots & \vdots & \vdots & \ddots & \vdots \\ 0 & 0 & 0 & \cdots & \lambda_n \end{pmatrix} = \mathbf{R}^T \mathbf{A} \mathbf{R}$$

where the rotation matrix $\mathbf{R}$ is an orthogonal matrix constructed by using the eigenvectors of $\mathbf{A}$ as successive columns.

Jacobi Rotations

A rotation by an angle θ in the plane of two coordinates, x_i and x_j, is effected by the matrix $\mathbf{R}_{ij}(\theta)$, which, except for elements $r_{ij} = -r_{ji} = -\sin\theta$, $r_{ii} = r_{jj} = \cos\theta$, is the same as the identity matrix $\mathbf{I}_n$. The angle is chosen so that the image of the off-diagonal element, a'_{ij}, is zero. One pass of an attempt to *diagonalize* a matrix by Jacobi rotations consists of the sequence of $\frac{1}{2}n(n+1)$ rotations needed to successively "zero" each off-diagonal element.

Spectral Decomposition

The eigenvectors of a real-symmetric matrix $\mathbf{A}$ can be used to construct a set of n projection operators, $\mathbf{P}_i = \hat{e}_i \hat{e}_i^T$, with $\sum_{i=1}^{n} \mathbf{P}_i = \mathbf{I}_n$. The matrix $\mathbf{A}$ may then be expanded in terms of the projection operators as

$$\mathbf{A} = \sum_{i=1}^{n} \lambda_i \mathbf{P}_i$$

where λ_i are the eigenvalues.

<table>
<tr><td>*Singular Value*
Decomposition</td><td>Any real $n \times m$ matrix $\mathbf{M}$ can be written as a product of three matrices, $\mathbf{M} = \mathbf{U}\mathbf{w}\mathbf{V}^T$, where $\mathbf{U}$ and $\mathbf{V}$ are orthogonal matrices of size $n \times n$ and $m \times m$, respectively, and $\mathbf{w}$ is a matrix of the same size as $\mathbf{M}$ with nonzero values only along the principal diagonal. The values on this diagonal, w_{ii}, are called the singular values of the matrix $\mathbf{M}$. This decomposition may be used to define a generalized inverse, $\mathcal{M}^{-1}$. This matrix then satisfies the relation $\mathcal{M}^{-1}\mathbf{M} = \mathbf{I}_m$, if $n < m$, and $\mathbf{M}\mathcal{M}^{-1} = \mathbf{I}_m$, if $n > m$ (but $\mathbf{M}\mathcal{M}^{-1} \neq \mathcal{M}^{-1}\mathbf{M}$), and is used by MATLAB to define backslash division of matrices.</td></tr>
</table>

MATLAB Functions Used

eig(A)   If $\mathbf{A}$ is a square matrix, the numerical solution of the eigenvalue equation, $\mathbf{A}\vec{x} = \lambda\vec{x}$, is computed by the function `eig(A)`. The use is `[V,D] = eig(A)`. The eigenvectors are returned as the columns of the matrix $\mathbf{V}$, and the corresponding eigenvalues are contained in the diagonal elements of the square matrix $\mathbf{D}$. If $\mathbf{A}$ is real-symmetric, the eigenvectors will be orthonormal and the eigenvalues will be real.

svd(A)   The singular value decomposition of an $n \times m$ matrix $\mathbf{M}$, with $n \geq m$, is computed by the function `svd`. The use is `[U,w,V] = svd(A)`. The three matrices returned, $\mathbf{U}$, an $n \times n$ matrix, $\mathbf{w}$, a diagonal matrix of the same size as $\mathbf{M}$, and $\mathbf{V}$, an $m \times m$ matrix, satisfy the equation

$$\mathbf{M} = \mathbf{U}\mathbf{w}\mathbf{V}^T$$

In MATLAB, this decomposition is particularly useful when defining a generalized inverse of a matrix. The situation with $n \leq m$ is defined similarly.

trace(A)   Returns the sum of the diagonal elements of the matrix $\mathbf{A}$. If $\mathbf{A}$ is square, the trace is equal to the sum of the eigenvalues.

sqrtm(A)   Computes the *matrix* square root of the square matrix $\mathbf{A}$. That is, a matrix $\mathbf{B}$ is returned that satisfies the relation $\mathbf{B}\mathbf{B} = \mathbf{A}$.

Bibliography for Part I

1. Forman S. Acton, *Numerical Methods That Work*, Harper & Row, New York, 1970.
2. B. Carnahan, H. Luther, and J. Wilkes, *Applied Numerical Methods*, John Wiley, New York, 1969.
3. T. J. Fletcher, *Linear Algebra Through Its Applications*, Van Nostrand Reinhold, New York, 1972.
4. Michael D. Greenberg, *Advanced Engineering Mathematics*, Prentice-Hall, Englewood Cliffs, NJ, 1988.
5. F. B. Hildebrand, *Methods of Applied Mathematics*, Prentice-Hall, Englewood Cliffs, NJ, 1952.

6. W. H. Press, B. P. Flannery, S. A. Teukolsky, and W. T. Vetterling, *Numerical Recipes, The Art of Scientific Computing*, Cambridge University Press, Cambridge, England, 1986.

7. B. T. Smith et al., *Matrix Eigensystem Routines—EISPACK Guide*, 2nd ed., vol. 6 of Lecture Notes in Computer Science, Springer-Verlag, Berlin and New York, 1976.

8. J. Stoer and R. Bulirsch, *Introduction to Numerical Analysis*, Springer-Verlag, Berlin and New York, 1980.

9. Gilbert Strang, *Linear Algebra and Its Applications*, 2nd ed., Academic Press, New York, 1980.

10. Samuel S. M. Wong, *Computational Methods in Physics and Engineering*, Prentice-Hall, Englewood Cliffs, NJ, 1992.

PART II / MATLAB
Applied to Functions of a Single Variable*

4 Series Expansions for Functions
5 Roots of Functions of a Single Variable
6 Minimum or Maximum of a Function
7 Numerical Integration Techniques

PREVIEW The theory of functions of a single variable encompasses such topics as continuity, limits, graphing, series expansions, roots, minima/maxima, differentiation, and integration. Each of these topics is treated in depth in a normal introductory calculus sequence. Our emphasis here is somewhat different in that we are attempting to both reinforce and extend these topics by providing numerical prescriptions for problems that are often treated only in an abstract analytical way in calculus. Our analysis begins with a review of what is meant by a function, classification of the types of functions that will be encountered, and a sampling of MATLAB techniques for constructing your own toolbox-like versions of functions. This is followed by the most fundamental of function definitions; the power series, and, in particular, the Taylor series. The procedures for testing and accelerating the convergence of a series make use of MATLAB programming techniques, such as the use of loops and simple logic structures.

The numerical analysis skills of a scientist or engineer are greatly enhanced by the use of a graph to visualize a problem or function. The ease with which all manner of functions can be plotted in MATLAB is among its most attractive features. Special attention is given to plotting one or more functions of a single variable using the toolbox function plot.

This part describes the standard topics of the numerical analysis of functions of a single variable, and illustrates the toolbox routines for solving these problems. These procedures include finding the roots and extrema of functions, and performing numerical integration of a function.

---

*This part makes use of MATLAB programming techniques described in Appendix Sections A.i through A.iv.

131

INTRODUCTION

A function is a mapping or correspondence between an independent variable, say x, and a dependent variable, say y. The idea is simple: *You give me an x; i.e., some number, and I'll do something to x, the same thing for all x's, and return a value y.* The operation on x could be defined in terms of algebraic operations as $3x^3 - x^2 + 1$ or could be: Given a value of x, I simply hit the e^x key on my calculator. The *function* then consists of the set of operations that will be performed on the arbitrary x. The only requirement on the function is that, for each and every x that it is given, there is only one outcome, one value of y, returned. That is, all functions must be *single valued*. This excludes functions like $\sqrt{x}$, unless we stipulate that, as a function, $\sqrt{x}$ returns only one of the two possible solutions of $y^2 = x$.

Another attribute of functions—but this one is not required—is continuity. The mathematical definitions of continuity are expressed in terms of limits, stating that, as we change the independent variable, x, infinitesimally, the resulting dependent variable, y, also changes infinitesimally; that is, there are no discontinuous jumps or gaps in the function. This feature of functions is adequately described by visualizing the function as "smooth," which has both a colloquial and a technical meaning in mathematics. The ideas associated with limits and continuity are the most important concepts in modern mathematics and are the basis for all of calculus. Unfortunately, digital computers are not adept at handling limit concepts; not only is the number set in a computer discrete, but it is a *finite* set. Nonetheless, procedures can be devised that will mimic the limit process and will be adequate for most applications. Thus, we will often let $\Delta x \to \epsilon \approx 0$, short of the infinitesimal dx. For our purposes, what is commonly meant by a smooth function will suffice, and this will be illustrated by graphs of the function over some range.

A mathematical definition of a function consists of prescribing a set of elementary arithmetic operations to be performed on the variable x. The combined number of these operations—multiplication, division, addition, subtraction—can be finite, in which case the function is a polynomial in x with positive or negative powers. The manipulation of polynomial functions using special MATLAB functions is described in Appendix Section A.ii. On the other hand, the set of required arithmetic operations can be infinite, meaning that the function can be expressed as a power series with an infinite number of terms. This is the topic of Chapter 4.

Once the function has been defined in terms of arithmetic operations, we can define a wide variety of operations that act on the function itself; i.e., functions of functions. Operations like integration and differentiation are operations on functions, as are root-solving techniques and procedures for finding the minimum or maximum of a function. These topics are addressed in Chapters 5 through 7.

4 / Series Expansions for Functions*

PREVIEW   A study of the formal properties of functions of a single variable begins with a discussion of the properties of infinite power series. The reason for this derives from the fact that a mathematical function must always be defined in terms of the elementary arithmetic operations. For a broad class of functions, the number of operations required in their definition is *infinite*; i.e., the function is defined in terms of an *infinite power series*. Of course, an infinite number of operations is, itself, undefined, and only becomes meaningful if terms in the series eventually become infinitesimal and the series *converges*. Under these circumstances, the infinite number of operations can be terminated, and the series is replaced by a *polynomial*.

However, the number of terms that need to be included is usually not known beforehand and must be determined by monitoring the size of each term during the evaluation of the series. That is, in addition to the summation structure used in evaluating a polynomial described in Appendix Section A.ii, we also require new MATLAB commands that compare quantities and execute a variety of tasks based on that comparison. It is this ability that is at the very heart of any computer language. Most problems that can be solved using a computer must be structured so as to continually monitor the behavior of a function or a procedure, and when a particular condition is satisfied, either terminate the calculation or move on to a different stage of the problem.

In this chapter, we investigate the properties of infinite power series, not only because of their importance to mathematical analysis, but also because they provide an excellent introduction to the decision structures of MATLAB.

Radius of Convergence of a Power Series

A power series expansion of a function is expressed by the infinite summation,[1]

$$f(x) = \sum_{k=0}^{\infty} a_k (x - x_0)^k \qquad (4.1)$$

---

*This chapter makes use of MATLAB programming techniques described in Appendix Sections A.i through A.iii.

[1]This definition can be generalized to include situations where x is complex and the summation is over negative integer values of k as well. In that case, it is known as a *Laurent series*. We limit the discussion to real x and nonnegative k, in which case the series is known as a *Taylor series*.

where the coefficients a_k are given as functions of the integers k, and x_0 is a constant value of the independent variable x. This is said to be an expansion of $f(x)$ about the point x_0. If the series on the right converges to a well-defined number for each value of x in the range $|x - x_0| < R$, it is said to have a radius of convergence R. (See any advanced calculus text for the standard tests of convergence.)

To simplify the expressions for the moment, we will take $x_0 = 0$.

Consider the well-known series for $\ln(1 + x)$,

$$\ln(1 + x) = x - \frac{x^2}{2} + \frac{x^3}{3} + \cdots + \frac{(-1)^{k+1}x^k}{k} + \cdots \quad \text{for } -1 < x \le +1 \tag{4.2}$$

$$\ln(0) = -\infty$$

$$\ln(2) = 0.693147\ldots$$

which has a radius of convergence given by x's in the range $-1 < x \le +1$. That is, the definition of $\ln(1 + x)$ is *defined* by this expansion starting with $x = 0$ out to $|x| < 1$. Because $\ln(1 + x)$ diverges as $x \to -1$, we expect the series to do likewise. However, this divergence is seen to render the series invalid for *all* x beyond the singularity, whereas the function $\ln(1 + x)$ remains well defined for $x > +1$. The conclusion is then that the definition of a function by means of a power series is valid only within the radius of convergence of the series, and further, that the limits of the radius of convergence are due to singularities of the function.[2]

ALGORITHM FOR SERIES SUMMATION

The general procedure for summing a power series consists of first obtaining a formal relationship between successive terms, making it possible to compute the next term in the series from the known value of the current term. Next, the size of each new term is monitored, and the summation is terminated when this size is less than some specified tolerance. Of course, it is assumed that the radius of convergence is known and that the current value of x is within that radius.

If r_k is the general form for the ratio of successive terms, $a_k x^k$, in the series

$$r_k = \frac{a_{k+1}x^{k+1}}{a_k x^k} = x\,\frac{a_{k+1}}{a_k}$$

---

[2]To obtain a series expansion for $\ln(1 + x)$ that is valid for $|x| > 1$, try writing

$$\ln(1 + x) = \ln x \left(1 + \frac{1}{x}\right)$$

$$= \ln x + \ln \left(1 + \frac{1}{x}\right)$$

$$= \ln(1 + (x - 1)) + \ln \left(1 + \frac{1}{x}\right)$$

The first term would have a series expansion valid for $-1 < (x - 1) \le +1$, whereas the second could be expanded in powers of $(\frac{1}{x})$ and would be valid for $-1 < \frac{1}{x} \le +1$. The values of x for which both expansions would be valid would be $+1 \le x \le +2$. Similar techniques can be used to obtain expansions valid for other radii of convergence.

then exponentiation can be avoided by computing the next term in the series from the current term by

$$(\text{term})_{k+1} = x \left(\frac{a_{k+1}}{a_k} \right) (\text{term})_k$$

For example, the series expansion for e^x is

$$e^x = 1 + x + \frac{x^2}{2!} + \frac{x^3}{3!} + \cdots + \frac{x^k}{k!} + \cdots$$

where $k! = k(k-1)(k-2)\cdots 2 \cdot 1$ is the factorial function, and the series is known to converge for all x. The ratio of successive terms is

$$r_k = \frac{x^{k+1}}{(k+1)!} \frac{k!}{x^k} = x \frac{k!}{(k+1)!} = x \frac{1}{k+1}$$

Because $\lim_{k \to \infty} r_k = 0$ for *any* finite value of x, the ratio test for the convergence of a series then guarantees that the series will converge for *all* x; i.e., the radius of convergence is ∞. The algorithm then for summing the first N terms of the series for an arbitrary value of x would be

```
sum  = 1;
term = x;
for k = 1:N
      sum  = sum + term;
      term = x*term/(k+1);
end
```

Of course, although the series is guaranteed to converge, we are not provided with any means of determining how many terms will be required to obtain a specified accuracy. Indeed, from the form of r_k in the expansion for e^x, successive terms will continue to *increase* in magnitude as long as $\frac{|x|}{k+1} > 1$.

What is needed is a set of MATLAB commands to compare the current value of term_k with some small quantity, say ϵ, and to terminate the summation when $|\text{term}_k| < \epsilon$. The MATLAB commands for performing such comparisons are called *relational* or *logical* operators. They are described in Appendix Section A.iii.

Acceleration of Convergence

An axiom of numerical analysis is that intelligence is mightier than computing power. There are numerous cases where small, but clever, changes to a computer program have resulted in a reduction of the execution times by 10- or a 100-fold, while simultaneously increasing the accuracy of the result. A good example of this relates to the summation of a series. The preceding algorithm, although valid, will often require inclusion of a great many terms to obtain a result of satisfactory accuracy. Of course, the more terms included in the summation, the more significant are the accumulating *round-off errors* resulting from each arithmetic operation. Using the series expansion for e^x to evaluate e^{-50} would prove to be extremely inefficient and, because of the round-off error, would very likely not yield meaningful results. Not only does the summation require that a great many terms be included before individual terms become small, but the alternating nature of the terms necessitates the subtraction of very large, nearly equal numbers, thus amplifying the round-off error. Even for more rapidly converging series, it is often worthwhile to seek some procedure for accelerating the convergence.

AITKEN'S METHOD

A method due to Aitken[3] is to write the sum in terms of its partial sums plus a remainder, R_k, as

$$s = \sum_{k=0}^{\infty} a_k = \sum_{k=0}^{n} a_k + R_{n+1}$$
$$= s_n + R_{n+1}$$

Of course, we do not know the value of the remainder, but if we assume that it satisfies the condition

$$R_n \approx c\alpha^n \tag{4.3}$$

i.e., R_n is proportional to some quantity, α^n, we can write

$$s - s_n = c\alpha^{n+1} = R_{n+1}$$
$$s - s_{n-1} = c\alpha^n \quad = R_{n+1}/\alpha$$
$$s - s_{n-2} = c\alpha^{n-1} = R_{n+1}/\alpha^2$$

three equations in the three unknown quantities, s, α, and R_n. Solving these equations for s, the value of the series, we obtain:

$$s = s_n - \frac{(s_n - s_{n-1})^2}{s_n - 2s_{n-1} + s_{n-2}} \tag{4.4}$$

Now, if the series satisfies the assumption in Equation 4.3, the exact value of the entire series is then given in terms of any three successive partial sums. For example, the binomial expansion of $(1-x)^{-1}$ for $|x| < 1$ is

$$(1-x)^{-1} = 1 + x + x^2 + x^3 + \cdots + x^k + R_k$$

For this particular series, the partial sums can be expressed as

$$s_n = \frac{1 - x^{n+1}}{1 - x}$$

Inserting this result into Equation 4.4 yields $s = (1-x)^{-1}$.

---

[3] Aitken, "On Bernoulli's Solution of Algebraic Equations," *Proc. Roy. Soc. Edinburgh*, vol. 46, 1926, pp. 289–305.

Even if the series in question does not satisfy the condition in Equation 4.3, insofar as it is *approximately* true, Equation 4.4 will return an approximate value for the complete sum. More importantly, this procedure suggests an algorithm: Use the partial sums s_1, s_2, and s_3 to compute an estimate for the sum s, and label it as $s_3^{(2)}$. Then use s_2, s_3, and s_4 to compute a second approximation, $s_4^{(2)}$, thus replacing the sequence of partial sums s_k by an *improved* sequence, $s_i^{(2)}$. Depending on the accuracy of the approximation in Equation 4.3, the second sequence should converge to a limit more quickly. In addition, if the procedure worked once, it should work again. Thus, use the new sequence, $s_i^{(2)}$, to compute a third set, $s_i^{(3)}$, by applying Equation 4.4 once again, and so on.

EXAMPLE 4.1

Acceleration of the Convergence of a Slowly Converging Series

We start with the partial sums for the slowly convergent series for $\ln(1 + x)$ with $x = 0.9$. These sums are computed and displayed in the first column of Table 4.1.

i	$s_i^{(1)} = s_i$	$s_i^{(2)}$	$s_i^{(3)}$	$s_i^{(4)}$	$s_i^{(5)}$	$s_i^{(6)}$
1	0.900000					
2	0.495000					
3	0.738000	0.646875				
4	0.573975	0.640075				
5	0.692073	0.642637	0.641936			
6	0.603500	0.641460	0.641830			
7	0.671828	0.642072	0.641862	0.641855		
8	0.618019	0.641725	0.641850	0.641854		
9	0.661066	0.641934	0.641855	0.641854	0.641855	
10	0.626198	0.641802	0.641853	0.641854	0.641854	
11	0.654726	0.641889	0.641854	0.641854	0.641854	0.641854

Table 4.1   *Aitken transformation of partial sums of the series for $\ln(1 + 0.9)$*

Notice that after the first two partial sums are computed, the progress of the calculation is to compute two more successive partial sums, s_k and s_{k+1}, and then use the Aitken transformation, Equation 4.4, to step across the triangular table computing higher-level approximations, s_k^m, for $m = 1, \ldots, \frac{k-1}{2}$ partial sums. The computed value for the series should become progressively more accurate as we move down the triangular table (include more and more terms) and as we move across (compute higher- and higher-level transformations).

The following MATLAB function implements this algorithm:

```
function [y,imax] = AitknSum(x,small)
%  AitknSum(x,small) will first produce a column
%  of partial sums for the series expansion of
%  ln(1+x).  The variable x must be a scalar.  The
```

```
%  Aitken transformation is then repeatedly applied
%  to this set until convergence is attained with-
%  in tolerance given by small.  The result is re-
%  turned in y and imax is the number of terms
%  included.
%==================================================
if  length(x)>1; error('x must be a scalar'); end
s(1,1) = x;   s(1,2) = x-x*x/2;
diff    = s(1,2);
term    = x^3/3;   i = 3;
while  abs(diff) > small
      s(1,i) = s(1,i-1) + term;
        term = -x*term*i/(i+1);
   s(1,i+1) = s(1,i) + term;
        term = -x*term*(i+1)/(i+2);
      kmax = (i+1)/2;
      for k = 2:kmax
         s(k,i)=s(k-1,i) - (s(k-1,i)-s(k-1,i-1))^2 ...
           /(s(k-1,i)+s(k-1,i-2)-2*s(k-1,i-1));
         s(k,i+1)=s(k-1,i+1) - (s(k-1,i+1)-s(k-1,i))^2 ...
           /(s(k-1,i+1)+s(k-1,i-1)-2*s(k-1,i));
      end
   i = i+2;
   y = s(kmax,2*kmax);
   diff = s(kmax,2*kmax) - s(kmax-1,2*kmax-2);
   imax = i;
   if imax > 20
     error('excessive terms in AitknSum')
   end
end
```

●

WHAT IF?   If $x \geq 1$, the series for $\ln(1+x)$ diverges. Reproduce the above example with $x = 1.1$ to see if Aitken's method can be applied to a divergent sequence. Experiment to find the largest x for which the method returns correct values for $\ln(1+x)$.

●

One of the first things you should notice concerning the preceding MATLAB function is that it has *not* been structured to accommodate input of a matrix as the independent variable, x. The reason for this relates to a limitation of MATLAB; namely, matrices in MATLAB *cannot* have more than *two* subscripts.[4] That is, if x were a column vector, then each element of the arrays $s_k^{(m)}$ would likewise be a column vector, requiring three indices.

---

[4]The reason for limiting the number of subscripts to merely two has to do with the fact that multiplication of three-dimensional or higher matrices is not defined.

Also, notice that very long statements in MATLAB can be broken into several lines by inserting an ellipsis (three or more periods followed by a carriage return) at a point in the long statement and continuing the remainder of the statement on the next line.

It is interesting to use this MATLAB function to explore some features of this particular series for $\ln(1 + x)$ or to adapt it to sum some other series. For example, the test for success in the function is based on the difference between the last two most accurate values for the series. The actual accuracy of the computation can be compared with the value returned by `log(1+x)`, and, in most cases, a very small number of terms is required to obtain a quite accurate value for the series.

Further experimentation with this MATLAB function reveals that the sum of the series for, say $x = 4$, accurately duplicates the value of $\ln(1 + 4)$. This is very surprising, as the series certainly does not even converge for $x > 1$. This is an example of a startling fact concerning the power series expansion of a function. If the power series converges for a finite range of x values, then the power series defines the function not only within its radius of convergence, but *everywhere*. That is, once we have a power series expansion for a function $f(x)$ that converges for $|x - x_0| < R$, transformations can be found that will yield a series representation that is valid for a different radius of convergence. The Aitken transformation is one such example; the *slight-of-hand* substitutions of footnote 2 on page 134 are another. This process of extending the region of validity of a power series is called *analytic continuation*. The proof of the statement and descriptions of analytic continuation methods require a study of the theory of complex variables.

Taylor Series

If a function is expanded as a power series in x that converges for $|x - x_0| < R$:

$$f(x) = \sum_{k=0}^{\infty} a_k(x - x_0)^k = a_0(x-x_0)^0 + a_1(x-x_0)^1 + a_2(x-x_0)^2 + \cdots$$

then certainly the series converges to $f(x_0)$ at $x = x_0$. In addition, within the radius of convergence, it can be shown that the derivative or integral of the series is the sum of the derivatives or integrals of the terms *and* they both converge with the same radius of convergence. Thus, the derivatives of $f(x)$ are

$$f(x) = f(x) \cdot \frac{1}{k}\frac{df(x)}{dx}$$

$$\frac{df(x)}{dx} = \sum_{k=1}^{\infty} ka_k(x - x_0)^{k-1} \implies \frac{1}{k}\frac{df(x)}{dx} = \sum_{k=1}^{\infty} a_k(x-x_0)^{k-1}$$

$$= a_1(x-x_0)^0 + a_2(x-x_0)^1 + a_3(x-x_0)^2 + \cdots$$

$$\frac{d^2 f(x)}{dx^2} = \sum_{k=2}^{\infty} ka_k(k - 1)(x - x_0)^{k-2}$$

$$\frac{d^3 f(x)}{dx^3} = \sum_{k=3}^{\infty} k(k - 1)(k - 2)a_k(x - x_0)^{k-3}$$

$$\vdots \quad \vdots \quad f(x) - \frac{1}{k}\frac{df(x)}{dx} = \sum_{k=0}^{\infty} a_k(x-x_0)^k - \sum_{k=1}^{\infty} a_k(x-x_0)^{k-1}$$

$$= a_0(x-x_0)^0 = \boxed{a_0}$$

Evaluating each of these equations at $x = x_0$, we obtain the relationship between the series coefficients and the derivatives of the function $f(x)$:

$$\frac{df^n(x)}{dx^n}\bigg|_{x=x_0} \equiv f^{[n]}(x_0) = n! \cdot a_n$$

Thus, if a function $f(x)$ has the property that *all* of its derivatives can be evaluated at $x = x_0$, then the Taylor series expansion of the function *about the point* $x = x_0$ is given by

$$f(x) = \sum_{k=0}^{\infty} \frac{1}{n!} \frac{df^n(x)}{dx^n}\bigg|_{x=x_0} (x - x_0)^n \qquad (4.5)$$

This equation, with $x_0 = 0$, is also called the *Maclaurin series*.

For example, the derivatives of the function $f(x) = \sin(x)$ have the values

$$\frac{d^n}{dx^n}\sin(x)\bigg|_{x=0} = \begin{array}{ll} (-1)^{(n-1)/2} & \text{for odd } n \\ 0 & \text{for even } n \end{array}$$

Thus, the Taylor/Maclaurin series expansion of $\sin(x)$ is

$$\sin(x) = \sum_{k=\text{odd}}^{\infty} \frac{(-1)^{(n-1)/2}}{n!} x^n$$

$$= x - \frac{x^3}{3!} + \frac{x^5}{5!} - \cdots$$

The Taylor/Maclaurin series expansions for several common functions are given in Table 4.2.

Series	For
$e^x = 1 + x + \frac{x^2}{2!} + \frac{x^3}{3!} + \cdots + \frac{x^n}{n!} + \cdots$	all x
$\sin x = x - \frac{x^3}{3!} + \frac{x^5}{5!} + \cdots + (-1)^n \frac{x^{2n+1}}{n!} + \cdots$	all x
$\cos x = 1 - \frac{x^2}{2!} + \frac{x^4}{4!} + \cdots + (-1)^n \frac{x^{2n}}{n!} + \cdots$	all x
$\ln(1 + x) = 1 - x + \frac{x^2}{2} + \cdots + (-1)^n \frac{x^n}{n} + \cdots$	$-1 \leq x < +1$
$\sinh x = x + \frac{x^3}{3!} + \frac{x^5}{5!} + \cdots + \frac{x^{2n+1}}{n!} + \cdots$	all x
$\cosh x = 1 + \frac{x^2}{2!} + \frac{x^4}{4!} + \cdots + \frac{x^{2n}}{n!} + \cdots$	all x
$(1 + x)^p = 1 + px + \frac{p(p-1)}{2!}x^2 + \cdots + \frac{p!}{(p-n)!\, n!}x^n + \cdots$	$\|x\| < 1$
$\dfrac{1}{1+x} = 1 - x + x^2 - x^3 + \cdots + (-1)^n x^n$	$\|x\| < 1$

Table 4.2   *Taylor/Maclaurin series of common functions*

The utility of the Taylor series expansion of a function in numerical analysis cannot be overstated. The Taylor series enables us to approximate a function to any desired level of accuracy and, thereby, to replace the function by a few simple algebraic terms. This idea permeates numerical analysis and will be used to find the roots of functions, to integrate functions, and to solve differential equations.

THE REMAINDER TERM IN TAYLOR SERIES

One last feature of the Taylor series has to do with the error introduced when the series is truncated after n terms. If the series is written in terms of its partial sums *plus a remainder term*, as

$$f(x) = \sum_{k=0}^{n-1} \frac{1}{k!} \frac{d^k f}{dx^k}\bigg|_{x=x_0} (x - x_0)^k + R_n(x) \tag{4.6}$$

the function $R_n(x)$ is the difference between the *actual* function $f(x)$ and the approximation to the function after including n terms. Of course, we cannot know this function precisely, or there would be no need for an infinite series. Nonetheless, it is possible to express the *form* of the remainder term as

$$R_n(x) = \frac{1}{n!} \frac{d^n f}{dx^n}\bigg|_{x=\xi} (x - x_0)^n, \tag{4.7}$$

where, in order for Equation 4.6 to be satisfied, the derivative is to be evaluated at some unspecified value ξ in the interval, $x_0 \leq \xi \leq x$. Although it is not possible to determine $R_n(x)$ precisely, it can be used to obtain *limits* on the size of the truncation error by determining the maximum of $R_n(x)$ over the interval. In addition, the form of the remainder states that, whatever the truncation error might be, at least its dependence on x is a constant times $(x - x_0)^n$. That is, the truncation of a Taylor series after the $(n-1)$ term yields an approximation that is said to be accurate *to order* $(x - x_0)^n$.

EXAMPLE 4.2

A Series Approximation to $x^{1/3}$

The Taylor series expansion for the function $x^{1/3}$ about the point $x_0 = 1$ is obtained by evaluating the first few coefficients using Equation 4.6.

$$f(x)|_1 = 1$$

$$\frac{df}{dx}\bigg|_1 = \left(\frac{1}{3} x^{-2/3}\right)\bigg|_1 = \frac{1}{3}$$

$$\frac{1}{2} \frac{d^2 f}{dx^2}\bigg|_1 = \frac{1}{2}\left(-\frac{2}{9} x^{-5/3}\right)\bigg|_1 = -\frac{1}{9}$$

$$\frac{1}{3!} \frac{d^3 f}{dx^3}\bigg|_1 = \frac{1}{6}\left(\frac{10}{27} x^{-8/3}\right)\bigg|_1 = \frac{5}{81}$$

so that the expansion of $x^{1/3}$ about $x_0 = 1$ to the *third* order is

$$x^{1/3} = 1 + \frac{1}{3}(x - 1) - \frac{1}{9}(x - 1)^2 + \frac{5}{81}(x - 1)^3 + R_4(x) \qquad \text{for } |x - 1| < 1$$

and the remainder term is given by

$$R_4(x) = \frac{1}{4!} \frac{d^4}{dx^4} x^{1/3} \bigg|_{x=\xi} (x-1)^4 = \frac{-10}{243} \xi^{-11/3} (x-1)^4$$

If we next wish to use this approximation to compute the cube roots of numbers somewhat *greater* than 1, the *maximum* value of $R_4(x)$ over the interval $1 \leq \xi \leq x$ is obtained at $\xi = 1$, or $[R_4(x)]_{\max} = \frac{-10}{243}(x-1)^4$. The accuracy of this series expansion for $x^{1/3}$ is illustrated in Table 4.3.

| | $f(x) = x^{1/3}$ | | | |
| | | | | |
x	$x^{1/3}$ (from series to third order)	$x^{1/3}$ (exact)	Actual error	Maximum value of remainder $|R_4(x)|_{\max}$
1.05	1.0163966...	1.0163964...	0.00000025...	0.00000026...
1.10	1.032284...	1.032280...	0.0000038...	0.0000041...
1.20	1.06271...	1.06265...	0.000057...	0.000066...
1.40	1.1195...	1.1187...	0.00082...	0.0011...
1.80	1.227...	1.216...	0.0107...	0.0169...

Table 4.3   *Comparison of actual error in the series for $x^{1/3}$ with remainder term*

You will notice from the table that the maximum value of the remainder term is always slightly larger than the actual error, the approximation improves significantly as $x \approx 1$, and the actual error is indeed proportional to $(x-1)^4$. (As x increases from $x = 1.1$ to $x = 1.2$ or as $(x-1)$ doubles, the error increases by a factor of $15 \approx 2^4$.)   ●

Padé Approximations

As we have seen, the radius of convergence of a power series expansion is determined by singularities in the function.[5] The radius of convergence and also the rate of convergence are, to a large extent, dependent on how close x is to a singularity of the function. Yet, approximations formulated in terms of Taylor series make no attempt to explicitly incorporate the features of these singularities. Not surprisingly, this defect may have seriously detrimental effects when

---

[5]It should be pointed out that these limiting singularities of a function can be at *complex*, as well as real, values of x. Thus, the Taylor series expansion of $(1 + x^2)^{-1} = 1 - x^2 + x^4 - \cdots$ is an alternating series that will certainly converge for $x < 1$, and diverge for $x > 1$, even though the function remains finite at ± 1. The singularity is found at $x = \pm i = \pm\sqrt{-1}$. The radius of convergence corresponds to $|x| < 1$, where, in this case, $|x|$ refers to the absolute magnitude of a complex quantity, $|x| = \sqrt{\bar{x} \cdot x}$.

approximating a function by a truncated power series. Thus, in the approximation

$$f(x) \approx \sum_{k=0}^{n} a_k(x - x_0)^k$$

if there are singularities in the function $f(x)$ on the left, these are apparently not reflected in the summation on the right, which for finite n has no singularities whatsoever.

Padé's approximation is an attempt to approximate a function and, simultaneously, incorporate some of the singular features of the function. The idea is quite simple. Instead of approximating the function as a truncated power series; i.e., a polynomial, a ratio of polynomials is tried:

$$f(x) \approx \frac{a_1 x^{n-1} + a_2 x^{n-2} + \cdots + a_{n-1} x + a_n}{b_1 x^{n-1} + b_2 x^{n-2} + \cdots + b_{n-1} x + b_n} \tag{4.8}$$

or if an approximation is desired over the range $x_0 \le x \le x_1$, we could write

$$f(x) \approx \frac{a_1 (x - x_0)^n + a_2 (x - x_0)^{n-1} + \cdots + a_n (x - x_0) + f_0}{b_1 (x - x_0)^n + b_2 (x - x_0)^{n-1} + \cdots + b_n (x - x_0) + 1} \tag{4.9}$$

Usually, both the numerator and the denominator polynomials are chosen to be of the same degree. The denominator polynomial can be designed to have roots at points where there are singularities in the function $f(x)$, so this formulation has the potential for application in regions where the Taylor series may be poorly convergent. As we shall also see, the Padé approximation is often surprisingly accurate.

The $2n$ polynomial coefficients, $a_i, b_i, i = 1, \ldots, n$, are determined as follows:

1. Evaluate the function at $2n$ values of x that are spaced over the interval of interest, $f_i \equiv f(x_i)$, $i = 1, \ldots, 2n$. (Do not include $f_0 = f(x_0)$ in this set.)

2. Rewrite Equation 4.9 as

$$a_1 (x_i - x_0)^n + \cdots + a_n (x_i - x_0) + f_0$$
$$= f_i[b_1 (x_i - x_0)^n + \cdots + b_n (x_i - x_0) + 1]$$

for each value of x_i.

3. Arrange these equations into a matrix equation of the form:

$$\begin{pmatrix} (x_1 - x_0)^{n-1} & \cdots & 1 & -f_1(x_1 - x_0)^{n-1} & \cdots & -f_1 \\ (x_2 - x_0)^{n-1} & \cdots & 1 & -f_2(x_2 - x_0)^{n-1} & \cdots & -f_2 \\ \vdots & \cdots & \vdots & \vdots & \cdots & \vdots \\ (x_n - x_0)^{n-1} & \cdots & 1 & -f_n(x_n - x_0)^{n-1} & \cdots & -f_n \\ (x_{n+1} - x_0)^{n-1} & \cdots & 1 & -f_{n+1}(x_{n+1} - x_0)^{n-1} & \cdots & -f_{n+1} \\ \vdots & \cdots & \vdots & \vdots & \cdots & \vdots \\ (x_{2n} - x_0)^{n-1} & \cdots & 1 & -f_{2n}(x_{2n} - x_0)^{n-1} & \cdots & -f_{2n} \end{pmatrix} \begin{pmatrix} a_1 \\ a_2 \\ \vdots \\ a_n \\ b_1 \\ \vdots \\ b_n \end{pmatrix} = \begin{pmatrix} \frac{f_1 - f_0}{x_1 - x_0} \\ \frac{f_2 - f_0}{x_2 - x_0} \\ \vdots \\ \frac{f_n - f_0}{x_n - x_0} \\ \frac{f_{n+1} - f_0}{x_{n+1} - x_0} \\ \vdots \\ \frac{f_{2n} - f_0}{x_{2n} - x_0} \end{pmatrix}$$

4. Solve this equation for the coefficients, a_i and b_i.

EXAMPLE 4.3

Example of a Padé
Approximation to
$\ln(1 + x)$

To approximate $\ln(1 + x)$ by a four-parameter Padé approximation over the range $0 \le x \le 2$, we begin with the equation

$$\ln(1 + x) \approx \frac{a_1 x^2 + a_2 x}{b_1 x^2 + b_2 x + 1}$$

evaluate $\ln(1 + x)$ for four values of x in the range $0 \to 2$, and construct the matrix

$$\begin{pmatrix} \frac{1}{2} & 1 & -\frac{1}{2}\ln(\frac{3}{2}) & -\ln(\frac{3}{2}) \\ 1 & 1 & -\ln(2) & -\ln(2) \\ \frac{3}{2} & 1 & -\frac{3}{2}\ln(\frac{5}{2}) & -\ln(\frac{5}{2}) \\ 2 & 1 & -1\ln(3) & -\ln(3) \end{pmatrix} \begin{pmatrix} a_1 \\ a_2 \\ b_1 \\ b_2 \end{pmatrix} = \begin{pmatrix} 2\ln(\frac{3}{2}) \\ \ln(2) \\ \frac{2}{3}\ln(\frac{5}{2}) \\ \frac{1}{2}\ln(3) \end{pmatrix}$$

which, when solved for a_i and b_i, yields

$$\begin{pmatrix} a_1 \\ a_2 \end{pmatrix} = \begin{pmatrix} 0.30303 \\ 0.99852 \end{pmatrix} \qquad \begin{pmatrix} b_1 \\ b_2 \end{pmatrix} = \begin{pmatrix} 0.08281 \\ 0.79492 \end{pmatrix}$$

A comparison of the accuracy of this four-parameter approximation to $\ln(1 + x)$ is given in Table 4.4.

x	$\ln(1+x)$ Approximate	$\ln(1+x)$ Exact	Difference
-0.80	$-1.45033\ldots$	$-1.60943\ldots$	$0.159\ldots$
-0.40	$-0.50472\ldots$	$-0.51082\ldots$	$0.00610\ldots$
0.00	0	0	0
0.40	$0.33645\ldots$	$0.33647\ldots$	$0.000020\ldots$
0.80	$0.58780\ldots$	$0.58778\ldots$	$0.000010\ldots$
1.20	$0.78845\ldots$	$0.78845\ldots$	$0.0000062\ldots$
1.60	$0.95552\ldots$	$0.95551\ldots$	$0.0000042\ldots$
2.00	$1.09861\ldots$	$1.09861\ldots$	0

Table 4.4   *Padé four-parameter approximation to* $\ln(1 + x)$

Clearly, the four-parameter approximation is reasonably accurate over the range $0 \le x \le 2$. As a curiosity, the denominator polynomial, $b_1 x^2 + b_2 x + 1$, has a root, and therefore, the approximation has a singularity, at $x = -1.387\ldots$, a reflection of the actual singularity in $\ln(1 + x)$. For a better approximation, more parameters are used, and you will find that the singularity will move closer to the actual singularity in $\ln(1 + x)$ at $x = -1$.

Constructing a MATLAB program to duplicate our procedures is surprisingly easy, as is illustrated here. (See also the demonstration code entitled PadeDemo on the programs disk.)

```
function = [a,b] = Padecoef(x0,x1,n)
   dx = (x1-x0)/(2*n);
   x = [x0+dx:dx:x1]';    X = x-x0;
```

```
f0 = log(1+x0);
f = log(1+x);    df = (f-f0)./X;
A = zeros(2*n,2*n);
for i = 1:n-1
        A(:,i)   = X.^(n-i);
        A(:,n+i) = -f.*(X.^(n-i));
end
A(:,n) = ones(2*n,1);
A(:,2*n) = -f;
B = inv(A);
a = B(1:n,:)*df;    b = B((n+1):2*n,:)*df;
if abs(det(A)) < 1.e-8
        disp('WARNING: Matrix is near singular')
        disp('determinant is)
        disp(det(A))
end
```

You should be aware that incorporating any extra information available concerning the function being approximated can further improve the accuracy of the Padé formula. For example, $\cos x$ is an *even* function of x; thus, the two polynomials should contain only even powers of x.

●

M-File Functions

Most of the toolbox functions in MATLAB that perform some action on an arbitrary function, such as integration, root finding, etc., expect a function that will accept a vector, $\vec{x}$, as input, and will return a vector $\vec{y}$ as output. That is, $\vec{y} = f(\vec{x})$. Thus, you should attempt to code your own M-file functions with this in mind. For example, the algorithm to sum a series can easily be generalized to simultaneously evaluate the series for a set of input values, x_i, by merely replacing the multiplication operations by element-by-element multiplications.

```
N = 10;
x = [1.0 1.1 1.2 1.3 1.4]';
term = x;
sum = zeros(size(x));
for k = 1:N
    sum = sum + term;
    term = x.*term/(k+1);
end
```

Also, many of the M-file functions you create will expect a *column* vector as input, causing the function to fail if you accidentally enter a *row* vector. This is most easily corrected by adding a statement of the form x = x(:) within the function. This statement will cause the contents of x, be it a matrix, column vector, or row vector, to be rewritten as a single column vector.

Problems

REINFORCEMENT EXERCISES

P4.1 Taylor Series.   For each of the following functions:

  (1) Determine the first few nonzero terms of the Taylor series expansion about the given point, $(x_0, f(x_0))$.
  (2) Determine the general form of the nth term.
  (3) Determine the radius of convergence of the series.
  (4) Evaluate the *maximum* value of the absolute magnitude of the remainder term, $R_4(x)$, over the interval $x_0 \longleftrightarrow x$.
  (5) Evaluate the function at the given point, x_1, first using the Taylor series including terms through $(x - x_0)^3$, and then using the exact expression for $f(x)$. Compare the difference with $R_4(x)$.
  a. $32x^5 - 80x^4 + 80x^3 - 40x^2 + 10x - 1$ $(x_0 = \frac{1}{2}, \qquad x_1 = 0)$
  b. e^{2x} $(x_0 = 1, \qquad x_1 = \frac{1}{2})$
  c. $\cos(1 + 2x)$ $(x_0 = 0, \qquad x_1 = -\frac{1}{2})$
  d. $\ln(1 + x)$ $(x_0 = 1, \qquad x_1 = 1.1)$
  e. $1/\sin(x)$ $(x_0 = \pi/4, \qquad x_1 = \pi/6)$

P4.2 Products of Series.   The MATLAB function fliplr(x) will resequence a *row* vector x in reverse order. That is, fliplr([1 2 3 4]) returns [4 3 2 1]. Thus, if c contains the coefficients in an expansion of, say e^x through fifth order; i.e., c = [1 1/2 1/6 1/24 1/125], the coefficients of the *polynomial* representing this truncated series are then a = fliplr(c). Use the MATLAB function conv for multiplying two polynomials to determine the coefficients of terms through x^5 in the Taylor series for $\sin x$ and $\cos x$. Compare with the series expansion for $\frac{1}{2} \sin 2x$.

P4.3 Recursive Functions.   In some computing languages it is not permitted for a function to reference itself. There is no such restriction in MATLAB. For example, write an M-file function to evaluate $\sin x$ and use the function to compute $\sin(\sin x)$. Compare the number of flops used by your function and compare with that used by the built-in MATLAB function sin.

P4.4 Experiments with Aitken Acceleration.   Compute and store the first 11 partial sums of the series expansion for e^x with $x = 5.1$. Use the code for the function AitknSum on page 137 to accelerate the convergence of the series, and compare the result with the exact value of $e^{5.1}$.

P4.5 Aitken Acceleration.   The Aitken acceleration method assumes that the remainder term in a series will have a dependence on n, approximately given as $c\alpha^n$, where c and α are constants. Doesn't it seem reasonable to assume that a more realistic n

dependence for the remainder term is perhaps $c\alpha^n/n$ or $c\alpha^n/n!$? Show that, in the first instance, Equation 4.4 is replaced by

$$(s - s_n)(s - s_{n-2}) = \frac{n^2}{n^2 - 1}(s - s_{n-1})^2$$

and in the second, it is replaced by

$$(s - s_n)(s - s_{n-2}) = \frac{n}{n + 1}(s - s_{n-1})^2$$

and that, in the limit $n \to \infty$, both equations are the same as Equation 4.4, suggesting that the simpler Aitken assumption is just as good.

P4.6 **Padé Approximation.** Obtain a four-parameter Padé approximation to $x^{1/3}$ over the range of x values, $1 \le x \le 2$, and compare the computed results with the Taylor series results in Table 4.3, as well as with the exact results.

P4.7 **Using the Function `feval`.** Adapt the MATLAB code on page 144 for an M-file function `[a,b] = PADE(`*name*`,x1,x2)` that will compute the two sets of coefficients of a Padé approximation to an arbitrary, four-parameter function *name* over the interval $x_1 \le x \le x_2$. Use this function to obtain Padé approximations to $\sin x$ over the interval $\pi/4 \le x \le \pi/2$.

EXPLORATION PROBLEMS

P4.8 **Great Circle Distances on the Earth.** A point on the earth's surface is specified in terms of latitude and longitude by angles (in degrees) and by specifying a particular hemisphere, e.g., $35°20'$N latitude would be in the Northern Hemisphere, about $35°$ above the equator, and $5°$W longitude would be $5°$ west of Greenwich, England. If we orient a coordinate system with the z axis through the North Pole and the x axis through Greenwich, then ordinary longitude and latitude are related to the common spherical angles, θ [polar angle, measured from the z-axis] and ϕ [azimuthal angle, measured in the x-y plane, from the x axis] as

$$\theta = 90° - \text{latitude} \quad \text{(if North)} \qquad \phi = \text{longitude} \quad \text{(if West)}$$
$$\theta = 90° + \text{latitude} \quad \text{(if South)} \qquad \phi = 360° - \text{longitude} \quad \text{(if East)}$$

Further, the angular separation between any two points on a sphere, Θ_{12}, can be expressed in terms of the coordinates of the two points, (θ_1, ϕ_1), and (θ_2, ϕ_2), as

$$\cos \Theta_{12} = \cos \theta_1 \cos \theta_2 + \sin \theta_1 \sin \theta_2 (\cos \phi_1 \cos \phi_2 + \sin \phi_1 \sin \phi_2)$$

and Θ_{12} is then obtained by taking the inverse cosine of this expression. Finally, the surface distance between the two points is given by $d = R\Theta_{12}$, where R is the radius of the earth.

    a. To designate the hemisphere, use a sign convention, such as (+) for North latitude or West longitude and (−) for South latitude or East longitude, and write an M-file function that assigns values to row vectors containing the locations of the following cities:

City	Latitude	Longitude
Chicago	$41°49'$N	$87°37'$W
Los Angeles	$35°12'$N	$118°02'$W
Montreal	$45°30'$N	$73°35'$W
London	$51°30'$N	$0°07'$W
Rio de Janeiro	$22°50'$S	$43°20'$W
Melbourne	$35°52'$S	$145°08'$W
Vladivostok	$43°06'$N	$131°47'$W
Johannesburg	$26°08'$S	$27°54'$W

b. Write an M-file function that:
  (i) Converts the latitude/longitude coordinates of each city to ordinary spherical angles, (θ, ϕ), *in radian measure.*
  (ii) Computes a square 8×8 table listing the distance in miles between each of the cities. (Use $R = 3958.89$ miles as the radius of the earth.) [Do not use for loops. Instead store the angles as column vectors and compute all the elements of $\cos\Theta_{12}$ as a matrix with a single line of MATLAB code.]

Summary

SERIES SUMMATION IN MATLAB

Structure of a Series Summation

The summation of an infinite series is accomplished by structuring the following steps in MATLAB:

Initialization	Start the variable that is to contain the value of the sum with an initial value of zero, and initialize the starting value of an individual term.	`sum = 0;`
		`term = 1;`
		`tol = 1e-6`

Loop	Set up a loop to repeatedly sum the first n terms of the sum.	`for i = 1:n`
Summation	The actual summation is accomplished by replacing sum by sum + term	`sum = sum+term;`
Redefine term	Compute the value of the next term in the series using the general form of the absolute value of ratio of terms, $r(n)$.	`term=r*term;`

Convergence Test	If the absolute magnitude of the terms approaches very small values, they will not appreciably alter the sum, and the loop is terminated.	`if abs(term)<tol;`
		`    break;`
		`end`

End   If the loop runs to completion without an   `end`
             indication of convergence, an insufficient   `if i==n`
             number of terms were included.   `    error('divergent')`
                                                           `end`

This structure can also be implemented using a `while` loop.

Definition   The Taylor series of a function, $f(x)$, expanded about the point, x_0, is

$$f(x) = f_0 + f_0'(x - x_0) + \frac{1}{2!}f_0''(x - x_0)^2 + \cdots + \frac{1}{n!}f_0^{[n]}(x - x_0)^n + \cdots$$

where all the derivatives are evaluated at $x = x_0$. If $x_0 = 0$, the series is also known as the Maclaurin series.

Remainder Term   The Taylor series can also be written as

$$f(x) = f_0 + f_0'(x - x_0) + \frac{1}{2!}f_0''(x - x_0)^2 + \cdots + \frac{1}{n!}f_0^{[n]}(x - x_0)^n + R_{n+1}(x)$$

where the remainder term, $R_{n+1}(x)$, is an *unknown* function of x of the form

$$R_n(x) = \frac{1}{n!}f^{[n]}(\xi)(x - x_0)^n$$

where $x_0 \le \xi \le x$.

Aitken
Acceleration of
Convergence   The value of an infinite series can be expressed in terms of its partial sums, plus a remainder term, as

$$s = \sum_{k=0}^{\infty} a_k = \sum_{k=0}^{n} a_k + R_{n+1}$$
$$= s_n + R_{n+1}$$

If it is then assumed that the remainder term is asymptotic to the expression

$$R_n \approx c\alpha^n$$

where c and α are constants, then the exact value of the sum can be determined from any three successive partial sums as

$$s = s_n - \frac{(s_n - s_{n-1})^2}{s_n - 2s_{n-1} + s_{n-2}}$$

This equation can be used successively to map a set of partial sums into a new, presumably more accurate, set and, depending on the accuracy of the assumption regarding R_n, will often yield sets of partial sums that are successively more rapidly convergent.

PADÉ APPROXIMATIONS

Definition   A function $f(x)$ is approximated by a ratio of two polynomials. Padé approximations are alternatives to Taylor series and are useful when the accuracy of the Taylor series is affected by the presence of nearby singularities in the function.

MATLAB Functions Used

feval

Used when a function F itself calls a second "dummy" function f. For example, the function F might find the root of an arbitrary function identified as a generic $f(x)$. Then, the name of the actual M-file function, say fname, is passed as a *character string* to the function F, either through its argument list or as a global variable, and the function is evaluated within F by means of feval. The use is feval(name,x1,x2,...,xn), where fname is a variable containing the name of the function as a character string; i.e., enclosed in single quotes, and x1,x2,...,xn are the variables needed in the argument list of function fname.

fliplr

Will "flip" the contents of a matrix or row vector left to right.

5 / *Roots of Functions of a Single Variable**

One of the most common tasks in science and engineering is finding roots of equations; that is, given a function $f(x)$, finding values of x such that $f(x) = 0$. This type of problem also includes determining the points of intersection of two curves. If the curves are represented by functions $f(x)$ and $g(x)$, the intersection points correspond to the roots of the function $F(x) \equiv f(x) - g(x)$.

Root-solving techniques are important for a number of reasons. They are useful, easy to understand, and usually easy to implement—with a minimum of instruction you are able to solve genuine problems in engineering. Root-solving techniques demonstrate, once again, that including a little extra information on the problem at hand usually yields significant rewards in increased accuracy and decreased computing time. In addition, these procedures provide an excellent illustration of many of the MATLAB techniques presented to this point. We now have an opportunity to introduce one of MATLAB's most important and useful features: the automatic plotting of functions. (Two-dimensional plotting procedures are described in Appendix Section A.iv.)

Very accurate and efficient MATLAB functions exist in the toolbox for finding roots of functions, but it is *essential* that you have firm understanding of how roots of functions are obtained by the computer before using the toolbox routines as *blackbox* functions. The techniques of finding the roots of a function are very dependent on the features of the particular function, much more so than techniques for solving matrix equations. Frequently you will write your own functions to find roots, taking into account special features of the particular function of interest, rather than rely on the MATLAB toolbox routines. (You would rarely do this with matrix equations.)

Essentially all the techniques for finding the roots of a function start with an initial guess of the root or with an interval known to contain a root; then we repeatedly refine the initial guess or shrink the initial interval. If the function is a polynomial, additional tricks can be employed that make use of the special features of polynomials. But first, let us review some of the well-known properties of roots of functions.

---

*This chapter makes use of MATLAB programming techniques described in Appendix Sections A.i through A.iv.

Roots of Polynomials

If $f(x)$ is a polynomial of degree n,

$$f(x) = a_1 x^n + a_2 x^{n-1} + \cdots + a_n x + a_{n+1}$$

then there are precisely n values of x for which $f(x) = 0$. This statement is due to Euler[1] and is well known from introductory algebra. If the roots of the polynomial are labeled as r_i, $i = 1, \ldots, n$, then the polynomial can be factored as

$$f(x) = a_1 x^n + a_2 x^{n-1} + \cdots + a_n x + a_{n+1}$$
$$= a_1 (x - r_1)(x - r_2) \cdots (x - r_n)$$

Note, however, that some of the n roots of the polynomial may have the same value (multiple roots) or they may be complex numbers. Examples are the polynomial $g(x) = x^2 - 2x + 1 = (x - 1)^2$, which has a multiple root (of multiplicity 2) at $x = 1$, whereas $h(x) = x^2 + x + 1$ has roots at $x = -\frac{1}{2}(1 \pm \sqrt{3}\, i)$. If the coefficients of the polynomial are real, then the complex roots of the polynomial always appear in conjugate pairs, $r = a \pm ib$.

Beginning several centuries ago, a popular area of research in mathematics was discovering various properties of the roots of polynomials. The most famous and important of the resulting theorems was Euler's; however, there are numerous lesser-known theorems describing the properties of polynomial roots that are occasionally helpful in finding the roots. One of these is due to Descartes[2] and is stated, without proof, as follows.

DESCARTES' RULE OF SIGNS

If the coefficients of the polynomial are real, and if the number of sign changes of the coefficients, reading left to right, is n, then

Positive real roots	The number of *positive* real roots is either n or n minus an even integer.
Negative real roots	The number of *negative* real roots is determined by rewriting the polynomial with $x \to -x$, counting the sign changes in the new polynomial, and applying the rule for positive roots.

For example, the polynomial

$$f(x) = x^4 - 5x^3 + 5x^2 + 5x - 6 = 0$$

---

[1] Leonhard Euler (1707–1783), one of the most prolific of all mathematicians, made substantial contributions to all areas of mathematics, despite the fact that he was blind for the last 17 years of his life.

[2] René Descartes (1596–1650), a French mathematician and philosopher, largely responsible for developing analytic geometry and famous for the philosophy contained in the statement, "I think, therefore I am," is also credited with many of the early advances in solutions of algebraic equations.

has three sign changes in the coefficients, $(\overset{1}{+1}, \overset{2}{-5}, \overset{3}{+5}, +5, -6)$, and thus will have either three or one real positive roots. The polynomial with x replaced by $-x$ has only one sign change, $(\overset{1}{+1}, +5, +5, -5, -6)$, and so there must be one negative real root. The roots of this polynomial are 1, 2, 3, and -1.

Often, much information is available concerning the properties of the roots of a polynomial, and numerous root-solving techniques that employ these special properties have been devised. These procedures can be counted on to accurately and efficiently compute the roots of a polynomial. However, they are often applicable to only a limited class of polynomials and are usually only used when extreme efficiency is required, say if the roots are to be computed thousands of times while varying the polynomial coefficients.

MATLAB, however, takes a different approach. If the size is not excessive, say less than 20×20, perhaps the thing that MATLAB does best is the evaluation of the eigenvectors and eigenvalues of a matrix. Therefore, the solution for the roots of a polynomial is recast in the form of an eigenvalue problem. Starting with a polynomial written in standard form, the equation for the roots is

$$f(x) = a_1 x^n + a_2 x^{n-1} + \cdots + a_n x + a_{n+1} = 0$$

Dividing through by a_1, this equation can be written as

$$x^n = -\frac{a_2}{a_1} x^{n-1} - \frac{a_3}{a_1} x^{n-2} - \cdots - \frac{a_n}{a_1} x - \frac{a_{n+1}}{a_1} \tag{5.1}$$

Next, we construct an $n \times n$ matrix with the coefficients on the right-hand side of Equation 5.1 as the first row and with 1s on a lower diagonal displaced by one from the main diagonal.

$$\begin{pmatrix} -a_2/a_1 & -a_3/a_1 & -a_4/a_1 & \cdots & -a_n/a_1 & -a_{n+1}/a_1 \\ 1 & 0 & 0 & \cdots & 0 & 0 \\ 0 & 1 & 0 & \cdots & 0 & 0 \\ \vdots & \vdots & \vdots & \ddots & \vdots & \vdots \\ 0 & 0 & 0 & \cdots & 1 & 0 \end{pmatrix}$$

The eigenvalues, λ_i, of this matrix are the *roots* of the polynomial, and the eigenvectors are of the form $[\lambda_i^{n-1}, \lambda_i^{n-2}, \ldots, \lambda_i]$. This can be seen by inspection of the eigenvalue equation

$$\begin{pmatrix} -a_2/a_1 & -a_3/a_1 & -a_4/a_1 & \cdots & -a_n/a_1 & -a_{n+1}/a_1 \\ 1 & 0 & 0 & \cdots & 0 & 0 \\ 0 & 1 & 0 & \cdots & 0 & 0 \\ \vdots & \vdots & \vdots & \ddots & \vdots & \vdots \\ 0 & 0 & 0 & \cdots & 1 & 0 \end{pmatrix} \begin{pmatrix} \lambda_i^{n-1} \\ \lambda_i^{n-2} \\ \lambda_i^{n-3} \\ \vdots \\ \lambda_i \end{pmatrix} = \lambda_i \begin{pmatrix} \lambda_i^{n-1} \\ \lambda_i^{n-2} \\ \lambda_i^{n-3} \\ \vdots \\ \lambda_i \end{pmatrix}$$

The first row of the matrix product duplicates Equation 5.1, and the remaining rows are trivially satisfied. Thus, to find the roots of the polynomial, this matrix is first constructed[3] and then the function `eig` is called to determine the eigenvalues.

The MATLAB toolbox function that executes these tasks is called `roots`. If the coefficients of a polynomial are contained in the vector a, then all of the roots of the polynomial will be computed by

```
r = roots(a)
```

For example, the roots of the polynomial on page 152 can be determined by

```
r = roots([1 -5 5 5 -6])

r =
      3.0000    2.0000    1.0000   -1.0000
```

Searching for Roots

A great many functions cannot be expressed as a polynomial. Functions such as the trigonometric functions $\sin x$ and $\cos x$, or the exponential function e^x and $\ln x$ are defined in terms of an infinite series (i.e., a polynomial of *infinite* degree). Other functions involving fractional powers are also often written in terms of elementary arithmetic operations by means of an infinite series. This implies that the number of real roots of these functions could be infinite.[4] (The function $\sin x = 0$ has an infinite number of solutions.) Thus, unlike polynomials, there are very few guides for starting a search for the roots of an arbitrary function. Some functions have many real, positive roots; others have no roots whatsoever. There is no substitute for first learning as much as you can about your particular function; there is no better way to do this than to plot the function over some range.

USING `plot` AND `ginput` TO SEARCH FOR A ROOT

Once you have plotted a function over the range where you expect to find a root, you can tell from the graph if there is indeed a root within this range and you can read from the graph the root's approximate value. To facilitate this step, you may wish to add a horizontal line on the graph corresponding to $y = 0$. This can be done by adding to the argument list of `plot` as follows:

---

[3]This matrix is called the *companion matrix* to the polynomial. There is a toolbox function named `compan` that will compute this matrix. The use is `compan(a)`, where a is the vector containing the coefficients of the polynomial.

[4]Of course, if the function contains only fractional powers of x and no transcendental functions like $\sin x$ or e^x, then the equation $f(x) = 0$ can repeatedly be written as $x^{a/b} = \ldots$ and then raised to the power b until there are no fractional powers left, leaving a polynomial with a well-defined number of roots.

```
plot(x,zeros(1,length(x)), ... )   % if x is a vector
or
plot(x,zeros(1,max(size(x)))), ...) % if x is a matrix
```

The function could then be replotted over a more limited range surrounding the root to obtain greater accuracy.

The MATLAB zoom *Command* An alternative is to use the MATLAB function zoom to more accurately determine the point on a plot where the function crosses the axis. The procedure is to first plot the graph, then enter the MATLAB statement

```
zoom on
```

Next, use the mouse to locate the cursor near the root and then click the left mouse button. Each time the left mouse button is clicked, the axes will be changed by a factor of 2, in effect "zooming in" on the point of interest. Clicking the right mouse button (shift-click on a Macintosh) will cause the plot to zoom out by a factor of 2. The command zoom off will turn off the zoom feature.

The MATLAB Function ginput Once the location of the root has been located to sufficient accuracy, the numerical value of the coordinates may then be read off the plot by use of the function ginput. The statement is

```
[xr,yr] = ginput(n)
```

where n is the number of points you wish to read from the graph. The coordinates of these points will be stored in the $n \times 1$ column vectors xr and yr. [*Note:* Use variable names different from the names (x,y) containing the graph points to avoid overwriting them.]

When the statement is entered, a pair of cross hairs will appear on the graph. (If running without windows, recall the plot with shg.) The cross hairs can be moved either with a mouse or (on most terminals) with the arrow keys. The coordinates of the cross hairs are then read by pressing a mouse button or by hitting any keyboard key. The reading of points continues until n points have been read. Of course, the accuracy of the values read will depend on the screen resolution and the steadiness of your hand.

STEPPING SEARCH FOR ROOTS

Frequently the search for a root of a function is only an incidental part of a larger problem, and it may be inconvenient to interrupt the calculation to plot the function to estimate the position of the root. In these situations a less reliable, but often satisfactory, alternative is to have the computer step through a range of x values and watch for a change of sign of the function indicating the position of a root. For example, if a root is thought to be in the range $x_0 < x < x_1$, then, assuming that the function $f(x)$ has been coded as an M-file function and the variables x_0 and x_1 have assigned values, the following code will accomplish this task:

```
f0 = f(x0);
if f(x1)*f0 > 0
    error('no root in interval')
end
x = x0+ dx = (x1-x0/50);
while x<x1
    if f(x)*f0 < 0
        xleft = x-dx;
        xrght = x;
        disp('interval containing a root is')
        disp([xleft xrght])
        break
    end
    x = x + dx
end
```

For example, the range of the ballistic trajectory of a projectile fired with a muzzle velocity of v_0 at an angle of θ with respect to the horizontal, ignoring air resistance, is

$$R(\theta) = \frac{2v_0^2 \sin\theta \cos\theta}{g}$$

where $g = 9.8$ m/s^2 is the acceleration of gravity. (The graph of such a trajectory is examined in Appendix Section A.iv.)

If we then wish to know the angle at which to fire the bullet in order to hit a target at a distance of 2200 m, we must solve the equation $R(\theta) - 2200 = 0$ for θ. Replacing $x_0, x_1, x,$ and $f(x)$ by $\theta_0 = 35°$, $\theta_1 = 55°$, θ, and $R(\theta) - 2200$ and adding statements defining v_0 and g yields the result:

```
interval containing a root is
36.6000        36.8000
```

The exact answer is

$$\theta = \frac{1}{2}\sin^{-1}\left(\frac{2200g}{v_0^2}\right) = 36.6899\ldots°$$

Refining an Interval by Interval-Halving

Once you have obtained a good idea of the vicinity of a root, either by plotting the function or by stepping, this interval can be systematically reduced in the following way:

1. Let x_0 and x_1 be the left and right ends of the interval known to contain a root; i.e., the function $f(x)$ changes sign between x_0 and x_1. Also, let $f_0 = f(x_0)$ and $f_1 = f(x_1)$.
2. Let x_m be the midpoint of the interval, $x_m = \frac{1}{2}(x_0 + x_1)$, and evaluate $f_m = f(x_m)$.
3. Determine in which half of the interval the function changes sign. That is, if $f_0 f_m < 0$, the root is on the left; if $f_1 f_m < 0$, it is on the right.
4. If the root is on the left, make the replacement

$$Left\ crossing \quad x_m \to x_1$$
$$f_m \to f_1$$

and if the root is on the right

$$Right\ crossing \quad x_m \to x_0$$
$$f_m \to f_0$$

5. While the interval $|(x_1 - x_0)| > \epsilon$, where ϵ is chosen to be a small positive convergence tolerance, return to Step 2.

This method is known as the *bisection method* for finding the roots of a function. It is popular because it is *guaranteed* to find a root of a function to any desired accuracy, provided the starting interval does indeed contain a root.

In addition, the method is suited to a variety of improvements. For example, if a root of the function $f(x) = 2xe^{-x^2} - \cos x$ is known to be in the interval $0 < x < 1$, we would begin by evaluating $f_0 = f(0) = -1$ and $f_1 = f(1) = 0.1955\dots$. However, because the value of $|f_1| \ll |f_0|$ we would expect the root to be closer to 1 than to 0. That is, using information about the *size* of the function as well as its sign, we should be able to improve the convergence rate of the algorithm. One approach is then to interpolate the function over this interval; i.e., connect the points (x_0, f_0) and (x_1, f_1) by a straight line and determine where the line intercepts the x axis. (See Figure 5.1.) This particular point is a better estimate of the root than is the midpoint of the interval. Thus, the improved procedure replaces the statement defining the midpoint, $x_m = \frac{1}{2}(x_0 + x_1)$, by the equation for the interpolated axis crossing:

$$x_m = x_0 - (x_1 - x_0)\frac{f_0}{f_1 - f_0}$$

in Step 2 of the bisection method. The remaining steps are unchanged.

You should try this method by hand calculation for a few steps to ensure that you have a grasp of the ideas (see Problem 5.3). Also, construct the code for a MATLAB function called `bisection`, which will take an arbitrary function, *fname*, and an interval known to contain a root, `x0` and `x1`, and will return a root to a tolerance, $|(x_1 - x_0)| <$ `small`.

THE MATLAB FUNCTION FOR ROOTS OF AN $f(x)$

Not surprisingly, there is a toolbox function that, after adding some further improvements, follows this procedure to obtain the roots of an arbitrary function

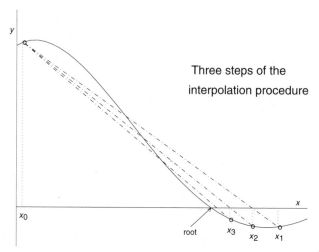

Figure 5.1   *Interpolation method for finding roots.*

of a single variable. The function is called `fzero`, and its use is

`r = fzero('`*fname*`',x0,small)`

That is, the root of the function `fname` is computed to a tolerance `small` starting from an *initial guess* for the root, `x0`. `fzero` starts at `x0` and searches in both directions until it finds a change in sign of the function, at which point the bisection algorithm is initiated. The improvements in the algorithm are that the type of interpolation employed can be either linear, like the preceding method, or quadratic, depending on the properties of the function near `x0`.

Users of bisection-type root-solving algorithms should be reminded that a function may change sign more than one time in a given interval, and if this is the case, the root returned may not be the root of interest. Furthermore, a function can have a root and never change sign; for example, $f(x) = (x-1)^2$. Neither `fzero` nor any bisection-type algorithm can find the roots of a function with multiple roots of even multiplicity.

Newton's Method for Finding Roots

The bisection method is based on first finding an interval known to contain a root and then systematically reducing the size of that interval. *Newton's method*, frequently referred to as the *Newton-Raphson method*, takes a bolder approach of starting instead with only a *guess* for the root, say x_0, rather than an interval. Then, using the value of the function at x_0 and the value of the slope of the tangent line to the function at x_0, the function is extrapolated to estimate the root. The process is then repeated using the new estimate for the root until the method either converges or fails.

Newton's method can be derived by starting with the Taylor series expansion for the function $f(x)$ about the point x_0:

$$f(x) = f(x_0) + \left.\frac{df}{dx}\right|_{x_0} (x - x_0) + \frac{1}{2!}\left.\frac{d^2 f}{dx^2}\right|_{x_0} (x - x_0)^2 + \cdots$$

We wish to find a value of x such that $f(x) = 0$. Assuming that x_0 is our guess for the root and that it is close to the actual root; i.e., $|(x - x_0)|$ is small, we need only retain a few terms in the series expansion. Indeed, if we terminate the series after the linear term, we have the following equation for the root x of $f(x)$:

$$f(x) = 0 \approx f(x_0) + \left.\frac{df}{dx}\right|_{x_0} (x - x_0)$$

Of course, if x_0 is not close to x, this equation may not even be approximately true. This equation can be rearranged to give an explicit expression for the root x:

$$x \approx x_0 - \frac{f(x_0)}{f'(x_0)} \tag{5.2}$$

where we have used the notation $f'(x_0)$ for $\left.\frac{df}{dx}\right|_{x_0}$. If our assumption of $|(x - x_0)|$ being small is valid, this equation should be a good estimate of the actual root of the function. But if $|(x - x_0)|$ is *not* small, what have we done? The answer is illustrated in Figure 5.2.

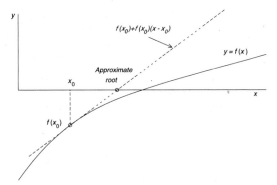

Figure 5.2 *Replacing a function by two terms of a Taylor series.*

Replacing the function $f(x)$ by the first two terms of the Taylor series is seen to be equivalent to approximating the function by a straight line through the point (x_0, f_0), which has the same slope as the tangent to the curve at that point. Setting this approximation to $f(x)$ equal to zero, we find the point where the line intersects the axis. As you can see in Figure 5.2, this procedure will not, in general, give the actual root of $f(x)$; however, the value generated by Equation 5.2 will, in practice, be *closer* to the actual root compared to the starting guess, x_0. Newton's method consists of repeating this process until the difference between successive values of the estimated root is less than some small tolerance.

Keep in mind, Newton's method attempts to find a root of a function by repeatedly approximating the function by straight lines. In each iteration, because we are monitoring not only the value of the function but also its slope, a substantial amount of information about the behavior of the function is used to construct the next step. In this sense, the information content of Newton's method is quite high, and we should expect that it will converge more rapidly to a root compared to the bisection-type algorithms. Indeed, it can be shown[5] that Newton's method is a *second-order* algorithm, meaning that if the error in the nth step is ϵ, the error in the next step will be proportional to ϵ^2. Bisection-type algorithms are *first-order* algorithms.

EXAMPLE 5.1

Newton's Method for Square Roots

The rate of convergence of Newton's method applied to most equations is quite impressive. For example, the square root of a number can be obtained in the following manner. Given a number N, we wish to determine a value of x such that $x = N^{1/2}$, or to phrase the problem in a different way, we wish to find a root of the equation, $f(x) = x^2 - N$. Applying Newton's method to this function, starting with an initial guess x_0 for the square root and using $f'(x) = 2x$ for the derivative, we obtain the following for the improved estimate of the square root:

$$x_1 = x_0 - \frac{f(x_0)}{f'(x_0)}$$

$$= x_0 - \frac{x_0^2 - N}{2x_0} \tag{5.3}$$

$$= \frac{1}{2}\left(x_0 + \frac{N}{x_0}\right)$$

To illustrate, we take $N = 111$ with $x_0 = 20$, an obviously poor first guess. Equation 5.3 then generates the following values for $\sqrt{111}$.

Step	x	x^2
0	20.0	400.0
1	12.8...	163.2...
2	10.73...	115.2...
3	10.537...	111.04...
4	10.53565...	111.0000032...
5	10.53565375...	111.0000000001...

Five iterations, beginning with a poor guess, have resulted in an answer that is correct to nine significant figures.

---

[5]Borse, *FORTRAN-77 and Numerical Methods*, 2nd ed., PWS Publishers, Boston, MA, 1991, p. 400.

If you have never done so, you should try this procedure to find the roots of several simple functions, using only a pocket calculator. (See Problems 5.6 and 5.7.)   ●

POTENTIAL PROBLEMS USING NEWTON'S METHOD

Newton's method is the most popular root-solving technique, and it can usually be relied upon to find a root quickly and accurately. However, in the following situations, the method will fail:

1. A poor initial guess may cause the method to fail. For example, if the initial guess x_0 is such that $f'(x_0)$ is small; i.e., the slope is nearly horizontal, then x_0 may be near a local maximum or minimum. A straight line approximation to the function will likely send the calculation out of the region of interest.
2. If the function has a vertical slope $(f' = \infty)$ at the root, successive improvements $\Delta x = -f(x_0)/f'(x_0)$ will be so small that Newton's method will never converge to the root.
3. As with bisection-type algorithms, Newton's method may have difficulty with multiple roots. For example, the function $f(x) = (x - r)^2$ has a multiple root at $x = r$, and both $f(r)$ and $f'(r)$ are identically zero at the root. However, as x approaches r, $f(x)$ approaches zero more quickly than does $f'(x)$, and Newton's method will still converge to the root, although the convergence rate is quite a bit slower than for a non-multiple root. If you know the root is multiple and have an idea of the multiplicity, Newton's method amended to

$$x_1 = x_0 - m\frac{f(x_0)}{f'(x)}$$

where m is the multiplicity of the root, will be found to converge at a rate characteristic of a second-order algorithm.

MATLAB CODE FOR NEWTON'S METHOD

Before constructing the MATLAB code to implement Newton's method for finding a root of an arbitrary function, you must have already created an M-file function for the function $f(x)$. Remember, to make the code as flexible as possible, the function $f(x)$ should be written so as to expect a *single* variable name in its argument list. If other parameters are required to evaluate $f(x)$, they should be given unique names and declared `global`.

In addition, Newton's method requires that the derivative of $f(x)$ be *separately* coded and stored in a second M-file, again written with a single variable name in the argument list. Once these tasks are accomplished, the MATLAB function shown in Figure 5.3 (p. 162) will use these two functions, starting with an initial guess x_0, and return a root of $f(x)$ to within a tolerance $|x - r| <$ `small`, where r is the actual value of the root. A demonstration of this

```
function r = Newton(x,fnc,dfdx,small)
%
% The starting guess for the root is x, fnc is a string
% that contains the name of an existing M-file function
% of a single variable, dfdx a string containing the
% name of a function of a single variable corresponding
% to the derivative of fnc, and small is the convergence
% tolerance.  Example fnc='sin', dfdx='cos', small=1.e-4
%-------------------------------------------------------
 r = x;              % Protect input value for x
 dx= 1;
 i = 0;             % A counter for iterations
 while abs(dx) > small
    i = i + 1;
    if i > 20       %excessive iterations
       error('excessive iterations, has not converged'}
    end
    df = feval(dfdx,r);
    if abs(df) < small
       error('small derivative, method diverging')
    end
    dx = -feval(fnc,r)/df;
    r = r + dx;
 end
```

Figure 5.3 *MATLAB code for Newton's method.*

code is contained in the script file named NewtDemo.m included on the applications disk.

Notice that in this program the value of the initial guess, x, that is passed to the function Newton is immediately assigned to a different variable upon entering the function. The reason for this replacement is that x will likely be altered within the function, and that altered value will be returned when the calculation is completed. If the value of x were needed later on, neglecting to protect the input variable in this way would cause this information to be lost. In all cases, functions should be written so as to leave all *input* quantities unchanged upon their completion.

In summary, Newton's method is characterized by a rapid convergence rate, an ability to handle multiple roots, and an easy-to-implement algorithm. These features are paid for by the method's potential for failure in certain circumstances and by the inconvenience of having to code a separate function for the derivative. The latter limitation is addressed in a final procedure called the secant method.

The Secant Method

The principal disadvantage in using Newton's method is that you must supply the code for *two* functions, $f(x)$, and $f'(x)$. Frequently, the task of finding the root of a function will be only an incidental part of a larger problem, and it may be inconvenient or distracting to take the time to code a separate function. In these cases you may wish to use some root-solving M-file function in your personal library that only requires the function $f(x)$ in its argument list. In most situations, the toolbox function `fzero` will suffice. If the function is very complicated or if roots must be computed for a large variety of parameter values, the faster convergence rate of Newton's method may be important. A variation on Newton's method is to use the basic equation defining the improvements to the guess in each step, $\Delta x = -f(x)/f'(x)$, but to let the computer attempt to compute the derivative numerically. Such a method is called the *secant method*.

We start with the basic ideas of Newton's method:

$$\Delta x_0 = -\frac{f(x_0)}{f'(x_0)}$$

$$x_1 = x_0 + \Delta x_0$$

and use an approximate expression for the derivative

$$f'(x) \approx \frac{f(x + \Delta x) - f(x)}{\Delta x}$$

where Δx is a known small quantity, say Δx_0. Thus, we begin the process with *two* input quantities: an initial guess, x_0, and a guess for the 0th-order correction, Δx_0. The starting values then correspond to knowing two points on the curve of f versus x, (x_0, f_0) and (x_1, f_1), where $x_1 = x_0 + \Delta x$ and $f_1 = f(x_1)$. We then use the preceding equations to compute the first correction:

$$\Delta x = -\frac{f(x_0)}{\left(\frac{f(x_0 + \Delta x_0) - f(x_0)}{\Delta x_0}\right)}$$

$$= -\Delta x_0 \frac{f_0}{f_1 - f_0}$$

$$= \Delta x_0 - \Delta x_0 \left(1 + \frac{f_0}{f_1 - f_0}\right)$$

$$= \Delta x_0 - \left(\Delta x_0 \frac{f_1}{f_1 - f_0}\right)$$

$$\equiv \Delta x_0 + \Delta x_1$$

where

$$\Delta x_1 = -\Delta x_0 \frac{f_1}{f_1 - f_0} \tag{5.4}$$

is the *improvement* in the estimate of the interval.

The algorithm is thus as follows:

1. Start with a guess for the root x_0 and a guess for an interval size Δx_0 so that $x_1 = x_0 + \Delta x_0$.
2. Compute the improvement to the interval, $\Delta x_1 = -\frac{f_1}{f_1 - f_0} \Delta x_0$.

3. Replace the pair of values $(x_0, \Delta x_0)$ by the pair $(x_1, \Delta x_1)$. The next pair of points is then $(x_1, x_2 = x_1 + \Delta x_1)$.

Whereas Newton's method is equivalent to repeatedly replacing the function by straight lines that are tangent to the function, the secant method can be shown (see Problem 5.9) to be equivalent to repeatedly replacing the function by straight lines drawn through the points (x_0, f_0) and (x_1, f_1)—that is, *secant* lines. The MATLAB code for an M-file function that implements the secant method is shown in Figure 5.4. A demonstration of this code is contained in the script file named ScntDemo.m included on the applications disk.

```
function r = Secant(x,dx,fnc,small)
%
% The initial guess is x, dx is a guess for the size of
% an interval containing the root, and fnc is a string
% containing the name of a function of a single variable.
% The convergence tolerance is small.  For example,
% x = 1.2, dx = 0.1, fnc='cos', small=1.e-5
%------------------------------------------------------------
 r = x;  dx0 = dx;  dx1 = dx0;  f0 = feval(fnc,r);
 i = 0;
 while abs(dx1) > small
     r = r + dx0;
     f1 = feval(fnc,r);
    dx1 = -dx0*f1/(f1-f0);
    if abs(dx1) > 1.e4; error('method diverging');end
    dx0 = dx1;    f0 = f1;
    i = i + 1;
    if i > 20;error('20 iterations is too many');end
 end
```

Figure 5.4   *MATLAB code for the secant method.*

Problems

REINFORCEMENT EXERCISES

P5.1 Properties of Roots of Polynomials.   In this problem you are to estimate the roots of a polynomial without using a calculator or a computer. You may use Descartes' rule, the values of the polynomial at $x = 0$ and ± 1, and "long division" of polynomials.
  a. Determine all four roots of $f(x) = x^4 + 6x^3 + 3x^2 - 10x$.
  b. Show that the function $g(x) = 8x^3 + 12x^2 + 14x + 9$ has only one real root, and that it is in the interval $-1 \le x \le 0$.
  c. Use the value of the derivative of the function, $h(x) = x^4 - 2x^3 + 2x^2 - 2x + 1$, at one of the real roots to determine the three remaining roots.

d. Show that the function

$$g(x) = x^4 + 7.7x^3 + 39.1x^2 + 14.4x - 13$$

has only one positive real root and one negative real root, and that the positive real root is less than 1. Also show that the real part of the complex roots is greater than 3.3.

P5.2 The Roots of Polynomials Using `roots`.
   a. Use the toolbox function `roots` to obtain the roots of the polynomials in Problem 5.1.
   b. Use the toolbox functions `compan` and `eig` to compute and diagonalize the companion matrix for each of the polynomials in Problem 5.1. The results should be the same as in part a.
   c. Do you think that the function `roots` will have any difficulty in finding multiple roots? Explain. The polynomial $f(x) = 25x^4 - 135x^3 + 256x^2 - 256x + 231 - 121$ has multiple roots. Use `roots` to find them.
   d. Use the toolbox function `conv` to multiply any two of the preceding polynomials together, then find the roots of the product and verify that the roots are simply the roots of both polynomials. Finally, use the toolbox function `deconv` to, in succession, divide the product by $(x - r_i)$, where r_i is one of the roots, to verify that the remainder is, in each case, zero.

P5.3 Roots by Interval Halving. Carry out six steps of the interval-halving process on the function $f(x) = x^3 - 2x - 5$, starting with the endpoints $x_0 = 1$ and $x_1 = 3$.

P5.4 Roots Found Using `fzero`. Find the roots of the following functions using the toolbox function `fzero`. You may have to first plot the functions to get an estimate of the initial guess.
   a. $xe^{-x^2} - \cos x$
   b. $\ln x - x/5$
   c. $(x/\sin x)^2 - \ln x/x^2 - 1.415$

P5.5 Multiple Roots. Show that the function `fzero` is not able to find multiple roots by attempting to find the root of $f(x) = (x - 3)^4(x^2 + 2x + 4)$. Next, use Newton's method with the multiplicity factor first equal to 2, and then 4, and compute a table of successive estimations of the root by this method.

P5.6 Newton's Method for Roots. Using Newton's method, show that if x_0 is a guess for the value of the nth root of a number N, then an improved guess is

$$x_1 = \frac{1}{n}\left[(n-1)x_0 + \frac{N}{x_0^{n-1}}\right]$$

[*Hint:* Show that this problem is the same as finding the root of the equation $f(x) = x^n - N$.]

P5.7 Newton's Method for the Inverse. In the early days of computing, calculators were mechanical, not electronic, and the earliest of these could multiply, but not divide. Use Newton's method to devise a scheme that does not employ division to iteratively calculate the inverse of a number, N, beginning with an initial guess for the inverse of x_0. [The condition on the initial guess is that $(2 - Nx_0) > 0$.]

P5.8 Convergence Rate.   Compare the execution times for the function `Secant` of Figure 5.4 with that of the toolbox function `fzero` when finding a root of the function in Problem 5.4a. Use the same convergence criteria for both functions, and start and stop the MATLAB stopwatch with the functions `tic` and `toc`.

P5.9 Graphical Interpretation of the Secant Method.   The function `Secant` of Figure 5.4 is used to find the roots of $f(x) = x^4 - 12x - 34$, starting with an initial guess of $x_0 = 2.0$ and $\Delta x_0 = 0.2$. The results of each step are displayed as:

Step	x	Δx
0	2.000	0.200
1	2.200	1.471
2	3.671	-1.085
3	2.587	0.177
4	2.764	0.136

Use `plot` to graph the function for $0.5 \leq x \leq 1.0$, and then on the hardcopy of the plot use the numbers in the table to graphically demonstrate how the secant method arrives at a root of the function.

P5.10 Complex Roots.   The function $f(x) = 3x^2 + 2x + 2$ has no real roots. If you attempt to find the roots of this function using the function `Secant` of Figure 5.4, the code will fail. However, if you add a small *imaginary* part to the initial guess, say $x_0 = 1 + 0.001i$, the code will quite quickly return the nearest complex root. Verify that this is true and explain why.

Summary

ROOT-SOLVING TECHNIQUES

Interval Halving

The most direct approach when searching for a root of a function of a single variable is to step along the independent variable axis with steps of ever-increasing size until the function is found to change sign. Once a sign change is observed between two points, x_0 and x_1, the function is then *interpolated* between these points to estimate position, x_m, of the axis crossing by the algorithm

$$x_m = x_0 - (x_1 - x_0)\frac{f(x_0)}{f(x_1) - f(x_0)}$$

The two subintervals $[x_0, x_m]$ and $[x_m, x_1]$ are then examined to determine in which segment the sign change occurs, and the step is then repeated for this smaller interval. Once a change in sign of the function has been located, this method will, without fail, progressively narrow in on the value of the root.

Newton's Method Starting with an estimate, x_0, of the root of a function, $f(x)$, an improved estimate is obtained by replacing the function by the first two terms of its Taylor series and extrapolating this straight line approximation to the point at which it crosses the axis. The algorithm is

$$x_1 = x_0 - \frac{f(x_0)}{f'(x_0)}$$

This point is then used as the starting point for the next iteration. Newton's method is more efficient than interval-halving techniques, but because it does not bracket a root, it is susceptible to divergences if started at points where $f'(x_0) \approx 0$.

The Secant Method If approximate difference expressions are substituted for the analytical expression for the derivative, Newton's method is known as the secant method. The starting point is, again, an initial guess for the root, x_0, and for an interval Δx_0 containing the actual root. The algorithm starts with the pair of points $[x_0, x_1 = x_0 + \Delta x_0]$, computes the new interval width Δx_1 by the equation

$$\Delta x_1 = \frac{-f(x_1)}{f(x_1) - f(x_0)} \Delta x_0$$

and replaces the original pair of points by $[x_1, x_2 = x_1 + \Delta x_1]$. This is repeated until $|\Delta x|$ is less than some prescribed convergence tolerance.

MATLAB Functions Used

roots If the $n + 1$ coefficients of an nth-degree polynomial are contained in the vector c, with c_1 being the coefficient of the highest power, then roots(c) will return the n roots of the polynomial. The roots may be real or complex.

compan The function roots finds the roots of a polynomial by finding the eigenvalues of the companion matrix to the polynomial. If the polynomial is of degree n, and has coefficients c_i, $i = 1, \ldots, n + 1$, the companion matrix is an $n \times n$ with terms $-c_{i+1}/c_1$, $i = 1, \ldots, n$ in the first row, 1s along the diagonal below the main diagonal, and 0s elsewhere. The companion matrix is computed by compan(c).

fzero If $f(x)$ is a function of a single variable and is coded in an M-file function with the name fname, and if x_0 is a guess for a root of this function, then fzero will attempt to find the actual root of the function to a tolerance, tol. The implementation is xroot = fzero('fname',x0,tol). If the third item in the argument list, tol, is omitted, a tolerance equal to eps, the machine accuracy, is used. If there is a nonzero fourth item in the list, intermediate steps in the calculation will be printed. The algorithm searches for a change in sign of the function and uses quadratic interpolation near the root.

ginput(n) Function ginput(n) is used to read the coordinates of n points directly from a two-dimensional plot. The cursor is positioned using a mouse and the coordinate values [x,y] are read by clicking the mouse. The process continues until n points are read, or until a carriage return is entered.

plot   The basic MATLAB program for plotting two-dimensional graphs. If x and y are two
                        vectors of the same size, then plot(x,y) will graph the elements of y (vertical) versus x
                        (horizontal). Multiple plots are possible by including more vector pairs in the argument
                        list. Also, the color and type of line used can be varied by including a third item along
                        with each vector pair. See Appendix Section A.iv for a more complete description of
                        plot.

title, xlabel,   Used to append a title and axis labels to a plot.
ylabel

6 / Minimum or Maximum of a Function

PREVIEW   In this chapter we will outline procedures for locating, and then refining, the estimation of a minimum of a function of a single variable. If you want to find a maximum, simply replace $f(x)$ by $-f(x)$ in what follows. This particular task, when generalized to functions of several variables, is the central ingredient in a very important and active area of research called *optimization*, and includes such topics as *variational methods* and *linear programming*, among others. We return to the multidimensional case in Chapter 10.

Mathematically, the minimum of a smooth function of one variable can be defined as follows:

Given a function $f(x)$ over an interval $a \leq x \leq b$, then $f(x)$ will have a *local* minimum at $x = x_0$ if $f'(x_0) = 0$ and $f''(x) > 0$.

Thus, finding the minimum appears to be a duplication of the root-solving problem, except that here we have to find the root of $f'(x)$. For simple functions, particularly functions of one variable, this method may indeed suffice. However, very often functions of a single variable are not easily differentiable. There is the added difficulty of ensuring that $f'(x) = 0$ returns a minimum, not a maximum or an *inflection point* (a point where both $f'(x)$ and $f''(x)$, and perhaps higher derivatives, are simultaneously zero). Resorting to numerical expressions for derivatives is not very profitable either. If the given function $f(x)$ can be evaluated to an accuracy ϵ (for example, $\epsilon = $ eps, the machine accuracy of MATLAB), the numerical derivative can be estimated only to an accuracy of $\sqrt{\epsilon}$, and the second derivative is even less accurate. The most accurate procedures will use the function itself to find the minimum by essentially riding the function *downhill* until a minimum is found.

The function $f(x)$ may have several minima in an interval. The *global* minimum is the overall minimum of the set of all *local* minima plus the values of $f(x)$ at both ends of the interval. There are no systematic procedures for finding global minima. The best we can do is assist you in finding a region containing a local minimum and then provide mechanisms to refine the location of the local minimum. If your function has a great many local minima, especially if they are close together, you are advised to spend the extra time to produce a graph of the function over the region of interest to help determine the global minimum. Needless to say, this can become extraordinarily difficult in multidimensional problems.

General Prescription for Finding a Minimum

The overall procedure for finding a single local minimum is as follows:

1. Follow the function *downhill*, with increasingly larger steps until the function is found to turn upward. There are several ways to accomplish this:

    (a) The stepping procedure can be done in the obvious manner, $b \Rightarrow b + \Delta x$, where Δx is a linearly increasing step size.

    (b) The last three values of the function can be used to fit a parabola and the position of the minimum of the parabola, if it is greater than b, is then used as the next step.

    (c) If the derivative of the function is available, it too may be used to fit an interpolating parabola to estimate the position of the turning point.

    (d) A combination of all these methods, along with higher-order fitting polynomials, can be used.

2. Once a minimum is *bracketed*, there are a number of popular methods for refining the result.

    (a) A method reminiscent of *interval halving*, called a *golden section search*, is used to decide which part of the current interval contains the minimum. Like interval halving, this procedure is *linearly* convergent.

    (b) Once again, over the bracketing interval, the function is fitted with a parabola, and the minimum of the parabola is used to obtain the next interval.

MATLAB contains a toolbox function **fmin** that will find the position of a local minimum of a dummy function. The use is

```
r = fmin(fname,a,b,small)
```

where **a** and **b** are points *known* to bracket a minimum, **fname** is a variable containing the name of a MATLAB M-file function of a single variable, and **small** is the convergence tolerance on the interval.[1] The procedures used in the function are a combination of the golden search method and polynomial interpolation. However, because this topic is so important in modern numerical analysis, we

---

[1]The best tolerance on the interval can be no better than $\sqrt{\epsilon}$, where ϵ is the floating-point accuracy of MATLAB; i.e, **eps**. This can be appreciated by considering the Taylor series expansion of the function near the minimum point x_0,

$$f(x_0 \pm \Delta x) \approx f_0 + \frac{f_0''}{2}\Delta x^2 \Rightarrow f_0 + \epsilon$$

where we have used the fact that $f_0' \approx 0$ near the minimum, and the smallest numerical distinction between $f(x_0 \pm \Delta x)$ and f_0 is given by the machine tolerance, ϵ.

feel that it is worth your while to better understand the methods, rather than rely totally on *black-box* routines. Indeed, you may be more comfortable writing your own codes, which can then frequently be tuned to fit the idiosyncrasies of a particular problem.

Stepping Search for an Interval Containing a Minimum

The function DOWNHILL will search for two points that bracket a minimum of the function $f(x)$. The basic features of the function are:

- The input quantities are two starting values of x, x_0, and x_1, the name of the function, and the maximum limit of the search (max), expressed as a multiple of $|x_1 - x_0|$. The output quantities are the limits of the smallest interval found that brackets the minimum, a and b.
- The first task is to determine whether or not the original interval already contains a local minimum or a local maximum. If there is a local minimum, the code simply returns. If a local maximum is within the interval, the direction of the search cannot be determined and the code displays a message and returns. Also, if a local maximum is encountered later in the calculation, the code again displays a message and returns.
- The next task is to determine which is the "downhill" direction and then to take a tentative step, $c = b + (b - a)$. We assume that $f(x_1) < f(x_0)$. If the reverse is true, the code "flips" the vector containing the x's and the f's, and continues. The downhill step moves the interval by an amount $\frac{1}{2}(x_1 - x_0)$ in the direction of decreasing $f(x)$.

The limit on the calculation is if the search goes beyond the specified range, or if more than 100 steps have been taken.

```
function [a,b] = DOWNHILL(x0,x1,max,fnc)
%DOWNHILL evaluates the function contained in the string
%    variable fnc at two points, x0, x1 (scalars) and then
%    follows the function downhill until a minimum is at-
%    tained.  It returns two x values in a,b that bracket
%    the minimum.  The range of the search is limited by
%    specifying the maximum distance as a multiple of the
%    distance abs(x1-x0) and is entered as max.  The func-
%    tion must accept column vector input.   The use is
%        [a,b] = DOWNHILL(x0,x1,max,fnc) , for example,
%        [a,b] = DOWNHILL(2.5,2.7,6,'cos')
%===========================================================

x = [x0 .5*(x0+x1) x1]'; f = feval(fnc,x);
if f(1)>f(2) & f(3)>f(2)
a=x(1); b=x(3); return
end
if f(1) < f(2)
x = flipud(x); f = flipud(f);
end
```

```
iter = 0;
while f(3) <= f(2)
   if f(1)<=f(2)
      disp('interval contains local max. between')
      disp([x(1) x(3)])
      a=x0; b=x0; return
   end
   b = 2*x(3)-x(2);
   x = [x(2) x(3) b]';
   f = [f(2) f(3) feval(fnc,b)]';
      iter = iter + 1;
      if abs(x(3)-x0)/abs(x1-x0) > max | iter > 100
   disp('out of range')
   disp('no minimum found between')
   disp([x0 x(3)]);
            a = x(3);b=x(3); return
      end
end
if x(3)<x(1);x=flipud(x);end
a = x(1); b = x(3);
```

Refining the Interval Containing a Minimum

At this stage, it is assumed that you know that a minimum of the function is within the interval $a \leq x \leq c$, of width $\Delta_0 = c - a$, and you have evaluated the function at one point within the interval, $x = b$, where $f_b < f_a$ and $f_b < f_c$. Notice the similarities and the differences of this method compared with those of the bisection method. In the bisection method, an interval is defined by *two* points (a, f_a) and (b, f_b), between which the function changes sign, $f_a f_b < 0$. To determine a minimum, we need at least *three* points: the left and right ends and a point between them where the value of the function is less than at either end. What is the best strategy for selecting the next value of x?

Consider such a line segment and assume, for sake of argument, that we select a point, x, in the larger subsegment:

$$\Delta_0 = c - a$$

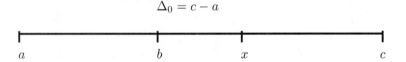

This defines two subsegments, $[abx]$ and $[bxc]$. Thus, if $f(b) < f(x)$, we would choose our next interval to be $[abx]$, whereas if $f(b) > f(x)$, it would be $[bxc]$. Clearly, with no other information available, we should choose x so that the size of the two segments are equal. In this way, no matter in which side the actual minimum resides, each step will reduce the bracketing interval by the same fraction. Consequently, we select x so that both subsegments are of equal length, $x - a = c - b$.

But how did we make the initial selection for point b? Obviously, by applying the same strategy in the *previous* step. Thus, from one step to the next, although the interval will shrink, the *relative* positions of the interior point to the ends will remain the same. That is, if the subsegment $[bxc]$ contains the minimum,

$$\frac{c-b}{c-a} = \frac{c-x}{c-b} \equiv \beta$$

and the condition that equates both subsegments can be written as

$$(c-x) = (c-a) - (c-b)$$

$$\frac{c-x}{c-b} = \frac{c-a}{c-b} - 1$$

$$\beta = \frac{1}{\beta} - 1$$

which is the equation for the *golden mean* of antiquity and has a solution $\beta = \frac{1}{2}(\sqrt{5}-1) = 0.61803\ldots$. Thus, the optimum position of the point within the first segment is a distance $\beta(c-a)$ from the right end or, defining $\gamma = 1 - \beta = 0.3820\ldots$, a distance $\gamma(c-a)$ from the left end. Each successive subsegment is then chosen to preserve the same relative distances.

Once the interval has been divided into two subsegments of equal size, we must determine in which subsegment the minimum lies. That is, we begin with the four points

$$\Delta_0 = x_4 - x_1$$

$$\Delta_1 = x_4 - x_2$$

with the points x_2 and x_3 located an equal distance from the left and right ends, respectively. In the next step, the interval is of width $\Delta_1 = \beta\Delta_0$.

If $f(x_3) < f(x_2)$ we replace the points

$$\begin{pmatrix} x_1 \\ x_2 \\ x_3 \\ x_4 \end{pmatrix} \rightarrow \begin{pmatrix} x_2 \\ x_3 \\ x_4 - \gamma\Delta_1 \\ x_4 \end{pmatrix}$$

If $f(x_2) < f(x_3)$ we replace the points

$$\begin{pmatrix} x_1 \\ x_2 \\ x_3 \\ x_4 \end{pmatrix} \rightarrow \begin{pmatrix} x_1 \\ x_1 + \gamma\Delta_1 \\ x_2 \\ x_3 \end{pmatrix}$$

The following MATLAB function, GOLDEN, which is included on the applications disk, will execute these operations on an arbitrary MATLAB function of a single variable and will return an estimate of the location of the minimum of the function. A demonstration program, entitled GoldDemo.m, is also included on the disk.

```
function r = GOLDEN(a,b,small,fnc)
%GOLDEN(a,b,small,fnc) will start with an interval, [a,b],
%  KNOWN to contain a minimum of the function fnc.  The
%  name of the function is contained in the string con-
%  stant fnc.  The algorithm is to successively reduce the
%  size of the interval by the golden search method until
%  the size is less than the specified tolerance, small.
%  The center of that interval is then returned as the
%  position of the minimum.  The quantities a,b,small, and
%  r are scalars.
%============================================================
beta  = .5*(sqrt(5) -1);       gamma = 1-beta;
x(1) = a;    x(4) = b;
if a>b
    x(1) = b;    x(4) = a;
end
d     = x(4) - x(1);
x(2) = x(1) + gamma*d;
x(3) = x(4) - gamma*d;
f     = feval(fnc,x);
while d > small
    d = beta*d;
    if f(3) < f(2)
        x = [x(2) x(3) x(4)-gamma*d x(4)];
        f = [f(2) f(3) feval(fnc,x(3)) f(4)];
    else
        x = [x(1) x(1)+gamma*d x(2) x(3)];
        f = [f(1) feval(fnc,x(2)) f(2) f(3)];
    end
end
r = .5*(x(2)+x(3));
```

For example, the function $y(x) = (x-1)^4$ has a minimum at $x = 1$. If the initial search interval is $[x_1, x_4] = [-2, 5]$, each iteration of GOLDEN will reduce this interval until the minimum is bracketed to within the specified tolerance. The logarithms of successive interval widths are plotted as a function of step number, i, in Figure 6.1. The straight line has a slope of -2 and indicates that the size of the interval decreases linearly as $(\Delta x)_{n+1} \approx e^{-2}\Delta x_n$.

It is not much more difficult to write a code that will fit a parabola through the last three points, and then use the parabola to estimate the position of the function's minimum, replacing the golden search positioning of the point x_4. These algorithms are quadratically convergent, in contrast to the linear convergence of the golden search method. Rather than constructing such a code, the reader is encouraged to read the listing[2] for the toolbox function fmin. This function combines the golden search method with a parabolic interpolation method

---

[2]The MATLAB function type can be used to display the contents of an M-file. For example, type fmin will display the entire contents of the MATLAB file fmin.m.

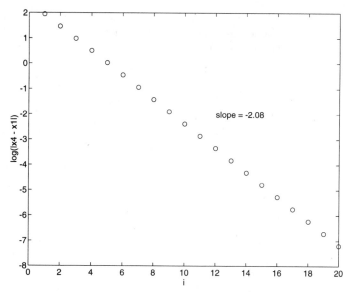

Figure 6.1 *The logarithms of successive interval widths, $\log \Delta x_i$, as a function of step number, i, for the golden search method.*

into a procedure known as *Brent's method*. The idea is to use the faster parabolic interpolation, if possible, and in steps where that fails, to resort back to the golden search method. The parabolic method fails in a particular step if the projected minimum point falls outside of the current interval, or if successive steps are becoming larger. The function also takes pains to never evaluate the function within a distance $\sqrt{\epsilon}$ from a previously computed point (because round-off error would negate any additional information from such points). Again, $\sqrt{\epsilon}$, where $\epsilon = $ **eps**, the MATLAB machine accuracy, is the limit of the accuracy of a search for the minimum of a function.

Applications of min/max computational methods can be found in a variety of problems that attempt to optimize one or a set of parameters. For those problems involving several simultaneous variables, we must wait until Chapter 10. However, there are numerous examples of single variable optimization problems.

ILLUSTRATIVE PROBLEM 6.1

Range of a Constant Thrust Rocket

Rather than following the ballistic trajectory of a simple projectile, it is interesting to compare ballistic trajectories with more realistic trajectories of rockets. Consider the simplified problem of a *constant thrust* rocket, where the rocket engine will supply a constant force to the rocket as long as the engine is operating. After burnout, the rocket will follow a ballistic trajectory.

The characteristics of the engine are given in terms of the following parameters:

Symbol	Description	Value
v_e	Exhaust velocity of propellant relative to the rocket	1500 m/s
Ω	Fuel-to-total mass ratio	0.125
t_c	Time of operation of engine (cutoff)	6 s
τ	Equal to t_c/Ω	48 s

Next, to make the problem tractable, we assume that the rocket always points in the same direction while the engine is operating. That is, the angle that the rocket makes with the horizontal is fixed, $\theta = $ constant. See Figure 6.2.

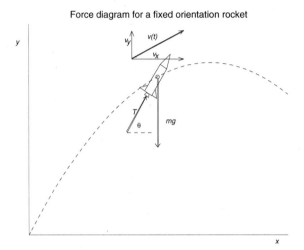

Force diagram for a fixed orientation rocket

Figure 6.2 *Force diagram for a rocket in flight.*

Under the assumption of fixed orientation angle, the equations of motion for the rocket can be solved,[3] and the results are:

Before cutoff: $0 \le t \le t_c$

$$v_x(t) = -\ln(1 - t/\tau)v_e \cos\theta$$
$$v_y(t) = -\ln(1 - t/\tau)v_e \sin\theta - gt$$
$$x(t) = v_e\tau \cos\theta f(t/\tau)$$
$$y(t) = v_e\tau \sin\theta f(t/\tau) - \frac{1}{2}gt^2$$

where

$$f(u) = u + (1 - u)\ln(1 - u)$$

The operative idea is, for a particular firing angle θ, to use the given parameters to compute the position and velocity components at the time of engine cutoff, (x_c, y_c) and (v_{xc}, v_{yc}). Thereafter, the normal ballistic equations are used for times $t > t_c$,

$$x(t) = x_c + v_{xc}(t - t_c)$$
$$y(t) = y_c + v_{yc}(t - t_c) - \frac{1}{2}g(t - t_c)^2$$

The *range* of the rocket is the x position when the rocket returns to the earth. That is, we solve the equation $y(t_{\text{hit}}) = 0$ for t_{hit}, and use it to compute $R = x(t_{\text{hit}})$.

---

[3]See Miele, *Flight Mechanics*, vol. 1, Addison-Wesley, Reading, MA, 1962.

The result is

$$R = x_c + \frac{v_{xc}v_{yc}}{g}\left[1 + \sqrt{1 + \frac{2gy_c}{v_{yc}^2}}\right] \qquad (6.1)$$

R is a function of the only parameter left unspecified, the orientation of the rocket, θ. Thus, we wish to find the maximum of $R(\theta)$.

The MATLAB code to solve this problem is as follows:

```
function R = THRUST(th)
global ve Omega tc g
%
% at engine cut-off
%
   tau = Omega*tc;
   u   = tc/tau;
   vxc = -log(1-u)*ve*cos(th);
   vyc = -log(1-u)*ve*sin(th) - g*tc;
    xc = ve*tau*sin(th)*(u + (1-u)*log(1-u));
    yc = ve*tau*cos(th)*(u + (1-u)*log(1-u))- .5*g*tc^2;
   arg = ones(length(th),1) - 2*g*yc.*vyc.^(-2);
     R = xc + (vxc.*vyc/g)*(1+sqrt(arg));
```

Finally, to find the *maximum* of this function using the procedures discussed thus far, we instead find the *minimum* of $-R$. That is, the line R = -R is added to the preceding function. Assuming that the maximum is for an angle θ in the range $\frac{\pi}{10} \le \theta \le \frac{\pi}{3}$, the statement

```
angle = fmin('THRUST',th1,th2,[0 1.e-7])*180/pi
```

yields the result $\theta_{\text{best}} = 49.45°$ as the optimum firing angle. Using this angle, the optimum range is $R(\theta_{\text{best}}) = 6010$ m.

Problems

REINFORCEMENT EXERCISES

P6.1 Minimum/Maximum of Simple Functions. Find the minima of the following functions both analytically and using the toolbox function, fmin.

    a. $f(x) = x^x$ with $0 \le x \le 1$ (Be careful when computing 0^0).

    b. $f(x) = x^3 - 3x^2 + x - 2$ with $0 \le x \le 2$

    c. $f(x) = \dfrac{\ln(1+x)}{(1+x)\sin^{-1}x} - \sqrt{\dfrac{1-x}{1+x}}$

    d. If the line $x + y = 1$ intersects the circle $x^2 + 3y^2 = r^2$, it does so at two points, (x_1, y_1) and (x_2, y_2). Find the value of r that minimizes the separation of the two intersection points.

***P6.2* Using the Downhill Search Method.**   The function

$$f(x) = 2500x^6 - 15000x^5 + 14875x^4 + 40500x^3 - 97124x^2 + 73248x - 19008$$

has several minima that may be difficult to find.

    a. Use the function DOWNHILL, starting with $x_0 = 0.8$ and $x_1 = 1.2$, to begin a search for a minimum. The function will return values (a, b) that bracket a minimum. Next, plot the function over the range $a \le x \le b$ to verify that a minimum is in the range. The graph should instead appear to have a *maximum!* Next, try plotting the function over a much narrower range surrounding the apparent maximum. Use this graph and the function fmin to determine the *local* minimum.

    b. The other minima are somewhat easier to find. Again, use DOWNHILL to determine a range both to the right and to the left of the minimum in part (a) that contains a minimum, and then use fmin to obtain the final positions.

***P6.3* Finding Minima/Maxima Using Derivatives.**   Obtain the expression for the first and second derivatives of the polynomial in Problem 6.2. Using the function roots, determine the five real roots, r_i, of the derivative of the polynomial. Then, using polyval, evaluate the second derivative at each of the values of r_i, thereby determining which of the values corresponds to the minima or maxima.

***P6.4* Function fmin and Brent's Method.**   Because the toolbox function fmin uses parabolic interpolation in place of the golden search method when the minimum is nearby, it should be faster than the function GOLDEN in most cases. Compare these two minimization functions when finding the minima of the following:

    a. $f(x) = \cos^2 x$; a minimum is at $x = \pi/2$.
    b. $f(x) = |\ln x|$; a minimum is at $x = 1$.

EXPLORATION PROBLEMS

***P6.5* Minimum of a Function of Two Variables.**   Although we will outline more efficient methods for finding the minimum of a function of two or more variables, it is a good exercise in writing MATLAB code to attempt to use only fmin to find a minimum of $f(x, y)$. The procedure might be structured as follows:

    a. Determine the intervals that bracket the overall minimum in both the x and y directions. That is, specify the search intervals as $x_a \le x \le x_b$ and $y_a \le y \le y_b$. The coordinates of the position of the minimum will be labeled as xmin, ymin.

    b. Write a dummy function that simply evaluates $f(x, y)$ as a function of a *single* variable y, with x and the name of the two-variable function $f(x, y)$ passed to the function as global variables called x0 and fname; i.e.,

```
function f = Fy(y)
  global x0 y0 fname
  f = feval(fname,x0,y);
```

and a similar function, Fx, that is a function of x only.

    c. Next, construct an M-file function to do the following:
        (i) Start with $x_0 = \frac{1}{2}(x_a + x_b)$ and $y_0 = \frac{1}{2}(y_a + y_b)$.

    (ii) Next, for this value of x_0, use `fmin` to find the y value corresponding to the minimum of `Fx`, and set this equal to `y0`.

    (iii) For this value of `y0`, use `fmin` to find the x value corresponding to the minimum of `Fx`, and set this equal to `x0`.

    (iv) The change in the values of `x0` and `y0` corresponds to the size of the step. Iterate the procedure until this step size becomes smaller than some tolerance, and stop the calculation.

    (v) Assemble all of this into an M-file function of the form

```
[xmin,ymin] = Fminmin(fname,xa,xb,ya,yb,tol)
```

Note: the quantities `x0`, `y0`, and `fname` will have to be `global` in the functions `Fx` and `Fy` and in the main MATLAB program. A graphical demonstration of the procedures used in this problem is contained in the file named `FmnDemo.m` on the applications disk.

P6.6 Applications of Fminmin. If a single strip of sheet metal is used to construct a water gutter for a house, it is desired that the carrying capacity of the gutter be *maximized*. The gutter is constructed from the metal strip of fixed width L by bending up both ends to an angle θ a distance d from both ends. See Figure 6.3.

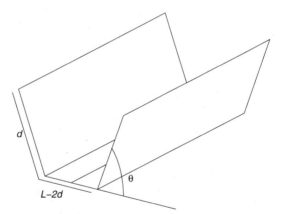

Figure 6.3 *Cross-section of a water gutter.*

The cross-sectional area is then

$$A(d, \theta) = d^2 \sin\theta \cos\theta + (L - d)d \sin\theta$$

a. Use the function `fminmin` of Problem 6.5 to find the values of d and θ that *maximize* $V(d, \theta)$, (or minimize $-V$). Use $L = 10$ cm.

b. The toolbox function `flops` is used to count the number of floating-point operations. Before executing your program, use `flops(0)` to reset the counter to zero. After the calculation is complete, `flops` will display the number of floating-point operations required to find the minimum. This will be of interest when we devise more efficient means for finding the minimum of $f(x, y)$ in Chapter 7. Also, be warned, this program will take several minutes if run on a PC. The values of d, θ that maximize the area are easily obtained using advanced calculus to be $d = 3\frac{1}{3}$ cm, $\theta = 60°$.

Summary

BRACKETING A MINIMUM OF A FUNCTION

Stepping Search
for Minimum

Although a root of a function can be bracketed by two points between which the function changes sign, the bracketing of a minimum requires at least three points. Thus, the calculation begins with three function evaluations, f_1, f_2, and f_3 at $x_1 > x_2 > x_3$, respectively. If f_2 is not less than *both* f_1 and f_3, the next point is taken in a "downhill" direction relative to the current set. This is repeated until a local minimum is bracketed.

Golden Search
Algorithm

Once it has been determined that the position of a local minimum is within the interval $[a, b]$, this interval is partitioned into three subsegments by evaluating the function at $x_1 = a + \gamma d$ and $x_2 = b - \gamma d$, where $d = b - a$, and $\gamma = (3 - \sqrt{5})/2$. The current interval is then reduced in size by the following algorithm:

$$\text{If } f(x_2) < f(x_1) \quad \text{we replace} \atop \text{the points} \quad \begin{pmatrix} a \\ x_1 \\ x_2 \\ b \end{pmatrix} \rightarrow \begin{pmatrix} x_1 \\ x_2 \\ b - \gamma d' \\ b \end{pmatrix}$$

$$\text{If } f(x_1) < f(x_2) \quad \text{we replace} \atop \text{the points} \quad \begin{pmatrix} a \\ x_1 \\ x_2 \\ b \end{pmatrix} \rightarrow \begin{pmatrix} a \\ a + \gamma d' \\ x_1 \\ x_2 \end{pmatrix}$$

with $d' = (1 - \gamma)d$.

MATLAB Functions Used

`fmin`

If the function $f(x)$ is coded in the form of an M-file function, for example, with the name *fname*, `fmin` will compute the minimum of this function over a specified interval $[x_1, x_2]$. The use is `xmin = fmin('`*fname*`',x1,x2)`. The name of the function to be minimized is included in the argument list as either a string constant enclosed in single quotes or a string variable containing the name of the function. An optional control vector labeled `OPTIONS` can be included as the fourth item in the argument list. If all the elements of this vector of length 18 are zero, default values of 10^{-4} are used for the convergence tolerance for x, and 500 for the maximum number of steps. Otherwise, `OPTIONS(2)`, if nonzero, will be the convergence tolerance, `OPTIONS(14)` will be the maximum number of steps, and `OPTIONS(1)`, if nonzero, will cause intermediate results to be displayed. Also, if the function $f(x)$ requires additional *parameters* (in addition to x) in its argument list, e.g., $f(x, \alpha, \beta)$, these can be included as items 5, 6, etc., in the argument list of `fmin`. Thus, to find the minimum of $f(x, \alpha, \beta)$, the reference to `fmin` would be of the form `fmin('`*fname*`',x1,x2,OPTIONS,alpha,beta)`.

`flops`

Returns the total number of floating-point operations executed during the current MATLAB session. `flops(0)` resets the counter to zero.

`type`

Is used to display the contents of MATLAB M-files. The use is `type` *name-of-function*.

7 / *Numerical Integration Techniques*

PREVIEW   The integral of a function of a single variable is defined in two equivalent ways: as either the opposite of differentiation, that is, as the *antiderivative*, or as the *area* under a curve. Whenever possible, we prefer to express the result of an integration of an elementary function, such as $\sin x$ or e^x, in terms of the antiderivative. Unfortunately, this is not possible for most of the integrals encountered in real problems. For example, an integral appearing as innocent as

$$I = \int_a^b e^{x^2}\, dx$$

cannot be expressed in terms of any elementary functions and must be evaluated numerically.

On the plus side, however, numerical procedures for evaluating integrals are usually easy to understand and, for the most part, are extremely stable; although they are limited by round-off error like any procedure, they will rarely fail completely. There are several varieties of algorithms that can be employed and tuned to fit the vagaries of most problems.

One important characteristic of a numerical method is whether it is a *closed-type* or *open-type* algorithm. Closed-type methods require the evaluation of the function at the endpoints of the interval; open-type methods do not. The most important and most basic of the closed-type methods is called *Romberg* integration, which is simply an extrapolation of ideas contained in the familiar *trapezoidal* rule. The trapezoidal rule estimates the area under a curve by replacing the integrand with a sequence of approximating trapezoids. This procedure is improved by replacing the function with a sequence of parabolic segments leading to Simpson's rule. The logical extension of the procedure leads to Romberg integration.

A very powerful *open-type* algorithm is *Gaussian quadrature*. In this technique, we estimate an integral by evaluating the function at *nonequally spaced* sampling points over the interval and by "weighting" the contribution of each point by a factor chosen to optimize the accuracy of integrals of polynomial integrands of varying degree.

Each of these algorithms is readily coded in MATLAB, and we discuss several examples of their applications. In addition, we describe and illustrate the use of the basic MATLAB integration functions quad and quad8.

The Trapezoidal Rule for Integration

Essentially all computational algorithms evaluate a definite integral by estimating the area under the curve of $f(x)$ by a variety of means, usually involving approximating the function by a polynomial over a limited range. For example, we are all familiar with the basic Riemann definition of the definite integral as a limit, described as follows.

RIEMANN DEFINITE INTEGRAL

The integral of $f(x)$ from $x = a$ to $x = b$; i.e., $\int_a^b f(x)\,dx$, is obtained by first partitioning the interval into n subsegments and evaluating the sum as

$$I_n = \sum_{i=1}^{n} f(x_i)\Delta x_i$$

where Δx_i is the width of the ith segment and $f(x_i)$ is a value of $f(x)$ for some x in the segment, and finally, taking the limit as $\Delta x_{max} \to 0$, where Δx_{max} is the width of the largest subsegment.

The Riemann integral has a clear geometrical interpretation as the area of a *histogram* of the function $f(x)$ over the interval $a \leq x \leq b$. See Figure 7.1.

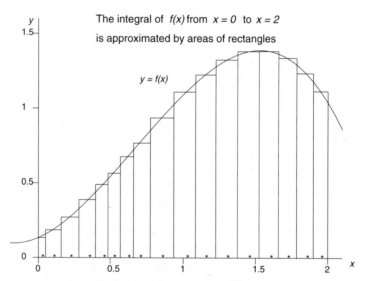

Figure 7.1   *Riemann integral as an area of a histogram.*

In this definition, the function $f(x)$ is replaced by a *constant* value over each segment; i.e., a polynomial of degree zero. Usually, a partition of equally sized segments is chosen so that

$$I = \int_a^b f(x)\,dx = \lim_{\Delta x \to 0} \sum_{i=0}^{n} f(x_i)\Delta x \qquad \left(\Delta x = \frac{b-a}{n}\right)$$

A slightly better estimate would be to approximate the function by straight lines over each segment; that is, to connect the points $x_i, f(x_i)$ and $x_{i+1}, f(x_{i+1})$. The area of the segment is then the area of a trapezoid, $\Delta A = \frac{1}{2}(f_i + f_{i+1})\Delta x$, where $\Delta x_n = (b-a)/n$ is the width of each of the segments. For a finite n, we then have the approximation

$$\int_a^b f(x)\,dx \approx \frac{(b-a)}{2n}\left[(f_a + f_1) + (f_1 + f_2) + \cdots + (f_{n-1} + f_b)\right]$$

$$= \frac{1}{2}\Delta x_n[f_a + 2f_1 + 2f_2 + \cdots + f_{n-1} + f_b]$$

$$= \frac{1}{2}\Delta x_n\left[f_a + f_b + 2\sum_{i=1}^{n-1} f(x_i)\right]$$

The trapezoidal rule is equivalent to replacing a function $f(x)$ by straight line segments at each point x_i, and it can be shown that the error in the method is proportional to $\Delta x^3 f''(\xi)$, where ξ is some (unknowable) point in the interval. Because this error estimate requires evaluating the second derivative of the function and then determining its maximum value over the interval, it is rarely used to evaluate the accuracy of the method. Furthermore, subsequent algorithms have much higher derivatives in their error estimates, making them substantially more difficult to utilize. Alternative accuracy checks will have to be devised.

The trapezoidal rule can be made quite amenable to machine computation and for later incorporation into higher-order algorithms by selecting a particular form for the partition of the interval. If the interval is subdivided into $n = 2^k$ segments, called panels, where k is a nonnegative integer, we have the following sequence of approximations to the integral:

$k = 0 \qquad n = 1 \qquad T_0 = \dfrac{1}{2}\Delta x_0(f_a + f_b) \qquad\qquad\qquad\qquad \Delta x_0 = (b - a)$

$k = 1 \qquad n = 2 \qquad T_1 = \dfrac{1}{2}\Delta x_1(f_a + 2f(a + \Delta x_1) + f_b) \qquad \Delta x_1 = \dfrac{(b-a)}{2}$

$k = 2 \qquad n = 4 \qquad T_2 = \dfrac{1}{2}\Delta x_2(f_a + 2f(a + \Delta x_2) + 2f(a + 2\Delta x_2)$

$\qquad\qquad\qquad\qquad\qquad + 2f(a + 3\Delta x_2) + f_b) \qquad\qquad\qquad \Delta x_2 = \dfrac{(b-a)}{4}$

$\vdots \qquad\qquad \vdots \qquad\qquad\qquad\qquad \vdots \qquad\qquad\qquad\qquad \vdots$

Notice that to proceed from k to $k+1$ the number of panels will double, but the only new information required is the value of the function at the midpoints of the previous panels. Furthermore, because $\Delta x_1 = \frac{1}{2}\Delta x_0$, $\Delta x_2 = \frac{1}{2}\Delta x_1$, etc., we can write the preceding expressions as follows:

$$T_0 = \frac{1}{2}\Delta x_0(f_a + f_b)$$

$$T_1 = \frac{1}{2}\Delta x_1[f_a + 2f(a + \Delta x_1) + f_b]$$

$$= \frac{1}{2}T_0 + \Delta x_1 f(a + \Delta x_1)$$

$$T_2 = \frac{1}{2}\Delta x_2[f_a + 2f(a + \Delta x_2) + 2f(a + 2\Delta x_2) + 2f(a + 3\Delta x_2) + f_b]$$

$$= \frac{1}{2}T_1 + \Delta x_2 \sum_{i=1,3} f(a + i\Delta x_2) \tag{7.1}$$

$$\vdots \qquad \qquad \vdots$$

$$T_k = \frac{1}{2}T_{k-1} + \Delta x_k \sum_{\substack{i=1 \\ \text{odd only}}}^{n-1} f(a + i\Delta x_k), \qquad \Delta x_k = \frac{(b-a)}{2^k}$$

The trapezoidal rule approximation to an integral consists of first computing $T_0 = \frac{1}{2}(b-a)(f_a + f_b)$ and then repeatedly applying Equation 7.1 until the
results change by less than some prespecified tolerance. To illustrate the trapezoidal rule, we will compute the value of the integral

$$I = \int_1^2 \left(\frac{1}{x}\right) dx$$

This integral is, of course, equal to $\ln 2 = 0.69314718\ldots$. The successive values
of the trapezoidal rule approximation with $a = 1$, $b = 2$, and $f(x) = 1/x$ are then

$$T_0 = \frac{1}{2}\left(\frac{1}{1} + \frac{1}{2}\right)(2 - 1) \qquad\qquad = 0.75$$

$$T_1 = \frac{T_0}{2} + \frac{1}{2}f\left(1 + \frac{1}{2}\right)$$

$$= \frac{0.75}{2} + \frac{1}{2}\left(\frac{1}{1.5}\right) \qquad\qquad = 0.708333\ldots$$

$$T_2 = \frac{T_1}{2} + \frac{1}{4}\left[f\left(1 + \frac{1}{4}\right) + f\left(1 + \frac{3}{4}\right)\right]$$

$$= \frac{0.70833}{2} + \frac{1}{4}\left(\frac{1}{1.25} + \frac{1}{1.75}\right) \qquad = 0.697024\ldots$$

Continuing the calculation through $k = 5$ yields

k	T_k
0	0.75
1	0.7083...
2	0.69702...
3	0.69412...
4	0.69339...
5	0.693208...
$\vdots$	$\vdots$
∞	0.693147...

The convergence of the trapezoidal method is not particularly fast, but the method is quite simple. A graphical demonstration of the trapezoidal method is contained in the script file named `TrapDemo.m` on the applications disk.

WHAT IF? The trapezoidal rule will give *exact* results for the integral of any polynomial of degree 2 or less. Use MATLAB to duplicate the preceding calculation for the quadratic $f(x) = 3x^2 - x + 5$ for $k = 1$, 2, and 3, and compare the results with the exact value of the integral. Can you prove this statement for an arbitrary quadratic? ●

Simpson's Rule

The trapezoidal rule was an improvement on the basic Riemann definition of a definite integral, where we replaced the $f = \text{const}$ approximation in each panel with a straight line approximation. Using a parabolic approximation should give even more accurate results.

The derivation of the trapezoidal rule consisted of dividing the interval into n subintervals or panels, replacing the function by a straight line connecting the endpoints, creating a trapezoid, and then adding the areas of the trapezoids. Simpson's rule simply duplicates these ideas using a parabolic replacement for the function. Three points are required to specify a parabola; therefore, the lowest-order Simpson's rule approximation will require two panels. In the trapezoid approximation we could simply write down the area of an individual trapezoid, but for Simpson's rule we must take a moment to derive the area beneath a parabola drawn through three points.

Area Under a Parabola Through Three Points We first wish to fit a parabola through three equidistant points, (x_1, f_1), $(x_1 + \Delta x, f_2)$, and $(x_1 + 2\Delta x, f_3)$, by writing the equation for the parabola as

$$y(x) = c_1(x - x_1)^2 + c_2(x - x_1) + c_3$$

The coefficients of the parabola can be determined by setting $y(x_1) = f_1$, $y(x_1 + \Delta x) = f_2$, and $y(x_1 + 2\Delta x) = f_3$, which yields the following matrix equation for the coefficients:

$$\begin{pmatrix} 1 & 0 & 0 \\ 1 & 1 & 1 \\ 1 & 2 & 4 \end{pmatrix} \begin{pmatrix} c_1 \Delta x^2 \\ c_2 \Delta x \\ c_3 \end{pmatrix} = \begin{pmatrix} f_1 \\ f_2 \\ f_3 \end{pmatrix}$$

or, after obtaining the inverse of the matrix,

$$\begin{pmatrix} c_1 \Delta x^2 \\ c_2 \Delta x \\ c_3 \end{pmatrix} = \frac{1}{2} \begin{pmatrix} 1 & -2 & 1 \\ -3 & 4 & -1 \\ 2 & 0 & 0 \end{pmatrix} \begin{pmatrix} f_1 \\ f_2 \\ f_3 \end{pmatrix}$$

Next, the area under the parabola from x_1 to $x_3 = x_1 + 2\Delta x$ is given by

$$(\text{area})_{13} = \int_{x_1}^{x_3} y(x)\, dx$$

$$= \frac{1}{3}c_1(x_3 - x_1)^3 + \frac{1}{2}c_2(x_3 - x_1)^2 + c_3(x_3 - x_1)$$

$$= \frac{8}{3}c_1\Delta x^3 + 4c_2\Delta x^2 + 2c_2\Delta x$$

$$= \frac{2}{3}\Delta x[4c_1\Delta x^2 + 3c_2\Delta x + 3c_3]$$

$$= \frac{2}{3}\Delta x(\, 4 \quad 3 \quad 3\,) \begin{pmatrix} c_1\Delta x^2 \\ c_2\Delta x \\ c_3 \end{pmatrix}$$

where the result is written as a product of a row vector times a column vector in the last line. If we now substitute the previous result for the column vector, we obtain

$$(\text{area})_{13} = \frac{1}{3}\Delta x[f_1 + 4f_2 + f_3] \tag{7.2}$$

Thus, the area under a general parabola over the interval a to b can be expressed in terms of the value of the parabola at the left and right ends, and at the midpoint of the interval.

This equation is the starting point of the derivation of Simpson's rule in the same way that the area of a single trapezoid is the basis of the trapezoidal rule.

DERIVATION OF SIMPSON'S RULE APPROXIMATION

The lowest-order Simpson's rule approximation to the integral of a function $f(x)$ over the interval $a \le x \le b$ begins by dividing the interval into two panels or three evaluation points. A parabola $y(x)$ is then drawn through the three points, and the area under the parabola is used as an approximation to the area under $f(x)$. Thus,

$$k = 1 \qquad n = 2^1 \text{ panels} \qquad \Delta x_1 = \frac{b - a}{2}$$

$$S_1 = \frac{1}{3}\Delta x_1[f_a + 4f(a + \Delta x_1) + f_b]$$

The next level of approximation is to halve the interval width and partition the interval into four panels, as shown in Figure 7.2. The area under the function $f(x)$ is then approximated as the area under the two parabolas drawn through the sets of points (f_a, f_1, f_2) and (f_2, f_3, f_b). The area under a single parabola is given by Equation 7.2, thus the area under the two parabolas is

$$S_2 = \frac{1}{3}\Delta x_2[(f_a + 4f_1 + f_2) + (f_2 + 4f_3 + f_b)]$$

$$= \frac{1}{3}\Delta x_2[f_a + 4(f_1 + f_3) + 2f_2 + f_b]$$

where $\Delta x_2 = (b - a)/2^2$ and $f_i = f(a + i\Delta x_2)$.

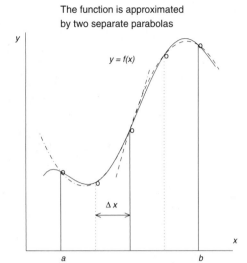

Figure 7.2   *Second-order Simpson's rule approximation is the area under two parabolas.*

Extending this method to $8, 16, 32, \dots$ panels, the result for S_k is easily obtained to be

Simpson's rule
$$S_k = \frac{1}{3}\Delta x_k \left[f_a + 4 \sum_{\substack{i=1 \\ \text{odd only}}}^{n-1} f(a + i\Delta x_k) + 2 \sum_{\substack{i=2 \\ \text{even only}}}^{n-2} f(a + i\Delta x_k) + f_b \right] \qquad (7.3)$$

Simpson's rule is justifiably an extremely popular method of evaluating integrals of smooth functions. It will converge quite nicely in most instances and is relatively easy to use. Also, Equation 7.3 can easily be adapted to handle an arbitrary odd number of unevenly spaced points, and it is the most common method for estimating the integral of experimentally obtained data.

WHAT IF   Simpson's rule will give the *exact* result for the integral of any polynomial of degree 3 or less. Use MATLAB to demonstrate this for a simple cubic. How many panels are required for the result to be exact?   ●

RELATIONSHIP BETWEEN SIMPSON'S RULE AND THE TRAPEZOIDAL RULE

Because both the trapezoidal rule (Equation 7.1) and Simpson's Rule (Equation 7.3) evaluate the function at precisely the same set of points, we might expect that the two expressions can be related. For example, the lowest Simpson's rule approximation, S_1, requires three values of the function, f_a, f_1, and f_b, and these values are the same ones entering into the trapezoidal approximations,

T_0 and T_1:

$$T_0 = \frac{1}{2}\Delta x_0(f_a + f_b), \qquad T_1 = \frac{T_0}{2} + \Delta x_1 f_1$$

$$S_1 = \frac{1}{3}\Delta x_1[f_a + 4f_1 + f_b]$$

Using $\Delta x_0 = 2\Delta x_1$, it is not difficult to show that

$$S_1 = T_1 + \frac{T_1 - T_0}{3}$$

Using the expressions for T_1, T_2, and S_2,

$$T_1 = \Delta x_2(f_a + 2f_2 + f_b)$$

$$T_2 = \frac{1}{2}\Delta x_2[f_a + 2(f_1 + f_2 + f_3) + f_b]$$

$$S_2 = \frac{1}{3}\Delta x_2[f_a + 4(f_1 + f_3) + 2f_2 + f_b]$$

where $f_i = f(a + i\Delta x_2)$ and $\Delta x_2 = (b - a)/4$, you can easily verify that

$$S_2 = T_2 + \frac{T_2 - T_1}{3}$$

Using the general expressions for T_k (Equation 7.1) and for S_k (Equation 7.3), this result can be further generalized to

$$S_k = T_k + \frac{T_k - T_{k-1}}{3} \tag{7.4}$$

This important result provides a convenient alternative method of obtaining the Simpson's rule results. First, compute the trapezoidal rule results through order k, then use Equation 7.4 to generate the Simpson's rule values likewise through order k. Notice that, by this method, once the trapezoidal rule values are obtained, the function itself is not evaluated at any additional points.

EXAMPLE OF SIMPSON'S RULE APPROXIMATION OF AN INTEGRAL

Again, consider the integral

$$I = \int_1^2 \frac{dx}{x}$$

First using Equation 7.3 for $k = 1$ yields

$$k = 1 \qquad n = 2^1 = 2 \qquad \Delta x_1 = \frac{b - a}{2} = \frac{1}{2}$$

$$S_1 = \frac{1}{3}\left(\frac{1}{2}\right)\left[1 + 4\left(\frac{1}{1.5}\right) + \frac{1}{2}\right]$$

$$= 0.694444\ldots$$

Continuing the calculation, we obtain the values listed in Table 7.1. For comparison, results are also included for the same integral obtained earlier by the trapezoidal method. Clearly, Simpson's rule converges much faster than the trapezoidal rule, at least for this example. The Simpson's rule values could also have been computed directly from the trapezoidal rule numbers by using Equation 7.4.

Order k	No. of panels n	T_k	S_k
0	1	0.75	
1	2	0.7083...	0.6944...
2	4	0.69702...	0.69325...
3	8	0.69412...	0.693154...
4	16	0.69339...	0.6931466...
5	32	0.693208...	0.6931473...
6	64	0.693162...	0.6931472...

Table 7.1 *Trapezoidal and Simpson's rule results for the integral $\int_1^2 dx/x$*

BEYOND SIMPSON'S RULE

Higher-level integration algorithms are next obtained by approximating the function $f(x)$ by interpolating polynomial segments with the degree of the polynomial now chosen to be 3,4,5, and so on. What results is an entire class of numerical integration techniques collectively called Newton-Cotes formulas. The trapezoidal rule and Simpson's rule are the first two of these formulas. To obtain the next formula in the set, a fourth-order polynomial, $y_4(x)$, is fit through the points associated with a four-panel partition of the interval, and the integral $\int y_4(x)\,dx$ is used as the next level approximation to $\int f(x)\,dx$. For eight panels, two fourth-degree segments are used, and so on. The formula that follows from this procedure is called *Boole's rule*.[1] Boole's rule is similar to, but more complicated than, the equations for the trapezoidal rule and Simpson's rule and is not in-

---

[1] The area beneath a quartic, from $x = a$ to $x = b$, is easily shown to be

$$\text{area} = \frac{2}{45}\Delta x[7f_a + 32f_1 + 12f_2 + 32f_3 + 7f_b]$$

where f_a, f_1, f_2, f_3, and f_b are the values of the quartic at the ends of the interval and at equally spaced points within the interval, and $\Delta x = (b - a)/4$. (See Borse, *FORTRAN-77 and Numerical Methods*, 2nd ed., PWS Publishers, Boston, MA, 1991, p. 577.) Boole's rule then approximates the area under an arbitrary function, $f(x)$, by subdividing the integration interval into $4n$ equal panels and summing the area beneath n separate quartics.

cluded here. Instead, let us consider how a higher-level algorithm such as Boole's rule relates to the earlier equations for the trapezoidal and Simpson's rules.

If the partitioning of the interval proceeds as before into $2, 4, 8, \ldots$ panels, the values of the function that enter Boole's rule will be the same set that was used in both Simpson's and the trapezoidal rule. It should then be possible to relate the Boole's rule calculation to the previous estimates in a manner similar to Equation 7.4. This is indeed possible, and, if the Boole's kth-order result is labeled as B_k, yields

$$B_k = S_k + \frac{S_k - S_{k-1}}{15}$$

Thus, once a table of Simpson's rule values has been determined, the improved Boole's rule results can easily be computed. For example, a third column can be added to Table 7.1 in the evaluation of the integral $\int_1^2 dx/x = 0.6931471806\ldots$, as in Table 7.2.

Order k	No. of panels n	T_k	S_k	B_k
0	1	0.75		
1	2	0.7083...	0.6944...	
2	4	0.69702...	0.69325...	0.69317...
3	8	0.69412...	0.693154...	0.693149...
4	16	0.69339...	0.6931466...	0.6931472...
5	32	0.693208...	0.6931473...	0.693147181...
6	64	0.693162...	0.6931472...	0.69314718054...
			exact =	0.6931471806...

Table 7.2   *A column of Boole's rule values is added to Table 7.1.*

You can no doubt see the pattern emerging in this analysis. The next step would be to fit the eight-panel partition with an eighth-degree polynomial, $y_8(x)$, approximate the area under the curve $f(x)$ by the area under a successively increasing number of polynomial segments, and label the kth-order such approximation as, say C_k. Finally, this higher-degree estimate is related to the previous area estimates B_k, S_k, and T_k. The result of this formidable, but straightforward, exercise is that the higher-degree approximations C_k are related to Boole's rule by

$$C_k = B_k + \frac{B_k - B_{k-1}}{63} \tag{7.5}$$

The limit suggested by the sequence of integration algorithms

$$T_k \rightarrow S_k \rightarrow B_k \rightarrow C_k \rightarrow \cdots$$

is known as *Romberg integration*[2] or Richardson extrapolation to the limit.

Romberg Integration

The first step in unifying the previous numerical integration algorithms is to define a new notation, T_k^m, where the subscript k labels the order of the approximation (n = number of panels = 2^k), and m identifies the *level* of the integration algorithm, that is

$$
\begin{aligned}
m &= 0 \quad \text{Trapezoidal rule} \quad && T_k^0 = T_k \\
m &= 1 \quad \text{Simpson's rule} \quad && T_k^1 = S_k \\
m &= 2 \quad \text{Boole's rule} \quad && T_k^2 = B_k \\
m &= 3 \quad && T_k^3 = C_k \\
&\ \ \vdots \quad && \quad \vdots
\end{aligned}
$$

Equations 7.1, 7.3, and 7.5, rewritten in terms of this new notation, become

$$T_k^1 = T_k^0 + \frac{1}{3}(T_l^0 - T_{k-1}^0)$$

$$T_k^2 = T_k^1 + \frac{1}{15}(T_l^1 - T_{k-1}^1)$$

$$T_k^3 = T_k^2 + \frac{1}{63}(T_l^2 - T_{k-1}^2)$$

The generalization of these results leads to the equation for the Romberg algorithm:

Romberg Integration

$$T_k^{m+1} = T_k^m + \frac{T_k^m - T_{k-1}^m}{4^{m+1} - 1}$$

The importance of this equation lies in the fact that, under normal conditions, quite accurate results may be obtained for the integral of a reasonably smooth[3] function with little more work than that required in obtaining the trapezoidal rule estimates.

---

[2] The proof of the Romberg algorithm, however, follows a different path. (See S. Kuo, *Computer Applications of Numerical Methods*, Addison-Wesley, Reading, MA, 1972.)
[3] *Smooth* means that the function must be continuous over the interval and all of its derivatives must be finite in that interval; that is, the function must be expandable as a Taylor series in this region.

CONSTRUCTING ROMBERG TABLES

The procedure is to start with the one-panel trapezoidal rule value,

$$T_0 \rightarrow T_0^0 = \frac{1}{2}(b-a)(f_a + f_b)$$

then increase the order of the calculation (increment k from 0 to 1) by using Equation 7.1:

$$T_k = \frac{1}{2}T_{k-1} + \Delta x_k \sum_{\substack{i=\text{odd only}}}^{n-1} f(a + i\Delta x_k)$$

Next, increase the *level* of the algorithm (i.e., from trapezoidal to Simpson's) by using Equation 7.4. These ideas can be combined in the form of a triangular table:

	m				
	0	1	2	3	4
$k = 0$	T_0^0				
1	T_1^0	T_1^1			
2	T_2^0	T_2^1	T_2^2		
3	T_3^0	T_3^1	T_3^2	T_3^3	
4	T_4^0	T_4^1	T_4^2	T_4^3	T_4^4

An increase in accuracy is achieved by increasing the number of panels (stepping down in the triangle) or by increasing the level of the algorithm (stepping across horizontally in the triangle). The chief advantage of the method is that the function is evaluated only to obtain the elements of the first column; i.e., the trapezoidal rule results.

The MATLAB code shown in Figure 7.3 implements the Romberg algorithm for the definite integral of a function of a single variable.

See Problems 7.1 through 7.4 for examples of trapezoidal, Simpson, and Romberg integration.

The MATLAB Toolbox Functions quad and quad8

There are two MATLAB toolbox functions, quad and quad8, that will evaluate the definite integral of a function $f(x)$ over a range $a \leq x \leq b$.

The operations of both of these functions are quite similar. The only significant difference is that quad is built around the basic Simpson's rule approximation, whereas quad8 employs an eight-panel Newton-Cotes formula. Both functions repeatedly divide the full interval in half and approximate the integral as a sum of parabolas (quad) or degree 8 polynomials (quad8). They each compare the most recently computed value with the previous one as a check on

```
function [t,T,dT] = ROMBRG(fnc,a,b,k)
%ROMBRG will integrate the function, fnc, from x=a to x=b
&    and produce a triangular Romberg table, T, of results.
&    The maximum order is k, and the number of panels for
%    this order is 2^k.  The last diagonal element of the
%    table, (the most accurate element) is returned as t.
%    The difference between the last two diagonal elements
%    is an indication of the error and is returned as dT.
%===========================================================
   n = 2^k+1;    N = k+1;    T = zeros(N,N);  dx = (b-a);
   T(1,1) = .5*dx*(feval(fnc,a)+feval(fnc,b));
   jn = 2^(k-1);
   for i = 1:k
       x   = a + (jn+1:2*jn:n)*dx;
       dx = dx/2;
       T(i+1,1) = .5*T(i,1) + dx*sum(feval(fnc,x));
       jn = jn/2;
   end
   for m = 1:k
     T(m+1:N,m+1) = T(m+1:N,m) + ...
              (T(m+1:N,m) - T(m:N-1,m))/(4^(m)-1);
   end
   t = T(k+1,k+1);   dT = abs(t-T(k,k));
```

Figure 7.3 *MATLAB code for Romberg integration of a function*

accuracy. However, both functions come with several added features that may be of interest to you. The complete forms of the definition lines are:

```
[A,count] =     quad(fname,a,b,tol,trace)
[A,count] =     quad8(fname,a,b,tol,trace)
```

where A is the value of the integral of the function referred to in the name fname from $x = a$ to $x = b$. The calculation is continued until successive values differ by less than tol. If this value is omitted from the argument list, a value of 10^{-3} is used. If you wish, both functions will return the total number of function evaluations required, count, in addition to the value of the integral. The fifth element in the argument list, trace, is a scalar flag used to control the display of intermediate results produced by quad and quad8. If this element is nonzero, both functions will display a plot of the integrand at the points used to evaluate the integral.

Finally, it must be emphasized that the function that is referenced by either quad or quad8 must be written using element-by-element operators. That is, it must expect a *vector* of input values and return a *vector* of output values.

Each of the methods described thus far, including the toolbox functions, requires the evaluation of the function at the endpoints of the interval; that is, these are *closed-type* algorithms. Quite frequently, you will need to have a value for an integral that you know converges, but in which the integrand is singular at one or both of the endpoints. For example, the integral

$$\int_0^1 \frac{dx}{\sqrt{x}} = 2\sqrt{x}\,\big|_0^1 = 2$$

is indeed finite, but, the integrand is singular at $x = 0$; therefore, none of the closed-type algorithms can be used in its evaluation. The most popular *open-type* integration algorithm, called *Gaussian quadrature*, is described below.

Gaussian Quadrature

All numerical integration techniques, and for that matter most numerical procedures of any kind, are based on the following concept: A complicated function may be approximated over sufficiently small intervals by polynomials of varying degree, and then the differentiation, integration, or other operation on the original function is replaced by the same operations on the simpler polynomial. The procedures of numerical integration that we have discussed so far—the trapezoidal rule, Simpson's rule, and Romberg integration—are examples of this concept. Romberg integration is the theoretical optimum procedure that can be applied to an arbitrary smooth function over a finite interval that contains no singularities, *provided* that the calculation is restricted to *evenly* spaced points. If the constraint of evenly spaced points is removed, a new class of numerical integration algorithms can be developed. These are known collectively as *Gaussian quadrature algorithms*. (Quadrature is the technical name for numerical integration.)

DERIVATION OF A SIMPLE GAUSSIAN QUADRATURE PROCEDURE

Each of the various numerical integration methods discussed thus far approximates the integral $\int f(x)\,dx$ by a sum of function evaluations at equally spaced points multiplied by so-called weight factors ω_i.

$$\int_a^b f(x)\,dx = \sum_{i=0}^n \omega_i f(x_i)$$

where $x_i = a + i\Delta x_k$, and the weights are specified by the particular type of integration algorithm. For example, the trapezoidal rule has a set of weights $\omega_i = \frac{1}{2}(1, 2, 2, \ldots, 2, 2, 1)\Delta x$, that is,

$$\int_a^b f(x)\,dx \approx \Delta x_k \frac{1}{2}(f_0 + 2f_1 + 2f_2 + \cdots + 2f_{n-1} + f_n)$$

The set of weights for Simpson's rule is $\omega_i = \frac{1}{3}(1, 4, 2, 4, \ldots, 4, 2, 4, 1)\Delta x$, that is,

$$\int_a^b f(x)\,dx \approx \Delta x_k \frac{1}{3}(f_0 + 4f_1 + 2f_2 + \cdots + 4f_{n-1} + f_n)$$

The set of multiplicative weight factors was determined by fitting successively higher-degree polynomials to segments of the function.

The great mathematician and scientist Karl Friedrich Gauss[4] suggested that the accuracy of the computational algorithm could be significantly improved if the positions of the function evaluations, *as well as* the set of weight factors, were left as parameters to be determined by optimizing the overall accuracy. By this statement he meant that the procedure should be *exact* when applied to polynomials of as high a degree as possible. If the approximation employs function evaluations at, say n points, the procedure has $2n$ parameters to be determined (the x_i and the factors ω_i). Because a general polynomial of degree N has $N + 1$ coefficients, the Gauss procedure with n points is required to be exact for any polynomial of degree $N = 2n - 1$ or less.

The Gaussian quadrature algorithms are usually stated in terms of integrals over the interval -1 to 1. For such an integral we would anticipate that, from symmetry, the sampling points and the weights would be the same for $\pm x$. Thus, the form of the algorithm is

$$\int_{-1}^{1} f(x)\, dx \approx \omega_0 f(0) + \sum_{i=1}^{k} \omega_i[f(x_i) + f(-x_i)] \qquad \text{for odd } n,\ k = \frac{(n-1)}{2}$$

$$\approx \sum_{i=1}^{k} \omega i[f(x_i) + f(-x_i)] \qquad \text{for even } n,\ k = \frac{n}{2} \tag{7.6}$$

The remaining task is to determine the parameters ω_i and x_i such that the approximation is of optimum accuracy in the sense discussed previously.

For example, in the case of two sampling points, the parameters are ω_1 and x_1, and these are determined by requiring that Equation 7.6 be *exact* whenever $f(x)$ is a polynomial of degree 3 or less. Applying this condition successively to the functions 1, x, x^2, and x^3 results in the following relations:

$$\int_{-1}^{1} 1\, dx = 2 = \omega_1[1 + 1]$$

$$\int_{-1}^{1} x\, dx = 0 = \omega_1[x_1 - x_1]$$

$$\int_{-1}^{1} x^2\, dx = \frac{2}{3} = \omega_1[x_1^2 + x_1^2]$$

$$\int_{-1}^{1} x^3\, dx = 0 = \omega_1[x_1^3 - x_1^3]$$

Thus, $\omega_1 = 1$ and $x_1^2 = \frac{1}{3}$. Equation 7.6 for $n = 2$ now reads

$$\int_{-1}^{1} f(x)\, dx \approx f\left(\frac{1}{\sqrt{3}}\right) + f\left(\frac{-1}{\sqrt{3}}\right) \tag{7.7}$$

You should verify that this remarkably simple approximation does indeed give the exact result for the integral of any polynomial of degree 3 or less.

---

[4] *Karl Friedrich Gauss* (1777–1855), was one of the most influential and prominent mathematicians of the nineteenth century.

If the function $f(x)$ is not a polynomial of degree 3 or less, the result of the $n = 2$ Gaussian quadrature expression will only be an approximation of the actual integral. The accuracy of the approximation will depend on how much the function $f(x)$ resembles polynomials of degree 3 or less. For example, consider the integral

$$I = \int_{-1}^{1} \cos(x)\, dx = \sin(x)|_{-1}^{1} = 1.6829\ldots$$

The Gaussian approximation to this integral is

$$I \approx \cos\left(\frac{-1}{\sqrt{3}}\right) + \cos\left(\frac{1}{\sqrt{3}}\right) = 1.6758\ldots$$

HIGHER-ORDER GAUSSIAN QUADRATURE PROCEDURES

To improve the accuracy of the approximation given in Equation 7.7, the number of sampling points is increased from 2 to $3, 4, \ldots$. For each choice of n, the number of points, the weight factors ω_i, and the position of the sampling points x_i must be determined by requiring that the approximation be exact for polynomials of degree $N \leq 2n - 1$. These have been determined and tabulated for $n = 2$ through 95.[5] An abbreviated table of the weights and sampling points for several values of n is given in Table 7.3.

APPLYING GAUSSIAN QUADRATURE TO INTEGRALS WITH A RANGE OTHER THAN −1 TO 1

If the integration interval is not −1 to 1, a change of variables must be effected before the Gaussian quadrature procedures of the previous section can be employed. For example, if the integral is from a to b,

$$I = \int_{a}^{b} f(x)\, dx$$

the change of variables is

$$x = \left(\frac{b - a}{2}\right)\xi + \left(\frac{b + a}{2}\right)$$

Thus, to evaluate the integral

$$I = \int_{0}^{2} \sin x\, dx$$

we would first change variables to $x = \xi + 1$, and evaluate the equivalent integral,

$$I = \int_{-1}^{1} \sin(\xi + 1)\, d\xi$$

---

[5]M. Abramowitz and I. Stegun, *Handbook of Mathematical Functions*, National Bureau of Standards, reprinted by Dover, New York, 1965.

n	i	x_i	ω_i
2	1	0.5773502692	1.0
3	0	0.0	0.8888888889
	1	0.7745966692	0.5555555556
5	0	0.0	0.5688888889
	1	0.5384693101	0.4786286705
	2	0.9061798459	0.2369268850
10	1	0.1488743390	0.2955242247
	2	0.4333953941	0.2692667193
	3	0.6794095683	0.2190863625
	4	0.8650633667	0.1494513492
	5	0.9739065285	0.0666713443
20	1	0.0765265211	0.1527533871
	2	0.2277858511	0.1471729865
	3	0.3737060887	0.1420961093
	4	0.5108670020	0.1316886384
	5	0.6360536807	0.1181945320
	6	0.7463319065	0.1019301198
	7	0.8391169718	0.0832767416
	8	0.9122344283	0.0626720483
	9	0.9639719273	0.0416014298
	10	0.9931285992	0.0176140071

Table 7.3 *Sampling points, x_i, and weights, ω_i, for Gaussian quadrature*

by Gaussian quadrature. The result for the $n = 5$ point calculation is

$$I \approx \omega_0 f(0) + \omega_1[f(x_1) + f(-x_1)] + \omega_2[f(x_2) + f(-x_2)]$$

$$= 0.569 \sin(1.0) + 0.237[\sin(0.906 + 1) + \sin(-0.906 + 1)]$$

$$+ 0.479[\sin(0.538 + 1) + \sin(-0.538 + 1)]$$

$$= 1.4161467 \quad [\text{Exact answer} = 1 - \cos(2) = 1.4161468\ldots]$$

Once again, the weights and sampling points of the Gaussian quadrature method of order n are $2n$ parameters that were determined by requiring that the algorithm be *exact* when applied to polynomials of degree $N \leq 2n - 1$. A very important generalization of this idea is to next require the procedure to be *exact* when applied to functions of the form $g(x)p_N(x)$, where $p_N(x)$ is an arbitrary polynomial of degree N as before, and $g(x)$ is any particular function you wish, as long as it has no free parameters. Thus, whereas the Gaussian quadrature method can be expected to give accurate results insofar as the integrand resembles a polynomial of degree N or less, the generalized algorithm will yield accurate results to the degree that the integrand resembles $g(x)p_N(x)$, even including possible singularities in $g(x)$. The most common of these generalizations of the Gaussian quadrature method are listed in Table 7.4. Also, in this context, the ordinary Gaussian quadrature is then called *Gauss-Legendre* quadrature. Each class of approximation has a different set of weights and sampling points, but all are evaluated as simply

$$\int_a^b g(x)f(x)\,dx = \sum_{i=1}^n \omega_i f(x_i)$$

Integral type	$g(x)$	$[a,b]$	Method name
$\displaystyle\int_{-1}^1 f(x)\,dx$	1	$(-1,1)$	Gauss-Legendre
$\displaystyle\int_{-1}^1 \frac{f(x)\,dx}{\sqrt{1-x^2}}$	$(1-x^2)^{-1/2}$	$(-1,1)$	Gauss-Chebyshev
$\displaystyle\int_0^\infty x^q e^{-x} f(x)\,dx$	$x^q e^{-x}$	$(0,\infty)$	Gauss-Laguerre
$\displaystyle\int_{-\infty}^\infty e^{-x^2} f(x)\,dx$	e^{-x^2}	$(-\infty,\infty)$	Gauss-Hermite

Table 7.4   *Common Gaussian quadrature algorithms*

It is no accident that the names associated with these procedures are also the names associated with some of the *special* polynomials—Legendre, Chebyshev, Laguerre, and Hermite polynomials. It turns out that the sampling points for each method are simply the zeros of the corresponding polynomials. (Thus, the sampling points for the $n = 2$ Gauss-Legendre method are $\pm 1/\sqrt{3}$, and these are likewise the roots of the second-order Legendre polynomial.)

You can find tables of the weights and sampling points for each of these classes of Gaussian quadrature, as well as others, in M. Abramowitz and I. Stegun, *Handbook of Mathematical Functions*, National Bureau of Standards, reprinted by Dover, New York, 1965. A script file that compares the various numerical integration methods of this chapter, including Gaussian quadrature, is included on the applications disk and is named `Ingl Demo.m`.

Improper Integrals

An integral is said to be *improper* if: (a) one or both of the limits are infinite, or (b) the integrand becomes unbounded ("blows up") for any values of the integration variable within the interval of integration. In both cases, the occurrence of infinity in either of the limits or the integrand may cause the integral to diverge. As with infinite summations, there are a variety of tests that can be used to determine whether or not an improper integral converges. And, just as with summations, even if you can prove that an improper integral does indeed converge and, therefore, represents a definite numerical value, it is often a difficult task to accurately obtain that value. Infinities in any form will cause problems in a computational algorithm; therefore, special techniques must be developed for these cases.

INTEGRALS WITH INFINITE LIMITS

In order for the integral

$$I = \int_a^\infty f(x)\,dx$$

to have meaning, that is to converge, clearly the function $f(x)$ must approach zero for large x. In fact, because

$$\int_a^\infty x^p = \left.\frac{x^{p+1}}{p+1}\right|_a^\infty$$

converges at the upper limit only if $p + 1 < 0$, we can conclude that the function $f(x)$ must approach zero *faster* than $1/x$. That is, $\lim_{x\to\infty} xf(x) = 0$. Similar consideration of the lower limit of the integral

$$I = \int_0^b f(x)\,dx$$

leads to the requirement that $f(x)$ must approach ∞ *more slowly* than $1/x$, or $\lim_{x\to 0} xf(x) = 0$. Once you have established that the integral is convergent, the most obvious method for its evaluation is similar to the technique used in evaluating an infinite summation. Simply integrate the function to a large, but finite, point chosen in such a way that you are sure that the remainder of the integral will only contribute an insignificant amount. If the integrand is largest at the beginning of the interval, one procedure is to break the interval $[a, \infty]$ into *four* parts, say

$$I = \int_a^\gamma f(x)\,dx + \int_\gamma^{3\gamma} f(x)\,dx + \int_{3\gamma}^{9\gamma} f(x)\,dx + \int_{9\gamma}^\infty f(x)\,dx$$

$$= I_1 + I_2 + I_3 + I_4$$

The value of γ is chosen so that the interval $[a, \gamma]$ contains the dominant part of the integral; that is, the region where $|f(x)|$ is the largest. If the sequence of inequalities

$$I_1 \gg I_2 \gg I_3$$

is satisfied, and I_3 is very small, we are probably justified in neglecting the last term, I_4, which cannot be integrated anyway. The first three integrals can be computed numerically by any of the methods discussed thus far.

TRANSFORMATIONS THAT REMOVE SINGULARITIES IN THE LIMITS

If you are uncomfortable with the indefiniteness of the partitioning method we have suggested, there is a better way; but it will require that you determine the asymptotic form of the integrand as it approaches the infinite limit. For example, if the integral is

$$\int_a^\infty f(x)\,dx$$

and $f(x) \to x^{-p}$ as $x \to \infty$, you would seek a transformation of variables that would map $x^{-p}\,dx \to d\xi$. That is, in order for the transformed integral to remain finite at the upper limit, $\int^c d\xi$, the upper limit itself must be finite. In our example, for $p = 2$, the transformation would then be $x = -1/\xi$, and

$$\int_a^\infty f(x)\,dx = \int_{-1/a}^0 f\left(\frac{-1}{\xi}\right)\frac{d\xi}{\xi}$$

Notice that we have traded the problem of an infinite upper limit for an integrand that is apparently indeterminate at $\xi = 0$. However, because the asymptotic form of f is known, the integrand can be explicitly evaluated at $\xi = 0$, or an open-type algorithm can be used.

SINGULARITIES WITHIN THE INTERVAL OF INTEGRATION

Even though a function $f(x)$ is infinite for some value of x either at the ends of the interval or within the interval, the integral may still be finite. The integral

$$I = \int_0^1 \frac{dx}{\sqrt{1-x^2}}$$

can be shown to be finite,[6] but clearly, the singularity at $x = 1$ will present problems. In addition to using an open-type algorithm, it is sometimes possible to find a change of variables that renders the singularity finite. For this integral, the transformation $x = \sin\theta$ will do the trick (and the integral is easily determined to be $\pi/2$). Trigonometric substitutions are usually useful when the singularity is due to terms like $(1-x^2)^{-p}$. However, it may not be possible to avoid the singularity by any simple change of variables. If the integral is indeed convergent, you might try to replace the integrand by its asymptotic form near the singularity, then explicitly (by hand!) integrate the function over a narrow interval containing the singularity, and then handle the rest of the interval numerically.

---

[6]Considering just the upper limit, the integral becomes

$$\int_*^1 \frac{dx}{\sqrt{1-x^2}} \Rightarrow \frac{1}{2}\int_*^1 \frac{dx}{\sqrt{1-x}} \to -\frac{1}{2}\int_*^0 \frac{d\xi}{\sqrt{\xi}} = \text{finite}$$

Problems

REINFORCEMENT EXERCISES

Code M-file functions for each of the functions in Table 7.5. These functions will then be used in the problems that follow.

No.	$f(x)$	Limits	Exact result
1	xe^{-x}	$0 \leq x \leq 1$	$1 - 2/e$
2	$x \sin x$	$0 \leq x \leq \pi/2$	1
3	$(1 + x^2)^{3/2}$	$0 \leq x \leq 1$	$1.56795196\ldots$
4	e^{-x^2}	$0 \leq x \leq 1$	$0.74684204\ldots$
5	$\sin x \tan x$	$0 \leq x \leq 1$	$0.38472019\ldots$

Table 7.5   *Test functions for closed-type integration*

P7.1 The Trapezoidal Rule. Write a MATLAB M-file function TRAP(a,b,k,fnc) that will return a *vector*, **T**, of the trapezoidal rule results for $\int_a^b fnc(x)\,dx$. The vector **T** will contain $k+1$ values obtained by using Equation 7.1. Test this routine by evaluating the integral of one or more of the given functions.

P7.2 Simpson's Rule. Write a MATLAB M-file function SIMP(a,b,k,fnc) that will return a *scalar*, **S**, containing the Simpson's rule results for $\int_a^b fnc(x)\,dx$. Use Equation 7.3 for $n = 2^n$ panels. Test this routine by evaluating the integral of one or more of the given functions.

P7.3 Using Trapezoidal Values to Compute Simpson's Rule. Use the values computed using the trapezoidal rule in Problem 7.1, along with Equation 7.4, to obtain the Simpson's rule values. Compare your results with those found in Problem 7.2.

P7.4 Romberg Integration. Construct a MATLAB function that will take a column vector, **T**, of elements representing the value of an integral computed using the trapezoidal rule with intervals successively reduced by half, as in Problem 7.1, and return a column vector, **r**, of the same length containing the elements along the diagonal of a Romberg table. Compare the accuracy of the trapezoidal rule calculation with the full Romberg integration for each of the test integrals.

P7.5 Integration Using quad or quad8. Evaluate each of the given functions using the toolbox functions quad or quad8, and compare your results with the "exact" results.

P7.6 Three-Point Gaussian Quadrature. Derive the values of the weights and sampling points for a *three-point* Gaussian quadrature.

P7.7 Singular Integrals.   The integral

$$\int_0^1 \ln(x)\, dx$$

has a singularity as $x \to 0$. Find the change of variable that allows the integral to be written as

$$-\int_0^\infty t e^{-t}\, dt$$

Describe how this integral might be evaluated numerically. Numerically evaluate the integral by implementing your proposed technique, and compare the result with the exact value of -1.

P7.8 Ten-Point Gaussian Quadrature.   Use the values in Table 7.3 to code an M-file function GAUSS10(a,b,fnc) that will perform a ten-point Gaussian quadrature evaluation of the integral of a function, $fnc(x)$, from $x = a$ to $x = b$. Test this function by evaluating each of the integrals in Table 7.6. Note that each integral has a singular point at one of the limits, and thus, closed-type integration algorithms cannot be used.

No.	$f(x)$	Limits	Exact result
1	$\sqrt{\ln \dfrac{1}{x}}$	$0 \le x \le 1$	$\dfrac{\sqrt{\pi}}{2}$
2	$x \log(\sin x)$	$0 \le x \le \pi$	$-\frac{1}{2}\pi^2 \ln 2$
3	$\dfrac{1}{x(1 - \ln x)\sqrt{\ln(1/x)}}$	$0 \le x \le 1$	π

Table 7.6   *Test functions for singular intervals*

P7.9 Infinite Limits.   The integrals of the functions in Table 7.7 can be estimated numerically by segmenting the integration interval as described on page 199. Use any appropriate closed-type integration algorithm and evaluate these integrals.

No.	$f(x)$	Limits	Exact result
1	$x e^{-x}$	$0 \le x < \infty$	$\dfrac{\sqrt{\pi}}{2}$
2	$x^2 e^{-x} \cos x$	$0 \le x < \infty$	$-\frac{1}{2}$
3	$\dfrac{1}{(1 + x)\sqrt{x}}$	$0 \le x \le 1$	π

Table 7.7   *Test functions for integrals with infinite limits*

Summary

A demonstration program that compares several of the methods discussed in this chapter is included on the program disk and is named `Ingl Demo.m`.

TRAPEZOIDAL RULE

Description

The area under a function $f(x)$ over the interval $[a, b]$ is approximated as the sum of areas of trapezoids of width $\Delta x_k = (b - a)/2^k$ and height of sides $f(x_i)$ and $f(x_{i+1})$.

Basic Equations

$$n = 2^k \qquad \Delta x_k = (b - a)/n$$

$$k = 0 \qquad T_0 = \tfrac{1}{2}\Delta x_0(f_a + f_b)$$

$$T_k = \tfrac{1}{2}T_{k-1} + \Delta x_k \sum_{\substack{i=1 \\ \text{odd only}}}^{n-1} f(a + i\Delta x_k)$$

SIMPSON'S RULE

Description

The area under a function $f(x)$ over the interval $[a, b]$ is approximated as the sum of the areas of the parabolas that are fitted to sets of three consecutive points.

Basic Equation

$$S_k = \frac{1}{3}\Delta x_2 \left[f_a + 4 \sum_{\substack{i=1 \\ \text{odd only}}}^{n-1} f(a + i\Delta x_k) + 2 \sum_{\substack{i=2 \\ \text{even only}}}^{n-2} f(a + i\Delta x_k) + f_b \right]$$

Relation to Trapezoidal Rule

$$S_k = T_k + \frac{T_k - T_{k-1}}{3}$$

ROMBERG INTEGRATION

Description

Fitting the integration points to higher- and higher-order polynomials leads to an extrapolation procedure that relates the trapezoidal rule to Simpson's rule, Simpson's rule to Boole's rule, etc., leading to an algorithm that is the optimum polynomial replacement for a given number of points. This is known as Romberg integration.

Basic Equation

$$T_k^{m+1} = T_k^m + \frac{T_k^m - T_{k-1}^m}{4^{m+1} - 1} \qquad m = 0, \ldots, k - 1$$

GAUSSIAN QUADRATURE

Description

An integral is estimated by evaluating the function at *nonequally spaced* sampling points, x_i, over the interval $[-1, 1]$ and by "weighting" the contribution of each point by a factor ω_i. The weights and sampling points are chosen to optimize the accuracy of the integrals of polynomial integrands of varying degree.

Basic Equation $\displaystyle\int_{-1}^{1} f(x)\,dx \approx \omega_0 f(0) + \sum_{i=1}^{(n-1)/2} \omega_i[f(x_i) + f(-x_i)]$ for odd n

$$\approx \sum_{i=1}^{n/2} \omega_i[f(x_i) + f(-x_i)]$$ for even n

IMPROPER INTEGRALS

Singularities in the Limits

The integral $\int_a^\infty f(x)\,dx$ is approximately evaluated by partitioning the infinite integration range into four or more segments in such a way that the sequence of contributions from the segments is decreasing, and suggestive that the final infinite segment contributes only an infinitesimal amount and can be neglected.

Singularities in the Integrand

First, it must be shown analytically that the integration of the singularity is indeed finite. The contribution from the region near the singular point is then usually estimated by replacing the integrand by its asymptotic form near the singularity.

MATLAB Functions Used

quad

The basic MATLAB numerical integration function, **quad**, uses a Simpson's rule algorithm and adjusts the step size until a specified accuracy is obtained. The use is **q = quad('*fname*',a,b)**, where *fname* is a string containing the name of the **M**-file function representing the integrand, and **a** and **b** are the limits. The default accuracy tolerance is 10^{-3}. A different tolerance can be used by including it as the fourth element in the argument list. If a nonzero fifth item is included in the argument list, a trace of the points used in evaluating the integral will be plotted. The function will also return a count of the function evaluations. **[q,count] = quad(...)**.

If, in addition to the independent variable, the integrand function requires additional parameters in its argument list, these may be included in the argument list of quad after the fifth position. Thus, the most general use would be

$$[\text{q}, \text{count}] = \text{quad}('\textit{fname}', \text{a}, \text{b}, \text{tol}, 1, \text{alpha}, \text{beta}, ...)$$

It is essential that the integrand function, *fname*, be structured so that it will accept a *vector* of input values, $\vec{x}$, and it will return a vector, $\vec{f}$, of output values.

quad8

This integration function uses a higher-order Newton-Cotes algorithm and it is used in precisely the same way as **quad**. It can be expected to more accurately handle integrands whose derivatives are singular within the range of the integral. Neither **quad** nor **quad8** will integrate singular integrals.

It is instructive to compare the execution times and number of function calls of these two functions for an integral like $\int_0^1 \sqrt{x}\,dx$.

Bibliography for Part II

1. M. Abramowitz and I. Stegun, *Handbook of Mathematical Functions*, National Bureau of Standards, reprinted by Dover, New York, 1965.

2. Forman S. Acton, *Numerical Methods That Work*, Harper & Row, New York, 1970.

3. A. C. Aitken, "On Bernoulli's Solution of Algebraic Equations," *Proc. Roy. Soc. Edinburgh*, vol. 46, 1926.

4. G. J. Borse, *Fortran-77 and Numerical Methods*, 2nd ed., PWS Publishers, Boston, MA, 1991.

5. B. Carnahan, H. Luther, and J. Wilkes, *Applied Numerical Methods*, John Wiley, New York, 1969.

6. S. D. Conte and C. de Boor, *Elementary Numerical Analysis*, McGraw-Hill, New York, 1972.

7. G. Dahlquist and A. Björck, *Numerical Methods*, Prentice-Hall, Englewood Cliffs, NJ, 1974.

8. P. J. Davis and P. Rabinowitz, *Methods of Numerical Integration*, Academic Press, New York, 1975.

9. R. W. Hamming, *Numerical Methods for Scientists and Engineers*, McGraw-Hill, New York, 1962.

10. F. B. Hildebrand, *Introduction to Numerical Analysis*, McGraw-Hill, New York, 1974.

11. S. Kuo, *Computer Applications of Numerical Methods*, Addison-Wesley, Reading, MA, 1971.

12. A. Miele, *Flight Mechanics*, vol. 1, Addison-Wesley, Reading, MA, 1962.

13. W. H. Press, B. P. Flannery, S. A. Teukolsky, and W. T. Vetterling, *Numerical Recipes, The Art of Scientific Computing*, Cambridge University Press, Cambridge, England, 1986.

14. J. Stoer and R. Bulirsch, *Introduction to Numerical Analysis*, Springer-Verlag, Berlin and New York, 1980.

15. D. M. Young and R. T. Gregory, *A Survey of Numerical Mathematics*, vols. I and II, Addison-Wesley, Reading, MA, 1972.

PART III / Functions of Two or More Variables*

PREVIEW   The addition of a second or a third independent variable into a problem dramatically complicates the analysis and lengthens the execution times of most algorithms by at least an order of magnitude. However, although the numerical procedures are more complicated, the ideas and the algorithms are usually rather transparent generalizations of the procedures of Part II. Therefore, as in Chapter 4, we will begin with the basic properties of multivariable functions, partial differentiation and series expansion, and make use of special MATLAB functions for producing contour graphs and three-dimensional perspective drawings of functions of two variables. These will be important for the heuristic descriptions of the application procedures that follow.

Next, we describe the topics of root solving and minimization of multivariable functions. The procedures we develop are reminiscent of the methods developed for single variable functions. That is, to evaluate a root of a function, $f(x, y)$, we begin by first finding a region in which the function changes sign, and then systematically reduce the size of the region. To locate a minimum, we again follow the function "downhill" until a valley is reached.

However, as we shall see, the difficulties introduced in going from one to two dimensions can be substantial. For example, in one dimension, two points will bracket a root or a minimum; in two dimensions, an *area* in the xy plane will be required. Once the boundary region for the root or minima is found, how is it systematically reduced in size? The capacity of MATLAB's easy-to-use three-dimensional graphics to aid in the visualization of these questions will prove to be invaluable.

The final topic is the numerical evaluation of multiple integrals. The procedures to evaluate multidimensional integrals are likewise derived from the various one-

---

*This part makes use of MATLAB programming techniques described in Appendix Sections A.i through A.v.

dimensional algorithms of Chapter 7, and they are particularly concise and transparent when written in MATLAB. A final chapter in this part discusses Monte Carlo integration, or the random number evaluation of integrals, particularly multidimensional integrals. This method is very popular for integrals with awkward boundaries, or if a relatively quick and approximate result is desired. Monte Carlo techniques will also be useful in Part IV in the analysis of data and in curve fitting.

INTRODUCTION

The mathematics of functions of a single variable is the main concern of most texts in numerical analysis, and understandably so. These functions can be represented as two-dimensional curves on a page, and very intuitive explanations relating to their derivatives, integrals, roots, minima/maxima, and other properties can be given. The properties of functions of two (or more) independent variables are much less intuitive. Two-dimensional curves are replaced by surfaces in three-dimensional space; concepts of tangent lines are replaced by tangent planes; area integrals are replaced by volume integrals; the intersection of two, three-dimensional functions becomes a two-dimensional curve, not a single point. Nevertheless, the calculus and numerical analysis of multivariable functions are extremely important areas of concern for the practicing engineer or scientist. The analysis and understanding of a great many real problems require at least two independent variables.

The thermodynamics of simple gases provides probably the simplest and most familiar examples of phenomena in which at least two independent variables are required. The workhorse of the theory is the simplified equation of state of an ideal gas, $PV = nRT$, which relates the three variables, pressure P (N/m^2), volume V (m^3), and temperature T (K). Any two of these variables may be considered as independent variables, and the third is then a dependent variable determined by the ideal gas equation, e.g., $P = P(V, T)$. Real systems of course have more complicated equations of state, which may include additional independent variables. Thermodynamics defines a variety of measurable quantities that relate to *processes*; such as expansion at constant pressure, or heating at constant volume, etc., in the form of *partial derivatives*; e.g., $(\partial V/\partial T)_P$, and seeks to find functions that represent quantities like the energy or the entropy as functions of several variables. Thermodynamics is usually classified as one of the most difficult of the basic sciences to master, and much of this difficulty is traceable to the fact that, to understand even the simplest phenomena, we are forced to think in more than two dimensions. Similar considerations apply to many other fields as well: fluid dynamics or heat flow in two or three dimensions, solid mechanics (stress-strain) in two or three dimensions, and many others. Proceeding from problems with one to two independent variables almost always requires a substantial expenditure of effort and a much deeper understanding of the problem.

Yet, to understand problems at the cutting edge of most areas of science and engineering, a proficiency in multivariable calculus and numerical analysis is essential. MATLAB can be used to great advantage to facilitate this understanding. The toolbox functions for instantly creating perspective three-dimensional graphs and contour plots will be helpful in providing the intuitive feel for these functions. We will generalize the procedures of Chapters 4 through 7 for single variable functions to multivariable functions in a mostly heuristic manner.

8 / *Derivatives of Multivariable Functions*

PREVIEW   A substantial part of the numerical analysis of single variable functions stems from the basic definition of the derivative:

$$f'(x) = \lim_{\Delta x \to 0} \left[\frac{f(x + \Delta x) - f(x)}{\Delta x} \right]$$

Newton's method for roots, most of the procedures we will see for solving differential equations, Taylor series expansions, and many other procedures begin by building approximation methods based on this simple equation. It is, therefore, reasonable to expect that similar numerical procedures for functions of two or more variables will likewise rely heavily on the definition of the derivative of $f(x, y)$.

The intuitive definition of the derivative of a function of a single variable at a point relates to the slope of the tangent line at that point. In multidimensions, tangent lines are replaced by tangent *planes*, and it becomes more difficult to visualize the derivative. If the function depends on two variables, $f(x, y)$, it can be represented by a surface. Then, the two-dimensional derivative at a point, (x, y), will be interpreted in terms of the tangent plane to the surface at that point.

The algebraic definition of the derivative of a function of a single variable relates to the changes in f as a result of infinitesimal changes in x. For a function of two variables, the generalized derivative will determine the changes in $f(x, y)$ resulting from infinitesimal changes in x, or in y, or both. These considerations lead to the definition of *partial derivatives*, which we discuss first.

The combination of two independent partial derivatives, one with respect to x, the other with respect to y, as the components of a *vector* is used to define the *vector* derivative of a function, called the *gradient* of f, designated by ∇f. In this chapter we also describe MATLAB functions for computing and for visualizing the gradient.

Partial Derivatives

In a straightforward generalization of the ordinary derivative, the definition of the derivative of a function of two or more variables is obtained by applying the definition of the ordinary derivative to one variable at a time. Thus, the *partial derivative* of the function $f(x, y)$ with respect to the variable x is defined to be equal to the ordinary derivative of $f(x, y)$ with respect to x, treating all other variables as if they were constant:

$$\frac{\partial}{\partial x} f(x,y) \equiv \left. \frac{df(x,y)}{dx} \right|_{y=\text{const}} \equiv \lim_{\Delta x \to 0} \frac{f(x + \Delta x, y) - f(x,y)}{\Delta x}$$

The partial derivative with respect to y is defined analogously. These two derivatives can then be used to determine the point at which a function of two variables has an extremum. For example, the maximum or minimum of a function, $f(x)$, of a single variable is obtained by finding the value of x at which the tangent line has zero slope, or where $df/dx = 0$. To find the maximum of a function of *two* variables, $f(x,y)$ (see Figure 8.1), we could proceed by *fixing* $y = y_c$ so that the function $f(x, y_c)$ is now a function of a single variable and corresponds to the "slice" of the function drawn at say, $y = 3$. The maximum of this curve in space is obtained by finding the point where $\partial f/\partial x|_{y_c} = 0$. We could proceed, similarly, with slices of the function along the x direction. That is, $\partial f/\partial y|_{x_c} = 0$ specifies the point along the slice at $x = x_c$ at which the function has zero slope in the x direction. Figure 8.2 illustrates two perpendicular slices of the function of Figure 8.1.

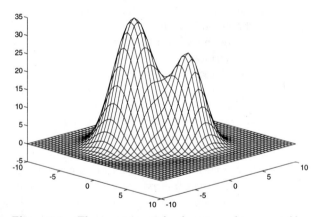

Figure 8.1   *The maximum of a function of two variables.*

Thus, the two *simultaneous* equations,

$$\frac{\partial f}{\partial x} = 0 \qquad \frac{\partial f}{\partial y} = 0$$

specify a pair of values (x, y) at which the slopes of tangent lines to the function are zero in both the x and y directions. Any combination of these two tangent lines will be a line lying in a *plane* that is parallel to the xy plane. The point where the tangent plane to the function is horizontal corresponds to a *critical* point of the function; that is, a *minimum*, a *maximum*, or a *saddle point*.[1] As

---

[1] A saddle point is a point where there is a minimum or maximum along one directional slice and the opposite, i.e., a maximum or minimum, in the perpendicular direction.

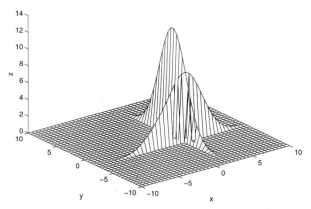

Figure 8.2 *Slices of the function of Figure 8.1 at $y = -2$ and $x = 2$.*

with single variable functions, higher derivatives can be used to distinguish the three possibilities. It is shown in calculus texts that if a quantity Δ is defined as

$$\Delta = \left[\left(\frac{\partial^2 f}{\partial x^2} \right) \left(\frac{\partial^2 f}{\partial y^2} \right) - \left(\frac{\partial^2 f}{\partial x \partial y} \right)^2 \right] \Bigg|_{(x_0, y_0)}$$

where (x_0, y_0) is the point where the first-order partial derivatives are zero, then the point (x_0, y_0) is a:

Maximum   if $\Delta > 0$ and either $\dfrac{\partial^2 f}{\partial x^2}\bigg|_{(x_0,y_0)} < 0$ or $\dfrac{\partial^2 f}{\partial y^2}\bigg|_{(x_0,y_0)} < 0$

Minimum   if $\Delta > 0$ and either $\dfrac{\partial^2 f}{\partial x^2}\bigg|_{(x_0,y_0)} > 0$ or $\dfrac{\partial^2 f}{\partial y^2}\bigg|_{(x_0,y_0)} > 0$

Saddle point   if $\Delta < 0$

If $\Delta = 0$, we do not have enough information, and higher-order derivatives must be used to determine the nature of the critical point.

If the partial derivatives, $\partial f / \partial x$ and $\partial f / \partial y$, exist in a neighborhood surrounding the point (x, y), they are, in turn, functions of x and y and can be differentiated, leading to second-order partial derivatives.

$$\frac{\partial}{\partial x}\left(\frac{\partial f}{\partial x}\right) = \frac{\partial^2 f}{\partial x^2} \qquad \frac{\partial}{\partial x}\left(\frac{\partial f}{\partial y}\right) = \frac{\partial^2 f}{\partial x \partial y}$$

$$\frac{\partial}{\partial y}\left(\frac{\partial f}{\partial x}\right) = \frac{\partial^2 f}{\partial y \partial x} \qquad \frac{\partial}{\partial y}\left(\frac{\partial f}{\partial y}\right) = \frac{\partial^2 f}{\partial y^2}$$

And again, if these partial derivatives exist in a neighborhood of (x, y), third-order derivatives can be defined. Further, it can be shown that if $\partial^2 f / \partial x \partial y$ and $\partial^2 f / \partial y \partial x$ are both *continuous*, the order of the differentiation is immaterial; i.e., $\partial^2 f / \partial x \partial y = \partial^2 f / \partial y \partial x$.

EXAMPLE 8.1

Partial Derivatives

The function

$$V(x,y) = \frac{q}{4\pi\epsilon_0}\frac{1}{|\vec{r}|}, \qquad \frac{1}{4\pi\epsilon_0} = 9 \times 10^9 \text{ N} \cdot \text{m}^2/\text{C}^2 \qquad (8.1)$$

represents the electric potential at a position $\vec{r} = x\hat{i} + y\hat{j}$ of a charge q located at the origin. Using $|\vec{r}| = \sqrt{x^2 + y^2}$, the derivatives of this function are

$$\frac{\partial V(x,y)}{\partial x} = -\frac{q}{4\pi\epsilon_0}\frac{x}{(x^2+y^2)^{3/2}} = -\frac{q}{4\pi\epsilon_0}\frac{x}{r^3}$$

$$\frac{\partial V(x,y)}{\partial y} = -\frac{q}{4\pi\epsilon_0}\frac{y}{(x^2+y^2)^{3/2}} = -\frac{q}{4\pi\epsilon_0}\frac{y}{r^3}$$

The electric field, $\vec{E}(x,y)$, is a vector defined at the position (x,y), and it is related to the electric potential by

$$E_x = -\frac{\partial V}{\partial x} \qquad E_y = -\frac{\partial V}{\partial y} \qquad (8.2)$$

Finally, the force on a second stationary charge, q_s, due to the presence of the first charge, q, is given by $\vec{F} = q_s\vec{E}$. Thus, knowing the functional form of the electric potential allows us to determine the force on stationary charges by means of the partial derivatives of the electric potential. (However, because a moving charge creates a *magnetic field* as well, problems involving the motion of charges are considerably more difficult.) ●

The Gradient

The infinitesimal change in a function $f(x,y)$ as a result of incrementing x by $x \rightarrow x + dx$, while holding y fixed, is

$$df_1 = \left.\frac{df(x,y)}{dx}\right|_{y=\text{const}} dx = \frac{\partial f(x,y)}{\partial x} dx \qquad [y = \text{constant}]$$

and the corresponding change in f caused by incrementing y by $y \rightarrow y + dy$ is

$$df_2 = \left.\frac{df(x,y)}{dy}\right|_{x=\text{const}} dy = \frac{\partial f(x,y)}{\partial y} dy \qquad [x = \text{constant}]$$

Thus, the change in the function as a result of an arbitrary *vector* displacement, $\vec{r} \rightarrow \vec{r} + d\vec{r}$, is

$$df = \frac{\partial f}{\partial x} dx + \frac{\partial f}{\partial y} dy$$

which can be interpreted as the *dot product* of two vectors:

$$df = \nabla f \cdot d\vec{r} \qquad (8.3)$$

where ∇f is a vector with components $(\partial f/\partial x, \partial f/\partial y)$. This combination is known as the *gradient* derivative of f, and it has very important physical significance.

For example, a function is constant along the contour curves of a contour map. Thus, if $d\vec{r}$ is nonzero, but along a contour line, $df = 0$, this implies that the direction of the gradient, ∇f, must be perpendicular to the contour line. Further, because the direction of $d\vec{r}$ that *maximizes* the quantity $|df|$ corresponds to $d\vec{r}$ parallel to ∇f, we conclude that the direction of the gradient points in the direction of greatest change in the function f. The gradient specifies the direction of the line of steepest ascent or descent of the function. Obviously, this interpretation of the gradient derivative will play a significant role in any algorithm to find the minimum/maximum of a function.

Using the definition of the gradient derivative, we can write the relationship of Example 8.1 between the electric potential $V(x, y)$ and the electric field $\vec{E}(x, y)$ as

$$\vec{E} = -\nabla V \tag{8.4}$$

The direction of the electric field will be perpendicular to the lines of constant potential, and the minus sign means that it will point "downhill."

THE MATLAB FUNCTIONS GRADIENT AND QUIVER

Using small, but finite, values for Δx and Δy, approximate values for the partial derivatives can be computed at each point. The MATLAB function to perform this task is called **gradient**. Its use in computing the electric field from the electric potential V is:

```
[Ex,Ey] = gradient(-V,dx,dy)
```

where V is a matrix of the values of the potential at specific grid points, and dx, dy are the spacings of the grid in the x and y directions, respectively. The function returns *two* matrices of the same size as V, representing the partial derivatives; in this case, $-\partial V/\partial x$ and $-\partial V/\partial y$. If the spacings are omitted, as in gradient(V), a spacing of $+1$ is assumed for both dx and dy.

The MATLAB function to represent this result on a graph is called quiver. Its use is

```
quiver(X,Y,Ex,Ey)
```

where X, Y are the matrices of the grid points and Ex and Ey are the x and y components of tiny arrows that are drawn at each grid point. A fourth parameter, S, can also be added to "scale" the size of the arrow. That is, $S = 2$ will double the length of the arrows, and $S = 0.5$ will reduce them by one-half. The function quiver is used most often in conjunction with a contour plot to indicate not only the lines of constant contour, but also the direction of steepest ascent.

EXAMPLE 8.2

The Electric Field Due to a Dipole of Charges

An electric dipole consists of two charges of equal but opposite charge separated by a distance d. The electric potential at a point is then simply the sum of the potentials from the individual charges. If the charges are positioned with $+q$ at (a, b) and $-q$ at $(-a, -b)$, the electric potential is

$$V(x,y) = \frac{q}{4\pi\epsilon_0}\left(\frac{1}{r_+} - \frac{1}{r_-}\right) \tag{8.5}$$

with

$$r_+ = \sqrt{(x-a)^2 + (y-b)^2}$$

$$r_- = \sqrt{(x+a)^2 + (y+b)^2}$$

The following MATLAB code is used to compute the electric potential for the dipole and, from it, the electric field by use of the function `gradient`. In addition, the potential function is plotted on a contour graph, and the electric field arrows are added using the function `quiver`.

```
%DIPOLE script file
%     The potential field from an electric dipole is plotted as
%     a contour graph, and the Electric Field, the gradient of
%     the electric potential, is represented using the function
%     quiver.
%===============================================================
  q   = 2e-6;   k = 9e9;   a = 1.5;   b = -1.5;
  x = -10:1:10;       y = x;
  [X,Y] = meshgrid(x,y);
  rp = sqrt((X-a).^2+(Y-b).^2);
  rm = sqrt((X+a).^2+(Y+b).^2);
   V = -q*k*(rp.^(-1) - rm.^(-1));
   i = find(V> 5000); V(i) =  5000*ones(length(i),1);
   i = find(V<-5000);0V(i) = -5000*ones(length(i),1);
  [Ex,Ey] = gradient(-V,1,1);
   E = sqrt(Ex.^2+Ey.^2);
    Ex = Ex./E;          % All E-vectors will
    Ey = Ey./E;          % be the same length.
  contour(X,Y,V)
  title('The field (arrors) and the potential of a dipole')
  axis('square')
  hold on
  plot(a,b,'o',-a,-b,'o');plot(a,b,'+',-a,-b,'-')
  quiver(X,Y,Ex,Ey)
  hold off
```

In this MATLAB code, we have truncated the values of the potential near the singularities at $\pm(x_0, y_0)$, so that $|V| < 5000$. This can be seen more clearly by plotting a perspective plot using `mesh(X,Y,V)`. Also, because the size of the arrows produced by `quiver` that represent the electric field will vary by several orders of magnitude, we have instead plotted only the direction of $\vec{E}$ by normalizing the $\vec{E}$ vectors at each point. The result is shown in Figure 8.3.

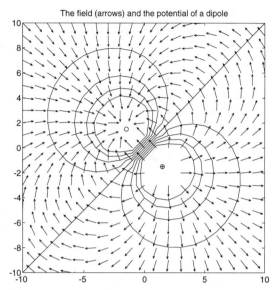

Figure 8.3 *The electric potential of a dipole. The direction of the electric field is indicated by the arrows.*

Notice how the direction of the electric field is always perpendicular to each contour line of the potential. ●

EXAMPLE 8.3

Heat Flow

The flow of heat is basically a *diffusion* process, wherein there is a transfer of energy from regions of high temperature to regions of low temperature. That is, if $T(x, y)$ is a surface representing a temperature distribution over some region expressed in terms of x and y, then heat flows downhill on this surface, meaning that the heat flow will be along the direction of $-\nabla \vec{T}$.

For example, if a solid metal bar of length ℓ and radius r has one end held at a fixed temperature T_{hot}, and the other end held at T_{cold}, there will be a constant flow of heat energy through the bar in a direction from the hot end to the cold end. The *rate* of heat flow in the bar, dQ/dt, is found experimentally to be proportional to the temperature difference $|T_{\text{hot}} - T_{\text{cold}}|$ and the cross-sectional area of the bar, A, and inversely proportional to the length of the bar, ℓ. The proportionality constant is called the *thermal conductivity*, k, and is characteristic of the particular metal of which the bar is made. These results are summarized in the *Fourier law of heat conduction*:

$$\frac{dQ}{dt} = -kA\frac{\Delta T}{\ell} \qquad (8.6)$$

The generalization of this result to an arbitrary geometry is

$$\frac{dQ}{dt} = -k(\delta A)\nabla T \cdot \hat{n} \qquad (8.7)$$

where dQ/dt is now the rate of heat energy flowing per second through an infinitesimal surface of area, δA, and $\hat{n}$ is a unit vector normal to that surface. Thus, $\nabla T \cdot \hat{n}$ is a measure of the temperature difference *across* that surface element.

Consider next the flow of heat in two dimensions. If a hot steam pipe of radius a and temperature $T_a = 200°C$ is surrounded by a sheath of insulation of thickness δ, and the outside radius, $b = a + \delta$, of the insulation is found to have a constant temperature of $T_b = 65°C$, the *rate* of heat flowing out of the pipe per second is a constant. If we then imagine a cylindrical surface drawn with a radius $a < r < b$ and length ℓ, the total heat flow through this surface must equal Q. The symmetry of the problem demands that the heat flow in a direction radially outward, so this determines the direction of the gradient of T. That is, $\nabla T = (dT/dr)\hat{r}$. And because the area of the cylinder is $2\pi r \ell$, and the normal to this surface points radially outward; i.e., is simply $\hat{r}$, Fourier's law states that

$$\frac{dQ}{dt} = -k(2\pi r\ell)\frac{dT}{dr}\hat{r} \cdot \hat{r}$$

or, because $\hat{r} \cdot \hat{r} = 1$,

$$dT = -\left(\frac{dQ/dt}{2\pi k\ell}\right)\frac{dr}{r}$$

so that

$$T(r) = T_a - \left(\frac{dQ/dt}{2\pi k\ell}\right)\ln(r/a)$$

Using the fact that $T(b) = T_b$, the constant term in parentheses can be eliminated to yield

$$T(r) = T_a\left[1 - \left(1 - \frac{T_b}{T_a}\right)\frac{\ln(r/a)}{\ln(b/a)}\right] \tag{8.8}$$

The distribution of the temperature in the insulation obtained from this equation is shown in Figure 8.4.

The components of the gradient of $T(x, y)$ are

$$\frac{\partial T}{\partial x} = \frac{dT}{dr}\frac{\partial r}{\partial x}$$

$$= -\left(\frac{T_a - T_b}{\ln(a/b)}\right)\frac{x}{r^2}$$

and

$$\frac{\partial T}{\partial y} = -\left(\frac{T_a - T_b}{\ln(a/b)}\right)\frac{y}{r^2}$$

where the term in parentheses is positive, indicating that the negative gradient of T points radially outward. Thus, the direction of the heat flowing out of the pipe is also radially outward. The temperature contours and the direction of the heat flow, as indicated by the gradient of the temperature, are illustrated in Figure 8.5. ●

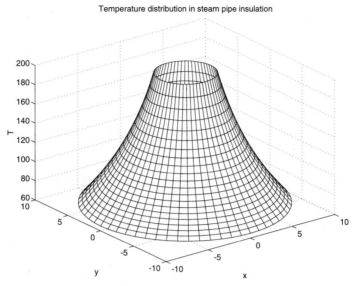

Figure 8.4 *The temperature in the insulation surrounding a steam pipe. The pipe radius is 3 cm and the insulation thickness is 7 cm.*

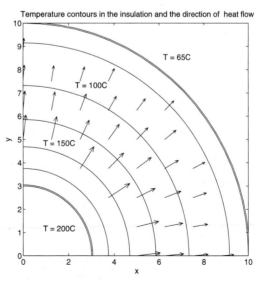

Figure 8.5 *Temperature contours in the insulation. The negative gradient of the temperature is indicated by the arrows and represents the direction of the heat flow.*

THE NORMAL TO A SURFACE

A very common structure in applied mathematics involves the flow of a quantity through a surface. In the heat flow example, the amount of heat flowing through an infinitesimal surface element δA was proportional to $\delta A \nabla T \cdot \hat{n}$ where $\hat{n}$ was the unit normal vector to the surface element. The total heat flowing through a finite surface is the integral of this quantity over the entire surface area.

The normal vector to a surface, $\hat{n}$, can itself be determined by using the gradient derivative as follows. A surface in three dimensions is specified by an equation of the form $z = f(x, y)$. If we think of this instead as $g(x, y, z) = f(x, y) - z = 0$, it can be thought of as the equation for the $g = 0$ contour in a four-dimensional space. Furthermore, the gradient of g is perpendicular to the contours of g, so that the direction of the normal to the surface $g(x, y, z) = 0$ is proportional to ∇g. Rewriting ∇g in terms of $z = f(x, y)$ and dividing by the length of the gradient, we obtain the expression for the normal to a surface $f(x, y)$ at a point:

$$\nabla g = \hat{i}\frac{\partial f}{\partial x} + \hat{j}\frac{\partial f}{\partial y} - \hat{k}$$

$$\hat{n} = \frac{\hat{i}\frac{\partial f}{\partial x} + \hat{j}\frac{\partial f}{\partial y} - \hat{k}}{\sqrt{\left(\frac{\partial f}{\partial x}\right)^2 + \left(\frac{\partial f}{\partial y}\right)^2 + 1}} \tag{8.9}$$

The Tangent Plane   Note that if the normal to the surface $z = f(x, y)$ at a point (x_0, y_0) is given by Equation 8.9, the expression for the *tangent* plane to the surface at the point (x_0, y_0) is simply the collection of points that satisfy the relation

$$\nabla g\big|_0 \cdot [(x - x_0)\hat{i} + (y - y_0)\hat{j} + (z - z_0)\hat{i}] = 0$$

or in terms of the function $f(x, y)$,

Equation for the tangent plane at a point (x_0, y_0)
$$z = z_0 + \frac{\partial f}{\partial x}\bigg|_0 (x - x_0) + \frac{\partial f}{\partial y}\bigg|_0 (y - y_0)$$

Taylor Series in Two Dimensions

The generalization of the one-dimensional Taylor series is obtained by replacing the operation $\left(\Delta x \frac{d}{dx}\right) f(x)$ by $\left(\Delta x \frac{\partial}{\partial x} + \Delta y \frac{\partial}{\partial y}\right) f(x, y)$ to obtain

$$f(x, y) = f(x_0, y_0) + \left(\Delta x \frac{\partial}{\partial x} + \Delta y \frac{\partial}{\partial y}\right) f(x, y)\bigg|_{x_0, y_0}$$

$$+ \frac{1}{2!} \left(\Delta x \frac{\partial}{\partial x} + \Delta y \frac{\partial}{\partial y}\right)^2 f(x, y)\bigg|_{x_0, y_0}$$

$$+ \cdots + \frac{1}{n!} \left(\Delta x \frac{\partial}{\partial x} + \Delta y \frac{\partial}{\partial y}\right)^n f(x, y)\bigg|_{x_0, y_0} + R_n \tag{8.10}$$

where

$$R_n = \frac{1}{(n+1)!} \left(\Delta x \frac{\partial}{\partial x} + \Delta y \frac{\partial}{\partial y} \right)^{n+1} f(x,y) \bigg|_{\xi,\eta} \qquad (8.11)$$

and (ξ, η) is an unknowable point somewhere in the intervals $x_0 \leq \xi \leq x_0 + \Delta x$, $y_0 \leq \eta \leq y_0 + \Delta y$, with $\Delta x = x - x_0$ and $\Delta y = y - y_0$. In this notation, the derivative in the second-order term corresponds to

$$\left(\Delta x \frac{\partial}{\partial x} + \Delta y \frac{\partial}{\partial y} \right)^2 f(x,y) \bigg|_{x_0,y_0} \equiv \left(\Delta x^2 \frac{\partial^2}{\partial x^2} + 2\Delta x \Delta y \frac{\partial^2}{\partial x \partial y} + \Delta y^2 \frac{\partial^2}{\partial y^2} \right) f(x,y) \bigg|_{x_0,y_0}$$

As an example of a two-dimensional Taylor series expansion, consider the function $V(x,y)$ for the potential of a dipole given in Equation 8.5. Expanding this function about the point $(0,0)$, we find that

$$V(0,0) = 0 \qquad \frac{\partial V}{\partial x}\bigg|_0 = \frac{Q}{4\pi\epsilon_0} \frac{2a}{(a^2+b^2)^{3/2}} \qquad \frac{\partial^2 V}{\partial x^2}\bigg|_0 = 0$$

$$\frac{\partial V}{\partial y}\bigg|_0 = \frac{Q}{4\pi\epsilon_0} \frac{2b}{(a^2+b^2)^{3/2}} \qquad \frac{\partial^2 V}{\partial y^2}\bigg|_0 = 0$$

$$\frac{\partial^2 V}{\partial y \partial x}\bigg|_0 = 0$$

Thus, the approximate expression for the potential near the origin is

$$V(x,y) \approx \frac{2Q}{4\pi\epsilon_0} \frac{(ax+by) - (a^2+b^2)}{(a^2+b^2)^{3/2}}$$

As with the one-dimensional Taylor series, it can be shown that, if the series converges at a point (x,y) *and* the remainder term $R_n \to 0$ as $n \to 0$, the series is a unique representation of the function.

For our purposes in numerical analysis, the Taylor series will be most useful as an approximation of the function in a small disk surrounding a point. That is, we will mostly be concerned with the linear terms in the expansion. In this regard, it is often illuminating to rephrase the expansion in a vector notation using the gradient derivative. If the location of the point (x,y) is characterized by the vector $\vec{r}$, then the first two terms of the Taylor series expansion about a point located by the vector $\vec{a}$ can be written as

$$f(\vec{r}) = f(\vec{a}) + (\vec{r} - \vec{a}) \cdot \nabla f(\vec{a}) + \cdots \qquad (8.12)$$

where $\nabla f(\vec{a})$ is the gradient of f evaluated at the position $\vec{a}$.

The Taylor series, in either one- or two-dimensional form, will prove to be very useful in the numerical techniques that follow.

Problems

REINFORCEMENT EXERCISES

P8.1 The Gradient Derivative. Show that if f is a function of r only, where $r = \sqrt{x^2 + y^2}$, then $\nabla f(r) = f'(r)\hat{r}$, where $\hat{r} = (x\hat{i} + y\hat{j})/r$.

P8.2 The Optimum Water Trough. A water trough is constructed from a single piece of sheet metal by bending the sides upward at an angle θ. The geometry is shown in Figure 8.6.

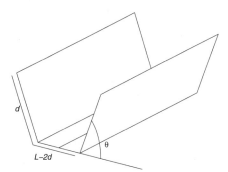

Figure 8.6 *A water-carrying trough formed from sheet metal.*

The width, L, of the sheet metal is 30 cm, and the variables d and θ are to be chosen so as to maximize the carrying capacity; i.e., to maximize the cross-sectional area. This area is given by

$$A(d, \theta) = (L - 2d)d\sin\theta + d^2\sin\theta\cos\theta$$

Determine the optimum values of d and θ.

P8.3 Normals to Surfaces. For each of the following functions, determine the general expression for the normal, $\hat{n}$, to the surface and the expression for the normal at the given point.
  a. $x^2 + y^2 + z = 0$ at $(x_0, y_0) = (0, 0)$
  b. $x^2 - y^2 + z = 0$ at $(x_0, y_0) = (0, 2)$
  c. A cone of height 2, upper radius of 2, and lower radius of zero; at the point $(x_0, y_0) = (1, 1)$
  d. A sphere of radius $r = 2$ at the point $(x_0, y_0) = (1, 1)$

P8.4 Tangent Planes. Show that the tangent planes to the following functions at the indicated points are parallel to the xy plane.
  a. $z^2 = (x^2 + 2y^2)^3$ at $(0, 0)$
  b. $z = (x - 1)^4 - 2(x - 1)(y + 1) + (y + 1)^4$ at $(1, -1)$
  c. $z = xy$, at $(0, 0)$

P8.5 Tangent Planes to the Function peaks. The toolbox function `peaks` can be used to plot a set of several hills and valleys. The statement, `[X,Y,Z] = peaks(N)` will produce $N \times N$ matrices for X, Y, Z which may then be plotted using `mesh`. The graph of this function has numerous minima and maxima. One of these is located at $(x_0, y_0) = (-0.0093196, +1.581368)$. Use each of the functions `contour`, `gradient`, and `quiver` to plot contour curves surrounding this point. Include on the graph a collection of arrows indicating the gradient near the point.

P8.6 Heat Flow in a Sphere. A solid spherical shell has its inner radius, $r = a$, held at a temperature T_a and its outer radius, $r = b$, held at a temperature $T_b < T_a$.
   a. From symmetry, the temperature in the sphere can only be a function of r. Thus, $\partial T / \partial x = (dT/dr)(\partial r / \partial x)\hat{r}$, etc. Show that $|\nabla T| = dT/dr$.
   b. Use the fact that the rate of heat (dQ/dt) flow through any sphere of radius $a \le r \le b$ is a constant, and the fact that the area of a sphere of radius r is $A = 4\pi r^2$, to determine $T(r)$.

P8.7 Taylor Series in Two Dimensions. Obtain the expansion of

$$f(x, y) = \sin(x - y) \cos(x + y)$$

through terms of order x^3 and y^3. Next, use the trigonometric identities

$$\sin(x - y) = \sin x \cos y + \cos x \sin y$$
$$\cos(x + y) = \cos x \cos y - \sin x \sin y$$

and the Taylor series expansions of $\sin \theta$ and $\cos \theta$ to verify your two-dimensional Taylor series.

P8.8 Electric Potential for a Quadrupole. A distribution of *four* charges is called a quadrupole. If the charges $+q$, $-q$, $+q$, and $-q$ are positioned at the corners of a square centered at the origin with sides of length $2a$, the electric potential at a point (x, y) is given by

$$V(x, y) = \frac{q}{4\pi\epsilon_0} \left[\frac{1}{\sqrt{(x - a)^2 + (y - a)^2}} - \frac{1}{\sqrt{(x + a)^2 + (y - a)^2}} \right.$$

$$\left. + \frac{1}{\sqrt{(x + a)^2 + (y + a)^2}} - \frac{1}{\sqrt{(x - a)^2 + (y + a)^2}} \right]$$

Use the values $q/4\pi\epsilon_0 = 1$ and plot the function $V(x, y)$ over the range $|x|, |y| \le 2$ first for $a = 1$. After computing the matrix containing the values of V, you should eliminate the singularities before plotting by using statements such as

```
i = find(V>=5);   V(i) = 5*ones(length(i),1);
i = find(V<=-5);  V(i) = -5*ones(length(i),1);
```

Use `meshc` to get a contour plot, as well as the graph of V. Next compute the gradient of V using `gradient`. Execute a contour plot with the gradients added by using the function `quiver`. Finally, execute the `quiver` plot alone to represent the direction of the electric field at each point on the plane. [Use `dx = dy = 0.2` in `gradient`, and a scaling factor of 2 in `quiver`.]

Summary

DERIVATIVES OF FUNCTIONS OF TWO VARIABLES

Partial Derivatives

By considering one variable at a time, the function $f(x, y)$ is viewed in terms of functional *slices* of a single variable. The ordinary one-dimensional derivative is then generalized to the partial derivative of the two-dimensional function as

$$\frac{\partial}{\partial x} f(x, y) \equiv \frac{df(x, y)}{dx}\bigg|_{y=\text{const}} \equiv \lim_{\Delta x \to 0} \frac{f(x + \Delta x, y) - f(x, y)}{\Delta x}$$

The Gradient

The total change in the function $f(x, y)$ as a result of infinitesimal changes in x and y is

$$df = \frac{\partial f}{\partial x}\, dx + \frac{\partial f}{\partial y}\, dy$$

which can be interpreted as the dot product of two vectors,

$$df = \nabla f \cdot \vec{dr}$$

where ∇f is a vector with components $(\partial f/\partial x, \partial f/\partial y)$ and is called the *gradient* of $f(x, y)$. The direction of the gradient indicates the path along which f is most rapidly increasing.

The Normal to a Surface

If $\hat{\imath}$, $\hat{\jmath}$, and $\hat{k}$ are the Cartesian unit vectors, the normal to a surface, $z = f(x, y)$, at a point (x_0, y_0) is given by Equation 8.9.

$$\hat{n} = \frac{\hat{\imath}\frac{\partial f}{\partial x} + \hat{\jmath}\frac{\partial f}{\partial y} - \hat{k}}{\sqrt{\left(\frac{\partial f}{\partial x}\right)^2 + \left(\frac{\partial f}{\partial y}\right)^2 + 1}}$$

where the derivatives are evaluated at (x_0, y_0).

The Tangent Plane

The plane tangent to a function $f(x, y)$ at a point $z_0 = f(x_0, y_0)$ is given by the equation

$$z = z_0 + \frac{\partial f}{\partial x}\bigg|_{x_0, y_0} (x - x_0) + \frac{\partial f}{\partial y}\bigg|_{x_0, y_0} (y - y_0)$$

The Two-Dimensional Taylor Series

The generalization of the Taylor series in two or more dimensions gives the following expression for the linear approximation to a function of two variables:

$$f(x, y) = f(x_0, y_0) + \frac{\partial f}{\partial x}\bigg|_{x_0, y_0} (x - x_0) + \frac{\partial f}{\partial y}\bigg|_{x_0, y_0} (y - y_0) + \cdots$$

MATLAB Functions Used

gradient

Will compute approximate partial derivatives of a function of two variables using central difference expressions. The use is [gx, gy] = gradient(z,dx,dy), where z is the matrix of function values over the grid defined by vectors x and y, and dx and dy are usually scalars characteristic of the spacing along those directions. Also, ordinary derivatives can be computed by Yderv = gradient(y,dy) if a single vector is used as the input along with the spacing of points.

`quiver`   Will draw small arrows on a two-dimensional grid, most often on a contour plot. The usual use is `quiver(x,y,dx,dy)`, where `x` and `y` are *matrices* designating the (x, y) positions at which to place the arrows on the grid, and `dx` and `dy` are matrices of the same size that determine the length and position of each arrow, and are often computed using `gradient`.

9 / Roots of Multivariable Functions

PREVIEW   Finding the zeros of a function in two or more dimensions is substantially more difficult than finding the roots of a single variable function. In addition, the problem can have different meanings. For example, the "zeros" of the function $f(x, y)$ will, if f is continuous, correspond to a closed trajectory in x and y and can be represented by the $f = 0$ contour on a contour plot. Or, the problem of finding the intersection of two functions, $f(x, y)$ and $g(x, y)$, is equivalent to finding the zeros of $F(x, y) = f(x, y) - g(x, y) = 0$ and will in turn yield a curve in space $y = y(x)$ as the zero contour of F.

A more common problem is that of finding the point (x_0, y_0) that represents the root of *two* simultaneous equations $f(x, y) = 0$ and $g(x, y) = 0$, with obvious generalizations to more than two dimensions. This problem may be viewed as a search for the intersection point or points of the two zero contours of f and g, and suggests the increased degree of difficulty in finding roots of simultaneous functions of several variables.

However, once the neighborhood of such a root has been established, a simple variation of Newton's method can be used to quickly converge to the value of the root.

To familiarize the reader with the techniques to be used, we first examine the features of a well-known function of two variables, the van der Waals equation of state for an imperfect gas.

Zero Contours of a Function of Two Variables

THE VAN DER WAALS EQUATION OF STATE

The van der Waals equation of state for one mole of an imperfect gas can be written in the form

$$p(v, t) = \frac{\frac{8}{3}t}{\left(v - \frac{1}{3}\right)} - \frac{3}{v^2} \tag{9.1}$$

where the pressure (P), volume (V), and temperature (T) have been written in terms of *scaled* variables, p, v and t, defined by

$$p = P/P_c \qquad v = V/V_c \qquad t = T/T_c$$

where P_c, V_c, T_c are the values of the variables at the critical point of the gas. The critical point is a unique set of the variables at which equal masses of the liquid and the vapor phase have equal densities. Critical point values are tabulated for all common gases.

Thus, the solution for the set of v and t values that will yield a pressure of $p = p_0$ is equivalent to finding the zeros of the function

$$f(v, t) = \frac{\frac{8}{3}t}{\left(v - \frac{1}{3}\right)} - \frac{3}{v^2} - p_0 = 0$$

or

$$t = \frac{3}{8}\left(p_0 + \frac{3}{v^2}\right)\left(v - \frac{1}{3}\right) \tag{9.2}$$

INVERSE FUNCTIONS

Quite frequently, using similar algebraic manipulations, we can solve the equation for the "zeros" of a function, $f(x, y) = 0$, to yield one variable as a function of the other, for example, $y = g(x)$. This completely solves the original problem, giving a curve in the xy plane on which the function f is zero. However, if you need the inverse of this function; i.e., $x = g^{-1}(y)$, you often face a formidable problem.[1] For example, in Equation 9.2, it is ordinarily more useful to be able to specify the temperature and compute the associated volume; i.e., have a relation of the form $v = v(t)$. One approach for evaluating $g^{-1}(y)$ that is effective for a limited range of y values is to utilize a Taylor series for g^{-1}; that is,

$$x = x_0 + (g_0^{-1})'(y - y_0) + \frac{1}{2}(g_0^{-1})''(y - y_0)^2 + \cdots \tag{9.3}$$

where the derivatives are evaluated at $x = x_0$. Then, because

$$(g_0^{-1})' = \frac{dx}{dy} = \frac{1}{y_0'} \qquad (g_0^{-1})'' = \frac{d}{dx}(y_0')^{-1} = -\frac{y_0''}{(y_0')^2}$$

Equation 9.3 can be rewritten as

$$x = x_0 + \frac{(y - y_0)}{y_0'} - \frac{1}{2}\frac{y_0''(y - y_0)^2}{(y_0')^2} + \cdots \tag{9.4}$$

For example, the function $y = g(x) = e^x$ can be inverted about the point $x = 0$ using $y_0 = y_0' = y_0'' = \cdots = 1$, to yield

$$x = (y - 1) - \frac{1}{2}(y - 1)^2 + \cdots$$

---

[1]Of course, if you plot the curve as y versus x, you need only rotate the graph by $90°$ to obtain x versus y. Even though this statement sounds a bit gratuitous, it does illustrate the fact that for a great many functions there is a difficulty in maintaining *single valuedness* in the inverse function. Thus, $y = \sin x$ is a well-behaved, single valued function for all x, whereas $x = \sin^{-1} y$ can only be defined for a range of y values such that $|y| \leq 1$.

which should match the expansion for the actual inverse function, $x = \ln(y)$ about the point $y = 1$. Of course, because

$$\ln(1 + \xi) = \xi - \frac{1}{2}\xi^2 + \frac{1}{3}\xi^3 + \cdots \pm \frac{1}{n}\xi^n + \cdots$$

by letting $y = 1 + \xi$, we see that at least the first few terms of the expansion are correct. Unfortunately, the higher derivatives of g^{-1} are progressively more complicated, limiting the utility of this approach to values of y very close to y_0.

NUMERICAL PROCEDURE FOR FINDING ZERO CONTOURS

Another, more numerical, approach is to track the function downhill by following the negative of the gradient derivative (or the reverse, if you start where $f(x, y) < 0$). Once you have found a single point on the $f = 0$ contour, move along the contour by taking steps perpendicular to the gradient, provided the gradient is not also zero. To construct the code to execute this algorithm, you will need to have separately coded, *three* distinct M-file functions for $f(x, y)$, $\partial f / \partial x$, and $\partial f / \partial y$.

Starting at a point (x_0, y_0), let the magnitude of the step be Δr. Then, if the function is positive, a step of this size is taken along the direction of $-\nabla f$. The components of the vector step are then determined as:

$$\Delta f = \sqrt{\left(\frac{\partial f}{\partial x}\right)^2 + \left(\frac{\partial f}{\partial y}\right)^2}$$

$$x_1 = x_0 - \left(\frac{\partial f}{\partial x}\right)\left(\frac{\Delta r}{\Delta f}\right)$$

$$y_1 = y_0 - \left(\frac{\partial f}{\partial y}\right)\left(\frac{\Delta r}{\Delta f}\right)$$

where the partial derivatives are evaluated at the point (x_0, y_0).

If the function is negative at (x_0, y_0), the step is along the direction of $+\nabla f$. This procedure is continued until the function changes sign. At that point, similar to the bisection method of Chapter 5, the stepping direction is reversed, and the step size is interpolated to approximate the position of the zero crossing.

Once the position of $f(x, y) = 0$ has been determined to a tolerance ϵ (i.e., $\Delta r < \epsilon$), you proceed to fill out the zero contour by now stepping *perpendicular* to the gradient (using a new, much larger step $\Delta \vec{c}$). The components of the perpendicular step vector satisfy the following equations:

$$(\Delta c)_x \frac{\partial f}{\partial y} - (\Delta c)_y \frac{\partial f}{\partial x} = 0$$

$$(\Delta c)_x^2 + (\Delta c)_y^2 = |\Delta \vec{c}|^2$$

or

$$\theta = \tan^{-1}\left[\left(\frac{\partial f}{\partial y}\right) \bigg/ \left(\frac{\partial f}{\partial x}\right)\right]$$

$$(\Delta c)_x = \Delta c \sin\theta$$

$$(\Delta c)_y = \Delta c \cos\theta$$

Every few steps you will have to rezero the function to make sure that you haven't stepped off the zero contour. In the simplest situations, this algorithm will successfully trace out the zero contour of a continuous function $f(x, y)$. However, the complete code must be prepared to handle a variety of potential problems, such as:

1. The function never reaches a zero value. It continues to climb or fall towards $\pm\infty$. Just as with functions of a single variable, the code must be written so as to avoid a runaway search.
2. If the original stepping search encounters a local minimum or maximum, it will be trapped in the depression or on the hump. So, in each step, the code will have to keep track of the direction of the search (either uphill or downhill), and if the direction changes and the function *does not* change sign, the search must be abandoned, and a new starting point used.

A MATLAB code that employs the preceding algorithm to trace out the zero contour of a function can easily be constructed. However, in most circumstances, we rarely need to know the values of a zero contour to such precision. Therefore, we do not provide in-depth examples of such applications, except to close this discussion with an important caveat regarding numerical methods in more than one dimension. As you can now see, when designing numerical procedures that apply to functions of several variables, it is frequently very difficult to visualize what precisely is going on or to select an optimum or even suitable starting point for the procedure. Intuition is even more important than in one-dimensional problems, and either a rough contour plot, a perspective graph, or both are usually essential to achieve a successful solution to the problem.

Simultaneous Nonlinear Functions

A different interpretation of "zeros" of multivariable functions involves the problem of finding the solutions of simultaneous equations of the form

$$f(x, y) = 0$$
$$g(x, y) = 0$$

If the functions f and g are linear in x and y, these are then simply matrix equations, and the methods of Chapter 2 can be used to easily obtain an answer. If, however, the equations are nonlinear, matrix methods are of no use, and, generally, the problem is very much more difficult to solve.

The fundamental problem is that the two functions f and g are, in general, totally independent. The two equations (or more in the generalized problem) define the zero contours of two independent functions, and we are looking for the individual points of intersection of these contour curves. The two curves may have many points of intersection, or they may have none. Unlike linear equations, where n independent equations are guaranteed to have n unique solutions, there is no corresponding statement for nonlinear functions that specifies how many, if any, intersection points there are. The only recourse is to first establish an approximate idea of the location of a solution by means of simultaneous contour plots of the functions. Once you know there is a solution and its approximate location, very efficient methods can then be used to converge on the result. These are based on a generalization of Newton's method in one dimension.

NEWTON'S METHOD IN TWO DIMENSIONS

After deciding on a starting point (x_0, y_0) known to be in the vicinity of the actual solution, both functions $f(x, y)$ and $g(x, y)$ are expanded in a Taylor series about this point. (See Equation 8.10.)

$$f(x, y) = f_0 + \left.\frac{\partial f}{\partial x}\right|_0 (x - x_0) + \left.\frac{\partial f}{\partial y}\right|_0 (y - y_0) + \cdots$$

$$g(x, y) = g_0 + \left.\frac{\partial g}{\partial x}\right|_0 (x - x_0) + \left.\frac{\partial g}{\partial y}\right|_0 (y - y_0) + \cdots$$

where the zero subscript indicates that the functions and all partial derivatives are to be evaluated at the point (x_0, y_0).

Then, by setting both f and g equal to zero, the two equations determine the values of x and y corresponding to the roots. As with Newton's method in one dimension, we truncate the series after the linear terms to obtain

$$\begin{pmatrix} \left.\frac{\partial f}{\partial x}\right|_0 & \left.\frac{\partial f}{\partial y}\right|_0 \\ \left.\frac{\partial g}{\partial x}\right|_0 & \left.\frac{\partial g}{\partial y}\right|_0 \end{pmatrix} \begin{pmatrix} x - x_0 \\ y - y_0 \end{pmatrix} \approx \begin{pmatrix} f_0 \\ g_0 \end{pmatrix}$$

Solving these equations for (x, y), we obtain

$$\begin{pmatrix} x \\ y \end{pmatrix} \approx \begin{pmatrix} x_0 \\ y_0 \end{pmatrix} - \begin{pmatrix} \left.\frac{\partial f}{\partial x}\right|_0 & \left.\frac{\partial f}{\partial y}\right|_0 \\ \left.\frac{\partial g}{\partial x}\right|_0 & \left.\frac{\partial g}{\partial y}\right|_0 \end{pmatrix}^{-1} \begin{pmatrix} f_0 \\ g_0 \end{pmatrix} \tag{9.5}$$

This equation can easily be generalized to n functions in n variables.

Next, in the spirit of the *secant method*, we approximate the partial derivatives as

$$\left.\frac{\partial f}{\partial x}\right|_0 \approx \frac{f(x_0 + \Delta x, y_0) - f(x_0, y_0)}{\Delta x} \equiv \frac{f_{\Delta x} - f_0}{\Delta x}$$

or

$$\begin{pmatrix} x \\ y \end{pmatrix} \approx \begin{pmatrix} x_0 \\ y_0 \end{pmatrix} + \begin{pmatrix} \Delta x' \\ \Delta y' \end{pmatrix}$$

where

$$\begin{pmatrix} \Delta x' \\ \Delta y' \end{pmatrix} = -\begin{pmatrix} \frac{f_{\Delta x} - f_0}{\Delta x} & \frac{f_{\Delta y} - f_0}{\Delta y} \\ \frac{g_{\Delta x} - g_0}{\Delta x} & \frac{g_{\Delta y} - g_0}{\Delta y} \end{pmatrix}^{-1} \begin{pmatrix} f_0 \\ g_0 \end{pmatrix}$$

One step of the algorithm consists of replacing the two starting points:

$$\begin{pmatrix} x_0 \\ y_0 \end{pmatrix} \begin{pmatrix} x_1 = x_0 + \Delta x_0 \\ y_1 = y_0 + \Delta y_0 \end{pmatrix} \Rightarrow \begin{pmatrix} x_1 \\ y_1 \end{pmatrix} \begin{pmatrix} x_2 = x_1 + \Delta x_1 \\ y_2 = y_1 + \Delta y_1 \end{pmatrix}$$

so that

$$\begin{pmatrix} \Delta x_1 \\ \Delta y_1 \end{pmatrix} = -\begin{pmatrix} \frac{f_{\Delta x} - f_0}{\Delta x} & \frac{f_{\Delta y} - f_0}{\Delta y} \\ \frac{g_{\Delta x} - g_0}{\Delta x} & \frac{g_{\Delta y} - g_0}{\Delta y} \end{pmatrix}^{-1} \begin{pmatrix} f_0 \\ g_0 \end{pmatrix} - \begin{pmatrix} \Delta x_0 \\ \Delta y_0 \end{pmatrix} \qquad (9.6)$$

The algorithm to implement this method is summarized as follows.

THE ALGORITHM FOR A TWO-DIMENSIONAL SECANT METHOD

1. Determine an initial guess, (x_0, y_0), for the root of the two simultaneous equations that is in the vicinity of a *known* root.
2. Select a *step*, $(\Delta x_0, \Delta y_0)$, that is likely to take you towards the root.
3. Specify a *tolerance*, ϵ, that will be used to terminate the search—when $|\Delta \vec{x}| < \epsilon$ the search will be ended.
4. Write M-file functions for the two functions $f(x, y)$ and $g(x, y)$.
5. `for iter = 1:imax`

   (a) Evaluate $(f_0, g_0), (f_{\Delta x}, g_{\Delta x})$, and $(f_{\Delta y}, g_{\Delta y})$.
   (b) Compute Δx_1 and Δy_1 using Equation 9.6.
   (c) `if` $|\Delta \vec{x}| < \epsilon$
         `return;`
      `else`
         replace $(x_0, y_0) \leftarrow (x_0 + \Delta x_0, y_0 + \Delta y_0)$
                  $(\Delta x_0, \Delta y_0) \leftarrow (\Delta x_1, \Delta y_1)$
      `end`

The MATLAB code for a two-dimensional secant method is given in Figure 9.1. For example, to find a root of the two functions

$$f(x, y) = x - \sin(x + y)$$
$$g(x, y) = y - \cos(x - y)$$

```
function x0 = SECNT2d(f,g,x,dx,tol)
%SECNT2d(f,g,x,dx,tol) uses a secant method generalized
%  to 2-d to find a root of the two equations f(x,y)=0
%  and g(x,y) = 0, starting from the initial guess, x.
%  Also needed is a guess, dx, for an interval contain-
%  ing the root. Both x and dx are 2-element column
%  vectors. The two functions names are contained in f
%  and g, and the maximum number of iterations is 20.
%  The calculation terminates if |dx| < tol.  The root
%  is returned in x0.
%==========================================================
for i = 1:20
    x0 = x;               f0 = [feval(f,x0) feval(g,x0)];
    x0 = x + [dx(1) 0]; fx = [feval(f,x0) feval(g,x0)];
                          fx =  (fx-f0)/dx(1);
    x0 = x + [0 dx(2)]; fy = [feval(f,x0) feval(g,x0)];
                          fy = (fy-f0)/dx(2);
    A = [fx;fy]+eps;
    x = x + dx;
    dx= (-A\f0'-dx')';
    if (norm(dx) < tol ); return; end;
end
disp('still outside tolerance after')
disp('20 iterations, abs(dx) =')
disp(norm(dx))
pause
```

Figure 9.1 *MATLAB code to implement the secant method in two dimensions.*

we might begin by using a contour plot to search for a starting point for the root. Contours of these two functions drawn at values of the functions at 0 and ± 0.1 are illustrated in Figure 9.2. From the graph, we estimate that a reasonable starting point for the search for the root of the functions is near the point $\vec{x}_0 = (x, y) = (0.9, 0.9)$. Using a step, $\vec{dx} = (dx, dy) = (0.1, 0.1)$, and a convergence tolerance tol = 1e-5, the MATLAB code of Figure 9.1 returns the root at $(x_r, y_r) = (0.935195\ldots, 0.997654\ldots)$ in 17 iterations. You will notice that the convergence is not nearly as rapid as it was for the one-dimensional secant method.

THREE OR MORE SIMULTANEOUS EQUATIONS

By far the most difficult step in solving three nonlinear equations in three unknowns is the search to find an appropriate starting point. Thus, the solution of the equations

$$f_1(x, y, z) = 0$$
$$f_2(x, y, z) = 0$$
$$f_3(x, y, z) = 0$$

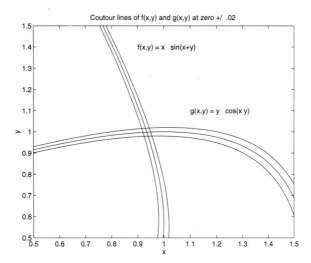

Figure 9.2 *Intersecting zero contours of two functions indicates a root.*

can usually be obtained smoothly and efficiently using a generalization of Newton's method; i.e.,

$$
\begin{pmatrix} x \\ y \\ z \end{pmatrix} \approx \begin{pmatrix} x_0 \\ y_0 \\ z_0 \end{pmatrix} - \begin{pmatrix} \frac{\partial f_1}{\partial x} & \frac{\partial f_1}{\partial y} & \frac{\partial f_1}{\partial z} \\ \frac{\partial f_2}{\partial x} & \frac{\partial f_2}{\partial y} & \frac{\partial f_2}{\partial z} \\ \frac{\partial f_3}{\partial x} & \frac{\partial f_3}{\partial y} & \frac{\partial f_3}{\partial z} \end{pmatrix}^{-1} \Bigg|_0 \begin{pmatrix} f_{1,0} \\ f_{2,0} \\ f_{3,0} \end{pmatrix}
$$

once you have a starting point.

Unfortunately, there is no general prescription for finding the vicinity of a root, or even the existence of a root for three or more simultaneous equations. To appreciate the problem, recall that an equation like $f(x, y, z) = 0$ defines a *surface* in space, and so in searching for a root of three simultaneous equations, you are looking for a single intersection point of *three* surfaces. Some less-than-satisfactory strategies might be:

- Fix one of the variables, say $z = z_c$, and draw simultaneous zero contours of the three functions. Concentrate on points where two of the curves intersect. Slightly alter z and see if the third curve moves towards or away from these points. Clearly, this method has the potential for great frustration.
- If any of the three functions can be rewritten to express one of the variables as a function of the other two; i.e., $f_1(x, y, z) = 0 \to z = z(x, y)$, then the remaining two equations can, at least formally, be written as functions of two variables: $f_2(x, y, z(x, y))$, $f_3(x, y, z(x, y))$, and the methods for two simultaneous equations can be used.
- If physical considerations suggest a solution to the problem should lie near a point (x_0, y_0, z_0), the search can be often greatly accelerated.

Without help from intuition or some other hint as to the vicinity of a root, the solution of three or more simultaneous nonlinear equations is very likely to be hopeless.

Problems

REINFORCEMENT EXERCISES

P9.1 Zero Contours. Find at least one of the zero contours of the following functions:

a. $f(x, y) = \dfrac{2 \sin x (\sin x^2 \cos y^2 - \cos x^2 \sin y^2)}{x^2 + y^2}$

b. $f(x, y) = \sin(x - y) \cos(x + y)$

c. $\dfrac{x^4 - y^4}{x^2 + y^2} + (y + 1)(x + y)$

P9.2 Intersection of Zero Contours. If V = [-0.2 0 0.2], contour(x,y,z,V) will draw contours at $z = 0 \pm 0.2$, thus banding the zero contour. Use this idea to find the zero contours of the first of the two functions given. Next, freeze this graph on the screen by using the **plot** command **hold on**, and duplicate the operation on the second function. Thus, determine the regions where the zero contours of the two functions intersect.

a. $f(x, y) = \sin(x - y);$ $g(x, y) = \cos(x + y)$

b. $f(x, y) = \dfrac{\sin(\sqrt{2x^2 + y^2})}{\sqrt{x^2 + 2y^2}}$

$g(x, y) = \ln\left(\dfrac{x^2 + y^4}{8}\right)$

P9.3 Roots of Two Simultaneous Functions. The roots of the functions in Problem 9.2 are near the points a. $(1, 1)$ and b. $(2.0, 1.5)$, respectively. Using these as the initial guess for the roots, find the actual values accurate to five significant figures using **SECNT2d**.

EXPLORATION PROBLEMS

P9.4 Temperature Increase in a Semi-infinite Slab. A semi-infinite medium is at a uniform temperature T_0, and at time $t = 0$ a constant heat flux, Q, begins to flow into the material at $x = 0$. The temperature will increase in the slab as a function of position, $x > 0$, and of time, $t > 0$, according to the equation

$$T(x, t) = T_0 + \frac{Qx}{k}\left[\frac{e^{-s^2}}{\sqrt{\pi}s} - \text{erfc}(s)\right]$$

where k is the material's thermal conductivity, $s^2 = x^2/4\alpha t$, α is the material's thermal diffusivity, and **erfc(s)** is a MATLAB toolbox function that corresponds to the integral

$$\text{erfc}(s) \equiv \frac{2}{\sqrt{\pi}} \int_s^\infty e^{-s^2}\, ds$$

Use the following parameters in this problem:

	Quantity	Value	Units
k	conductivity	0.015	$J/(s \cdot m°C)$
α	diffusivity	2.5×10^{-5}	m^2/s
x	position in slab	$0 \le x \le .25$	m
t	time	$0 \le t \le 120$	s
T_0	initial temperature	20	$°C$
Q	heat flux	200	$J/m^2 s$

Now suppose that a chemical reaction will take place when the temperature reaches $50°C$. Find the set of points (x, t) for which $T(x, t) = 50°C$. From the contour at $T = 50°C$ determine the speed at which the $T = 50°C$ isotherm is moving through the slab.

P9.5 Quantum Mechanics: Particle Confined in a Spherical Well. The root of the two equations

$$\frac{x}{\tan x} = -y$$

$$x^2 + y^2 = s^2$$

where $s^2 = 2mV_0 a^2/\hbar^2$, determines the energy of the ground state for a particle of mass m confined in a spherical well of radius a and potential strength V_0. (The constant $\hbar$ is Planck's constant divided by 2π, $\hbar = 1.055 \times 10^{-34}$ J·s.) The energy of the state is $E = -(y/s)^2 V_0$. Use $s = 3.5$, $V_0 = 30$ MeV. (The atomic unit of energy, MeV, corresponds to one million electron volts.) Solve for the energy of the ground state. To obtain the first excited state of this system, replace the first of the preceding two equations by

$$\frac{1}{x \tan x} - \frac{1}{x^2} = \frac{1}{y} + \frac{1}{y^2}$$

and once again find the root and the energy.

Summary

ZERO CONTOURS

Inverse Functions   If the equation $f(x, y) = 0$ can be written as $y = g(x)$, then the curve of y versus x is the curve of the zero contour of $f(x, y)$. If instead, the equation is expressed as $x = h(y)$, it is frequently necessary to invert the function h to obtain $y = h^{-1}(x)$ if a y versus x plot of the zero contour is needed.

Stepping Procedures for Zero Contours   The function $f(x, y)$ is followed downhill; i.e., along the direction of the negative gradient, $-\nabla f$, until it changes sign. The next step is taken perpendicular to the direction of the gradient and along the contour lines, thereby tracing out a zero contour.

ZEROS OF SIMULTANEOUS FUNCTIONS

The Initial Guess   The zero of two functions of two variables is a single point corresponding to the intersection of the two zero contour lines. For three functions it is the single point of intersection of three surfaces. Without suggestions from physical considerations or from other sources, the estimation of the starting point of a search for a common root is often impossible. For two functions, the choice of an initial guess can be greatly aided by simultaneous contour plots of the two functions, and the use of `ginput` to determine the point of intersection.

Newton's Method   Once an initial guess for a root, (x_0, y_0), is obtained, Newton's method for improving a guess for a root in one dimension is easily generalized to two or more dimensions, as

$$\begin{pmatrix} x_1 \\ y_1 \end{pmatrix} \approx \begin{pmatrix} x_0 \\ y_0 \end{pmatrix} - \begin{pmatrix} \frac{\partial f}{\partial x}\big|_0 & \frac{\partial f}{\partial y}\big|_0 \\ \frac{\partial g}{\partial x}\big|_0 & \frac{\partial g}{\partial y}\big|_0 \end{pmatrix}^{-1} \begin{pmatrix} f_0 \\ g_0 \end{pmatrix}$$

where the function f and its derivatives are to be evaluated at the guess (x_0, y_0). This step is then iterated until sufficient accuracy is obtained, or until the method diverges.

Two-Dimensional   Replacing the partial derivatives by approximate difference expressions enables the two-
Secant Method   dimensional Newton's method to be structured so that only external M-file functions for $f(x, y)$ and $g(x, y)$ need to be coded. The algorithm is initiated with a starting guess, (x_0, y_0), and an estimate of the step sizes required in the x and y directions, Δx and Δy. The improvements in these quantities are obtained by the replacements:

$$\begin{pmatrix} x_0 \\ y_0 \end{pmatrix} \begin{pmatrix} x_1 = x_0 + \Delta x_0 \\ y_1 = y_0 + \Delta y_0 \end{pmatrix} \Rightarrow \begin{pmatrix} x_1 \\ y_1 \end{pmatrix} \begin{pmatrix} x_2 = x_1 + \Delta x_1 \\ y_2 = y_1 + \Delta y_1 \end{pmatrix}$$

where

$$\begin{pmatrix} \Delta x_1 \\ \Delta y_1 \end{pmatrix} = - \begin{pmatrix} \frac{f_{\Delta x} - f_0}{\Delta x} & \frac{f_{\Delta y} - f_0}{\Delta y} \\ \frac{g_{\Delta x} - g_0}{\Delta x} & \frac{g_{\Delta y} - g_0}{\Delta y} \end{pmatrix}^{-1} \begin{pmatrix} f_0 \\ g_0 \end{pmatrix} - \begin{pmatrix} \Delta x_0 \\ \Delta y_0 \end{pmatrix}$$

MATLAB Functions Used

`erf, erfc`   Toolbox functions to evaluate error function integrals. The function `erf(x)` evaluates $(2/\sqrt{\pi}) \int_0^x e^{-t^2} dt$. The function `erfc` evaluates the *complementary* error function, defined as $(2/\sqrt{\pi}) \int_x^\infty e^{-t^2} dt = 1 - \text{erf(x)}$.

10 / *Minimization of a Multivariable Function*

PREVIEW   A very common and important problem in all branches of science and engineering is that of minimizing or maximizing a single function of several variables. This problem occurs naturally as part of many optimization codes that are used to find the "best" combination of variables to maximize the profit of a business, minimize the energy in an atomic configuration, find the most favorable or shortest path along a surface, or optimize the fit of a multiparameter function to data. One usually seeks the minimum of [1] a nonlinear function of many variables. To ease the graphical interpretation, much of the analysis that follows deals with nonlinear functions of just two variables, $f(x, y)$; we generalize the two-dimensional results to more variables when appropriate.

Basically two types of algorithms are used to find the minimum of a multivariable function. The first is to attempt to *bracket* the point of the minimum with the hope of successively reducing the volume of the bracketing region until sufficient accuracy is obtained. This method, known as a *downhill simplex method*, is the one used by the MATLAB function fmins. Simplex methods are not highly efficient and are appropriate only if the number of variables is not too large, say less than ≈ 6. We describe the algorithm early in the chapter. The second, more complicated, and usually more efficient, class of methods attempts to use information contained in the derivative of f to follow the function downhill, perhaps along a curving path, until the function reaches a local minimum. These types of procedures are called *directional set methods* or *quasi-Newton methods*, and they have numerous variations. We describe these methods later in this chapter.

Simplex Methods for Finding Minima

In Chapter 6 we saw that when searching for the minimum of a function of one variable, the most important task was to first *bracket* the minimum; i.e., to find an interval that is known to contain a local minimum. The remaining task is to simply, systematically reduce the interval, thus "trapping" the minimum. In one dimension, three function evaluations are required to establish that an

---

[1]Again, if a maximum is desired, simply replace the function f by $-f$.

interval contains a minimum. However, trying to extend this bracketing idea to multidimensions reveals a dilemma. In two dimensions, five points (the point (x_0, y_0) and its four nearest neighbors on a two-dimensional grid) should suffice to establish that the point (x_0, y_0) is near a minimum contained in a region bounded by the four nearest neighbors. However, in this case, the minimum is *not* trapped. The function could fall towards a minimum along a line between two of the neighboring points. Thus, in two or more dimensions, it is *impossible* to firmly bracket a minimum. For this reason, multidimensional minimization is much more difficult and much less efficient than one-dimensional minimization.

Because escape-proof bracketing is no longer possible, we abandon the hope of first finding and systematically shrinking a volume down to the point of the minimum. Nevertheless, the geometrical ideas contained in the preceding discussion are useful for establishing a method of search that should be able to find a path downhill on the surface to a local minimum. We will not define volumes in terms of rectangular parallelepipeds in multidimensions; instead, the volume elements will be those contained in the simplest possible volume-containing structures, called *simplexes*. Thus, in two dimensions, a *volume* corresponds to an area in the xy plane, and the structure would be a triangle defined by any three non-collinear points. In n dimensions, the simplex is the region contained within $n + 1$ points. Another way to visualize the simplex method is to imagine the region bounded by the tips of the $n + 1$ vectors $\vec{x}_0$, $\vec{x}_i = \vec{x}_0 + \lambda_i \hat{e}_i$, where $\vec{x}_0$ is an arbitrary vector, say the initial guess for the minimum point, and $\hat{e}_i$, $i = 1, \ldots, n$, are orthogonal unit vectors in the n-dimensional space. The λ_i's are usually chosen to be the same and equal to λ, which then defines the initial volume presumed to contain the minimum. The procedure involves moving the simplex downhill or, if that is not possible, shrinking its size. The algorithm for doing this is due to Nelder and Mead,[2] and it is the procedure followed by the toolbox function, `fmins`.

DOWNHILL SIMPLEX METHODS

For simplicity, consider a function of two variables, $f(x, y)$. The initial guess for the point of the minimum is $\vec{x}_0$, and the two-dimensional simplex is then a triangle with vertices located by the vectors $\vec{x}_0$, $\vec{x}_1 = \vec{x}_0 + \lambda \hat{\imath}$ and $\vec{x}_2 = \vec{x}_0 + \lambda \hat{\jmath}$. Or,

$$\vec{x}_0 = \begin{pmatrix} a \\ b \end{pmatrix} \qquad \vec{x}_1 = \begin{pmatrix} a + \lambda \\ b \end{pmatrix} \qquad \vec{x}_2 = \begin{pmatrix} a \\ b + \lambda \end{pmatrix}$$

The function is then evaluated at these points, and the points are ordered so that $\vec{x}_{\max}$, $\vec{x}_{\min}$, and $\vec{x}_{\text{next}}$ correspond, respectively, to the points where the function is largest, smallest, and next to largest in value. These three points, even in the many-dimensional case, are used to determine the appropriate strategy for the next step in the algorithm, which seeks to replace $\vec{x}_{\max}$ by a carefully chosen $\vec{x}$ in the direction of decreasing f. The algorithm is performed as follows:

---

[2]J. A. Nelder and R. Mead, *Computer Journal*, vol.7, 1965, p. 308

1. Determine the points $\vec{x}_{max}$, $\vec{x}_{min}$, and $\vec{x}_{next}$ corresponding to the maximum, minimum, and next-to-largest values of the function; i.e., f_{max}, f_{min}, and f_{next}.

2. Reflect the point $\vec{x}_{max}$ along a line from $\vec{x}_{max}$ through the midpoint of the opposite face of the triangle. This point is the average of all vectors, excluding $\vec{x}_{max}$. That is, in n dimensions, $\vec{x}_{avg} = \frac{1}{n} \sum_{i \neq i_{max}} \vec{x}_i$. The reflected point will be called $\vec{x}_{try1} = \vec{x}_{avg} - \vec{x}_{max}$. See Figure 10.1.

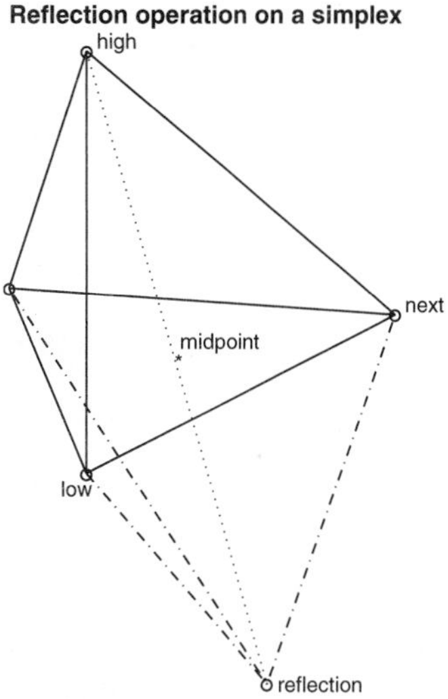

Reflection operation on a simplex

Figure 10.1 *Reflection of the maximum point of the simplex.*

3. If $f(\vec{x}_{try1}) < f_{min}$, the reflection has found a lower point, so try an even larger move along that line: $\vec{x}_{try2} = \gamma \vec{x}_{try1} + (1 - \gamma)\vec{x}_{avg}$, where γ is a scale value greater than one.

   (a) Now if $f(\vec{x}_{try2}) < f_{min}$, replace $\vec{x}_{max} = \vec{x}_{try2}$.
   (b) Otherwise, you've gone too far. However, the point $\vec{x}_{try1}$ can still be used; $\vec{x}_{max} = \vec{x}_{try1}$.

4. Otherwise, the reflected point is not an overall lowest point. However, if it is between f_{next} and f_{max}:

   (a) It can still be used to replace $\vec{x}_{max} = \vec{x}_{try1}$.
   (b) Also, try a point along the line, but on the same side as $\vec{x}_{max}$, say $\vec{x}_{try2} = \beta \vec{x}_{max} - (1 - \beta)\vec{x}_{avg}$, where β is a contraction factor (< 1). That is, seek to shrink the simplex along the reflection line. If successful; i.e., if $f(\vec{x}_{try2}) < f_{max}$, then $\vec{x}_{max} \leftarrow \vec{x}_{try2}$.

5. If nothing along the reflection line seems to work, try a contraction of the simplex towards the lowest point. That is, for all $\vec{x}_i$ except the minimum point, make the replacement, $\vec{x}_i \rightarrow \frac{1}{2}(\vec{x}_i + \vec{x}_{min})$. (See Figure 10.2.)

6. Test for completion and cycle back to start over.

Contraction of a simplex towards the low point

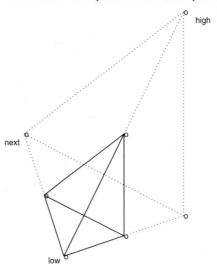

Figure 10.2 *Contraction of the simplex towards the minimum.*

The preceding algorithm can be readily translated into a MATLAB routine and used to find the minimum of an arbitrary function starting from an initial guess for the position of the minimum, and for the characteristic length scale, λ, of the starting simplex volume. Actually, the people at MATLAB have already done this for you. The toolbox function implementing a downhill simplex method in multidimensions is called `fmins`, and its use is demonstrated in Illustrative Problem 10.1.

ILLUSTRATIVE PROBLEM 10.1

Use of the Downhill Simplex Method of fmins to Find the Minimum of a Function

To illustrate this procedure, we find the minimum of the function

$$f(x, y, z) = (x - 2)^4 + (y - 1)^2 + 3(z - 1)^6$$

which clearly has a minimum value of zero at the point $(2, 1, 1)$. We first code the M-function file for this function.

```
function f = Ftest(x)
%Ftest: The input to the function is a vector, while
%       the output is a single value; i.e., a scalar.
%
  f = (x(1)-2)^4 + (x(2)-1)^2 + 3*(x(3)-1)^6;
```

We start with an initial guess of $\vec{x}_0 = (1.6, 1.2, 0.8)$. Also, the search will be successful if $|\Delta \vec{x}/\vec{x}| < \epsilon_x$, or if $|\Delta f(\vec{x})| < \epsilon_f$. As with minimizing single variable functions, the smallest values of these convergence tolerances must be greater than $\sqrt{\text{eps}}$ and eps, respectively, where eps is the computer accuracy. These quantities are delivered to the function `fmins` by means of the vector **options** as

$$\text{options} = [I_{\text{display}}, \epsilon_x, \epsilon_f];$$

where $I_{\text{display}} = 1$ (or nonzero) results in a display of intermediate results by the function fmins. The use of **fmins** to solve for the minimum might then be:

```
x = [1.6, 1.2, 0.8];
options = [1, 1.e-5, 1.e-9];
xmin = fmins('ftest',x,options)
```

which will generate a display like

```
cnt =
   358
contract

v =
    2.0000e+00      2.0000e+00      2.0000e+00      2.0000e+00
    1.0000e+00      1.0000e+00      1.0000e+00      1.0000e+00
    1.0000e+00      1.0000e+00      1.0000e+00      1.0000e+00
fv =
    4.1573e-26      1.0583e-25      1.3228e-25      2.7155e-25

xmin =
    2.0000    1.0000    1.0000
```

The first three vectors in the display are those defining the vertices of the simplex. The fourth is the test point obtained by either a reflection or contraction operation. Note that the correct minimum was found, but that it required 358 steps. The method is not particularly efficient, but it is extremely easy to use and will be found to be especially useful in optimizing the fit of a multiparameter test function to data, as will be illustrated in Chapter 15.

WHAT IF? The computational effort required to find the minimum of a function of several variables grows dramatically with the number of variables. Reduce the number of variables in Illustrative Problem 10.1 to two, and rerun the minimization code. By what approximate factor does the speed of the calculation increase? What if the number of variables in the test function is increased to four? ●

Minimization Using Directional Methods

In one-dimensional problems, the usual choice is between (a) methods that are easy to understand and easy to implement, but are rather inefficient; i.e., methods that only use values of the function at a variety of points to proceed to a minimum or a root, or (b) generally more efficient methods having a higher information content, such as using the slope of the function as well. The latter are usually more complicated to code. The situation in multidimensions is no different. To improve on the downhill simplex method of starting at a point and then simply "groping" for the way downhill, we might next try to use properties of the function itself to lead the way.

Multidimensional directional methods are often highly efficient compared with the downhill simplex method, particularly when the path downhill contains many curves and/or shallow slopes. The price to be paid is that the methods are usually quite complicated and difficult to code. In addition, there are several different directional methods from which to choose, all of which fall into two broad categories: *directional set methods* and *quasi-Newton methods*. All directional set methods start with an initial guess for the position of the minimum, $\vec{x}_0$, *carefully* select a direction in which to step, and then use *one-dimensional* methods to find the minimum of the function *along that direction*. The methods differ only in their selection of the optimum direction in which to proceed. We describe only the simplest (and least efficient) directional method, the *steepest descent method*, and direct our more detailed analysis towards an understanding of the *quasi-Newton method*, which uses ideas familiar to the reader from earlier encounters with the elementary Newton-Raphson algorithm for finding roots.

THE METHOD OF STEEPEST DESCENT

The most obvious choice for the best direction downhill is to use the negative gradient of the function, $-\nabla f$. The statement of the problem is quite simple. At a point where a function $f(x,y)$ is a minimum, *all* of the components of the gradient derivative will be equal to zero; i.e., if there are n independent variables, x_i, $i = 1, \ldots, n$, then we have to solve n independent equations of the form

$$\nabla f(\vec{x}) = 0 \quad \Rightarrow \quad \frac{\partial f}{\partial x_1} = 0$$

$$\frac{\partial f}{\partial x_2} = 0$$

$$\vdots$$

$$\frac{\partial f}{\partial x_n} = 0$$

This problem seems to be strikingly similar to the problem of Chapter 9, that of solving for the zeros of n simultaneous nonlinear equations, and thus appears to carry with it the formidable, and even hopeless, problem of finding a suitable starting point at which to begin the numerical search. However, there is a profound difference. Now our equations are *not* a random collection of disconnected equations, they are all related in that they are all directional derivatives of the same function in a set of n orthogonal directions. That is, even if $\vec{x}$ is not at a minimum, $-\nabla f(\vec{x})$ can be used to characterize the steepest direction downhill on the multidimensional surface represented by $f(\vec{x})$. Thus, it would seem that all we have to do is follow the function down the negative gradient until a minimum is found. The procedure is as follows: (a) Evaluate ∇f at a trial point and use the direction of $-\nabla f$ as the direction of steepest descent of the function. (b) Find the minimum of the function f along this direction. (*Note:* This step is strictly a one-dimensional problem and could readily be accomplished using the methods of Chapter 6.) (c) At this point, reevaluate ∇f; if it is not zero (or less than the convergence tolerance), repeat the process. This procedure, called the *method of steepest descent*, almost always succeeds in finding a local minimum, but it is rarely used for two reasons:

1. If the function $f(x, y)$ is quite complicated, its n derivatives contained in $\nabla f(\vec{x})$ are likely to be much worse, and you will need to separately code each and every one of them. However, in all cases, algorithms that make use of the analytic expressions for the derivatives in place of simpler differences in the function will be substantially more efficient as you converge to the answer.

2. Unless the negative gradient at the initial guess points directly at the ultimate minimum, and the function has no hills or valleys along the route—clearly a very rare situation—the method of steepest descent will be found to be quite inefficient. The method usually takes a great many small steps as it follows the function downhill along a curving path. This is illustrated in Figure 10.3 for a function of two variables that is represented by contour lines. Remember, ∇f at a point is always perpendicular to a contour line through that point. Thus, after the first step, wherein a minimum is found at point 1, by starting at point 0 and moving along the line parallel to ∇f, the next step *must* be perpendicular to the previous direction. Unless you are extremely lucky in your selection of the initial point, the path to a minimum is apt to be a zigzag path of numerous steps.

Each step of the steepest descent method
is perpendicular to the previous step

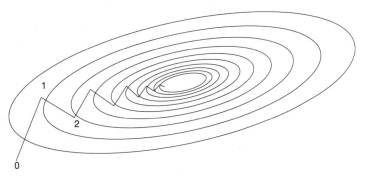

Figure 10.3 *The steepest descent method may take a zigzag path to the minimum.*

Still, the negative gradient does provide a starting direction. How can we make use of the directional information available in the gradient to proceed more efficiently to the ultimate minimum? Of the many algorithms that have been devised in response to this question, the most popular, and perhaps the easiest to understand, is called the *quasi-Newton method* and is described in the next section.

Quasi-Newton Method for Minimization of Multivariable Functions

The method is known as *quasi-Newton* because it starts from a Taylor series expansion of the function about a point as did the one-dimensional Newton-Raphson method for finding roots. It is also known as the *variable metric optimization method.*

The multidimensional Taylor series expansion for a function $f(\vec{x})$ expanded about the point $\vec{x}_0$ is

$$f(\vec{x}) = f(\vec{x}_0) + (\vec{x} - \vec{x}_0) \cdot \nabla f_0 + \frac{1}{2}(\vec{x} - \vec{x}_0) \cdot \mathbf{H}_0 \cdot (\vec{x} - \vec{x}_0) + \cdots \qquad (10.1)$$

where the function and its derivatives on the right-hand side are to be evaluated at $\vec{x}_0$. The matrix $\mathbf{H}_0$ is the set of second partials,

$$(\mathbf{H}_0)_{ij} = \left.\frac{\partial^2 f}{\partial x_i \partial x_j}\right|_0 \qquad (10.2)$$

and is known as the *Hessian* of f at $\vec{x}_0$.

If f is a function of a single variable, a minimum corresponds to a point where $\frac{df}{dx} = 0$ *and* $\frac{d^2 f}{dx^2} > 0$. (A maximum is characterized by $\frac{d^2 f}{dx^2} < 0$.) Similarly, if f is a multivariable function, a minimum is characterized by $\nabla f = 0$, *and* its Hessian is *positive definite* and *symmetric*.[3] If we were able to evaluate at a trial point, not just ∇f, but the complete Hessian matrix of second partials, the operation of minimization would be greatly simplified. To see why, consider the Taylor series expansion for ∇f to the same order as that for $f(\vec{x})$:

$$\nabla f = \nabla f_0 + \mathbf{H}_0 \cdot (\vec{x} - \vec{x}_0) + \cdots$$

Forcing $\nabla f = 0$ at the minimum point, $\vec{x}_m$, and truncating the expansion after the linear terms, we can rewrite this equation as

$$\mathbf{H}_0(\vec{x}_m - \vec{x}_0) \approx -\nabla f_0$$

which is simply a matrix equation that can easily be solved for the position of the minimum,

$$\vec{x}_m = \vec{x}_0 - \mathbf{H}_0^{-1}\nabla f_0 \qquad (10.3)$$

If f were no worse than quadratic in the variables, this equation would be *exact*, and the minimum would be found in one step. As in Newton's method of finding roots, with more complicated functions this step will have to be iterated until changes in the projected minimum differ by less than a convergence tolerance.

---

[3]A symmetric positive definite matrix has all its eigenvalues positive.

For example, in Illustrative Problem 10.1, the minimum of the function $f(x, y, z) = (x - 2)^4 + (y - 1)^2 + 3(z - 1)^6$ was found by the downhill simplex method. Because both the gradient and the Hessian of this function are easily computed, they can be used to find the minimum again, this time by iteration of the preceding Newton-like formula. The gradient and the Hessian evaluated at the initial guess, $\vec{x}_0 = (1.6, 1.2, 0.8)^T$, are

$$\nabla f|_0 = \begin{pmatrix} 4(x - 2)^3 \\ 2(y - 1) \\ 18(z - 1)^5 \end{pmatrix}\Bigg|_0 = \begin{pmatrix} -0.256 \\ 0.4 \\ -0.0058 \end{pmatrix}$$

and

$$\mathbf{H}_0 = \begin{pmatrix} 12(x - 2)^2 & 0 & 0 \\ 0 & 2 & 0 \\ 0 & 0 & 90(z - 1)^4 \end{pmatrix}\Bigg|_0 = \begin{pmatrix} 1.9200 & 0 & 0 \\ 0 & 2 & 0 \\ 0 & 0 & 0.1444 \end{pmatrix}$$

so that

$$\mathbf{H}_0^{-1} = \begin{pmatrix} 0.5208 & 0 & 0 \\ 0 & 0.5 & 0 \\ 0 & 0 & 6.9444 \end{pmatrix}$$

The first iteration of the algorithm predicts the position of the minimum to be at

$$\vec{x}_{\min}^{(1)} \approx \vec{x}_0 - \mathbf{H}_0^{-1} \cdot \nabla f_0 = \begin{pmatrix} 1.7333 \\ 1.0000 \\ 0.8400 \end{pmatrix}$$

In the next iteration, ∇f and $\mathbf{H}^{-1}$ are first computed at the new value of $\vec{x}$ and then used to again estimate the position of the minimum. After 20 iterations, the position of the minimum is approximately $x_{\min} = [1.9999 \ 1.0000 \ 0.9977]^T$.

Of course, in typical problems, the function f will be substantially more complicated and the chore of coding all $\frac{1}{2}n(n - 1)$ independent elements of the Hessian matrix may prove to be impossible, or at least unfeasible.

In quasi-Newton methods, each iteration begins with an attempt to build up successively better and better approximations to $\mathbf{H}_0^{-1}$ using only computed values of f and ∇f. Once obtained, one iteration of the method then computes an estimate of the minimum point at, say $\vec{x}_1$. The quasi-Newton method is then repeated to build up an approximation to $\mathbf{H}_1^{-1}$ at the new point. The method for computing $\mathbf{H}_0^{-1}$ is due to Davidon,[4] was improved by Fletcher and Powell,[5] and is known as the DFP method. It has been further improved several times since, but we present here only the basic DFP algorithm.

---

[4] W. C. Davidon, *Variable Metric Method for Minimization*, AEC Report No. ANL-5990, Argonne National Laboratory, Argonne, IL, 1959.
[5] R. Fletcher and M. J. D. Powell, "A Rapidly Convergent Descent Method for Minimization," *Computer J.* **6**, 1963, pp. 164–168.

THE DFP ALGORITHM TO CALCULATE
THE INVERSE HESSIAN

First of all, we assume that f is quadratic in the variables $\vec{x}$ and establish procedures to compute the exact inverse of $\mathbf{H}_0$. For nonquadratic functions, these expressions will then serve as the first iteration of an iterative algorithm. In this context, notice that if $\mathbf{H}_0^{-1}$ is the exact inverse Hessian at the point $\vec{x}_0$, then by writing down the expansions of ∇f about the point $\vec{x}_0$ for two distinct points $\vec{x}_a$ and $\vec{x}_b$, and subtracting, we obtain

$$(\vec{x}_b - \vec{x}_a) = \mathbf{H}_0^{-1} \cdot (\nabla f_b - \nabla f_a) \tag{10.4}$$

That is, the difference between any two points is related to the difference in the gradients at those two points multiplied by the inverse Hessian at $\vec{x}_0$. In particular, the relation between successive computed points, $\vec{x}_i$ and $\vec{x}_{i+1}$, along the path to the minimum must satisfy

$$\delta\vec{x}_i = \mathbf{H}_0^{-1} \cdot \delta(\nabla f_i) \tag{10.5}$$

where

$$\delta\vec{x}_i = (\vec{x}_{i+1} - \vec{x}_i) \qquad \delta(\nabla f_i) = (\nabla f_{i+1} - \nabla f_i)$$

Next, labeling the current ith-order approximation to the inverse of the Hessian as $\mathbf{A}_i$ ($\mathbf{A}_i \approx \mathbf{H}_0^{-1}$), the DFP procedure first requires that the next order approximation, $\mathbf{A}_{i+1}$, whatever it might be, satisfy Equation 10.5 as well. That is,

$$\delta\vec{x}_i = \mathbf{A}_{i+1} \cdot \delta(\nabla f_i) \tag{10.6}$$

where

$$\mathbf{A}_{i+1} \equiv \mathbf{A}_i + \delta\mathbf{A}_i$$

This is a rather weak restriction, requiring only that the change in the approximate inverse Hessian, $\delta\mathbf{A}_i$, in this step satisfy the equation:

$$\delta\mathbf{A}_i \cdot \delta(\nabla f_i) = \delta\vec{x}_i - \mathbf{A}_i \cdot \delta(\nabla f_i) \tag{10.7}$$

The various forms of DFP algorithms then correspond to differing choices for the improvement in $\mathbf{A}_i$ that will satisfy Equation 10.7. There is a broad class of matrices that can potentially satisfy this requirement, but the original DFP method chooses the following:

$$\delta\mathbf{A}_i = \frac{\delta\vec{x}_i \delta\vec{x}_i^T}{\delta\vec{x}_i^T \delta(\nabla f_i)} - \frac{\mathbf{A}_i \delta(\nabla f_i)\delta(\nabla f_i)^T \mathbf{A}_i^T}{\delta(\nabla f_i)^T \mathbf{A}_i \delta(\nabla f_i)} \tag{10.8}$$

For this choice, they have shown that if f is quadratic, the sequence of matrices $\mathbf{A}_i$ will converge to the exact inverse Hessian in no more than n steps, where n is the number of independent variables.

One procedure to implement this algorithm is to start with initial guesses for the inverse Hessian, $\mathbf{A}_0$, and the starting position, $\vec{x}_0$, and to compute the next position vector using

$$\vec{x}_1 \approx \mathbf{A}_0 \cdot \vec{x}_0 \tag{10.9}$$

The initial guess for $\mathbf{A}_0$ is usually chosen to be the identity matrix. The improvement to $\mathbf{A}_0$ is computed using Equation 10.8 and the process is then iterated. This is illustrated in Example 10.1.

EXAMPLE 10.1

The DFP Algorithm Applied to a Test Function

The function used in Illustrative Problem 10.1 was

$$f(x, y, z) = (x - 2)^4 + (y - 1)^2 + 3(z - 1)^6$$

The gradient of this function is

$$\nabla f = \begin{pmatrix} 4(x-2)^3 \\ 2(y-1) \\ 18(z-1)^5 \end{pmatrix}$$

Once again, we start with the initial guess, $\vec{x}_0 = [1.60, 1.20, 0.80]$. The DFP algorithm, beginning with $\mathbf{A}_0$ equal to the identity matrix, yields the following sequence of results:

$$\vec{x}_1 = \vec{x}_0 - \mathbf{A}_0 \cdot (\nabla f)_0$$

$$= \begin{pmatrix} 1.60 \\ 1.20 \\ 0.80 \end{pmatrix} - \begin{pmatrix} -0.2560 \\ 0.4000 \\ -0.0058 \end{pmatrix}$$

$$= \begin{pmatrix} 1.856 \\ 0.8000 \\ 0.8058 \end{pmatrix}$$

and at this value of $\vec{x}$,

$$\nabla f_1 = - \begin{pmatrix} 0.0119 \\ 0.4000 \\ 0.0050 \end{pmatrix}$$

so that

$$\delta \vec{x}_0 = \vec{x}_1 - \vec{x}_0 = \begin{pmatrix} 0.2560 \\ -0.4000 \\ 0.0058 \end{pmatrix} \qquad \delta(\nabla f)_0 = \nabla f_1 - \nabla f_0 = \begin{pmatrix} 0.2441 \\ -0.8000 \\ 0.0008 \end{pmatrix}$$

Thus, using Equation 10.8, we obtain

$$\delta \mathbf{A}_0 = \begin{pmatrix} 0.0862 & 0.0114 & 0.0036 \\ 0.0114 & -0.4965 & -0.0051 \\ 0.0036 & -0.0051 & 0.0001 \end{pmatrix}$$

resulting in

$$\mathbf{A}_1 = \begin{pmatrix} 1.0861 & 0.0114 & 0.0036 \\ 0.0114 & 0.5035 & -0.0051 \\ 0.0036 & -0.0051 & 1.0001 \end{pmatrix}$$

and

$$\vec{x}_2 = \begin{pmatrix} 1.8735 \\ 1.0015 \\ 0.8087 \end{pmatrix}$$

Continuing this for the required two additional steps yields

$$\vec{x}_3 = \begin{pmatrix} 1.9207 \\ 0.9982 \\ 0.8349 \end{pmatrix}$$

Because the test function was not quadratic, the procedure has not converged within the first three steps. However, continuing the process will yield results that converge to the exact result.   ●

QUASI-NEWTON METHODS WITH LINEAR MINIMIZATION

It should be pointed out that the most common implementations of this algorithm add one slight twist that can often dramatically improve the efficiency, particularly if the function is nearly quadratic as it approaches its minimum. The idea is to use Equation 10.9, not to supply the next x value, but merely to indicate a direction in which to search for the minimum. That is, if we define a vector, $\vec{d}$, by the equation

$$\vec{d} = \mathbf{A}_i \cdot \delta(\nabla f_i)$$

and interpret both $\vec{x}_i$ and $\vec{d}$ to be "constants," then the function $f(\vec{x}_i + \lambda \vec{d})$ is a function of a *single* variable, λ. In that case, highly efficient one-dimensional minimization codes, such as fmin, can be used to find the value of λ that causes the function $f(\vec{x}_i + \lambda \vec{d})$ to be a minimum. The next point in the multidimensional space is then given by $\vec{x}_{i+1} = \vec{x}_i + \lambda \vec{d}$.

To implement this refinement in MATLAB, you will have to write a simple two-line M-file function that finesses the multivariable function in question, $f(\vec{x})$, to appear to be a function of a single variable λ. This can be accomplished by the following code:

```
function y = F1dmin(c,x,d)
   z = x + c*d;
   y = feval(fname,z);
```

where *fname* is a string constant containing the name of the multivariable function $f(\vec{x})$. The necessary MATLAB statement to determine the optimum λ to use in each step is then

```
fmin('F1dmin',0,3,[0 1.e-3],x,d);
```

In this statement, it is expected that $0 < \lambda \leq 3$. The elements of the row vector, [0 1.e-3], are part of an optional OPTIONS vector that is used to control the output and operations of the function fmin. (See help fmin for details.) The last two entries, x and d, are the additional parameters in the argument list for the function F1dmin.

The improvements resulting from this change vary from quite significant to merely a factor of 2 or so in the reduction of the number of iterations required to obtain a solution. Because each step now requires a separate linear minimization, you will notice a measurable degradation of the speed of the overall method. In short, the details of the function itself determine whether or not this additional feature will improve the multivariable minimization code.

MATLAB CODE FOR THE QUASI-NEWTON METHOD

We are now at a point where we could assemble a MATLAB routine to implement the DFP algorithm, but first, let us try one additional alteration. If we replace the exact expressions for the components of the gradient by simple difference expressions, such as

$$\left.\frac{\partial f}{\partial x_i}\right|_0 \approx \frac{1}{h}\left(f\left(x_1, x_2, \ldots, x_i + \frac{1}{2}h, \ldots\right) - f\left(x_1, x_2, \ldots, x_i - \frac{1}{2}h, \ldots\right)\right)$$

where h is an appropriately chosen small value, perhaps we can get away with a procedure that only requires evaluations of the function itself, thereby avoiding the complications of coding all the elements of the gradient. Of course, what we give up in doing this is the guarantee that the inverse Hessian can be obtained in just n steps. However, the slower convergence may be more than compensated by ease in coding. The MATLAB code to implement the DFP algorithm using approximate gradients is given in Figure 10.4.

The first task is to construct the code for the approximate gradient. Note that the function of Figure 10.4 is named Gradynt, so as not to conflict with the toolbox function gradient. Also, note that the $n \times 1$ column vector $\vec{x}$ is first mapped into an $n \times n$ matrix of identical columns by using the function meshgrid, and finally, the diagonal matrix h*eye(n,n) is added or subtracted to x to produce a set of column vectors with increments h in each of the n directions.

```
function g = Gradynt(f,x,h)
%Gradynt computes the approximate gradient of the function
%    whose name is contained in the string variable f.
%    Central differences are used, and f is a function of n
%    variables, where n is the length of x.  The interval
%    used in the central differences is h, and the deriva-
%    tives are returned as a column vector in g.
%            Use: g = Gradynt(f,x,h)
%================================================================
```

(Figure 10.4 continued)

```
if isstr(f)~=1;error('name must be a string');end
if min(size(x))>1; error('x must be a vector or scalar');end
x = x(:);
n = length(x);
X = meshgrid(x)';
xp = X + 0.5*h*eye(n,n);
xm = X - 0.5*h*eye(n,n);
g = zeros(n,1);
for i = 1:n
    g(i) = (feval(f,xp(:,i)) - feval(f,xm(:,i)))/h;
end
```

Next, the quasi-Newton algorithm is easily assembled and reads:

```
function x1 = QNewtn(fnc,x0,h,tol)
% QNewtn will find the minimum of an existing function of
%      several variables starting from the initial guess
%      vector, x0, and the initial step size h.  It uses
%      the approximate gradients of the function which are
%      computed by the function Gradynt.  Also, the function
%      f1dmin is simply the existing function evaluated at
%      x+c*Dir, which, for a given x and Dir, is considered
%      to be a function of a single variable, c.  The pro-
%      cedure is to use the gradient information to select
%      a direction to search for the overall minima.  A one
%      dimensional search is done along this direction using
%      the toolbox function fmin with c as the variable.  To
%      avoid this step, and simply use a ''full'' step in the
%      indicated direction, replace the line lambda = ...,
%      by lambda = 1.  The process iterates until the frac-
%      tional changes in the step size is less than tol. The
%      use is:  x1 = QNewtn(fnc,x0,h,tol).
%===========================================================
n = length(x0);
A = eye(n,n);   x = 2*x0;  h = 1.e-5;
g0= Gradynt(fnc,x0,h);    d = -g;
while (norm(x-x0)/norm(x0)) > tol
    lambda = fmin('F1dmin',0,3,[0 0 1.e-3],fnc,x,d);
    x  = x0 + lambda*d;
    g = Gradynt(fnc,x,h);
    dx = x - x0;   dg = g - g0;
    dX = A*dg;
    dA = d*d'/(d'*dg) - dX*dX'/(dg'*dX);
    A  = A + dA;   x0= x;   g0= g;   d = -A*g
end
```

Figure 10.4 *A Quasi-Newton algorithm employing approximate derivatives for minimizing a multivariable function*

After 50 iterations of this secant-like adaptation of the DFP method applied to the function $f(x, y, z) = (x - 2)^4 + (y - 1)^2 + 3(z - 1)^6$, starting at $x_0 = [1.6\ 1.2\ 0.8]'$, with $h = 0.05$, we obtain the following results:

$$x_{\text{DFP}}^{(50)} = \begin{pmatrix} 1.99070 \\ 0.9999995 \\ 0.94409 \end{pmatrix}$$

and the exact same results are obtained using the actual gradients in place of the difference expressions.

You should compare this with an "exact" Newton-like algorithm using the analytic expression for the inverse of the Hessian matrix; that is, an iterative algorithm based on Equation 10.3. The result is:

$$x_{\text{Newton}}^{(50)} = \begin{pmatrix} 1.9999999994 \\ 1.0000000000 \\ 0.9999971455 \end{pmatrix}$$

As a final comparison, we give the results of the previous downhill simplex calculation using the toolbox function `fmins`, also after 50 iterations:

$$x_{\text{simplex}}^{(50)} = \begin{pmatrix} 2.02994 \\ 1.00015 \\ 0.92317 \end{pmatrix}$$

Of course, the closer the function f resembles a quadratic form in the variables, the more the calculation using the DFP algorithm will approach the exact Newton-like results (and of course the reverse is also true).

Finally, the number of iterations required for a satisfactory solution will be approximately the number of independent variables squared. Thus, if there are numerous variables in the problem, it likely is worth your time to spend considerable effort in first finding a good starting point for the algorithm. You could, for example, start with a downhill simplex method, hoping that it can navigate its way along a curving path to the vicinity of the minimum point. Then, once near the point of the minimum where the function is more likely to be near-parabolic, switch to quasi-Newton methods. Or, if *some* of the elements of the gradient are quite simple, you may wish to code a separate M-file function for the exact expressions for the derivatives. Finally, if possible, a contour plot of the function can aid immeasurably in any search for minima or maxima. A graphical demonstration of the quasi-Newton method is contained in the file `QNwtDemo.m` on the applications disk.

Problems

REINFORCEMENT EXERCISES

P10.1 **Numerical Examples.** Find the minima of the following functions using both the function `fmin` and the function `QNEWTN`. Start the search at the indicated point.

    a. $z = 3x^2 - 2xy + y^2 + 3x - 4y + 2$, at $(x, y) = (0, 1)$
    b. $z = 3x^2 + xy^2 + y^4 - 5x + 3y^2$, at $(x, y) = (\frac{1}{2}, \frac{1}{2})$

P10.2 Tracing the Path to the Minimum. The function $z = 3(x-1)^2 + 2(x-1)(y-2) + (y-2)^4$ has a minimum at $(x,y) = (1,2)$. First, draw a contour plot centered about the minimum point. Next, use the function QNEWTN to search for the minimum starting from the point $(0,0)$ and adapt the code so that each of the intermediate points, (x_i, y_i), is *saved* in separate vectors. With hold on, plot the trajectory of the solution as it converges on the minimum point.

P10.3 Rosenbrock's "Banana" Function. The function

$$f(x,y) = 100(y - x^2)^2 + (1-x)^2$$

is often given as a good test function of any algorithm for finding minima. The function has a very shallow valley, and many algorithms will find it difficult to traverse through this valley. Use the toolbox function fmins to find the minima starting at $(x,y) = (-1.2, 1)$. Because the region of near-zero slope is so extensive, the function QNEWTN cannot be used unless the approximate gradients computed in Gradynt are replaced by an M-file function that computes the two components of the gradient analytically. Even with this change, you will find that the shallow slopes cause the quasi-Newton method with one-dimensional minimization to take approximately ten times the number of iterations that fmins does.

P10.4 Minima in Five Dimensions. A test function, due to Powell, that will also cause difficulties in the search for minima is

$$f(x_1, x_2, x_3, x_4) = (x_1 + 10x_2)^2 + 5(x_3 - x_4)^2 + (x_2 - 2x_3)^4 + 10(x_1 - x_4)^4$$

Starting at the point $(1, -1, 0, 1)$, search for the minima using fmins or QNEWTN. (Once again, you will have to use analytic derivatives in place of the approximate derivatives supplied by Gradynt.) The minimum is at $(0,0,0,0)$, but it is extremely difficult to converge to this point. Change variables to $\alpha = x_1 + 10x_2$, $\beta = x_3 - x_4$, $\gamma = x_2 - 2x_3$, $\delta = x_1 - x_4$ and prove that the *only* minimum of this function is at $(x_1, x_2, x_3, x_4) = (0,0,0,0)$.

P10.5 Zeros of Simultaneous Functions. In Chapter 9, it was stated that unless there is some additional information, the search for a root of several simultaneous nonlinear functions is frequently hopeless. Yet the search for a minimum of a single multivariable function usually leads to a solution. What then is wrong with the proposal to find the *root* of three functions in three variables by, instead, searching for the *minimum* of $|f_1| + |f_2| + |f_3|$?

P10.6 The DFP Algorithm. Prove that the matrix $\delta \mathbf{A}_i$ of Equation 10.8 is indeed a solution to Equation 10.7.

P10.7 Quadratic Functions. Construct an M-file function that is quadratic in three variables, x, y, and z, and has an obvious minimum at, for example, $(1,2,3)$. Starting with an initial guess of $x_0 = (0,0,0)$ and using the function QNEWTN for 5 iterations, find the minimum. Compare the speed of the solution with a similar problem wherein the exponent 2 on one of the terms is replaced by a 4.

P10.8 Minimum Distances. Four towns, labeled A, B, C, and D, are located at the points $(0,0)$, $(1,5)$, $(5,2)$, and $(7,3)$. They are to be connected by roads leading out of the towns and meeting at a single intersection at point, P, located at (x,y). Find the position of P that minimizes the total lengths of the roads.

P10.9 Paths of Light Rays. If the speed of light in a transparent medium is v, and the speed in a vacuum (or air) is c, the ratio $n = \frac{c}{v}$ is called the index of refraction of the medium. Consider then the path from point P to Q of the light beam traveling through a transparent material of thickness d and index of refraction n. The actual path followed by the light ray is the one that takes the *least* amount of time to get from point P to point Q.

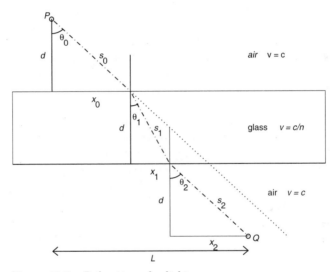

Figure 10.5 *Refraction of a light ray.*

From Figure 10.5, the travel time from P to Q is $T = (s_0 + ns_1 + s_2)/c$. The length elements, s_0, s_1, s_2, must satisfy the equations:

$$s_0 \sin \theta_0 + s_1 \sin \theta_1 + s_2 \sin \theta_2 = L$$

$$s_i \cos \theta_i = d, \qquad \text{for } i = 0, \ldots, 3$$

Thus, once the quantities L, d, and n are specified, the travel time, T, can be expressed as a function of (θ_0, θ_1):

$$\frac{c}{d} T(\theta_0, \theta_1) = \frac{1}{\cos \theta_0} + \frac{n}{\cos \theta_1} + \frac{1}{\cos \theta_2}$$

with θ_2 determined from

$$\tan \theta_2 = \frac{L}{d} - \tan \theta_0 - \tan \theta_1$$

Write an M-file function, `PathTime(theta)`, that will determine the values of (θ_0, θ_1) that correspond to a minimum of the travel time from point P to point Q. Show that the values obtained satisfy $\theta_0 = \theta_2$ and Snell's law: $\sin \theta_0 = n \sin \theta_1$.

EXPLORATION PROBLEMS

P10.10 **Binding Energies of Nuclei.** The total binding energy; i.e., the amount of energy required to completely disassemble a nucleus, is approximately given by the equation:

$$B(Z, N) = b_v A \left[1 - k_v \left(\frac{N - Z}{A} \right)^2 \right] - b_s A^{2/3} - b_C \frac{Z^2}{A^{1/3}}$$

where N is the number of neutrons and Z the number of protons in the nucleus. The *atomic number* is $A = N + Z$. The parameters have the values:

$$b_v = 15.6 \text{ MeV} \qquad k_v = 1.5$$
$$b_s = 17.2 \text{ MeV}$$
$$b_C = 0.7 \text{ MeV}$$

The quantity that will, in turn, indicate the stability of the nucleus to radioactive decay is the binding energy per particle, B/A. The larger this quantity, the more stable the nucleus. Use the expression for B/A to determine the region of the most stable nuclei. (Only investigate nuclei with $N \geq 6$ and $Z \geq 6$.) Finally, determine the values of Z and N that correspond to maximum binding.

P10.11 **The Variational Principle.** Consider an $n \times n$ eigenvalue problem, $\mathbf{A}\hat{e}_i = \lambda_i \hat{e}_i$, where $\hat{e}_i$ and λ_i are, respectively, the n eigenvectors and eigenvalues of the matrix A. (Assume that A is real-symmetric, and perhaps review the discussion in Chapter 3.) The eigenvectors $\hat{e}_i$ are of unit length and are orthogonal. They also satisfy the completeness condition, $\sum_{i=1}^{n} \hat{e}_i \hat{e}_i^T = \mathbf{I}_n$. Next, an arbitrary vector, $\vec{v}$, can be expanded in terms of the $\hat{e}_i$'s as $\vec{v} = \sum_{i=1}^{n} \hat{e}_i \hat{e}_i^T \vec{v}$. The effect of inserting $\vec{v}$ into the eigenvalue equation is

$$\mathbf{A}\vec{v} = \sum_{i=1}^{n} \mathbf{A}\hat{e}_i \hat{e}_i^T \vec{v}$$

$$= \sum_{i=1}^{n} \lambda_i \hat{e}_i \hat{e}_i^T \vec{v}$$

$$\geq \lambda_{\min} \sum_{i=1}^{n} \hat{e}_i \hat{e}_i^T \vec{v}$$

$$\geq \lambda_{\min} \vec{v}$$

Multiplying both sides of this inequality by $\vec{v}^T$, we obtain

$$\lambda_{\min} \leq \left(\frac{\vec{v}^T \mathbf{A} \vec{v}}{\vec{v}^T \vec{v}} \right) \equiv R_A(\vec{v})$$

Thus, we have a prescription for estimating the smallest eigenvalue of a matrix by choosing an arbitrary vector, $\vec{v}$, and then finding the minimum of the quantity, $R_A(\vec{v})$.

For example, let

$$\mathbf{A} = \begin{pmatrix} 4 & -2 & 1 & 0 \\ -2 & 4 & -2 & 1 \\ 1 & -2 & 4 & -2 \\ 0 & 1 & -2 & 4 \end{pmatrix} ; \qquad \vec{v} = \alpha \begin{pmatrix} 1 \\ 2 \\ -2 \\ 1 \end{pmatrix} + \beta \begin{pmatrix} 2 \\ 1 \\ 1 \\ -2 \end{pmatrix}$$

and find the minimum value of $R_A(\alpha, \beta)$. Compare this value with the actual smallest eigenvalue, $\lambda_{\min}$, and, using the computed values of α and β, compare $\vec{v}$ with $\hat{e}_{\min}$.

Try adding a third vector to the set comprising $\vec{v}$ with an additional parameter γ. If this third vector is not a linear combination of the first two, the results will improve. Adding a fourth linearly independent vector will generate the *exact* minimum eigenvalue.

Summary

SIMPLEX METHODS

A Simplex

A simplex in n dimensions is the simplest geometrical shape that can enclose a *volume* in n dimensions. In two dimensions, a simplex is a triangle.

Downhill Simplex Method

A set of operations designed to either progress towards the minimum point of a function or to shrink an enclosing simplex towards the minimum point. The operations consist of *reflections* of the simplex away from the high point, or *contractions* of the simplex away from the high point or towards the low point.

DIRECTIONAL METHODS

Steepest Descent

The negative gradient of a function defines the direction of steepest descent, which may then be followed to find the minimum. The procedure is to determine a direction by evaluating $-\nabla f$ at a point and then to minimize the function (of now one fewer variable) along this direction. The gradient is again computed, and the process repeated. The difficulty is that the next step will, by definition, be perpendicular to the original direction and may result in a needlessly zigzag path to the minimum.

QUASI-NEWTON METHODS

Definition

The Quasi-Newton methods are based on the multidimensional Taylor series expansion for a function and the fact that, at the minimum point, $\nabla f = 0$. This leads to the Newton-like algorithm

$$\vec{x}_m = \vec{x}_0 - \mathbf{H}_0^{-1} \nabla f_0$$

to obtain an improved estimate, $\vec{x}_m$, of an initial guess, $\vec{x}_0$, for the point of the minimum. The quantity, $\mathbf{H}_0$, is the Hessian matrix evaluated at $\vec{x}_0$:

$$(\mathbf{H}_0)_{ij} = \left. \frac{\partial^2 f}{\partial x_i \partial x_j} \right|_0$$

The DFP Algorithm

The inverse to the $n \times n$ Hessian matrix is needed to execute one step of the quasi-Newton method. The DFP algorithm is an iterative procedure to approximate this matrix. The matrix $\mathbf{H}_0^{-1}$ is first guessed to be the identity matrix; then after n iterations, it is of the form $\mathbf{H}_0^{-1} \rightarrow \mathbf{H}_0^{-1} + \delta\mathbf{A}$, where

$$\delta\mathbf{A}_i = \frac{\delta\vec{x}_i\delta\vec{x}_i^T}{\delta\vec{x}_i^T\delta(\nabla f_i)} - \frac{\mathbf{A}_i\delta(\nabla f_i)\delta(\nabla f_i)^T\mathbf{A}_i^T}{\delta(\nabla f_i)^T\mathbf{A}_i\delta(\nabla f_i)}$$

are used to approximately compute $\mathbf{H}_0^{-1}$.

Quasi-Newton Plus Linear Minimization

The DFP algorithm is used to merely supply a *direction*, $\vec{d} = \mathbf{A}_i \cdot \delta(\nabla f_i)$, where $\mathbf{A}$ is computed by the DFP iterative relation, and the function $f(\vec{x}_i + \lambda\vec{d})$ is then minimized along this direction by minimizing with respect to λ. This new point is then used as the input for the next step.

MATLAB Functions Used

fmins

The downhill simplex method is implemented in MATLAB by means of the function fmins. The use is xmin = fmins('*fname*',x0), where '*fname*' is a string constant that contains the name of an M-file function that will return a *scalar* corresponding to the value of the function. The quantity x0 is a vector containing the initial guess for the values of the variables that minimize the function.

The default convergence criterion is 10^{-4} for both $|\Delta\vec{x}|$ and $|\Delta f|$ and the default limit on the number of iterations is 500. These may be changed by including them as elements 2, 3, and 14 of the optional OPTIONS control array, which is then included as the third item in the argument list of fmins.

Also, if OPTIONS(1) is nonzero, intermediate steps in the calculation will be displayed. If the function '*fname*' requires additional parameters in its argument list, they may be included in the argument list of fmin after the OPTIONS array.

11 / *Multiple Integrals*

PREVIEW   Multiple integrals appear frequently in all areas of science and engineering. At first glance, the construction of algorithms for their numerical evaluation appears to present little difficulty. However, usually as a result of idiosyncrasies within the programming language itself, subtle, and sometimes annoying, problems are encountered. In this chapter, we describe these problems and construct the MATLAB code to address them. Each of the prescriptions can be adapted to employ any of the one-dimensional integration algorithms described in Chapter 7.

However, the principal difficulty in extending integration algorithms from one to several dimensions is the dramatic increase in the number of sampling points required. As a result, most real-life situations entail a mix of analytical and numerical evaluations of integrals. If any part of the integral can be handled analytically, that part most certainly should be done before attempting a purely numerical solution.

Finally, improper integrals, integrals with singularities within the region of integration, or highly oscillatory integrals are even more trouble in multidimensions; often precluding any reliable numerical result. There are no general prescriptions for handling improper or pathological multidimensional integrals, but usually successful procedures, if they exist, are simply adaptations of the methods used to evaluate similarly difficult one-dimensional integrals.

MATLAB Code for Iterated Integrals

A multiple integral can be written as a sequence of one-dimensional integrations over each of the independent variables.[1] For example, the three-dimensional integral

$$I = \int f(x, y, z) \, dx \, dy \, dz$$

is usually written successively as

$$I = \int_{z_1}^{z_2} dz \int_{y_1(z)}^{y_2(z)} dy \int_{x_1(y,z)}^{x_2(y,z)} f(x, y, z) \, dx$$

---

[1] This form of the multiple integral is known as an *iterated integral*.

255

The result of the first integration over x results in a function of y and z only; the second integration results in a function of z only. Thus, the three integration operations can be segmented as

$$I = \int_{z_1}^{z_2} G(z)\,dz$$

$$G(z) = \int_{y_1(z)}^{y_2(z)} H(y,z)\,dy$$

$$H(y,z) = \int_{x_1(y,z)}^{x_2(y,z)} f(x,y,z)\,dx$$

which can easily be generalized to more dimensions.

Each of the integrations is now one dimensional and should be easy to evaluate using any of the one-dimensional procedures of Chapter 7. However, this takes a bit of fine tuning of the existing one-dimensional codes. Basically, the problem is that each of these codes was written to expect a function of a *single* variable, whereas the integrand is now a function of two or more independent variables. Either the integration codes will have to be rewritten, or the functions somehow *finessed* to *appear* to be functions of a single variable. To make the resulting code as portable as possible, we will follow the latter alternative.

Fortunately, the basic integration functions in MATLAB allow more than one parameter in the argument list of the function representing the integrand. That is, to evaluate the integral

$$I = \int_0^1 e^{-\alpha x}\,dx$$

we could either pass the parameter α to the integrand function, $f(x) = e^{-\alpha x}$, via a global statement:

```
global alpha
alpha = 0.5;
I = quad('arg',0,1);
```

```
function f = arg(x);
   global alpha
   f = exp(-alpha*x);
```

or include the parameter within the arguments of both arg and quad.

```
alpha = 0.5;
I = quad('arg',0,1,[ ],[ ],alpha);
```

```
function f = arg(x,alpha);
  f = exp(-alpha*x);
```

Blank matrices have been entered in the normal positions for the quad parameters tol and TRACE, and cause the function to use default values, $(10^{-3}$, and 0, respectively). This procedure avoids the use of global variables.

To simplify the discussion, we will consider in detail only the evaluation of a double integral. Triple integrals are then an obvious, but complicated, further generalization.

A double integral, when written as an iterated integral, takes the form

$$I = \int f(x,y)\, dx\, dy$$

$$= \int_{y_1}^{y_2} dy \int_{x_1(y)}^{x_2(y)} f(x,y)\, dx$$

$$= \int_{y_1}^{y_2} G(y)\, dy$$

with

$$G(y) \equiv \int_{x_1(y)}^{x_2(y)} f(x,y)\, dx$$

Thus, in MATLAB code, the first step in evaluating the integral is to construct the function $G(y)$, which, for an arbitrary value of y, evaluates an integral over x from $x = x_1$ to $x = x_2$. The input to this function, namely y, will be a vector of values, and for *each* element of y, a separate integration is performed. That is, if y_i is the ith element of the vector y, the corresponding integration is

$$G(y_i) = \int_{x_1(y_i)}^{x_2(y_i)} f(x,y_i)\, dx$$

or, in MATLAB code:

```
function f = G(y,X1,X2,fnc)
  x1 = X1(y);      x2 = X2(y);
  n = length(y);
  for i = 1:n
    f(i) = quad8(fnc,x1(i),x2(i),[ ],[ ],y(i));
  end
```

Here we have assumed that the functions that specify the upper and lower limits already exist and are called X2 and X1, respectively.

Also, special care must be employed in writing the actual integrand function of the double integral, fnc(x,y). The function G interprets the function fnc as a function of a *single* variable, x, with the values of y being passed as *parameters*. Thus, this function must be written to expect vector input for x, but *scalar* input for y. The function should return a vector of the same size as x.

The value of the double integral is now easily obtained by simply integrating $G(y)$ over the range defined by the scalar values y_1 and y_2.

```
I = quad8('G',y1,y2,[ ],[ ],X1,X2,fnc)
```

In some programming languages, particularly FORTRAN, a procedure is not permitted to "call" itself. Thus, in the preceding code, the function quad is used to integrate $G(y)$, which, in turn, calls quad to integrate fnc(x). This is not permitted in FORTRAN and would require that two copies of the code for quad

---

```
y1 = 1;   y2 = 4;
Xhi= 'Xhi';
Xlo= 'Xlo';
fnc = 'Arg2d';
I = DblIntg(fnc,Xlo,Xhi,y1,y2);
```

The limits of the integration are over the domain $y_1 \leq y \leq y_2$, and $X_1(y) \leq x \leq X_2(y)$, where X_1 and X_2 are functions defined later. Arg2d is the name of the integrand function. Function DblIntg then evaluates the integral

$$I = \int_{y_1}^{y_2} \left(\int_{X_1(y)}^{X_2(y)} f(x,y)dx \right) dy$$

```
%======================================
function a = DblIntg(fnc,Xlo,Xhi,y1,y2)

  a = quad8('G2d',y1,y2,[ ],[ ],Xlo,Xhi,fnc)
%======================================
```

The inner integral over x is evaluated by the function G2d.

```
function f = G2d(y,X1,X2,fnc)
 y=y(:);  n = length(y);
 if isstr(X1)==1;
    x1 = feval(X1,y);
 else
    x1 = ones(size(y))*X1;
 end
 if isstr(X2)==1;x2=feval(X2,y);else;x2=ones(size(y))*X2;end
 for i = 1:n
   f(i) = quad8(fnc,x1(i),x2(i),[ ],[ ],y(i))
 end
 f = f(:);
```

G2d is the integral within the parentheses.

$X1$, $X2$ are either functions of y or are simply scalars.

For each value of y, do a separate integration.

Return a column vector.

```
======================================
function f = Xlo(y)
  f = ...
```

The lower limit is a function of y.

```
======================================
function f = Xhi(y)
  f = ...
```

The upper limit is a function of y.

```
======================================
function f = Arg2d(x,y)
  f = ...
```

The integrand of the double integral

Figure 11.1 *MATLAB code for double integrals*

be used and given different names. In MATLAB, there is no such prohibition, and there is no need for including renamed copies of the integration code.

Notice that only in the last integral of an iterated integral are the limits simple numbers. All inner integrals will, in general, have functions that define their limits. However, in those situations in which some or all of the limits of an inner integral are simply scalar numbers, it is convenient to adapt the code to recognize the difference and to accept either functions or scalars for the specification of inner limits. This is accomplished by means of the built-in MATLAB function `isstr`.[2] An example of the MATLAB code for evaluating a double integral is summarized in Figure 11.1.

To apply these ideas, consider the integral of the function $f(x, y) = x^2 + y^2$ bounded by the lines $x = 2$, $y = 1$, and the curve $y = x^2$. The two-dimensional integration region, $\mathcal{R}$, is indicated in Figure 11.2. If the integral is segmented as

$$I = \int_{y=1}^{4} \left(\int_{x=\sqrt{y}}^{x=2} (x^2 + y^2)\, dx \right) dy$$

the integration can be performed analytically, resulting in

$$I = \int_{y=1}^{4} \left(y^2 x + \frac{1}{3} x^3 \right) \Big|_{x=\sqrt{y}}^{4} dx$$

$$= \int_{y=1}^{4} \left(\frac{8}{3} + 2y^2 - \frac{1}{3} y^{\frac{3}{2}} - y^{\frac{5}{2}} \right) dy$$

$$= \frac{1006}{105} = 9.580952\ldots$$

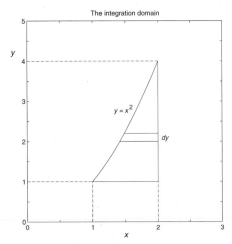

Figure 11.2 *The boundaries of the integration region* $\mathcal{R}$.

---

[2]The function `isstr(s)` returns a value of 1 if `s` is a string constant; if not, zero is returned.

The function `DblIntg` was used to evaluate this integral. The computed value of the integral was accurate to 13 digits (compared with the exact value of 1006/105), and the number of flops was approximately 12,000.

Romberg Techniques for Double Integrals

Numerically evaluating a one-dimensional integral by the methods of Romberg integration is one of the most efficient integration procedures available, particularly if the integrand is relatively smooth. The method consists of first approximating the integrand function by straight line segments over a partition of the integration interval into 2^k panels, for $k = 0, \ldots, n$. This operation generates a sequence of trapezoidal approximations to the integral, T_k. Next, without ever evaluating the function again, this set of trapezoidal numbers is combined to yield the same values that would have been obtained from a Simpson's rule approximation over the same partitions: $S_k = T_k + \frac{1}{4}(T_k - T_{k-1})$. Simpson's rule is equivalent to approximating the integrand by quadratic segments, so it is suggestive to attempt to combine the Simpson numbers to obtain a set of approximations that are equivalent to fitting the integrand by quartic segments; i.e., Boole's rule. The result is $B_k = S_k + \frac{1}{16}(S_k - S_{k-1})$. Extending these ideas results in the Romberg technique. The details are given in Chapter 7.

The same approach can be applied to two-dimensional integrals. For the moment, consider such an integral over a rectangular region:

$$I = \int_c^d \int_a^b f(x, y) \, dx \, dy$$

To simplify the notation later, we next rewrite this integral centered about the point (x_0, y_0):

$$I = \int_{y_0 - \Delta y}^{y_0 + \Delta y} \int_{x_0 - \Delta x}^{x_0 + \Delta x} f(x, y) \, dx \, dy$$

and expand the function $f(x, y)$ in a Taylor series about this point:

$$f(x, y) \approx f(x_0, y_0) + \left.\frac{\partial f}{\partial x}\right|_0 (x - x_0) + \left.\frac{\partial f}{\partial y}\right|_0 (y - y_0) + \cdots$$

By restricting the approximation to linear terms, we obtain the first approximation to the integral:

$$I \approx 4f_0 \Delta x \Delta y$$

That is, a linear approximation to the function yields an integral that is equal to the area of the rectangle times the value of the function at the *center* of the rectangle. The rectangle is next partitioned into four equal sections, and the integral approximated by the sum of the values of the function at the centers of each rectangle multiplied by their area. The next step involves halving both Δx and Δy, resulting in 16 rectangles.

Thus, the procedure averages the function over a succession of finer and finer grids. The grids are specified by square matrices **X** and **Y** containing the x and y coordinates of each point. Next, the integrand function is coded in the M-file function `fnc(X,Y)` to expect square matrices **X** and **Y** and to return a matrix of values. The MATLAB code to accomplish this generalization of the trapezoidal rule is as follows:

```
x = (a+b)/2;   y = (c+d)/2;
dx= (b-a);     dy= (d-c);     Area = dx*dy;
T(1) = feval('fnc',x,y);
for i = 2:k
   dx= dx/2;        dy= dy/2
   x = [x+dx/2;x-dx/2];   y = [y+dy/2;y-dy/2];
   X = ones(size(x))*x'; Y = y*ones(size(y'));
   T(i+1) = sum(sum(feval('fnc',X,Y)))/4^(i-1)*Area;
end
```

The next level of approximation is to retain the quadratic terms in the Taylor series expansion. The value for the integral is then

$$I \approx 4\Delta x \Delta y \left[f_0 + \frac{1}{6} \left(f_{xx}\Delta x^2 + f_{yy}\Delta y^2 \right) \right]$$

where we have used the shortened notation $f_{xx} \equiv \partial^2 f/\partial x^2\big|_0$, etc.

To obtain expressions for the quantities f_{xx} and f_{yy}, we next use the Taylor expansion to evaluate the function at the four points specified by $x_\pm = x_0 \pm \frac{1}{2}\Delta x$ and $y_\pm = y_0 \pm \frac{1}{2}\Delta y$. The values of the function at these points are designated as $f_{++}, f_{+-}, f_{-+}, f_{--}$. The result is easily found to be

$$\begin{pmatrix} f_{++} - f_0 \\ f_{+-} - f_0 \\ f_{-+} - f_0 \\ f_{--} - f_0 \end{pmatrix} = \frac{1}{2} \begin{pmatrix} 1 & 1 & \frac{1}{4} & \frac{1}{4} \\ 1 & -1 & -\frac{1}{4} & \frac{1}{4} \\ -1 & 1 & -\frac{1}{4} & \frac{1}{4} \\ -1 & -1 & \frac{1}{4} & \frac{1}{4} \end{pmatrix} \begin{pmatrix} f_x \Delta x \\ f_y \Delta y \\ f_{xy}\Delta x \Delta y \\ f_{xx}\Delta x^2 + f_{yy}\Delta y^2 \end{pmatrix}$$

This matrix equation is solved to obtain

$$f_{xx}\Delta x^2 + f_{yy}\Delta y^2 = 2(f_{++} + f_{+-} + f_{-+} + f_{--}) - 8f_0$$

Inserting this into the expression for the integral, we obtain

$$I \approx \frac{4}{3}\Delta x \Delta y \left[(f_{++} + f_{+-} + f_{-+} + f_{--}) - f_0 \right]$$

$$= \frac{1}{3}(4T_1 - T_0)$$

where $T_0 = 4\Delta x_0 \Delta y_0 f_0$ and $T_1 = 4\Delta x_1 \Delta y_1 (f_{++} + f_{+-} + f_{-+} + f_{--})$ are the first two trapezoidal approximations to the integral. Notice that this relation is exactly the same as the connection between Simpson's rule and the trapezoidal rule for one-dimensional integrals.

It is not difficult to extend this result to higher orders. If the two-dimensional grid is divided into n^2 rectangles, with $n = 2^k$, then the kth-order trapezoidal rule result for T_k is simply the mean of the values of the integrand function at the midpoints of these rectangles. Furthermore, you can show that the Simpson's rule result of the same order can be obtained directly from the trapezoidal results by the Romberg equation:

$$S_k = \frac{1}{3}(4T_k - T_{k-1})$$

To extend the level of the algorithm requires considerable algebra. However, it is possible to show that the next level of approximation; i.e., replacing the integrand by a quartic function of x and y, will generate a result, B_2, similar to Boole's rule, and that this result can be expressed as a combination of Simpson's rule results,

$$B_k = \frac{1}{15}(16S_k - S_{k-1})$$

which is identical to the associated Romberg equation. From these suggestive results it is tempting to assume that the remainder of the Romberg algorithm follows and to apply the Romberg method to the evaluation of a double integral. That is, once the set of trapezoidal rule approximations for a double integral has been computed by our code, this set is refined by the normal Romberg algorithm developed for one-dimensional integrals. Thus, we alter our code by inserting the following just before the **end** statement:

```
for m = 1:i-1
  T(m+1:i,m+1) = T(m+1:i,m) + (T(m+1:i,m)-T(m:i-1,m))/(4^m-1);
end
t = T(i,i);
dT= abs(t-T(i-1,i-1));
```

The quantity t is the final, most accurate, element of the Romberg table. dT is the difference between the last two diagonal elements and is an indication of the accuracy of the computed result. This code is quite efficient when integrating a function that is relatively smooth over a *rectangular* region; i.e., one in which the limits of the inner integral are constants. If this method is compared with the function DblIntg, which uses the function quad8 to perform the integration, it is found that both methods require approximately the same number of floating-point operations to obtain the same accuracy. However, by minimizing the use of **for** loops, the Romberg procedure has a clock time, as measured by tic-toc, that is about 50 times faster.

To adapt the code to the more common situation, where the inner limits are either constants *or* arbitrary functions of the second integration variable, requires a fair amount of dexterity when constructing the nonrectangular grid on which to evaluate the trapezoidal approximations. The interested reader will find one approach to constructing this grid in the listing of the function `Rmbg2d` included on the program disk.

EXAMPLE 11.1

Multidimensional Romberg Integration

The fact that the simple Romberg extrapolation procedure, which was developed for one-dimensional integrals, can easily be adapted to multidimensional integrals can frequently be used to great advantage, particularly when such integrals are evaluated using *vectorized* or partially vectorized arithmetic.

As an example of the application of the Romberg technique in three dimensions, consider the following test integral:

$$I = \int_0^2 dz \int_0^2 dy \int_0^2 dx \ \sin(xy)e^{-zy}$$

The *exact* value of the integral is $I = 1.81365676\ldots$. Next, the volume is partitioned into "cubes" by halving the step sizes in each of the three directions. The function is evaluated at the center of each cube, and the average of these values is determined. This is repeated for three more iterations. The trapezoidal rule values obtained in this manner are:

n	T
1	$2.476479\ldots$
8	$2.005760\ldots$
64	$1.862897\ldots$
512	$1.826036\ldots$
4096	$1.816756\ldots$

The ordinary Romberg extrapolation procedure is then applied to this set of numbers to obtain:

$2.47647901\ldots$
$2.00575952\ldots$ $1.84885302\ldots$
$1.86289726\ldots$ $1.81527651\ldots$ $1.81303807\ldots$
$1.82603598\ldots$ $1.81374882\ldots$ $1.81364697\ldots$ $1.81365664\ldots$
$1.81675574\ldots$ $1.81366235\ldots$ $1.81365658\ldots$ $1.81365674\ldots$ $1.81365674\ldots$

The most accurate values in each order of the calculation are those along the diagonal. Thus, this calculation has yielded a final value for the integral that is accurate to about $8\frac{1}{2}$ significant figures. The number of flops was 28,609 and the elapsed time was $\approx \frac{1}{4}$ second. Performing the same integration the standard way, as iterated integrals, requires slightly fewer flops ($\approx 24,000$) to attain the same accuracy, indicating that the standard method is slightly more efficient. However, the actual clock time for the latter calculation was, once again, longer by a factor of approximately 50. This is a consequence of the partial *vectorization* of the calculation. ●

Special Considerations Regarding Multidimensional Integrals

COMPUTING TIMES

The MATLAB code on page 258 can easily be generalized to handle three-dimensional integrals (see Problem 11.2) or integrals of even higher dimensions. Of course, the amount of computation, as measured by function evaluations, will increase as N^d, where N is the number of function evaluations needed for a comparable one-dimensional integral, and d is the dimensionality. Thus, to achieve satisfactory accuracy in one dimension for an integral whose integrand is quite smooth might require 50 or more function evaluations and perhaps 5 seconds of computer time. A comparable integral in three dimensions then requires approximately 125,000 function evaluations and would consume about $3\frac{1}{2}$ hours. You will readily appreciate the importance of carrying out any possible analytical integration steps before turning to the computer to evaluate multiple integrals.

Very frequently, a physical problem that requires three-dimensional integrals for its solution can be greatly simplified by invoking some fundamental symmetry of the problem. For example, when some problems are expressed in cylindrical coordinates, the underlying symmetry of the problem will suggest an *invariance* with respect to rotations about the cylinder axis. That is, the solution of the problem cannot depend on the angle measured about the cylinder axis. Then, by all means, we "integrate out" the dependence on this angle by analytically performing the angle integration first. Even in more complicated cylindrical problems, the angle integrals can frequently be done analytically; saving you a factor of 50 or more in computing time.

Finally, the resulting accuracy of a multidimensional integral can be proportionately improved by reducing the number of integration dimensions. Round-off error can easily render the computed value of an integral meaningless if the number of function evaluations becomes extremely large.

LIMITATIONS ON ACCURACY

Because multidimensional integration is simply a sequence of one-dimensional integrals embedded one within another, the same considerations apply if any of the integrals are *improper*. That is, if any of the limits are infinite, or if there are any infinite singularities within the integration region, you must first convince yourself that the integral itself remains finite, and this must be done analytically. Then, knowing that the integral does indeed represent a well-defined number, you must devise means whereby the computer can accurately estimate the value of the integral. These schemes were described in Chapter 7. Incorporating them in a multidimensional integration algorithm can be quite difficult, but is basically straightforward. You should be prepared to handle the following problems:

1. One or more limits of the integration are infinite. You must demonstrate that the integrand falls off significantly faster than one over the integration variable. The integral is typically evaluated by replacing infinite limits by appropriately chosen cutoffs.

2. The integrand contains a singularity, and the singularity is at the edge of the region. In this case, it must be shown that the integrand diverges slower than one over the integration variable in the vicinity of the singularity. The contribution to the integral from the neighborhood surrounding the singularity is evaluated analytically using the asymptotic form of the integrand. The remainder of the integral is done numerically.

3. The integrand diverges within the integration region. Often, the integrand will "change sign" across the singularity. That is, if for each point in which the integrand approaches $+\infty$ there is a corresponding point approaching $-\infty$ at the same rate, it may be possible to define a value for the integral known as the *principal value*, corresponding to the subtraction of the infinities.

Also, because one-dimensional integration algorithms were designed to be most accurate if the integrand resembled a polynomial, the same is true for multidimensional integrals. The multidimensional numerical integration process can be expected to return an accurate result if the integrand is "smooth" and resembles a polynomial of modest degree. If the integrand contains numerous "wiggles" of unknown size and frequency, or if it contains a few huge "bumps" at unknown locations, then, unfortunately, the numerical evaluation of the integral will probably be impossible.

Problems

REINFORCEMENT EXERCISES

P11.1 Two-Dimensional Integrals.

a. Reverse the order of the integrations on page 259. That is, show that the integral over the same region $\mathcal{R}$ can be written as

$$\int_1^2 dx \int_1^{x^2} (x^2 + y^2)\, dy$$

Evaluate the integral numerically.

b. The circle $x^2 + y^2 = 5^2$ and the ellipse $x^2/8^2 + y^2/3^2 = 1$ intersect at the point (x_c, y_c). Determine this point and use it to evaluate the area that is within both the circle and the ellipse. See Figure 11.3.

c. Sketch the integration region for the following integral.

$$\int_{x=1}^2 dx \int_{y=\sqrt{x}}^x \ln\left(\frac{x}{y}\right) dy + \int_{x=2}^4 dx \int_{y=\sqrt{x}}^2 \ln\left(\frac{x}{y}\right) dy$$

Switch the order of integration and adapt the code for part a to evaluate the integral.

P11.2 Romberg Multidimensional Integration.
Evaluate the integral of Problem 11.1b by the Romberg procedure by using function Rmbg2d on the applications disk.

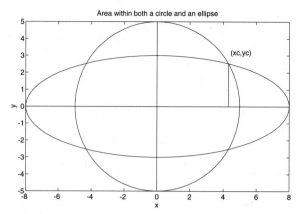

Figure 11.3   *The area within both a circle and an ellipse.*

P11.3 **Three-Dimensional Trapezoidal Rule.**   Write a short MATLAB function that will evaluate a three-dimensional integral over a cubical volume by repeatedly partitioning the volume into subcubes obtained by successive halving of the step sizes and then averaging the integrand function evaluated at the centers of these cubes. Test your code by using the integrand function of Example 11.1 to duplicate the trapezoidal rule numbers that are quoted there.

EXPLORATION PROBLEMS

P11.4 **Three- and Four-Dimensional Integrals.**
   a. Express the three-dimensional volume of the region bounded by the four planes: (i) $z = 8$, (ii) $z = x + 2y$, (iii) $y = 0$, (iv) $x = 0$ in terms of an integral over x, y, and z. Analytically evaluate the integral for the volume.
   b. What is the volume of the four-dimensional sphere of radius r?

P11.5 **Triple Integrals.**   Develop a MATLAB code to evaluate a *triple* integral of the form:

$$\int_{z_1}^{z_2} dz \int_{y_1(z)}^{y_2(z)} dy \int_{x_1(yz)}^{x_2(yz)} F(x, y, z)\, dx$$

   a. First test the code by evaluating the integral in Problem 11.2.
   b. If the region bounded by the planes in Problem 11.4a has a density given by $e^{-(x^2+y^2+z^2)/4}$, numerically determine the mass of the region. This triple integral will take considerable time for MATLAB to evaluate using our previous code, and it illustrates the primary reason that multiple integrals are so difficult for computers to evaluate. The sheer number of function evaluations for even the simplest of triple integrals can be enormous. When you run this code, keep track of the number of `flops` required.
   c. Rewrite the first integral over x in terms of an error function, thereby reducing it to a two-dimensional integral.

Summary

Iterated Integrals   A multidimensional integral such as $\int f(x, y, z)dx\,dy\,dz$ can be written as successive one-dimensional integrals

$$I = \int_{z_1}^{z_2} dz \int_{y_1(z)}^{y_2(z)} dy \int_{x_1(y,z)}^{x_2(y,z)} f(x, y, z)\, dx$$

Because the basic MATLAB integration functions are constructed to integrate a function of a *single* variable, the challenge is to structure the MATLAB code so that each step of the iterated integral is computed as an integral of functions of a single variable over well-defined limits. This can be accomplished by starting with the innermost integral and replacing, say $f(x, y, z)$, by a newly defined function, $F_{yz}(x)$. The new function is merely the same function as f, but is made to *appear* to be a function of a single variable with the other independent variables passed via **global** statements. The integral $G(y, z) = \int F_{yz}(x)\, dx$ is itself a function of y and z only. The next integral (over y) is handled in a similar manner.

Multidimensional   The integrand is first evaluated at the midpoint of the integration region and multiplied
Romberg   by the area of the region. This value is labeled T_0^0. Next, the integration region is
Integration   subdivided into subregions by halving the lengths along each of the integration axes; i.e., four rectangles in two dimensions, eight parallelepipeds in three dimensions, etc. The integrand, evaluated at the centers of each region, is multiplied by the volume of the region and the average of these values is labeled T_1^0. The regions are again halved, and the process is repeated. The Romberg equation,

$$T_k^{m+1} = T_k^m + \frac{T_k^m - T_{k-1}^m}{4^{m+1} - 1} \qquad m = 0, k - 1$$

is then used to compute the limit of this sequence of results.

MATLAB Functions Used

`isstr(s)`   Tests whether or not the value contained in the variable, **s**, is a *string*. A value of 1 is returned if **s** is a string constant, if not, zero is returned.

12 / *Monte Carlo Integration*

PREVIEW   In each of the one-dimensional integration algorithms developed in Chapter 7 and subsequently applied to multidimensional integrals in Chapter 11, the function was evaluated at a relatively small number of carefully selected sampling points, chosen so that the algorithm would yield exact results if the integrand were a polynomial of modest degree. If instead, we abandon completely any selection criteria for the sampling points in favor of taking a huge number of them, perhaps we can achieve, by sheer numbers, results that were earlier obtained by intelligence.

Of course, as computers inexorably become faster and faster, and as their storage capacity becomes ever larger, it becomes more and more tempting to simply use a relatively "dumb" algorithm and let the computer work harder. The old argument that each computer problem must be very carefully tuned so that the efficiency of the computation is near optimum before the problem is submitted to the computer, is now often ignored by *experienced* computational scientists and engineers in favor of brute force methods, like Monte-Carlo integration. I emphasize *experienced*, because even if this approach is to be used, the programmer must be aware of the potential costs to be incurred, and this is possible only if he or she is familiar with the alternatives.

However, as we shall see, Monte-Carlo integration *is* often the preferred approach. Some multidimensional integration problems involve limits that are so awkward that determining the sequence of limits in an iterated integral is nearly impossible. We describe problems of this nature at the end of this chapter.

Integrals as an Average of a Function

The one-dimensional integral

$$\frac{1}{L} \int_0^L f(x)\,dx \equiv \langle f \rangle \tag{12.1}$$

is also defined as the average of the function over the interval $0 \leq x \leq L$. This means that, if we sample the function at a large number, N, of randomly selected points in the interval, we can obtain an approximation to the average of f, expressed as

$$\langle f \rangle \approx \frac{1}{N} \sum_{i=1}^{N} f(x_i) \tag{12.2}$$

The approximate error in this expression for $\langle f \rangle$, after including N sampling points, depends on two quantities:

1. The *spread* of f values. This is usually expressed in terms of the standard deviation of f, which is defined as

$$\sigma = \sqrt{\langle f^2 \rangle - \langle f \rangle^2} \tag{12.3}$$

   that is, the *average of f^2* minus the (*average of f*)^2. Thus, if f is a constant for all x, then $\sigma = 0$, and a single point will be sufficient to determine $\langle f \rangle$.
2. The number of terms included in the summation, N. Clearly, the more sampling points we include, the more accurate the computed value of $\langle f \rangle$ should be.

If we can assume that errors are distributed in a normal distribution; i.e., described by a Gaussian, or bell-shaped, curve, then an approximate expression for the error in $\langle f \rangle$ is given, for large N, by the *standard error*[1]

$$\langle f \rangle \approx \frac{1}{N} \sum_{i=1}^{N} f(x_i) \pm \frac{\sigma}{\sqrt{N}} \tag{12.4}$$

Thus, to reduce the error, we attempt to either reduce σ or increase N, or both. Reducing σ, as we shall soon see, will usually correspond to a change of variables that causes f to be as close to "flat" as possible when expressed in the new variables. Also, with a fixed σ, to achieve one additional significant figure in accuracy will require approximately 100 times the current number of points. These two features are the principal guides we use in fashioning random sampling procedures to evaluate integrals.

Returning to Equations 12.1 through 12.3, we have the following equation for estimating one-dimensional integrals:

$$I = \int_a^b f(x)\, dx \approx (b-a) \left[\frac{1}{N} \sum_{i=1}^{N} f(x_i) \pm \frac{\sigma}{\sqrt{N}} \right] \tag{12.5}$$

$$= (b-a)\langle f \rangle \pm \delta I$$

where

$$\delta I = (b-a) \frac{\sigma}{\sqrt{N}} \tag{12.6}$$

is the standard error associated with this particular estimation of the value of the integral.

These equations generalize in an obvious manner to multidimensions,

$$\int dx_1 \int dx_2 \cdots \int f(\vec{x})\, dx_n \approx V \left[\frac{1}{N} \sum_{i=1}^{N} f(\vec{x}_i) \pm \frac{\sigma}{\sqrt{N}} \right] \tag{12.7}$$

---

[1]See any standard text on statistics; for example, D. Schiff and R. B. D'Agostino, *Practical Engineering Statistics*, John Wiley, New York, 1996.

where V is now the *volume* of the n-dimensional integration region; i.e., $dV = dx_1 dx_2 \cdots dx_n$, and $\vec{x}_i$ is a vector in that space.

Monte Carlo Integration Using Random Sampling

The method of integration using a random sampling of function evaluations is called the *Monte Carlo method*, and it is named after one of the temples of probability theory, the casino at Monte Carlo. The basic element of the procedure is the toolbox function $\texttt{rand(m,n)}$,[2] which will return an $m \times n$ matrix of random numbers, all in the range $0 \rightarrow 1$.

In order to understand the simplicity of the method and the limitations on its accuracy, it is best to start with a simple one-dimensional integral. The integral

$$I = \int_0^\pi \sin x\, dx = 2$$

represents the area under the curve $y = \sin x$ from $x = 0$ to $x = \pi$. The following MATLAB code integrates $\sin x$ over this interval by repeatedly selecting 1000 random points at a time, averaging $\sin x$ over these thousand points, and then updating the current expression for the average of $\sin x$ as well as $\sin^2 x$, which is used to compute the standard deviation. Notice that, in this case, we can compute σ exactly because $\langle \sin x \rangle = \frac{2}{\pi}$ and $\langle \sin^2 x \rangle = \frac{1}{2}$; thus, we have $\sigma = \sqrt{\frac{1}{2} - \frac{4}{\pi^2}} = 0.3078\ldots$.

```
function [z,err] = MCsinx(a,b)
exact= cos(a) - cos(b);
 avg = [0 0];                % Will contain <f>, <f^2>
   for N = 1:100
            x = a + (b-a)*rand(1000,1);
          term = sum([sin(x) sin(x).^2])/1000;
          avg = ((N-1)*avg + term)/N;
        err(N) = (b-a)*sqrt((avg(2)-avg(1)^2)/1000/N);
         z(N) = (b-a)*avg(1);
   end
N = 1000*(1:100)';
V = [z' exact+err' exact-err' exact*ones(100,1)];
plot(N,V)
title('Monte Carlo Integration of sin(x)');
xlabel('N')
```

[2] The function $\texttt{rand}$ with no argument returns a single scalar value in the range $0 \rightarrow 1$. Also, $\texttt{rand}$ will automatically return a different set of random numbers each time it is called. Be cautioned that $\texttt{rand}$ with a single argument, such as $\texttt{rand(1000)}$, does not return 1000 random numbers, but rather a 1000×1000 square matrix of random numbers. Finally, to compute, say 1000 random numbers, it is much faster to use the code $\texttt{x = rand(1000,1)}$, rather than $\texttt{for i = 1:1000; x(i) = rand; end}$.

The results of this program are shown in Figure 12.1.

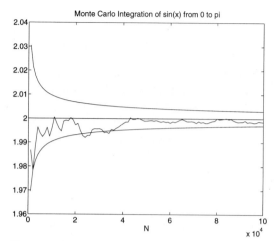

Figure 12.1 *A Monte Carlo approximation to $\int_0^\pi \sin x \, dx$. The smooth curves represent the exact result ± the standard error term.*

The procedure seems to be converging to the exact value of 2.0, but extremely slowly. From Equation 12.5, we see that the convergence rate is proportional to $\sqrt{N}$. Every algorithm we have encountered to this point has had a convergence rate that was at least linear, and several were quadratic. This example clearly illustrates the significant impact that the convergence rate can have on the solution of a problem: 100,000 points to achieve three-figure accuracy for a one-dimensional integral of a smooth function. How could a procedure this crude ever be the preferred method in any problem? We will attempt to answer this question momentarily, but first, let us examine more carefully the factors that affect the accuracy of Monte Carlo calculations.

Equation 12.5 can be used to estimate the approximate size of the error in a one-dimensional Monte Carlo integration. For example, in the integral that we have just computed, we have $\sigma \approx 0.308$, so that

$$\int_0^L f(x) \, dx = \int_0^\pi \sin x \, dx = 2 = \pi \frac{\sum_{i=1}^N f(x_i)}{N} \pm \frac{0.967}{\sqrt{N}} \qquad (12.8)$$

Then, to achieve an accuracy of four significant figures, $0.967/2\sqrt{N} < 10^{-4}$, requires $N > 23 \times 10^6$; a huge calculation for so simple an integral.

There is a way to somewhat improve the convergence rate of such integrals. The error term is proportional to the standard deviation, which, as we have seen, is zero if the function is constant. Therefore, a change of variables that results in an integrand that is more nearly constant over the interval will reduce σ and, thereby, the error term. Note, however, that the convergence *rate* will not be changed. Of course, in our example, the variable change $\xi = \cos x$ renders the integral $\int_{-1}^{1} d\xi$, which will yield the value 2 in one step. A somewhat more challenging integral is represented by

$$ I = \int_0^\pi e^{-x^2} \sin x \, dx $$

The value of this integral (as obtained by `quad`) is $0.424437\ldots$. The graph of the integrand is shown in Figure 12.2a. The Monte Carlo code adapted to this integral yields the results shown in Figure 12.2b. For this integrand, the ratio $\sigma/\langle f \rangle = 0.342$, whereas for the integral of $\sin x$, $\sigma/\langle f \rangle = 0.154$. Thus, the results should be roughly half as accurate for the same number of points.

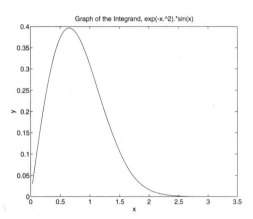

Figure 12.2a *The function $e^{-x^2} \sin x$.*

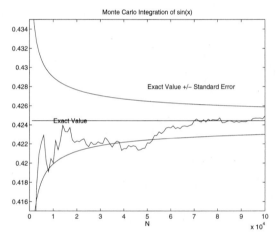

Figure 12.2b *A Monte Carlo approximation to $\int_0^\pi e^{-x^2} \sin x \, dx$.*

Next, experimenting with the transformation $t = e^{-x^2}$, the integral becomes

$$ I = \frac{1}{2} \int_{e^{-\pi^2}}^{1} \frac{\sin x(t)}{x(t)} \, dt $$

where $x(t) = \sqrt{-\ln(t)}$. The integrand of this much more complicated appearing integral is shown in Figure 12.3a.

Thus, it appears that this function will be more amenable to a Monte Carlo calculation. The results of the Monte Carlo code applied to this function are shown in Figure 12.3b. You can see that the results, as indicated by the size of the standard errors, are substantially more accurate.

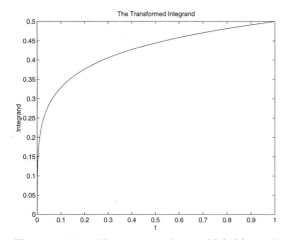

Figure 12.3a *The integrand* $\sin x(t)/x(t)$ *with* $x(t) = \sqrt{-\ln t}$.

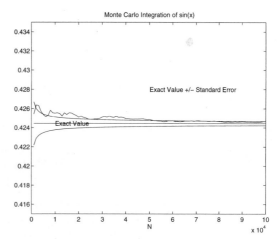

Figure 12.3b *Monte Carlo integration of the transformed function.*

MULTIDIMENSIONAL MONTE CARLO INTEGRATION ALGORITHM

The multidimensional code to implement the Monte Carlo algorithm in several dimensions is an easy generalization of the one-dimensional program. If the integration is over a rectangular Cartesian basis with limits $a_1 \leq x_1 \leq b_1$, $a_2 \leq x_2 \leq b_2, \ldots$, and $a_n \leq x_n \leq b_n$, the statement selecting a set of, say, 100 random points is

```
x = a + rand(100,1).*(b-a)
```

where **a** and **b** are *vectors* containing the limits of the individual variables.

If the integration is over variables expressed in spherical or cylindrical coordinates, then special care must be taken that the points are equally distributed throughout the volume. For example, the spherical coordinates of a point, (r, θ, ϕ), are related to ordinary Cartesian coordinates by the equations:

$$\begin{aligned}
x &= r\cos\phi\sin\theta \\
y &= r\sin\phi\sin\theta \\
z &= r\cos\theta \\
d\tau &= dx\,dy\,dz = r^2\sin\theta\,dr\,d\theta\,d\phi
\end{aligned} \qquad (12.9)$$

with limits specified as $0 \leq \phi \leq 2\pi$, $0 \leq \theta \leq \pi$, and $0 \leq r \leq a$. A random selection of 100 points over this domain is

```
phi     = rand(100,1)*2*pi
theta   = rand(100,1)*pi
r       = rand(100,1)*a
```

In addition, the extra term in the spherical volume element, $r^2 \sin\theta$, *must* be included in the integrand. This term is called the *Jacobian*[3] of the transformation from Cartesian coordinates to spherical coordinates.

The cylindrical coordinates of a point, (ρ, ϕ, z), are related to the Cartesian coordinates by the equations:

$$x = \rho \cos\phi$$
$$y = \rho \sin\phi$$
$$z = z \tag{12.10}$$
$$d\tau = dx\,dy\,dz = \rho\,d\rho\,d\phi\,dz$$

with the limits specified as $0 \le \phi \le 2\pi$, $0 \le z \le h$, and $0 \le \rho \le a$. In this case, the extra term added to the integrand is simply ρ.

As we have seen, obtaining satisfactory accuracy using Monte Carlo methods, even in one dimension, requires a large number of evaluation points. As with any multidimensional integral, if satisfactory accuracy requires N function evaluations for a one-dimensional integral, the multidimensional integral will require on the order of N^d function evaluations, where d is the dimension of the space.

Before attempting any major multidimensional Monte Carlo calculation, you should estimate the total computing time. The function **rand** in MATLAB is not particularly fast, and you may find that to obtain sufficient accuracy, this function may have to be called tens of millions of times. If that is the case, your program may have to run overnight, and it is then *essential* that you make provisions for catastrophic failure of the computing system. Thus, your programs should write intermediate results to a file every few minutes. The MATLAB statement for writing to a data file is **fprintf** and is described in Appendix Section A.vi.

Monte Carlo Methods for Integrals with Complex Boundaries

Quite often, the integrand of a multidimensional integral is reasonably simple and smooth, but the boundaries of the integration region are very awkward or difficult to express quantitatively. For example, the integral to obtain the volume

---

[3]The Jacobian is a function that relates volume elements in two different coordinate representations. For example, if the two coordinate systems are Cartesian (x, y, z) and spherical (r, θ, ϕ), the notation and definition of J is

$$J = \frac{\partial(x, y, z)}{\partial(r, \theta, \phi)} = \begin{vmatrix} \frac{\partial x}{\partial r} & \frac{\partial x}{\partial \theta} & \frac{\partial x}{\partial \phi} \\ \frac{\partial y}{\partial r} & \frac{\partial y}{\partial \theta} & \frac{\partial y}{\partial \phi} \\ \frac{\partial z}{\partial r} & \frac{\partial z}{\partial \theta} & \frac{\partial z}{\partial \phi} \end{vmatrix} = r^2 \sin\theta$$

so that if $F(r, \theta, \phi) \equiv f[x(r, \theta, \phi), y(r, \theta, \phi), z(r, \theta, \phi)]$, then

$$\int_{\mathcal{R}} f(x, y, z)\,dx\,dy\,dz = \int_{\mathcal{R}} F(r, \theta, \phi)\frac{\partial(x, y, z)}{\partial(r, \theta, \phi)}\,dr\,d\theta\,d\phi$$

of a cylinder of height h and radius a is quite trivial if the cylinder is centered at the origin and the problem is expressed in cylindrical coordinates:

$$V = \int_{z=0}^{z=h} \int_{\phi=0}^{\phi=2\pi} \int_{\rho=0}^{\rho=a} d\rho \, d\phi \, dz$$

$$= \pi a^2 h$$

To express the same integral in Cartesian coordinates is more difficult and becomes extremely arduous if the cylinder is not centered at the origin and not positioned parallel to one of the axes. The basic integral has not become more difficult, but determining the limits on the x, y, and z integrations is now a formidable task. Of course, for this problem, the correct procedure is to transform the variables to a body-centered coordinate system and introduce cylindrical coordinates. Unfortunately, this is not always possible. Thus, we have no alternative but to express the limits of each of the integrals in terms of the dependent variables at hand.

However, although the limits of the integrals may turn out to be unacceptably complicated, it usually is much easier to determine whether or not a randomly chosen point is inside the volume of integration. Thus, a point (x, y, z) will be inside a cylinder centered at the origin if

$$x^2 + y^2 \le a^2$$

$$|z| \le \frac{1}{2}h$$

Similar conditions can be devised for the displaced cylinder.

Next, imagine a regularly shaped volume surrounding the cylinder. The volume could be a sphere, a cube, or some other such structure. If we construct a large set of points randomly chosen throughout the regular volume, we expect the points to be more or less evenly distributed throughout. Thus, the fraction of those points that fall within the cylinder should be proportional to the *volume* of the cylinder itself. This provides an alternate procedure for integrating over the volume of the cylinder. Of course, some care must be used in choosing the enclosing volume. Because all the points that fall outside the integration volume are discarded, to optimize the accuracy of the calculation, the enclosing volume should match, as closely as possible, the actual volume of the integral.

ONE-DIMENSIONAL INTEGRALS AS RATIOS OF AREAS

As a first illustration of this method, we again return to the integral $\int_0^\pi \sin x \, dx$. This time, we will interpret the integral as a *surface* integral. Random values for both x and y will be chosen in the range $0 \le x \le \pi$ and $0 \le y \le 1$, and we will find the fraction of those points that fall under the sine curve. This is equivalent to a two-dimensional integral of a function that is *one* if $y \le \sin x$, and zero otherwise. The integrand is represented by Figure 12.4.

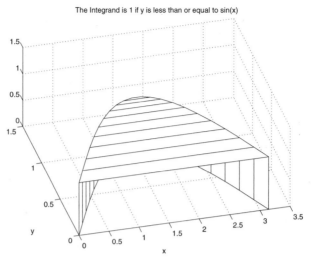

Figure 12.4   *The function is one if $y \leq \sin x$ and zero elsewhere.*

The code to evaluate this integral is as follows:

```
function [z,err] = MC2sinx(a,b)
 avg = 0;     exact = cos(a)-cos(b);
   for N = 1:100
            x = a + (b-a)*rand(1000,1)+eps;
            y = rand(1000,1);
         term = length(find(y <= sin(x)))/1000;
         avg = ((N-1)*avg + term)/N;
       err(N) = (b-a)*sqrt(avg*(1-avg)/1000/N);
         z(N) = (b-a)*avg(1);
   end
sigma = err(100)*sqrt(1e5)/(b-a)/exact;
N = 1000*(1:100)';
V = [z' exact+err' exact-err' exact*ones(100,1)];
plot(N,V)
```

The results of the calculation are shown in Figure 12.5, where you can see that, although the result does indeed converge to the exact result of 2, the error estimates are somewhat larger than those of the earlier calculation of the same integral (compare Figure 12.1).

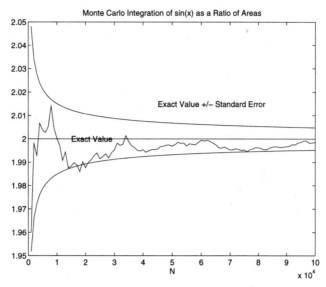

Figure 12.5 *A Monte Carlo integration of $\int_0^\pi \sin x\, dx$ as a ratio of areas.*

MONTE CARLO MULTIDIMENSIONAL INTEGRALS WITH COMPLEX BOUNDARIES

Finally, we attempt to illustrate those situations when a Monte Carlo calculation is the best we can do. Consider the relatively simple-appearing integral representing the volume enclosed by three identical cylinders intersecting at right angles. (See Figure 12.6.)

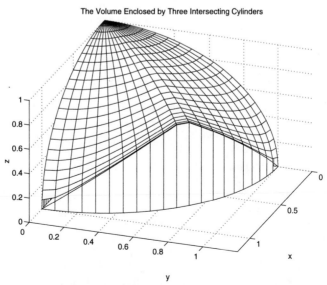

Figure 12.6 *Volume enclosed by three intersecting cylinders.*

The difficulty is in expressing the limits of the integration region. Try it! On the other hand, the condition that a random point lie within the region is simply:

$$(x^2 + y^2 \le a^2), \qquad \text{and } (x^2 + z^2 \le a^2), \qquad \text{and } (y^2 + z^2 \le a^2)$$

The points that are furthest from the origin are the points that are common to all three cylindrical surfaces; i.e., $x = y = z = (1/\sqrt{2})a$. Thus, a sphere of radius $r = \sqrt{3/2}\,a$ will completely enclose the volume. Alternatively, we could have chosen Cartesian coordinates and a cube whose sides are of length $2a$. The choice of spherical coordinates is suggested because enclosing the volume by a sphere, rather than a cube, will greatly reduce the amount of wasted volume where the integrand is zero.

In this example, the integral is $\int_{\mathcal{R}} r^2\, dr \sin\theta\, d\theta\, d\phi$, where $\mathcal{R}$ is the volume of the region enclosed by the three intersecting cylinders. The Monte Carlo integration is over a sphere of radius a with an integrand that is one if inside the volume of the cylinders, and zero elsewhere. The enclosing spherical volume closely approximates the volume $\mathcal{R}$; thus, we can expect that σ is relatively small. Nevertheless, this is a three-dimensional integral and we can also expect that a large number of points will be required to obtain a rough estimate of the integral's value. The MATLAB code to evaluate the integral is as follows:

```
function [V,err] = MC3cylsp(a)
%MC3cylsp is a Monte Carlo integration to obtain the volume
%     enclosed by three similar cylinders intersecting at right
%     angles.  The radius of the cylinders is a, and the sampling
%     is over a sphere of radius sqrt(1.5)*a.  The code performs
%     the integration in spherical coordinates, and is equivalent
%     to averaging the Jacobian [r^2 sin(theta)] over the points
%     that lie within the cylinder's boundaries.  The ''phase
%     space'' factor for averaging is C = (R)(pi/2)(pi/2) for the
%     three variables r, theta, and phi in the first octant of a
%     sphere.
%
%     The progress of the integration is then plotted vs N^(1/6),
%     where N is the number of sampling points, and its value is
%     compared with the exact value.  This problem is more easily
%     done in Cartesian coordinates, but this code illustrates
%     the use of the Jacobian for non-rectangular coordinates.
%================================================================
 R = sqrt(3/2)*a;
 Vsphere = pi*R^3/6; Vexact = 2*(1-1/sqrt(2))*a^3;
 C = R*pi^2/4;                  % Phase space volume
 avg  = 0;    avg2 = 0;   NN = 10;  n = 10000;
 for N = 1:NN
     r    =    R*rand(n,1);
     theta= 0.5*pi*rand(n,1);
     phi  = 0.5*pi*rand(n,1);
     x    = r.*cos(phi).*sin(theta);
     y    = r.*sin(phi).*sin(theta);
     z    = r.*cos(theta);
```

```
                    test = (x.^2+y.^2<=a^2)&(x.^2+z.^2<=a^2)&(y.^2+z.^2<=a^2);
                    i    = find(test);
                    term = sum(r(i).^2.*sin(theta(i)))/n;
                    avg  = ((N-1)*avg + term)/N;
                    avg2 = ((N-1)*avg2+ term^2)/N;
                    V(N) = C*avg;
                 err(N)= C*sqrt((avg2-avg^2)/N) ;
               end
             N = n*(1:NN);
             N = N.^(1/6);
             Q = [V' Vexact+err' Vexact-err' Vexact*ones(NN,1)];
             plot(N,Q);
             title('Monte Carlo Integration of a Volume within 3 cylinders')
             ylabel('Volume'); xlabel('N^(1/6)')
             text(4,Vexact,'Exact Value')
```

As in the one-dimensional case, if the integrand is a steeply peaked function, it will be to your advantage to attempt to change variables to make the new integrand as close to a constant as possible. The code above, as well as others in this chapter, is contained on the applications disk.

It is not possible to write a general-purpose Monte Carlo code to handle an arbitrary multidimensional integral, but the procedure should be quite clear.

Summary of the Monte Carlo Integration Method

1. If the function is sharply peaked, find a change of variables that will result in an integrand that is as close to a constant as possible. This will reduce the standard deviation of the function and will increase the rate of convergence.
2. If the method has been chosen because of difficulty in determining the intermediate limits of the iterated integral, replace the volume by an enclosing volume, V, with simpler limits, *but* take care to match the enclosing volume, as closely as possible, to the volume of the original integral. The enclosing volume will usually be rectangular in Cartesian coordinates (x, y, z), but could be part of a cylinder in cylindrical coordinates, or part of a sphere in spherical coordinates.
3. Estimate the number of points, N, you will need to obtain a first estimate of the integral and, from this, the computing time for the calculation. If need be, provide for saving intermediate results in your program.
4. Code the statement(s) that will determine whether or not a random point is inside the integration region, $\mathcal{R}$.
5. `for i = 1:N` use the function **rand** to compute an n-dimensional random vector that has its components scaled to fit within the enclosing volume. If the point is within the region $\mathcal{R}$, evaluate the integrand at this point and add the value to the accumulating sum, S, for the integral.
6. The value of the integral is then approximately SV/N.

7. Execute the entire program at least one more time, using a different seed for the function **rand**.

The Monte Carlo method was a method of last resort for our problem, forced on us by the fact that the limits of intermediate integrals in the iterated integration were unavailable. Unfortunately, this situation occurs very frequently with multidimensional integrals. This accounts for the popularity of the Monte Carlo method; a popularity that is certainly not a result of the cleverness, accuracy, or efficiency of the method.

Finally, there is an entire field of investigation called *simulation* that is closely related to the discussion of random numbers and Monte Carlo integration of this chapter. In simulation we attempt to mimic extremely complicated phenomena by using the computer's random number generator to simulate normal fluctuations in nature. Thus, a program could be constructed that simulates the diffusion of gas particles by a "random walk" of one or several particles with a slight drift in one direction. This type of analysis is growing rapidly in popularity, but is outside the scope of this text.

Problems

REINFORCEMENT EXERCISES

P12.1 Uniform Distributions of Random Numbers. A histogram of, say 1000 random numbers, can be graphed by the statements:

```
z = rand(1000,1); hist(z)
```

If the numbers contained in **z** are truly random and uniformly distributed, the histogram should be almost "flat." Flat, that is, with a fluctuation about a constant value, where the size of the fluctuation decreases with the square root of the number of terms. The statement **[n,x] = hist(z)** will return the heights of the individual columns in the histogram in the vector n. Compute the standard deviation of n for 100, 1000, and 10,000 random points and test the hypothesis that the fluctuations should be proportional to $1/\sqrt{n}$.

P12.2 Normal Distribution of Random Numbers. The function **randn** produces a normal, or bell-shaped, distribution of random numbers with a variance of +1. That is, the numbers are distributed like e^{-x^2}. Does this mean that if **z = randn(1000,1)** and **Z = exp(-z.^2)**; that a histogram of Z would be *flat*? How about the distribution of $\ln(Z)$? First, predict the answer; then produce the histogram to see if your prediction is correct.

P12.3 Random Distribution of Numbers in an Ellipse. Produce vectors x and y containing 1000 random numbers evenly distributed over the range $-3 \le x \le +3$ and $-1 \le y \le +1$. Discard all the values that fall outside of the ellipse $(x^2/3^2) + y^2 = 1$, and plot the points using lower case "o" for each point. Do the points appear to be evenly distributed? Next, produce a uniform distribution of 1000 values of θ and ρ in the range $0 \leq \theta \leq 2\pi$ and $0 \leq \rho \leq 1$. From these, compute x and y values from $x = 3\rho\cos\theta$ and $y = \rho\sin\theta$. Again, discard all points outside the ellipse and plot the remaining points. Are the points evenly distributed? Explain the difference.

P12.4 Two-Dimensional Monte Carlo Integrals. Evaluate the following integrals over the specified domains, $\mathcal{R}$, by using a Monte Carlo integral employing first 1000, and then 10,000, random points. Next, after sketching the domain, $\mathcal{R}$, of the integral, use the results of Chapter 11 to evaluate each of the double integrals as iterated integrals. Compare the results and any difficulties in the two methods.

    a. $\int\int_{\mathcal{R}} \sin(\ln(\sqrt{x^2 + y^2}))\, dx\, dy$ with $\mathcal{R}$ defined as the region within the two circles, $(x - 1)^2 + y^2 \leq 4$ and $x^2 + y^2 \geq 1$, that satisfies $x \leq 1$.

    b. $\int\int_{\mathcal{R}} [e^{-x^2} \sin y + e^{-y^2} \sin x]\, dx\, dy$ with $\mathcal{R}$ given by the region $y \geq \sqrt{x} - 2$, $y \leq \frac{1}{2}x + 1$, and $y \leq 4 - x$.

P12.5 Three-Dimensional Monte Carlo Integrals. The volume of the ellipsoid, $x^2/a^2 + y^2/b^2 + z^2/c^2 = 1$, is $V = \frac{4}{3}\pi abc$. Obtain an estimate of the volume of one octant of the ellipsoid by means of a Monte Carlo integral employing 1000 random values each for x, y, and z in the range $0 \leq x \leq a$, $0 \leq y \leq b$, and $0 \leq z \leq c$, for a variety of choices of a, b, and c. Can you *scale* this problem so that the integral is independent of a, b, and c? Rewrite your code as a one-line MATLAB statement.

P12.6 Integrand with Three Cylinders. Evaluate a Monte Carlo integration of the function $f(x, y, z) = e^{-(x^2+y^2+z^2)}$ over a region contained within the three intersecting cylinders of the example shown in Figure 12.6.

Summary

INTEGRALS AS AVERAGES

Basic Equation

Relating the definition of the average of a function over an interval in terms of the integral

$$\frac{1}{b - a} \int_a^b f(x)\, dx \equiv \langle f \rangle$$

to the definition of an average over a *sampling*,

$$\langle f \rangle \approx \frac{1}{N} \sum_{i=1}^{N} f(x_i)$$

yields the equation fundamental to the Monte Carlo method:

$$\int_a^b f(x)\, dx = (b - a) \left[\frac{1}{N} \sum_{i=1}^{N} f(x_i) \pm \frac{\sigma}{\sqrt{N}} \right]$$

which suggests that an integral may be evaluated by sampling methods. The quantity $\sigma = \sqrt{\langle f^2 \rangle - \langle f \rangle^2}$ is the standard deviation of the spread of f values.

Multidimensional Integrals

The generalization to multidimensional integrals is

$$\int dx_1 \int dx_2 \cdots \int f(\vec{x})\, dx_n = V \left[\frac{1}{N} \sum_{i=1}^{N} f(\vec{x}_i) \pm \frac{\sigma}{\sqrt{N}} \right]$$

where V is the volume of the integration region.

INTEGRALS AS RATIOS OF AREAS

Complex Boundaries

If a principal difficulty in the evaluation of an integral is associated with complex integration boundaries, the integral can be replaced by a second integral whose integration boundaries *enclose* the original, but are simpler, *and* the integrand, f, is replaced with a function that is equal to f if a point is within the original boundaries, and *zero*, if outside. The average of this function is then related to the original integral divided by the enlarged volume.

MATLAB Functions Used

rand	The basic random number generator in MATLAB. The various options available are as follows:
rand(n)	An $n \times n$ matrix with random entries uniformly distributed over the interval 0 to 1.
rand(m,n)	An $m \times n$ matrix with random entries.
rand(size(A))	Is the same size as A.
rand	With no arguments is a scalar whose value changes each time it is referenced.
randn	Switches to a normal distribution with mean 0.0 and variance 1.0.
hist	Used to plot histograms. If X is a vector, hist(X) will arrange the elements of X by size, from smallest to largest, and will then plot a histogram of 10 equally spaced bins over the range of X values. hist(X,N) uses N bins.

Bibliography for Part III

1. Forman S. Acton, *Numerical Methods That Work*, Harper & Row, New York, 1970.
2. K. W. Brodlie, "Unconstrained Minimization," in *The State of the Art in Numerical Analysis*, Chapter 3, D. A. H. Jacobs, ed., Academic Press, London, 1977.
3. G. Dahlquist and A. Björck, *Numerical Methods*, Prentice-Hall, Englewood Cliffs, NJ, 1974.

4. W. C. Davidon, *Variable Metric Method for Minimization*, AEC Report No. ANL-5990, Argonne National Laboratory, Argonne, IL, 1959.

5. D. Schiff and R. B. D'Agostino, *Practical Engineering Statistics*, John Wiley, New York, 1996.

6. R. Fletcher and M. J. D. Powell, "A Rapidly Convergent Descent Method for Minimization," *Computer J.* **6**, 1963.

7. J. M. Hammersley and D. C. Handscomb, *Monte Carlo Methods*, Methuen, London, 1964.

8. D. E. Knuth, "Seminumerical Algorithms," in *The Art of Computer Programming*, 2nd ed., vol. 2, Addison-Wesley, Reading, MA, 1981.

9. J. A. Nelder and R. Mead, *Computer Journal*, vol. 7, 1965.

10. H. Niederreiter, "Quasi-Monte Carlo Methods," *Bull. Amer. Math. Soc.* **84**, p. 957, 1978.

11. J. Ortega and W. Rheinbolt, *Iterative Solution of Nonlinear Equations in Several Variables*, Academic Press, New York, 1970.

12. W. H. Press, B. P. Flannery, S. A. Teukolsky, and W. T. Vetterling, *Numerical Recipes, The Art of Scientific Computing*, Cambridge University Press, Cambridge, England, 1986.

13. A. Ralston and P. Rabinowitz, *A First Course in Numerical Analysis*, McGraw-Hill, New York, 1978.

14. J. Stoer and R. Bulirsch, *Introduction to Numerical Analysis*, Springer-Verlag, Berlin and New York, 1980.

15. J. F. Traub, *Iterative Methods for the Solution of Equations*, Prentice-Hall, Englewood Cliffs, NJ, 1964.

PART IV /
Data Analysis
and Modeling*

PREVIEW   Interpreting raw data is one of the most creative activities of a scientist. When faced with a few hundred or several million values of some measured quantities, the scientist must somehow distill this mass of numbers into a concise and elegant statement about nature. Most often, measurements are made under preconceived ideas about some phenomena, in which case the experimenter must determine, with some certainty, whether or not the results support the hypothesis or invalidate it. This may mean that a proposed relationship between two or more variables can be tested by graphing the data in a particular form, or that a specific functional form is to be adjusted to best represent the data. Frequently, the summary of the data will consist of a few simple quantities, such as the average or the standard deviation. An experimenter might then need to determine the "best" value for a quantity, along with an estimate of the experimental uncertainties. One point, of which all those involved in data analysis are keenly aware, is that secure knowledge of the accuracy of a measured quantity is often more important than the actual value of the quantity. For example, the measured mass of a particular type of elementary particle, the electron neutrino, ν_e, is thought to be zero,[1] or, within experimental error, $m_\nu < 10^{-35}$ kg. If its true mass is as large as 10^{-35} kg, it is possible that the enormous number of these particles distributed throughout the universe could

---

*This part makes use of MATLAB programming techniques described in Appendix Sections A.i through A.vi.

[1] It may strike you as odd that an elementary particle could have precisely *zero* mass. However, you may be familiar with the quantum mechanical description of electromagnetism, in which one realization of the fields is in terms of radiation, such as light, and can be represented in terms of *particles* called *photons*, which have a mass of precisely zero.

contribute sufficient mass to eventually slow and even reverse the expansion of the universe. If the true mass is less, the expansion may continue unabated forever. Experimental uncertainties can be extremely important.

In this part, we review several techniques for summarizing and characterizing data. Data will either consist of a set of values representing a single quantity, for example, the responses to a questionnaire, or represent pairs, triplets, etc., of values that have a relationship to one another; e.g., the (x_i, y_i) coordinates of the trajectory of a projectile, or the pressure, volume, and temperature values of a gas measured at a succession of time values (P_i, V_i, T_i, t_i).

The first step in summarizing the data usually consists of characterizing the entire set by a few statistics, such as the average and the variance. MATLAB will enable you to then make statements like, "The measured value is $x = \ldots$ with an uncertainty $\Delta x = \ldots$."

Next, we review a variety of schemes for summarizing data in terms of functional forms; i.e., curve fitting. We will impose our ideas concerning the relationship between two or more experimental control variables onto the data and attempt to determine how well the data comply. Ordinarily, the functions will contain adjustable parameters that are tuned to optimize the fit to the data. In situations like this, the data summary might consist of statements like, "Within confidence limits, $\chi^2 = \ldots$, the data are described by the function $y_i = h\,e^{-\lambda x_i}$ with $h = \ldots \pm \ldots$ and $\lambda = \ldots \pm \ldots$." This type of analysis will progress from fitting lines, parabolas, or polynomials, to fitting exponentials or other functions, and finally to the lifeblood of the electrical engineer, fitting to expansions in terms of sines and cosines. This is Fourier analysis, and it is perhaps the primary impetus for the creation of MATLAB itself.

Again, our task is to extract from a collection of numbers an underlying statement of truth concerning nature, and to phrase this statement in the simplest and most precise form possible. The "handling" of data usually involves either the computation of the quantities that characterize the data; for example, the *mean*, the *standard deviation*, etc; or summarizing the data by some manner of curve fitting. In either task, the goal is to reduce the complications associated with hundreds, thousands, or perhaps millions of data points to a handful of items that accurately reflect what the data are trying to tell us.

13 / *Statistical Description of Data*

PREVIEW   This chapter is primarily concerned with characterizing a large set of data by a few quantities called *statistics* and *moments* of the data.

We begin with a discussion of numbers that result from some form of measurement, and that are presumed to bear a relationship to an underlying actual value of a quantity. If the set of numbers corresponds to repeated measurements of the same (or similar) quantities, it is to be expected that the results will likely clump around the actual value being measured. In such situations it is useful to characterize the data by the *moments* of the distribution.

The most common statistical functions used to characterize a data set are the *mean*, the *median*, and the *standard deviation*. We briefly describe and illustrate the MATLAB toolbox functions that easily calculate these and many other statistical quantities.

In addition, measurements are always accompanied by measurement errors. The results of a measurement are always a "cloudy image" of reality, with the observed values distributed about the unknowable underlying reality. This situation can be described by associating error bars with the results, or by using some other means of describing the scatter of the measured results about the actual values.

The "normal" spread of measurements about the actual values is given by the *binomial distribution*, which can be approximated in different regimes by *Poisson* or *Gaussian* distribution functions. Thus, the comparison of two measurements is equivalent to determining the overlap between two distribution functions. This idea, in turn, leads to a discussion of *likelihood* functions. These are a measure of the probability of occurrence of a particular set of experimental results scattered about known actual values, with their scatter characterized by a set of standard deviations. The *quality* of a data set may then be quantified in terms of a distribution about the maximum of the *likelihood* function, and characterized by a measure called *chi square*, χ^2. Using χ^2 we may then compute *confidence* levels that quantify the representation of the data as a faithful image of the actual values.

Moments of a Distribution

Distribution moments attempt to characterize a set of data that has a strong tendency to fall near a central value. We want to determine that central value and to describe the nature of the spread of the values around it.

THE MEAN OR AVERAGE OF A DISTRIBUTION

The best known of the distribution moments is the *mean* or *average*, which is defined as

$$\bar{x} \equiv \frac{1}{N} \sum_{i=1}^{N} x_i$$

where N is the total number of data values, x_i, and the overbars are used to denote the mean or average. This same quantity is often denoted with angle brackets as $\langle x \rangle$.

The MATLAB evaluation of $\bar{x}$ is obtained by the function `mean(x)`. If x is a vector quantity, `mean(x)` returns the mean of its components computed as `sum(x)/length(x)`. If x is an $m \times n$ matrix, `mean(x)` will return n values, a row vector, corresponding to the average value of the elements of the n *columns* of the matrix.

THE MEDIAN OF A DISTRIBUTION

The mean is not the only measure of the location of the center of the distribution. Another commonly used quantity is the *median*, or central value, of the distribution. If the data consist of an odd number of values, the median, designated as x_{med}, corresponds to that data value that has exactly the same number of points above it as below. If the data are an even number of points, the median is defined as the average of the two central values.

Clearly, calculating the median is much more difficult than computing the mean. The data set must be sorted so that the data fall in ascending or descending order; the median is associated with the middle of the ordered distribution. The calculation of the median in MATLAB is achieved by the function `median(x)`. As with `mean(x)`, a single value is returned if x is a vector; if x is a matrix, a row vector containing the median of each column of x is returned.

The difference between the median and the mean can give some indication of the distribution of the x values. For example, in a company with a large number of moderately paid employees and a handful of sumptuously paid executives, the median would be fairly close to the average of just the lowly workers; whereas in computing the overall average, the high salaries of a few might be sufficient to give a false impression of the actual pay of a typical employee. This distribution would have a large peak centered about the typical worker's pay plus a long tail extending out to the executive salaries. That is, the distribution would be asymmetrical.

There are additional statistical constructions to measure both the spread of the distribution, its asymmetry, and even its shape. These correspond to higher moments of the distribution.

THE STANDARD DEVIATION AND THE VARIANCE

The best known measure of the *spread* of a distribution is the *sample variance*, and it is defined as:

$$\text{Var} = \frac{1}{N-1} \sum_{i=1}^{N} (x_i - \overline{x})^2 \tag{13.1}$$

where $\overline{x}$ is the average of the distribution. Thus, to compute the variance, two passes through the data are required; the first to compute $\overline{x}$, and the second pass to sum the square of the deviations from the average. The reason that the factor multiplying the sum is $1/(N-1)$ rather than $1/N$ arises from the fact that, starting with N independent data values, we first compute $\overline{x} = \frac{1}{N}\sum x_i$. Thus, in subsequent computations the N data are no longer independent; e.g., $x_N = N\overline{x} - \sum_{i=1}^{N-1} x_i$, and so there are only $N-1$ independent terms entering into the variance calculation. If, however, the mean is known to be a certain value and is not computed from the data, the factor $1/(N-1)$ is then replaced by $1/N$.

THE STANDARD DEVIATION

The *standard deviation* is a well-known measure of the spread of a distribution and is defined to be the square root of the variance:

$$\sigma \equiv \sqrt{\text{Var}} \tag{13.2}$$

For large N, the difference between the factors $1/(N-1)$ and $1/N$ is small, and the variance can be simplified to read

$$\text{Var} \approx \frac{1}{N} \sum_{i=1}^{N} (x_i - \overline{x})^2$$

$$= \frac{1}{N} \left[\sum_{i=1}^{N} x_i^2 - 2\overline{x} \sum_{i=1}^{N} x_i + \overline{x}^2 \sum_{i=1}^{N} 1 \right]$$

$$= \overline{x^2} - 2\overline{x} \cdot \overline{x} + \overline{x}^2$$

$$= \overline{x^2} - \overline{x}^2$$

That is, the variance is the average of x^2 minus the (average of x)^2, and the standard deviation becomes for large N,

$$\sigma \approx \sqrt{\overline{x^2} - \overline{x}^2}$$

This expression for the standard deviation can be obtained with a *single* pass through the data.

The MATLAB function **std(x)** operates much like **mean** and **median**, and it returns the *sample* standard deviation[2] of the elements of the vector x or returns

---

[2]For a set of elements x_i, $i = 1, \ldots, n$, the sample standard deviation, s, is equal to the square root of the sample variance; i.e.,

$$s = \left[\frac{1}{n-1} \sum_{i=1}^{n} (x_i - \langle x \rangle)^2 \right]^{1/2}$$

a row vector of the sample standard deviations of the elements of the columns of a matrix x.

The use of the standard deviation in estimating the spread of a distribution is aided by *Chebyshev's theorem*, which states that if $k > 1$, the fraction of all measurements between $\overline{x} + k\sigma$ and $\overline{x} - k\sigma$ is greater than $\left(1 - \frac{1}{k^2}\right)$. Thus, at least 75% of all measurements are within $k = 2$ standard deviations of the mean.

THE MEAN ABSOLUTE DEVIATION

A second, less common, measure of the spread of a distribution is the average of the absolute deviations from the average, defined as

$$\delta = \frac{1}{N} \sum_{i=1}^{N} |x_i - \overline{x}| \tag{13.3}$$

This quantity is less popular as a measure of spread than the standard deviation because the absolute magnitude function has awkward mathematical properties, making it difficult to use in theoretical arguments. However, when compared with the standard deviation, it is less susceptible to being dominated by one or two terms with a large deviation, and indeed is a useful alternative for distributions that are quite broad with a substantial number of points far from the mean.

There is no toolbox function to directly compute δ, but one is easily constructed and illustrated as follows:

```
function delta = absdev(x)
avg = mean(abs(x - mean(x)));
```

Modes of a Distribution

The *mode* of a distribution corresponds to the most likely value in the distribution. If we think of the data as representing a single underlying value clouded by experimental errors and measurement fluctuations, we can picture the data as a continuous function, $\omega(x)$, corresponding to the probability that a particular measurement will result in the value x. Because a measurement must result in some value, $\int_{-\infty}^{+\infty} \omega(x)\,dx = 1$, with 1 representing certainty. The most likely value, or mode, of the distribution corresponds to the maximum of the function $\omega(x)$. Of course, the probability distribution could have more than one local maximum. For example, if there are two local maxima, the distribution is termed *bimodal*. Characterizing a distribution in terms of modes is clearly going to be more useful than using means, medians, and standard deviations if the distribution of the data contains two or more clearly defined maxima.

All of this is perfectly straightforward if the function $\omega(x)$ is continuous and differentiable. But we were only imagining the function $\omega(x)$. All we actually have is a set of numbers that appear clustered about some central value; that is, the density of observations increases as we near the most likely value. In elementary statistics, this idea is used to estimate the mode by a process called *binning*, counting the number of values that fall between equally spaced values of x. These numbers are then plotted in the form of a histogram, and the tallest bin of the histogram is expected to contain the most likely value. Of course, scientists are usually averse to binning data, because it necessarily reduces the information content of the data.

USING HISTOGRAMS TO LOCATE DISTRIBUTION MODES

A histogram counts the number of points that fall within a selection of intervals all of the same width. First, let us summarize the properties of the function used in constructing a histogram. The toolbox function `hist(x,n)` will draw a histogram of the values contained in the vector x using n equally sized bins spaced from x_{min} to x_{max}. If n is omitted, a value of $n = 10$ is assumed. Of course, the function proceeds by first sorting the x values, determining the minimum and maximum, counting the number of values in each bin, and finally, plotting these numbers using the function for bar graphs, `bar`. The tallest bin should contain the mode of the distribution.

A related operation is to fix the number of points per bin, and then determine the width of each bin. The narrower the bin, the denser the data near that point. The idea is to first sort the data into N values x_i, next to select an integer value, for example, greater than or equal to 5, and finally to scan through the data to determine the location of the minimum width bin. We will assume that the mode is then at the central point of this bin. The code is as follows:

```
x = sort(x);
m = 2;
N = length(x);
for i = m+1:N-m
  w(i-m) = (x(i+m)-x(i-m))/(2*m+1);
end
[xmax mmax] = max(w)
```

The problem with this method is that there is nothing unique about the choice of the number of points, $2m + 1$, that are to be included in each bin. The code will have to be rerun several times with different choices for m, and not surprisingly, significantly different results may be obtained in each run. What is happening is that a choice of small m may lead to the selection of a narrow spike in the data as the maximum, whereas using a larger value causes the spike to be washed out, and a completely different value of x suggested as the position of the most likely point. Usually, this observation indicates that the data cannot support an interpretation of the spike as a genuine maximum. Either the data must be improved (more measurements taken to fill in the spike), or the interpretation discarded.

Higher Moments of a Distribution: The Skewness and the Kurtosis

The *mean, median, standard deviation,* and *mode* are intended to characterize the central peak and spread of a distribution. Higher moments can be used to help classify the actual shape of the distribution function. Higher moments will require calculation of quantities like $\overline{x^3}$ and $\overline{x^4}$. Because these involve raising x to rather high powers, you can expect that they will be strongly dominated by a few *outlier* points far from the central peak. If there are many such points, the utility of higher moments is greatly diminished or invalidated.

THE SKEWNESS

The *skewness,* or *third moment,* of a distribution is defined as

$$Skew \equiv \frac{1}{N} \sum_{i=1}^{N} \left[\frac{x_i - \overline{x}}{\sigma} \right]^3 \tag{13.4}$$

and is a measure of the *asymmetry* of the distribution. A positive value for *Skew* reflects a distribution that stretches out farther to the right (values greater than $\overline{x}$) than to the left; the converse is true for a negative *Skew*.

THE KURTOSIS

The kurtosis, or fourth moment, is defined as

$$Kurt \equiv \frac{1}{N} \sum_{i=1}^{N} \left[\frac{x_i - \overline{x}}{\sigma} \right]^4 - 3 \tag{13.5}$$

and is a measure of the "pointyness" of the distribution. A large positive value of *Kurt* suggests a very sharp, even cusplike, peak to the distribution; a large negative value represents a distribution with a broad, relatively flat, peak. The negative three is included in the definition so that a *normal,* or *Gaussian,* distribution will have zero kurtosis, as we will see later in the chapter.

Again, because both *Skew* and *Kurt* can so easily be dominated by a few outlier points, you should use caution when employing these functions to interpret data.

Characterizing a distribution by its moments is clearly related to the Taylor series expansion of a function, and it is not surprising that an expansion in terms of moments, if it converges, will uniquely define a distribution function. Also, if the higher moments, like the kurtosis, are defined to be zero if the distribution is Gaussian-like, the expansion can be expected to rapidly converge whenever the distribution is Gaussian-like.

ILLUSTRATIVE PROBLEM 13.1

Moments of Random Distributions

The function **rand** should produce a set of numbers evenly distributed over the interval $0 \to 1$. Thus, the average, median, and mode will approach $\frac{1}{2}$, and the standard deviation should approach $\sigma \approx 1/2\sqrt{3}$. (See Problem 13.2.) The skewness should be approximately zero, and the kurtosis should be negative. The following tests these assumptions. (See the script file **Moments.m**.)

```
x    = rand(1,500);
avg = mean(x);      mid = median(x);
absdev = mean(abs(x-avg));
sigma = std(x);
X   = sort(x);      N  = length(x);
for k = 1:N-4
    w(k) = X(k+4)-X(k);
end
[pmin,imode] = min(w);
mode = X(imode);
disp('mean   median   mode   absdev  sigma')
disp([avg mid mode absdev sigma])
skew = mean((x-avg).^3);
kurt = mean((x-avg).^4)-3;
disp('skewness  kurtosis')
disp([skew kurt])
```

The output of this code is:

mean	median	mode	absdev	sigma
0.5106	0.5115	0.9629	0.2524	0.2931

skewness	kurtosis
−0.0013	−2.9876

Because the distribution is "flat," the mode or maximum of the distribution is of little meaning. The code above is contained in a script file named `moments.m` on the applications disk. Next, to obtain a more interesting distribution, replace the first line of the preceding code by

```
s  = 2.0;        % The standard deviation
x0 = 6.0;        % The median of the distribution
x1 = -s*log(rand(1,500));
x2 =  s*pi*rand(1,500);
x = x0 + [x1.*cos(x2) x1.*sin(x2)];
```

This will now produce a set of random numbers with a distribution function

$$\omega(x) = \frac{1}{\sqrt{2\pi}s}e^{-(1/2)(x-x_0)^2/s^2}$$

called the *Gaussian distribution*, which has a standard deviation equal to s and a mean equal to x_0. This code employs the *Box-Muller* method for generating Gaussian distributions of random numbers. See Press et al. (1992) for details. The output of this code is given in the following paragraph, and the distribution of the x's is shown as a histogram in Figure 13.1 (page 294).

In addition, MATLAB has a special random number generator that will return random numbers distributed in a Gaussian fashion with a mean of 0 and a standard deviation of 1. The function is `randn(m,n)`. It will return an $m \times n$ matrix of Gaussian distributed numbers, and in most respects, is similar to `rand`.

mean	median	mode	absdev	sigma
6.0304	6.0438	6.0504	1.5585	1.9605

skewness	kurtosis
0.3973	44.9011

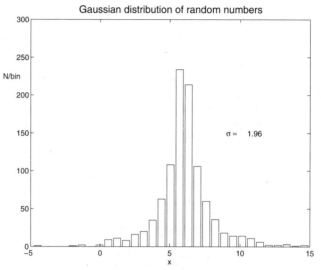

Figure 13.1 *Random numbers distributed as* $(1/\sqrt{2\pi}s)e^{-(1/2)(x-x_0)^2/s^2}$

Distribution Functions

THE BINOMIAL DISTRIBUTION

Probability theory is an attempt to predict the occurrence of *events*. The definition of an event can vary, but for our purposes, will correspond to one measurement of a particular quantity. The basic assumption of the theory is that successive events, or measurements, are entirely independent of one another. For example, the measurement could be the number of cars that pass in one minute, the number of radioactive atoms that decay in one second, the pressure in a gas container, how many heads are up in tossing N coins, etc. Measurements of each of these quantities would be expected to return a particular value, for example, the most likely value—the mode, plus or minus some degree of fluctuation. The probability distribution of expected results is called a distribution function.

Consider repeated measurements of the same quantity; for example, the number of coins that are heads up when tossing 10 coins. The *event* is the flipping of the 10 coins and observing how many are heads up. The *value* of the measurement can be any integer between 0 and 10. If $0 \le p \le 1$ is the probability of observing "heads" on one coin flip, then $q = 1 - p$ is the probability of measuring "tails." Ideally, both are equal to $\frac{1}{2}$. Then, in one coin flip, the probability of *either* heads or tails is $(p + q) = 1$, and in N coin flips, $(p + q)^N = 1$. Expanding this latter expression by the binomial expansion, we have

$$(p + q)^N = 1 = \sum_{i=0}^{N} \frac{N!}{(N - i)!\, i!} q^{N-i} p^i$$

The factorials, $n!$, are defined for nonnegative integers as $n! = n(n - 1) \cdot (n - 2) \cdots 2 \cdot 1$, and $0! \equiv 1$. The interpretation of this summation is that it represents all possible outcomes of N flips of a coin, and a particular term, for example, $i = m$, is the probability of m heads in N coin flips. The distribution of heads in N coin flips is then

Binomial
Distribution $$\omega_N(m) = \frac{N!}{(N - m)!\, m!} (1 - p)^{N-m} p^m$$
Function

where p is the probability of heads on one toss, and m is the number of heads up out of the N coins. The number of separate elements in a measurement is called the number of *degrees of freedom*; in this case the number of degrees of freedom is N. This distribution is known as the *binomial probability distribution function* for N degrees of freedom.

If $\bar{x}$ is the average number of heads in N tosses, clearly,

$$\bar{x} = Np = N(1 - q)$$

Next, we use these relations to replace the probabilities p and q in the earlier expression to obtain the *distribution* function for heads. Also, because our intent is to obtain a continuous distribution function, we replace the integer m by the continuous[3] variable x. Thus, making the replacements $m \to x$ and $p \to \bar{x}/N$, the binomial distribution function can be written as

$$\omega_N(x) = \frac{N!}{(N - x)!\, N^x\, x!} \bar{x}^x \left(1 - \frac{\bar{x}}{N}\right)^{N-x} \tag{13.6}$$

Let $\omega(x)\, dx$ represent the probability of obtaining a value of x in the interval x to $x + dx$ in a single measurement, and let both ω and x be continuous variables. Assuming the normal conditions of probability theory; i.e., the measurements are independent of one another, this function can be obtained from the binomial distribution by permitting the number of possibilities to approach infinity. That is, in the coin toss example, increasing the number of coins in each toss will

---

[3]In the event that x is not a nonnegative integer, the factorials are generalized to the *gamma* function,

$$\Gamma(x) = \int_0^\infty t^{x-1} e^{-t}\, dt$$

The gamma function duplicates the properties of the ordinary factorial, shifted by one:

$$\Gamma(n + 1) = n!$$

(See also Problem 13.6).

increase the number of possible outcomes, and as the number of possibilities approaches infinity, $\lim_{N \to \infty} \omega_B(m) \Rightarrow \omega(x)$.

There are two ways to take the limit of Equation 13.6 as $N \to \infty$:

1. Assume that $\overline{x}$ approaches a fixed *value* independent of N; i.e., the probability $p = \overline{x}/N \to 0$.
2. Assume that $\overline{x}$ approaches a fixed *fraction* of N, and the probability $p = \overline{x}/N \not\to 0$.

The first category includes measurements such as the sizes of grains of sand. If x is the size of a sand grain, and N is the number of sand particles on the beach, the value of $\overline{x}$ will remain finite as $N \to \infty$, and $p = \overline{x}/N \to 0$.

On the other hand, if $\overline{x}$ represents the number of heads in a single toss of N coins, as $N \to \infty$, $\overline{x}$ will approach $\frac{1}{2}N$, so $p \to \frac{1}{2}$; thus this falls under the second category.

A slightly different example is found in the distribution of radioactive decays observed in a macroscopic sample of atoms. If $\overline{x}$ represents the average number of atoms that radioactively decay per second out of a sample of N atoms, typical values might be $N \approx 10^{23}$ atoms, and the number of decays will usually be of the order of hundreds per second. Thus, $p \approx 10^{-21}$ decays/atom/s, not zero. However, because p is extremely small, this phenomenon can be quite accurately described using the limit of the binomial distribution as $p \to 0$.

THE POISSON DISTRIBUTION

The *Poisson distribution function* is the limit of the binomial expansion, Equation 13.6, as $N \to \infty$ while $\overline{x}$ remains finite; i.e., $p = \overline{x}/N \to 0$. In taking the limit of Equation 13.6, we expect $\omega(x)$ to be approximately zero for values of x far from $\overline{x}$. Thus, we may consider x to remain finite (and near $\overline{x}$), so that

$$\lim_{N \to \infty} \left[\frac{N!}{(N-x)!\, N^x} \right] \Rightarrow 1$$

and

$$\lim_{N \to \infty} \left(\left(1 - \frac{\overline{x}}{N}\right)^{N-x} \right) \Rightarrow \lim_{N \to \infty} \left(\left(1 - \frac{\overline{x}}{N}\right)^{N} \right) = e^{-\overline{x}}$$

and in the limit of large N, the binomial distribution function can be replaced by the Poisson distribution function.

Poisson Distribution Function
$$\omega_P(x) = \lim_{N \to \infty} \omega_N(x) = \frac{\overline{x}^x\, e^{-\overline{x}}}{x!} = \frac{\overline{x}^x\, e^{-\overline{x}}}{\Gamma(x+1)}$$

THE NORMAL OR GAUSSIAN DISTRIBUTION FUNCTION

The second limiting case, that of $N \to \infty$, $p \not\to 0$, is more difficult to obtain. The procedure is to first replace x by a variable measured from the average, $\overline{x}$,

$$\xi = x - \overline{x}$$

so that ω will be nonzero only for "small" values of ξ:

$$\omega_N(\xi) = \frac{N!}{(N - \overline{x} + \xi)! \, N^{\overline{x}+\xi}} \frac{\overline{x}^{\overline{x}+\xi}}{(\overline{x} + \xi)!} \left(1 - \frac{\overline{x}}{N}\right)^{N-\overline{x}-\xi} \tag{13.7}$$

Next, *Stirling's approximation* for large factorials,

$$n! \approx \sqrt{2\pi} \, n^{n+(1/2)} e^{-n} \tag{13.8}$$

is used to replace each of the factorials in Equation 13.7. The result is

$$\omega(\xi) = \sqrt{\frac{N}{2\pi \overline{x}(N - \overline{x})}} \left(1 - \frac{\xi}{N - \overline{x}}\right)^{-(N-\overline{x})+\xi+(1/2)} \left(1 + \frac{\xi}{\overline{x}}\right)^{-\overline{x}-\xi+(1/2)}$$

The $\frac{1}{2}$'s in the exponents can be neglected in the limit $N \to \infty$.

To proceed further, we use the trick $C = e^{\ln C}$ to write the product of the two terms as a sum of logarithms of the form $\ln(1 + \epsilon)$ with $\epsilon \ll 1$. Finally, using the series expansion of the logarithm,

$$\ln(1 + \epsilon) \approx \epsilon - \frac{1}{2}\epsilon^2 + \cdots \qquad (\epsilon \ll 1)$$

the limit of the binomial distribution function for $p \not\to 0$, known as the *normal*, or *Gaussian*, distribution, is then found to be

Gaussian or Normal Distribution

$$\omega_G(\xi) = \frac{e^{-(1/2)(\xi/\sigma)^2}}{\sqrt{2\pi} \, \sigma}$$

where

$$\sigma = \sqrt{\frac{\overline{x}(N - \overline{x})}{N}} = \sqrt{Np(1 - p)}$$

is the standard deviation of the distribution.

For example, the presidential election of 1960 was so close that some pundits suggested that the result was indistinguishable from a random selection wherein each voter simply flipped a coin. Well, if that were the case, we would expect the final vote tally to be within a few standard deviations from the 50/50 mode of the distribution. For this election the number of voters was $N \approx 60,000,000$, $p = \frac{1}{2}$, and President Kennedy received about 200,000 more votes than candidate Nixon, or $\approx 0.20\%$. However, this value is still 18 standard deviations from a tie—perhaps not a landslide, but outside the limits of a random selection.

In Figures 13.2a and 13.2b are plotted (smooth curves) of a Gaussian distribution with $p = \frac{2}{3}$, and a Poisson distribution with $\overline{x} = 10$. These are then compared (+'s) with comparable binomial distributions for $N = 100$ points.

The area under the Gaussian distribution for a range of ξ represents the probability of a measurement returning a value of ξ in that range. Thus, the probability of a measurement being within δ of the mean, $\overline{x}$, is

$$P(\pm\delta) = \frac{1}{\sqrt{2\pi}\sigma} \int_{-\delta}^{\delta} e^{-(1/2)(\overline{x}/\sigma)^2} d\xi = \frac{2}{\sqrt{\pi}} \int_0^{\sqrt{2}\sigma\delta} e^{-t^2} dt$$

where, in the last step, we have made a change of variables, $t = \xi/\sqrt{2}\sigma$, and used the fact that the integral is even in ξ. This last integral is a common mathematical function known as the *error function*. The error function appears frequently in the solution of differential equations. It is extensively tabulated, and it is available on most computing systems. In MATLAB, this integral is evaluated by the toolbox function `erf(k)`; i.e.,

$$\mathtt{erf(k)} = \frac{2}{\sqrt{\pi}} \int_0^k e^{-t^2} \, dt$$

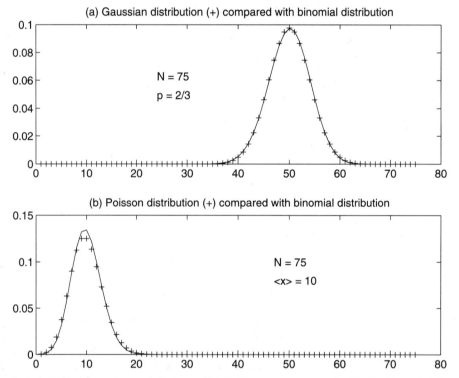

Figure 13.2 *Gaussian and Poisson distributions are compared with the binomial distribution function.*

Properties of Distribution Functions

If x is continuous, the distribution function $w(x)$ has the meaning that $w(x)\,dx$ is the probability of obtaining a value of x in the interval $x \rightarrow x + dx$ in a single measurement. Therefore, it must satisfy $\int w(x)\,dx = 1$, where the limits of the integral are over all possible values of x.

If m is an integer, the discrete distribution function $w(m)$ must satisfy $\sum_{m=0}^{\infty} w(m) = 1$. Thus, for the Gaussian distribution,

$$\int_{-\infty}^{\infty} \omega_G(x)\, dx = \frac{1}{\sqrt{2\pi}\sigma} \int_{-\infty}^{\infty} e^{-(1/2)(x-(\bar{x}/\sigma))^2}\, d\xi$$

$$= \frac{2}{\sqrt{\pi}} \int_0^{\infty} e^{-t^2}\, dt = 1$$

where we have used the identity

$$\int_0^{\infty} x^{2n} e^{-ax^2}\, dx = \frac{1 \cdot 3 \cdot 5 \cdots (2n-1)}{2^{n+1}a^n} \sqrt{\frac{\pi}{a}} \tag{13.9}$$

to evaluate the last integral.

Similarly, for the Poisson distribution,

$$\sum_{m=0}^{\infty} \omega_P(m) = e^{-\bar{x}} \left(\sum_{m=0}^{\infty} \frac{\bar{x}^m}{m!} \right)$$

$$= e^{-\bar{x}} e^{\bar{x}}$$

$$= 1$$

The following identities can be used along with Equation 13.9 to evaluate the moments of the Poisson and Gaussian distribution functions:

$$\sum_{m=1}^{\infty} m \frac{\bar{x}^m}{m!} = \bar{x} e^{\bar{x}}$$

$$\sum_{m=1}^{\infty} m^2 \frac{\bar{x}^m}{m!} = (\bar{x}^2 + \bar{x}) e^{\bar{x}}$$

$$\sum_{m=1}^{\infty} m^3 \frac{\bar{x}^m}{m!} = (\bar{x}^3 + 3\bar{x}^2 + \bar{x}) e^{\bar{x}}$$

$$\sum_{m=1}^{\infty} m^4 \frac{\bar{x}^m}{m!} = (\bar{x}^4 + 6\bar{x}^3 + 7\bar{x}^2 + \bar{x}) e^{\bar{x}}$$

and give the following results:

	Mean	Standard Deviation	Skewness	Kurtosis
Poisson	$\bar{x}$	$\sqrt{\bar{x}}$	$\frac{1}{\sqrt{\bar{x}}}$	$\frac{1}{\bar{x}}$
Gaussian	$\bar{x}$	σ	0	0

Combinations of Distribution Functions

If x and y are two random variables with continuous distribution functions, $\omega_1(x)$ and $\omega_2(y)$, it would be interesting to determine the distribution function for combinations of x and y, as in a function $z = f(x, y)$. It is clearly possible to calculate the moments of z as

$$\overline{z^n} = \int \int [f(x,y)]^n \omega_1(x)\omega_2(y)\,dx\,dy$$

and use these to characterize the distribution of z's. In the following sections we consider a few simple examples of such functions.

THE AVERAGE OF THE AVERAGES

A particularly important combination of x and y is the sum, or average, of the two

$$z = \frac{1}{2}(x+y)$$

So that

$$\overline{z} = \frac{1}{2} \int \int (x+y)\omega_1(x)\omega_2(y)dx\,dy$$

$$= \frac{1}{2}(\overline{x} + \overline{y})$$

(13.10)

Thus, if X is a measure of the average of x's obtained by sampling, for example, n points,

$$X = \frac{1}{n}(x_1 + x_2 + \cdots + x_n)$$

then, repeating this process over and over again, we obtain a distribution of X values. The average of these averages is then:

$$\overline{X} = \frac{1}{n}(\overline{x_1} + \overline{x_2} + \cdots + \overline{x_n})$$

(13.11)

THE STANDARD DEVIATION OF THE AVERAGES

Certainly, the more times we measure $\overline{X}$, the more accurate should be our determination of $\overline{x}$ itself. That is, the spread of the X's should decrease with repeated measurements.

The standard deviation of the sum of two variables can be obtained from the preceding discussion to be

$$\sigma^2(ax+by) = \int \int [(ax+by) - (a\overline{x} + b\overline{y})]^2 \omega_1(x)\omega_2(y)\,dx\,dy$$

$$= a^2 \int (x-\overline{x})^2 \omega_1(x)\,dx + b^2 \int (y-\overline{y})^2 \omega_2(y)\,dy$$

$$+ 2ab \int \int (x-\overline{x})(y-\overline{y})\omega_1(x)\,dx\,\omega_2(y)\,dy$$

The first two integrals are $\sigma^2(x)$ and $\sigma^2(y)$, respectively. The third integral should be zero if x and y are independent events; i.e., there should be equal amounts of positive and negative contributions.

Thus,

$$\sigma^2(ax + by) \approx a^2\sigma^2(x) + b^2\sigma^2(y) \tag{13.12}$$

The generalization to the deviations of the averages of Equation 13.11 is obviously

$$\sigma^2(X) = \frac{1}{n^2}(\sigma^2(x) + \sigma^2(x) + \cdots + \sigma^2(x))$$

$$= \frac{1}{n}\sigma^2(x)$$

or if X is the average of N measurements of x,

$$\sigma^2(X) = \frac{1}{N}\sigma^2(x) \tag{13.13}$$

Note that this relation is true, independent of the particular distribution function describing the x measurements. This result is known as the *central limit theorem*. It states that, although the spread of x values, characterized by σ, may be large, the uncertainty in the average, $\overline{X}$, decreases as $1/\sqrt{N}$, where N is the number of separate measurements of the average and that, independent of the underlying distribution of x, the distribution of X will approach a Gaussian form.

EXAMPLE 13.1

The Central Limit Theorem and Coin Flips

The number and standard deviation of heads up in tossing 10 coins could be duplicated by:

```
coin = fix(2*rand(10,1));% coin = either 0's or 1's
up   = sum(coin);
sig2 = std(coin)^2;
```

The total number of arrangements of the 10 coins is 2^{10}, and the number of ways of getting i heads up in a single toss of 10 coins is $C_i = 10!/(10 - i)!\,i!$. The probability of getting i heads up is

$$p_i = \frac{10!}{(10 - i)!\,i!}2^{-10}$$

The averages of this distribution are

$$\overline{x} = \sum_{i=0}^{10} ip_i = 5$$

$$\overline{x^2} = \sum_{i=0}^{10} i^2 p_i = 27.5$$

and so, $\sigma = \sqrt{\overline{x^2} - \overline{x}^2} = \sqrt{2.5}$. The mean and standard deviation of a single measurement should then be 5 ± 1.58. Next, the same measurement is repeated

N times and averaged by:

```
for k = 1:N
   coin = fix(2*rand(10,1));
   up(k) = sum(coin);
   sig2(k)= std(coin)^2;
end
M   = mean(up);
Sig = sqrt(sum(sig2)/N);
```

The code above is contained in a script file named `CentLimit.m` on the applications disk. The result of the average of several measurements should approach 5 coins with a decreasing standard deviation. That is, even though the standard deviation associated with each measurement is quite large, repeated measurements will produce a much more accurate value of the measurement. This is illustrated in Table 13.1, where the "measurement" is repeated N times for $N = 10$, 10^2, 10^3, and 10^4.

N	$\bar{x}$	$\sigma(\bar{x})$	$\sqrt{2.5/N}$
0	4.3000	0.5088	0.5000
10^2	4.9900	0.1583	0.1581
10^3	5.0300	0.0500	0.0500
10^4	5.0092	0.0158	0.0158

Table 13.1 *Repeated measurements of a quantity reduce the standard deviation* ●

Finally, for the more general combination, $z = f(x, y)$, if $\sigma(x)$ and $\sigma(y)$ are "small," it can be shown that, similar to a Taylor series in two dimensions,

$$(z - \bar{z}) \approx \left(\frac{\partial f}{\partial x}\right)(x - \bar{x}) + \left(\frac{\partial f}{\partial y}\right)(y - \bar{y})$$

and it is then not difficult to show that

$$\sigma^2(z) = \overline{(z - \bar{z})^2} \approx \left(\frac{\partial f}{\partial x}\right)^2 \sigma^2(x) + \left(\frac{\partial f}{\partial y}\right)^2 \sigma^2(y) \qquad (13.14)$$

Comparison of Distributions

In this section we will assume that we have measured N pairs, (x_i, y_i), and we are looking for a functional relationship of the form $y = f(x)$. Further, the independent variable is a control parameter, and for simplicity, we will assume that there are no errors or uncertainties in the x values. Thus, for each x we measure a y. The experimental arrangement is such that there is an uncertainty in the y's. We, therefore, expect that for a constant x, numerous measurements of

y will yield a distribution of y's centered about a mean $\bar{y}$. Finally, we will assume that the distribution of each of the y values can be described by a Gaussian distribution centered about $\bar{y}_i$, with a standard deviation, σ_i.

THE MAXIMUM LIKELIHOOD FUNCTION

The probability of obtaining a particular y_i in the interval $y_i \leftrightarrow y_i + dy$ in a single measurement is, as before,

$$P(y_i)\,dy_i = \omega_G(y_i)\,dy_i$$

The probability of duplicating the entire set of data, all N measurements, is called the *likelihood function*, $\mathcal{L}$, defined as

$$\mathcal{L}(y_1, y_2, \cdots y_N)\,dy_1 dy_2 \cdots dy_N \equiv \omega_G(y_1)\omega_G(y_2)\cdots\omega_G(y_N)\,dy_1 dy_2 \cdots dy_N$$

so that

$$\mathcal{L}(y_1, y_2, \cdots y_N) = \frac{1}{(2\pi)^{N/2}\prod_{i=1}^{N}\sigma_i}\exp\left[-\frac{1}{2}\sum_{i=1}^{N}\left(\frac{y_i - \bar{y}_i}{\sigma_i}\right)^2\right] \tag{13.15}$$

CHI-SQUARE

The expression in the exponent of Equation 13.15 is called *chi-square*,

$$\text{\textit{Chi-square}} \qquad \chi^2 \equiv \sum_{i=1}^{N}\left(\frac{y_i - \bar{y}_i}{\sigma_i}\right)^2$$

and we expect that the value of $\chi^2 \approx N$, the number of data points. That is, the average spread of each of the y_i values about their underlying means should be approximately equal to σ_i, or each term in the sum should be of order one. Roughly speaking, if the computed value of χ^2 is *too large*, the indications are that the data are not consistent with the *expectation values*, $\bar{y}_i$. If χ^2 is *too small*, usually the spread of the data, the σ_i's, have been underestimated, or the data are bogus; i.e., the experiment is too good to be true. We will next determine, more precisely, the actual distribution of χ^2 for data containing underlying random errors.

THE DISTRIBUTION OF χ^2

The most likely outcome of the experiment corresponds to the maximum of $\mathcal{L}$, or a minimum of χ^2. To find the distribution of χ^2, make the change of variables, $(y_i - \bar{y}_i)/\sigma_i = \eta_i$, and assume that all the σ_i's are the same and equal to σ; in terms of the η_i's, the likelihood function takes the form

$$\mathcal{L}(y_1, y_2, \ldots, y_N)\,dy_1 dy_2 \cdots dy_N = (2\pi)^{-N/2}e^{-(1/2)\chi^2}\,d\eta_1 d\eta_2 \cdots d\eta_N \tag{13.16}$$

with $d\eta_1 d\eta_2 \cdots d\eta_N = d(vol)$, the infinitesimal volume element in an N-dimensional space, and $r^2 = \eta_1^2 + \eta_2^2 + \cdots + \eta_N^2 = \chi^2\sigma^2$, the square of the distance of a particular point from the origin in that space.

Because the probability only depends on r^2, all points that fall within a spherical shell of radius r and thickness dr will have the same likelihood. Thus, if the expression is rewritten in terms of spherical coordinates in this N-dimensional space, the volume of the spherical shell, $d(vol)$, is of the form

$$d(vol) = C r^{N-1} \, dr$$

where C is a constant. To determine C, we require that the integral of the probabilities given by Equation 13.16 be equal to 1:

$$\int \mathcal{L} \, d(vol) = 1$$

Equation 13.9 is used to evaluate the integral. Inserting this value into the expression for $\mathcal{L}$, we obtain the χ^2 distribution function.

Chi-square Distribution Function
$$\mathcal{L}(\chi^2) d\chi^2 = \frac{(\chi^2)^{((N/2)-1)}}{2^{N/2} \Gamma(N/2)} e^{-(1/2)\chi^2}$$

This distribution has a mean of N and a standard deviation of $\sqrt{2N}$.

THE χ^2 PROBABILITY FUNCTIONS $P(\chi^2, N)$ AND $Q(\chi^2, N)$

The integral $\int_0^{\chi^2} \mathcal{L}(t) \, dt$ represents the probability of N experimental data generating a chi-square equal to or better (i.e., *smaller*) than the limit χ^2. This integral is known as an *incomplete gamma function*. In terms of MATLAB functions, it is evaluated as

$$P(\chi^2, N) = \texttt{gammainc(chisqd/2, N/2)}$$

A somewhat more common criterion is to use, instead, $Q(\chi^2, N) \equiv 1 - P(\chi^2, N)$, which represents the probability that the observed chi-square could be as large or larger than χ^2.

EXAMPLE 13.2

Is the Roulette Wheel Honest?

A roulette wheel has 38 supposedly equal slots, each numbered $00, 0, 1, \ldots, 36$. A small ball is dropped on the spinning wheel and if the player's number comes up, the payoff is at 35 to 1. The probability of getting a particular number on one spin should then be $1/38$. Let y_k be the number of times the number k comes up in N spins of the wheel. If the wheel is honest, the occurrence of each number will be approximately $N/38$ with a standard deviation of $\sqrt{N/38}$. The χ^2 for the observed numbers is

$$\chi^2 = \sum_{k=00}^{36} \frac{(y_i - \frac{N}{38})^2}{N/38} \tag{13.17}$$

We expect that χ^2 will be approximately equal to the number of degrees of freedom in the experiment; i.e., 38, and that the standard deviation of χ^2 will be approximately $\sqrt{2} \cdot 38$. If χ^2 is much larger, the wheel is likely biased. How large a value are we willing to accept before calling the authorities? Well, if $\chi^2 = 45$, evaluating $Q(\frac{45}{2}, \frac{38}{2}) \Rightarrow 1 - \texttt{gammainc(45/2, 38/2)} = 0.2022$, we could conclude with $\approx 80\%$ certainty that the wheel is rigged. A more conservative person might require that the observed chi-square be more than one or two standard deviations beyond the mean. That is,

$$\sigma = \sqrt{2 \cdot 38} \approx 8.7$$

$$\chi_1^2 = \overline{\chi^2} + \sigma \approx 46.7$$

$$\chi_2^2 = \overline{\chi^2} + 2\sigma \approx 55.4$$

$$Q\left(\frac{\chi_1^2}{2}, \frac{38}{2}\right) = 1 - \texttt{gammainc}(46.7/2, 38/2) = 0.1573$$

$$Q\left(\frac{\chi_2^2}{2}, \frac{38}{2}\right) = 1 - \texttt{gammainc}(55.4/2, 38/2) = 0.0338$$

indicating that, with very high certainty, we can conclude that a χ^2 as large as 50 would indicate that the wheel is indeed biased.

Or, put another way, say the wheel is indeed biased, but only slightly. For the honest wheel, the expected probabilities are $p = \frac{1}{38}$ for each number, and in a thousand spins of the wheel, each number should occur approximately $\frac{1000}{38}$ times, plus or minus $\sqrt{\frac{1000}{38}}$. If instead, the probability of double zero is $p_{00} = \frac{1}{33}$, and the probabilities of all other numbers are equal to $p_k = \frac{1}{37}(1 - \frac{1}{33})$, a gambler aware of this bias, who consistently bets double zero, could expect to make approximately a 6% profit (i.e., $\frac{35}{33} - 1$) over an extended period. How would this bias affect the observed χ^2?

To compare the observed distribution of this particular wheel with the expected results of an ideal wheel, we use Equation 13.17 to compute χ^2. However, in this case, the distribution of numbers will no longer be equal. The frequency of double zero will be centered about Np_{00} rather than $N/38$. All other numbers will occur at a reduced frequency centered about Np_k. Defining the change in the probabilities, $\delta p_{00} = p_{00} - \frac{1}{38}$ and $\delta p_k = p_k - \frac{1}{38}$, Equation 13.17 can be rearranged as follows:

$$\chi^2 = \frac{38}{N} \sum_{i=00}^{36} [(y_i - Np_i) + N\delta p_i]^2$$

$$= \frac{38}{N} \sum_{i=00}^{36} \left[(y_i - Np_i)^2 + 2(y_i - Np_i)N\delta p_i + (N\delta p_i)^2\right]$$

The first summation can be estimated to be

$$\sum_{i=00}^{36} (y_i - Np_i)^2 \approx \sigma_{00}^2 \frac{(y_{00} - Np_{00})^2}{\sigma_{00}^2} + 37\sigma_k^2 \frac{(y_k - Np_k)^2}{\sigma_k^2}$$

$$\approx \sigma_{00}^2 + 37\sigma_k^2$$

$$= N(p_{00} + 37p_k)$$

$$= N$$

The second term in the sum is approximately zero if the y_i's are evenly distributed about the actual probabilities, p_{00} and p_k. The third term has a value $C \approx N^2(\delta p_{00}^2 + 37\delta p_k^2) = 1.63 \times 10^{-5}N^2$. Thus, the likely value for the observed χ^2 is

$$\chi^2 \approx 38(1 + 1.63 \times 10^{-5}N)$$

If, as above, we require a $\chi^2 > 50$ before we conclude with confidence that the wheel is biased, we will have to observe approximately $N = 19{,}400$ spins or, at 3 spins a minute, about 108 hours of observation.   ●

Problems

REINFORCEMENT EXERCISES

P13.1 Chebyshev's Theorem. Use the Box-Muller method on page 293 to produce a Gaussian distribution of 1000 values of x centered about $x_0 = 10$ with a standard deviation of $\sigma = 5$. Next, use the toolbox function `find` to determine what fraction of the points are in the range $x_0 - k\sigma$ to $x_0 + k\sigma$ for $k = 1.5, 2, 2.5, 3$ and compare with the prediction of Chebyshev's theorem. Repeat for a *flat* distribution; i.e., `x = x0*(rand(1,1000)`, using the value returned by `std(x)` for σ.

P13.2 Standard Deviation of a Flat Distribution. A perfectly flat distribution of x's over the range $a \le x \le b$ is $x_i = a + i\Delta x$ with $\Delta x = (b-a)/N$ and $i = 0, 1, \ldots, N$. Prove that the standard deviation of such a distribution is exactly

$$\sigma = \frac{b-a}{2\sqrt{3}}\sqrt{1 + \frac{2}{N}}$$

You will need the following identities:

$$\sum_{i=0}^{N} i = \frac{N(N+1)}{2} \qquad \sum_{i=0}^{N} i^2 = \frac{N(N+1)(2N+1)}{6}$$

Compare this expression with `std(rand(N,1))` for $N = 100$, 500, and 1000.

P13.3 The Central Limit Theorem. The central limit theorem states that the spread of the distribution of repeated measurements of the same quantity will ultimately follow a Gaussian distribution, and it will narrow proportional to $1/\sqrt{N}$, where N is the number of measurements. For $N = 100, 500, 1000, 5000, \ldots, 10^5$ compute $\Delta\bar{x} = |\bar{x} - \frac{1}{2}|$ and $\Delta\sigma = |\sigma - \frac{1}{2\sqrt{3}}|$ for a random distribution of N points over the range 0 to 1. Graph $\Delta\bar{x}\sqrt{N}$ and $\Delta\sigma\sqrt{N}$ versus $\log N$. Do your results support the central limit theorem?

P13.4 Combinations of Distributions: Standard Deviations.
    a. Assume that for values of x and y near $\bar{x}$ and $\bar{y}$, the function $z = f(x, y)$ can be adequately approximated by

$$(z - \bar{z}) \approx \left(\frac{\partial f}{\partial x}\right)\bigg|_{\bar{x},\bar{y}}(x - \bar{x}) + \left(\frac{\partial f}{\partial y}\right)\bigg|_{\bar{x},\bar{y}}(y - \bar{y})$$

and prove that the standard deviation of the z values is given by

$$\sigma^2(z) = \overline{(z - \bar{z})^2} \approx \left(\frac{\partial f}{\partial x}\right)^2\bigg|_{\bar{x},\bar{y}}\sigma^2(x) + \left(\frac{\partial f}{\partial y}\right)^2\bigg|_{\bar{x},\bar{y}}\sigma^2(y)$$

   b. Let x = 1+ rand(1,1000); y = 2+rand(1,1000), and use MATLAB to compute $\sigma(z)$ for: (i) $z = 3x - y$ and (ii) $z = \sqrt{x^2 + y^2}$ and compare with the result using the preceding equation.

P13.5 Bimodal Distributions. The procedure for finding a single *mode* of a distribution simply scanned the data and looked for the point where the distribution was most dense. If the distribution is bimodal, the problem is more complex. A general-purpose program that scans for a second local minimum in the density of points is much more difficult to write. Usually it is easier and quite satisfactory to locate a potential second peak by eye on a histogram plot, and then use the original code to search for the mode over a restricted range including just the second peak. The following code will produce a distribution with two peaks, one smaller than the other. Using MATLAB functions, find the location of both peaks.

```
x1 = -2*log(rand(1,200));
x2 =  2*pi*rand(1,200);
x  =  6 + [x1.*cos(x2) x1.*sin(x2)];
x1 = -3*log(rand(1,400));
x2 =  3*pi*rand(1,400);
y  =  12 + [x1.*cos(x2) x1.*sin(x2)];
x = [x y];
[n,t] = hist(x,50);
plot(t,n);
```

P13.6 The Gamma Function.
   a. Prove that the gamma function, $\Gamma(x)$, defined in footnote 3 on page 295, duplicates the properties of the factorial; i.e., for $n = 0, 1$, and 2, show explicitly that $\Gamma(n + 1) = n!$.
   b. For *noninteger p*, integrate by parts to show that

$$\Gamma(p) = (p - 1)\Gamma(p - 1)$$

   c. Use the definite integral $\int_0^\infty e^{-\xi^2} d\xi = \sqrt{\pi}/2$ to show that $\Gamma(\frac{1}{2}) = \sqrt{\pi}$. [*Hint:* Use the change of variables $t = \xi^2$.]
   d. Determine the values of $\Gamma(5\frac{1}{2}), \Gamma(-\frac{1}{2})$.

P13.7 Derivation of the Gaussian Distribution.
   a. Fill in the algebra in the derivation of the Gaussian distribution on page 296.
   b. A common, but very slow way to generate a Gaussian distribution is to sum a uniform distribution of random numbers. Thus, the code

```
x = rand(1,1000);
for i = 1:900
    X(i) = mean(x([i:i+100]));
end
```

should produce a set of 900 random numbers distributed about the value 0.5. What will be the standard deviation of these numbers? Compare this value with the value computed by std(X).

P13.8 Chi-Square Distribution Function.

a. Convert the integral

$$\int \mathcal{L}\, d(vol) = \frac{C}{(2\pi)^{N/2}} \int_0^\infty e^{-(1/2)r^2}\, r^{N-1}\, dr = 1$$

to a gamma function with the change of variables, $t = r^2/2$. Evaluate the constant C.

b. Show that the confidence level function, Q, defined as

$$Q(\chi^2, N) = 1 - \int_0^{\chi^2} \mathcal{L}(t)\, dt \qquad (13.18)$$

can be written as

$$Q(\chi^2, N) = 1 - \frac{1}{\Gamma(N/2)} \int_0^{(1/2)\chi^2} t^{(N/2)-1} e^{-t}\, dt$$

$$= 1 - \texttt{gammainc}\left(\frac{\chi^2}{2}, \frac{N}{2}\right)$$

c. For $N = 10$, 100, and 1000, produce a graph of $Q(\chi^2, N)$, as a function of χ^2/N, over the range $0 \le \chi^2/N \le 1.5$. From the graph estimate the value of χ^2 corresponding to a confidence level of 90%.

P13.9 Biased Roulette Wheel Revisited.   Suppose you suspect that the roulette wheel is biased, *and* you believe you know which number is favored. We will assume that the favorable position is again double zero. In this case, there are only *two* possibilities: Either double zero comes up or not. The number of degrees of freedom is thus reduced to two. Assuming the probability of each number on an honest wheel is $p = \frac{1}{38}$ and $q = 1 - p$, the χ^2 for N spins of the wheel is now

$$\chi^2 = \frac{(N_{00} - pN)^2}{\sigma_{00}^2} + \frac{[(N - N_{00}) - qN]^2}{\sigma_q^2}$$

where N_{00} is the number of times double zero comes up. The standard deviations are taken to be $\sigma_{00}^2 = \sigma_q^2 \approx N/33$. Assuming that the actual probability of double zero is again $p_{00} = \frac{1}{33}$, and defining $\delta p_{00} = \frac{1}{33} - \frac{1}{38}$ as the change in the probability for double zero, and $\delta q_{00} = \frac{32}{33} - \frac{37}{38}$ as the change in the probability of *not* double zero,

a. Show that χ^2 is approximately

$$\chi^2 \approx 2 + 33N[(\delta p_{00})^2 + (\delta q_{00})^2]$$

b. Use the toolbox function `gammainc` to show that the likelihood of a χ^2 as large as 6 is only 0.05, and that to obtain this value of χ^2, we need only make 3800 observations; much fewer than the amount in the example in the text. Explain why this is the case.

***P13.10* The Lorentzian Probability Distribution.** The distribution of energies of particles emitted in radioactive decay, or the energy lost in a resistive element in an ac circuit containing inductors and capacitors, is given by a function called the Lorentzian, and is defined as

$$\omega_L(E) = \frac{L_0}{(E - E_0)^2 + (\Gamma/2)^2}$$

where E_0 is the central energy and Γ is approximately the full width of the distribution measured at half the maximum. The following MATLAB code will duplicate this distribution:

```
function x = lorentz(e0,g,N)
 y = (g/2)*tan(pi*rand(1,N)) + e0;
 i = find((y < e0+5*g) & (y > e0-5*g));
 x = y(i);
```

This distribution produces very long tails in both directions, and has therefore been truncated to values of **x** in the range $|x - e_0| \leq g/2$.

    a. Compute a set of x's using this code with $e_0 = 5$, $g = 2$, and $N = 1000$, and produce a histogram with approximately 40 bins.

    b. On the same graph, (i.e., **hold on**), graph the Lorentzian function with the constant L_0 chosen so that the maximum of the curve will match the height of the tallest bin on the histogram.

    c. Compute $\bar{x}$ and $\sigma(x)$ theoretically, after restricting the range to $|x - e_0| \leq g/2$ and normalizing the expression for $\omega_L(x)$. Compare with the MATLAB computed quantities, **mean(x)** and **std(x)** and explain any large discrepancies.

***P13.11* Poisson Distribution of Random Numbers.** The following MATLAB code is based on an algorithm described in Knuth for obtaining a Poisson distribution of random numbers. The number of random numbers produced will be approximately $\frac{2}{3}N$, and the average will be approximately m, which must be an integer.

```
function x = poisdist(m,N);
 y = tan(pi*rand(1,N);
 Y = 1 + sqrt(2*m-1)*y/(m-1);
 R = (1+y.^2).*exp((m-1)*(log(Y)-Y+1));
 i = find(Y>0 & Y<3.3 & R >= rand(1,N));
 x = (m-1)*Y(i);
```

    a. For $m = 7$ and $N = 1500$, use this code to compute roughly 1000 x's that are distributed as a Poisson distribution. From these data compute $\bar{x}$, $\sigma(x)$, and the skewness. Compare with the results expected for a Poisson distribution. [*Note:* The code will produce a distribution with a very long tail that has been truncated. This will artificially reduce the skewness.]

    b. Use the data to produce a histogram with approximately 40 bins. On the same graph, plot $C\omega_p(x) = C(m^x e^{-m}/x!)$ for $x = 1:22$, choosing C so that the maximum of the curve matches the height of the tallest bin.

Summary

MOMENTS OF A DISTRIBUTION

Mean

The mean or average of a distribution is defined by

$$\overline{x} \equiv \frac{1}{N} \sum_{i=1}^{N} x_i$$

Median

The median, x_{med}, satisfies the condition that the number of x values less than x_{med} is equal to the number of x values greater than x_{med}.

Variance

The variance is a measure of the spread of x values from the median and is defined by

$$\text{Var} = \frac{1}{N-1} \sum_{i=1}^{N} (x_i - \overline{x})^2$$

Standard Deviation

The standard deviation, $\sigma = \sqrt{\text{Var}}$, is given by the equation

$$\sigma^2 = \frac{1}{N-1} \sum_{i=1}^{N} (x_i - \overline{x})^2$$

$$\approx \frac{1}{N} \sum_{i=1}^{N} (x_i - \overline{x})^2$$

$$= \overline{x^2} - \overline{x}^2$$

Skewness

The skewness is a measure of the asymmetry of a distribution, and it is defined as

$$Skew \equiv \frac{1}{N} \sum_{i=1}^{N} \left[\frac{x_i - \overline{x}}{\sigma} \right]^3$$

Kurtosis

The kurtosis, or fourth moment, of a distribution is defined by

$$Kurt \equiv \frac{1}{N} \sum_{i=1}^{N} \left[\frac{x_i - \overline{x}}{\sigma} \right]^4 - 3$$

DISTRIBUTION FUNCTIONS

Binomial Distribution

If p is the probability of occurrence of an event in a single measurement, the probability of m occurrences in N separate measurements is given by

$$\omega_N(m) = \frac{N!}{(N-m)!\,m!} (1-p)^{N-m} p^m$$

Poisson Distribution

The limit of the binomial expansion, as $N \to \infty$ while $\overline{x}$ remains finite; i.e., $p = \overline{x}/N \to 0$:

$$\omega_P(x) = \lim_{N \to \infty} \omega_N(x) = \frac{\overline{x}^x\, e^{-\overline{x}}}{x!} = \frac{\overline{x}^x\, e^{-\overline{x}}}{\Gamma(x+1)}$$

Normal or
Gaussian
Distribution

The limit of the binomial distribution as $N \to \infty, p \not\to 0$:

$$\omega_G(x) = \frac{e^{-(1/2)((x-\overline{x})/\sigma)^2}}{\sqrt{2\pi}\,\sigma}$$

COMBINATIONS OF DISTRIBUTIONS

Standard
Deviation of Two
Independent
Variables

The standard deviation of the sum of two independent measurements is given by

$$\sigma^2(ax + by) = a^2\sigma^2(x) + b^2\sigma^2(y) \tag{13.19}$$

COMPARISONS OF DISTRIBUTIONS

Chi-square

The likelihood of obtaining a particular set of N pairs of values (x_i, y_i) in one complete measurement of N points is related to the quantity chi-square, which is defined as

$$\chi^2 \equiv \sum_{i=1}^{N} \left(\frac{y_i - \overline{y}_i}{\sigma_i} \right)^2$$

Distribution of χ^2

The maximum value of χ^2 determines the most likely result of the measurements. The distribution of χ^2 is given by the function

$$\mathcal{L}(\chi^2)\, d\chi^2 = \frac{(\chi^2)^{((N/2)-1)}}{2^{N/2}\Gamma(N/2)} e^{-(1/2)\chi^2}$$

Confidence Level

The likelihood that an observed chi-square could be as large as the value χ^2 is given in terms of the incomplete gamma function as

$$P(\chi^2, N) = \texttt{gammainc(chisqd/2, N/2)}$$

where N is the number of data pairs.

MATLAB Functions Used

`mean`

If x is a vector, `mean(x)` returns a single value equal to the average or mean of the values in x. If x is a matrix, the mean of each of the columns of x is returned as a row vector.

`median`

If x is a vector, its elements are sorted and the element midway between $x_{\min}$ and $x_{\max}$ is returned. If x is a matrix, the mean of each column is returned as a row vector.

`std`

If x is a vector, `std(x)` computes the sample standard deviation, s, defined as

$$s = \left[\frac{1}{N-1} \sum_{i=1}^{N} (x_i - \overline{x})^2 \right]^{1/2}$$

If x is a matrix, the standard deviation of each column is returned as a row vector.

gamma, gammainc   The *gamma* function,

$$\Gamma(x) = \int_0^\infty e^{-t} t^{x-1} \, dt$$

is a generalization of the factorial function defined so that $\Gamma(n+1) = n!$ if n is a non-negative integer. The function **gamma(x)** computes this function at each value contained in **x**.

The *incomplete* gamma function is defined as

$$P(x, a) = \frac{1}{\Gamma(a)} \int_0^x e^{-t} t^{a-1} \, dt$$

and is computed by the function **gamminc(x,a)**.

hist   Will produce a histogram plot. The use is **hist(y,n)**, where **y** is a vector containing the data (unsorted) and **n** is the number of bins to be used. If **n** is omitted, a value of 10 is used.

bar   Used by **hist** to draw a bar graph. The use is **bar(x,y)**, which, for vectors of the same size, will draw a bar graph of the elements of **y** versus **x**. If **x, y** are matrices of the same size, one bar graph is drawn for each column.

14 / *Linear Least Squares Analysis*

An experiment will usually generate data in the following way. For each value of x, the value of y is measured. It is assumed that there are *no* errors in the independent variable x, and all of the experimental errors associated with the measurements can be attributed to errors in y. Thus, the output of the experiment is a data set of N points, (x_i, y_i), *and* an estimate of the error associated with each y measurement; i.e., Δy_i.

The *likelihood* function of Chapter 13 represented the probability of occurrence of an entire set of data, (x_i, y_i), and the maximum likelihood corresponded to the *minimum* of the quantity χ^2, which was defined as

$$\chi^2 = \sum_{i=1}^{N} \left(\frac{y_i - \overline{y}_i}{\sigma_i} \right)^2$$

where σ_i is the *known* standard deviation of the measurements of the ith data point, and $\overline{y}_i$ is the *expectation* value of the ith measurement. That is, $\overline{y}_i$ is the underlying *actual* value of y. It is assumed that the measured values, y_i, will be distributed in a Gaussian manner about $\overline{y}_i$.

In this chapter we will assume that the *actual* values of the quantity y can be represented by some form of *hypothesis*. A typical form of the hypothesis is that the actual values of y are related to the independent variable x by some functional form containing one or more parameters. We then seek the *optimum* set of these parameters in fitting the functional form to the data *and* some quantitative statement concerning the "goodness" of the fit. These ideas lead to a procedure for obtaining an optimum fit of a functional form to data known as *least squares curve fitting*. Numerous MATLAB toolbox functions exist for facilitating a wide variety of curve fits to data.

We illustrate and extend the very powerful and easy-to-use MATLAB procedures available for fitting straight lines and polynomials to data. Finally, we discuss the very important estimation of the accuracy of the computed parameters.

The most important consideration in evaluating the fit of a curve to data is the determination of whether or not the curve is *within the expected error* associated with the measurements y_i. Therefore, we begin with a discussion of the underlying error in experimental data.

Systematic Versus Random Errors

If the experimental errors, Δy_i, associated with each measurement are known, we may make the reasonable replacement $\sigma_i \approx \Delta y_i$, where σ_i is the standard deviation of the measurements about the ith point. Of course, this assumes that there are no *systematic* or *nonrandom* measurement errors in the experiment. Examples of systematic errors might be an unnoticed change in air temperature while measuring several trajectories of a projectile or ignoring gravity while measuring the motion of small dust particles in turbulent air, or simply a wrong measurement of mass. As we know, random errors will diminish as $1/\sqrt{N}$ with repeated measurements, but systematic errors are much more perverse; no manner of repeated measurements will correct data that are simply wrong. However, if we have an idea about what the actual expectation values of the measurement should be; i.e., the $\overline{y}_i$, and have an accurate estimate of the size of the random errors, Δy_i, the computed χ^2 will likely be too large if there are also significant systematic errors.[1] This may cause the experimenter to adjust the error quantities, Δy_i, or to search for the source of the systematic errors.

Alternatively, if the objective is to *determine* the expectation values by minimizing χ^2, as is done in the *least squares curve fitting* procedures of this chapter, the indicators of systematic error may be more difficult to detect. A perfectly acceptable value of χ^2 is not a guarantee that systematic errors will not invalidate your conclusions concerning the results of an experiment. Particularly because systematic errors are frequently caused by phenomena of which we are unaware or have decided to ignore; they are the basic reason why experiments can never *"prove"* the validity of any theory. One conflicting measurement is sufficient to invalidate, for example, $E = mc^2$, but a million confirming measurements do not ensure that the next measurement will not invalidate everything. You cannot use statistics and probability theory to compute a "confidence" level for hidden systematic errors. Even if something has worked successfully for centuries, there is still no way to predict, with certainty, that it will work successfully the next time it is tried. How many times was Newton's law, $F = ma$, found to be acceptable within experimental errors before Einstein showed that Newton's laws were invalid in the face of large velocities or large gravitational fields? That is, $F = ma$ was always wrong. The data before this century never *proved* that $F = ma$, they were simply not sufficiently precise to suggest an inconsistency.

MINIMIZING χ^2 TO OBTAIN AN OPTIMUM FIT

Again, the likelihood function, $\mathcal{L}(\chi^2)$, represents the probability distribution of complete sets of measurements about the *known* expectation values $\overline{y}_i$, assuming a Gaussian spread of the data about the points, $\overline{y}_i$, characterized by *known*

---

[1] A large data set will frequently contain a few incorrect points, either from being recorded incorrectly, or from other accidental errors. These points will usually appear as *outlier* points, and they can significantly distort the calculation of χ^2. For this reason, when using statistical analysis to decide whether or not to accept or reject a hypothesis, it is best to use very conservative interpretations based on χ^2. Better yet, the experimenter should seek out the reasons for outlier points and correct them.

standard deviations, σ_i. The idea behind *least squares curve fitting* is to turn this definition around to treat the expectation values, in some sense, as *parameters*, and assert that the maximization of $\mathcal{L}(\chi^2)$ will yield the true value of the expectation value.

For example, if the hypothesis that we are attempting to verify from the data is that there is a functional relationship between the variables x and y of the form $y = f(x)$, then we can use this relation to compute the underlying expectation values for each value of x; i.e., $\overline{y}_i = f(x_i)$. In addition, we will assume that the standard deviations, characterizing the spread of each data about the underlying expection value, can be replaced by the known experimental error in the measurement of that point; i.e., $\sigma_i \approx \Delta y_i$. Finally, if the nature of the relation is made somewhat imprecise by specifying only the *form* of the function that might contain several adjustable parameters, $\alpha_1, \alpha_2, \ldots \alpha_n$,

$$y = f(\vec{\alpha}, x)$$

the expression for χ^2 becomes a *function* of the α's:

$$\chi^2(\vec{\alpha}) = \sum_{i=1}^{N} \left(\frac{y_i - f(\vec{\alpha}, x_i)}{\Delta y_i} \right)^2 \tag{14.1}$$

Minimizing this expression will yield the optimum set of parameters for this particular choice of functional form. Then, if by using these optimal α's the computed value of χ^2 is now within acceptable limits, we may conclude that the data are consistent with our hypothesis relating x and y.

Keep in mind that all curve-fitting procedures will determine the optimum fit of a particular functional form to the data, and they will compute the best-fit values for the various parameters in the model. These parameters will correspond to a *local* minimum of χ^2. There is rarely a guarantee that there is not a totally different set of parameters that may lead to an even better fit. That is, the solutions of the equations for the minimum will usually not be *unique*. Next, any worthwhile procedure will also compute the uncertainty in the best-fit parameters. Again, computed results are next to useless without an estimate of their accuracy. Finally, the "goodness of fit" should be stated in a quantitative manner by computing the variance of the data from the curve; i.e., the minimum value of χ^2.

DEGREES OF FREEDOM IN PARAMETER FITTING

There is one more consideration before we can test these ideas. We have said that the number of degrees of freedom in a set of N independent data points, (x_i, y_i), is, N, which is then the most likely value of χ^2. When fitting a function with n parameters to the data, the number of degrees of freedom must be reduced by n. For example, if we were to fit a straight line with two parameters; i.e., $y(x) = \alpha_1 + \alpha_2 x$, to just two points, the fit would always be exact, and χ^2 would be zero, not 2. Similarly, a quadratic with three parameters can be drawn through any three points. It would thus seem that the most likely value of χ^2, when fitting the data to a function with n parameters, is $\nu = N - n$. The reason for this revised relationship is quite clear. Minimizing the equation for χ^2 will add to the analysis n additional equations of the form

$$\frac{\partial \chi^2}{\partial \alpha_i} = 0$$

each of which will then be a relationship between the data points, thereby reducing the number of independent points by n.

It can then be shown that the resulting distribution of χ^2 is the same as obtained earlier, except that N is replaced by $\nu = N - n$. Thus, the *confidence level* of a fit will now be evaluated by using the functions $P(\chi^2, \nu) = $ gammainc(chisqd/2, (N − n)/2) and $Q(\chi^2, \nu) = 1 - P(\chi^2, \nu)$.

Least Squares Fitting of a Straight Line

As an easy illustration of the ideas of least squares curve fitting, consider the model function to be a simple straight line,

$$y = f(x) = \alpha_1 x + \alpha_2$$

so that

$$\chi^2(\alpha_1, \alpha_2) = \sum_{i=1}^{N} \left(\frac{y_i - \alpha_1 x_i - \alpha_2}{\Delta y_i} \right)^2 \tag{14.2}$$

The equations for the minimum of this function of two variables are then

$$\frac{\partial \chi^2}{\partial \alpha_2} = -2 \sum_{i=1}^{N} \left(\frac{y_i - \alpha_1 x_i - \alpha_2}{(\Delta y_i)^2} \right)$$

$$\frac{\partial \chi^2}{\partial \alpha_1} = -2 \sum_{i=1}^{N} x_i \left(\frac{y_i - \alpha_1 x_i - \alpha_2}{(\Delta y_i)^2} \right)$$

If we assume for simplicity that the errors, Δy_i are all the same[2] and equal to Δy, these equations can be simplified by writing them in terms of the averages: $\overline{x} = (1/N) \sum x_i$, $\overline{y} = (1/N) \sum y_i$, $\overline{xy} = (1/N) \sum x_i y_i$, etc.[3] The result is

$$\overline{y} - \alpha_1 \overline{x} - \alpha_2 = 0$$

$$\overline{xy} - \alpha_1 \overline{x^2} - \alpha_2 \overline{x} = 0$$

or in matrix notation

$$\begin{pmatrix} \overline{x^2} & \overline{x} \\ \overline{x} & 1 \end{pmatrix} \begin{pmatrix} \alpha_1 \\ \alpha_2 \end{pmatrix} = \begin{pmatrix} \overline{xy} \\ \overline{y} \end{pmatrix} \tag{14.3}$$

which is of the form $\mathbf{A}\vec{\alpha} = \vec{b}$ and has a solution $\vec{\alpha} = \mathbf{A}^{-1}\vec{b}$.

---

[2]It is not much more difficult to retain the differences in Δy_i. The approximation is made primarily to simplify the notation.

[3]Note that $\frac{1}{N} \sum_{i=1}^{N} 1 = \frac{1}{N}(1 + 1 + \cdots + 1) = 1$.

THE EQUATIONS FOR THE BEST-FIT LINE

The equations for the slope and intercept are merely 2×2 matrices, and it is easier to solve them directly for the solution. Defining $D_{xx} = \overline{x^2} - \overline{x}^2$ (i.e., the *variance* of x) and $D_{xy} = \overline{xy} - \overline{x} \cdot \overline{y}$, the solution for the slope, α_2, and the intercept, α_1, of the best-fit line can be written as

$$\alpha_1 = \frac{D_{xy}}{D_{xx}}$$

$$\alpha_2 = \overline{y} - \alpha_1 \overline{x}$$

THE UNCERTAINTIES IN THE SLOPE AND INTERCEPT

To obtain an estimate of the uncertainties in the computed values of α_1 and α_2, we make use of the relationship between the standard deviation of a function $f(x, y)$, and the standard deviations of its variables x and y; i.e., σ_x and σ_y,

$$\sigma_f^2 = \left(\frac{\partial f}{\partial x}\right)^2 \sigma_x^2 + \left(\frac{\partial f}{\partial y}\right)^2 \sigma_y^2$$

In this case, the errors in α_i are caused by the errors in y_i, so that each α_k is a function of the y_i's. Thus

$$\sigma^2(\alpha_k) = (\Delta y)^2 \sum_{i=1}^{N} \left(\frac{\partial \alpha_k}{\partial y_i}\right)^2 \tag{14.4}$$

The partial derivatives can now be obtained from the solutions for α_k,

$$\frac{\partial \alpha_1}{\partial y_i} = \frac{1}{D_{xx}} \frac{\partial}{\partial y_i}(\overline{xy} - \overline{x} \cdot \overline{y})$$

$$= \frac{1}{N} \frac{x_i - \overline{x}}{D_{xx}}$$

$$\frac{\partial \alpha_2}{\partial y_i} = \frac{\partial}{\partial y_i}(\overline{y} - \alpha_1 \overline{x})$$

$$= \frac{1}{N} \frac{\overline{x^2} - x_i \overline{x}}{D_{xx}}$$

Inserting these expressions into the equation for $\sigma(\alpha_i)$, we obtain the following equations for the uncertainties in the intercept and slope of the best-fit straight line:

$$\sigma^2(\alpha_1) = \frac{(\Delta y)^2}{N} \frac{1}{D_{xx}}$$

$$\sigma^2(\alpha_2) = \frac{(\Delta y)^2}{N} \frac{\overline{x^2}}{D_{xx}} \tag{14.5}$$

EVALUATING THE FIT OF THE LINE TO THE DATA

Finally, we compute χ^2 using Equation 14.2, inserting these values for α_1 and α_2. With this value of χ^2, we next evaluate the probability function $Q(\chi^2, N-2)$. If Q is larger than, say $\frac{1}{5}$, you may be content with, if not absolutely convinced of, the fit of the data to a line. However, if $Q < 0.1$ or so, there are problems with the fit. There are three possible interpretations for a low probability Q.

1. The data are inconsistent with a linear fit. The model is demonstrated to be invalid. However, before reaching this conclusion, you must check the next two possibilities.

2. The experimental errors, Δy, have been *underestimated*. Reassess the errors. Are the errors for all points indeed the same? If need be, redo the calculation using different Δy's for each point. If χ^2 is still too large, you have one more hurdle to clear before the model can be discarded.

3. The error distribution may be non-Gaussian. Least squares analysis is based on the assumption of a normal; i.e., Gaussian distribution of random errors. Unfortunately, it is more common than many experimenters care to admit for data to have a nicely shaped Gaussian distribution as random error should, *but* with much longer "tails" than would a Gaussian. That is, most of the data look OK, but there are a small number of *outliers*, which, as we know, will dominate any least squares analysis. If this is the case with your data, before abandoning the model, you should investigate a curve-fitting procedure known as *robust* analysis, which attempts to reduce the dominating effect of a few outlier points.[4] Often, this means minimizing the mean *absolute* difference, $|y_i - \overline{y}|$, rather than the mean square difference, $(y_i - \overline{y})^2$. See also Problem 14.2.

EXAMPLE 14.1

Least Squares Fit of a Line to Random Data

Rather than laboriously typing in large amounts of data to test the preceding idea, we can take advantage of MATLAB's random number generating function to create any amount and variety of pseudodata. For example, the function rand produces a flat distribution of random numbers, so a subsequent sorting of the numbers and then a plotting of them in sequence should produce a straight, if somewhat ragged, line. This is illustrated in Figure 14.1 for $N = 101$ points. This graph was produced by the following code:

```
N = 50;
y = sort(rand(1,N));
dx= 1/(N-1);      x = 0:dx:1;
plot(x,y,'+')
```

---

[4]For a detailed explanation of the techniques of robust analysis, see Press et al., *Numerical Recipes, The Art of Scientific Computing*, Cambridge University Press, Cambridge, England, 1986, p. 539.

Next the matrices in the equation for the coefficients of the line are computed. Recall, this equation reads

$$\begin{pmatrix} \overline{x^2} & \overline{x} \\ \overline{x} & 1 \end{pmatrix} \begin{pmatrix} \alpha_1 \\ \alpha_2 \end{pmatrix} = \begin{pmatrix} \overline{xy} \\ \overline{y} \end{pmatrix}$$

or $\mathbf{A}\vec{\alpha} = \vec{b}$. The MATLAB code then is

```
A = [mean(x.^2 mean1x);mean(x) 1];
b = [mean(x.*y) mean(y)];
alpha = A\b;
```

To plot the line superimposed on the data, we add

```
Y = alpha(2)*x + alpha(1)
plot(x,Y,x,y,'+')
```

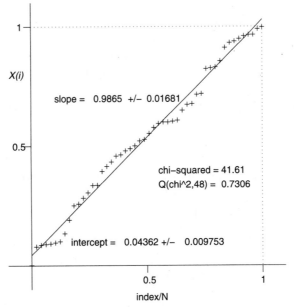

Least squares fit of a line to 50 sorted random numbers

slope = 0.9865 +/- 0.01681

chi–squared = 41.61
Q(chi^2,48) = 0.7306

intercept = 0.04362 +/- 0.009753

index/N

Figure 14.1 *Least squares fit to a sorted list of random numbers.*

Finally, we compute the errors in the slope and intercept, and calculate the *goodness of fit*. To do this, we will need an estimate of the spread of the y values from the best-fit line. We expect that line to be approximately $y = x$; therefore, the spread of the y values from the line should be equal to the standard deviation of the x values from their *mean*; i.e., 0.5. This quantity is, in turn, given by the standard deviations of the x values themselves, divided by $\sqrt{N}$. (See Equation 13.13.) Because we know that the standard deviation of random numbers in the range 0 to 1 is $\sigma_x \approx \frac{1}{2\sqrt{3}} \approx \frac{1}{4}$, we will use $\Delta y = \frac{1}{4\sqrt{N}}$ as the standard deviations of the y's and add the code

```
sy = 0.25/sqrt(N);
sx = std(x);
salpha = sy*[sqrt(mean(x.^2)) 1]/sqrt(N)/sx;
chi2    = sum((y-Y).^2)/sy^2;
Q       = 1 - gammainc(chi2/2,(N-2)/2);
```

The values of these quantities were added to the graph using `gtext`.

You will not be surprised to learn that MATLAB has a toolbox function that will fit a straight line (or indeed a polynomial of degree n) to data. The name of the function is `polyfit` and it returns the coefficients of the best-fit polynomial. The use is

```
a = polyfit(x,y,n);
```

where n is the degree of the fitting polynomial. For a straight line, $n = 1$. The coefficients of the best-fit polynomial are returned in the vector, $\vec{a}$, and the evaluation of the fitted polynomial is achieved with the function `polyval(a,x)`; i.e.,

$$y(x) = a_1 x^n + a_2 x^{n-1} + \cdots + a_{n-1} x + a_n$$
$$= \text{polyval}(\text{a}, \text{x})$$

There is an additional option available with the toolbox functions `polyval` and `polyfit` that will determine the size of the error in y such that 50% of the points match the curve within this error. (See Problem 14.7.) See the help file for more information concerning these two functions. There is no toolbox function to calculate χ^2.

Finally, consider writing the model equation for each of the N points individually:

$$y_1 = \alpha_1 + \alpha_2 x_1$$
$$y_2 = \alpha_1 + \alpha_2 x_2$$
$$y_3 = \alpha_1 + \alpha_2 x_3$$
$$\vdots \quad \vdots \quad \vdots$$
$$y_N = \alpha_1 + \alpha_2 x_N$$

which represents N equations in two unknowns, α_1 and α_2. These can be written in MATLAB matrix form as $\mathbf{M}\vec{\alpha} = \vec{y}$, where $\mathbf{M}$ is an $N \times 2$ rectangular matrix, $\vec{\alpha}$ is a two-element column vector, and $\vec{y}$ is an N-element column vector. This equation can be solved directly using the "backslash" division operator, as shown here:

```
M = [ones(N,1) x'];
alpha = M\y';
```

Compare these results with the previous two calculations of the coefficients. ●

Linear Least Squares Fitting of Polynomials

The generalization of the method of finding a best-fit line to finding a best-fit polynomial is very straightforward. To see the general pattern of the results, we try next the model function,

$$y(x) = \alpha_1 x^2 + \alpha_2 x + \alpha_3$$

and form the χ^2 function,

$$\chi^2(\alpha_1, \alpha_2, \alpha_3) = \sum_{i=1}^{N} \left(\frac{y_i - (\alpha_1 x^2 + \alpha_2 x + \alpha_3)}{\Delta y_i} \right)^2$$

Again assuming that the Δy_i's are all the same, and using the same ideas as in the two-parameter fit, the three equations for the minimum of χ^2,

$$\frac{\partial \chi^2}{\partial \alpha_1} = \frac{\partial \chi^2}{\partial \alpha_2} = \frac{\partial \chi^2}{\partial \alpha_3} = 0$$

are easily shown to yield

$$\begin{pmatrix} \overline{x^4} & \overline{x^3} & \overline{x^2} \\ \overline{x^3} & \overline{x^2} & \overline{x} \\ \overline{x^2} & \overline{x} & 1 \end{pmatrix} \begin{pmatrix} \alpha_1 \\ \alpha_2 \\ \alpha_3 \end{pmatrix} = \begin{pmatrix} \overline{x^2 y} \\ \overline{xy} \\ \overline{y} \end{pmatrix} \tag{14.6}$$

The solution to this matrix equation will return the values of the coefficients, α_k, of the best-fit quadratic to the data. The generalization of the method to polynomials of arbitrary degree is obvious.

The same values are returned by the function `polyval`.

The backslash operator, \, also solves an overdetermined matrix equation by a least squares procedure, and it is the preferred method to use in MATLAB. That is, if $\vec{x}$ and $\vec{y}$ are column vectors containing the data points, to find the $n+1$ parameters of the best-fit polynomial of degree n, we first form the rectangular $N \times n$ matrix M in MATLAB[5] as

---

[5]In the case where the matrix **M** is *square*, it is known as the *Vandermonde* matrix, and it can be computed using the MATLAB function `vander(x)`. In the present nonsquare generalization of the Vandermonde matrix, however, you should use the code we have shown.

```
N = length(x);
M = ones(N,1);
xk = x;
for k = 1:n;
    M = [xk M];
    xk = x.*xk;
end
```

$$M = \begin{pmatrix} x_1^n & x_1^{n-1} & \cdots & x_1 & 1 \\ x_2^n & x_2^{n-1} & \cdots & x_2 & 1 \\ x_3^n & x_3^{n-1} & \cdots & x_3 & 1 \\ \vdots & \vdots & \ddots & \vdots & \vdots \\ x_N^n & x_N^{n-1} & \cdots & x_N & 1 \end{pmatrix}$$

If the number of data points is N, the solution of the N equations in n unknowns is obtained in a least squares sense by the simple statement

```
a = M\y;
```

OBTAINING THE UNCERTAINTIES IN THE PARAMETERS

To obtain the uncertainties in the elements of $\vec{a}$, we need to compute $\partial a_k / \partial y_i$ and evaluate $\sigma^2(a_k)$ using

$$\sigma^2(a_k) = (\Delta y)^2 \sum_{i=1}^{N} \left(\frac{\partial a_k}{\partial y_i} \right)^2$$

First, by multiplying the overdetermined set of equations, $\mathbf{M}\vec{a} = \vec{y}$, by $\mathbf{M}^T$:

$$\mathbf{M}^T \mathbf{M} \vec{a} = \mathbf{M}^T \vec{y} \tag{14.7}$$

the equation becomes more tractable, involving a square $n \times n$ matrix, $\mathbf{C} = \mathbf{M}^T \mathbf{M}$ (that is, $\mathbf{M}^T_{[n \times N]} \cdot \mathbf{M}_{[N \times n]} = \mathbf{C}_{[n \times n]}$). Equation 14.7 is, in fact, identical to the degree n generalization of Equation 14.6 with the coefficient matrix given by $\mathbf{C} = \mathbf{M}^T \mathbf{M}$. Thus, an alternative equation for the solutions, $\vec{a}$, is

$$\vec{a} = \mathbf{C}^{-1} \mathbf{M}^T \vec{y} \tag{14.8}$$

or, in terms of the individual components,

$$a_k = \sum_{j=1}^{n} \sum_{i=1}^{N} \mathbf{C}_{kj}^{-1} \mathbf{M}_{ji}^T y_i$$

Noting that $\mathbf{M}$ is a function of the x's only, we obtain

$$\frac{\partial a_k}{\partial y_i} = \sum_{j=1}^{n} \mathbf{C}_{kj}^{-1} \mathbf{M}_{ji}^T$$

so that

$$\sigma^2(a_k) = (\Delta y)^2 \sum_{i=1}^{N} \left(\frac{\partial a_k}{\partial y_i} \right)^2$$

$$= (\Delta y)^2 \sum_{i=1}^{N} \sum_{j=1}^{n} \mathbf{C}_{kj}^{-1} \mathbf{M}_{ji}^T \sum_{\ell=1}^{n} \mathbf{C}_{k\ell}^{-1} \mathbf{M}_{\ell i}^T$$

$$= (\Delta y)^2 \sum_{j=1}^{n} \sum_{\ell=1}^{n} \mathbf{C}_{kj}^{-1} \mathbf{C}_{k\ell}^{-1} \sum_{i=1}^{N} \mathbf{M}_{\ell i}^T \mathbf{M}_{ij}$$

and using the fact that $\mathbf{C} = \mathbf{M}^T\mathbf{M}$ in the last line, we obtain

$$\sigma^2(a_k) = (\Delta y)^2 \mathbf{C}_{kk}^{-1} \tag{14.9}$$

THE MATLAB CODE FOR POLYNOMIAL LEAST SQUARES FITTING

Thus, the complete MATLAB code to obtain an nth-degree polynomial fit to N data points, assuming an error of dy in the y values, is

```
function [a,da] = LstSqr(x,y,dy,n);
N = length(x);
M  = ones(N,1);
xk = x;
for k = 1:n;
        M = [xk M];
        xk = x.*xk;
end
a = M\y;
da = dy*sqrt(diag(inv(M'*M)));
```

A cautionary word regarding least squares polynomial fitting: The coefficient matrix has a tendency to become near singular for relatively modest values of n.[6] This will have the effect of strongly amplifying any round-off or other errors in the problem. You should, therefore, never attempt a polynomial least squares fit for n greater than 4 or 5.

General Linear Least Squares

If you think of the least squares fitting of a polynomial to data as simply finding the optimum set of coefficients α_k in the *linear* combination of the functions, $1, x, x^2, \ldots, x^n$,

$$y(x)\alpha_1 x^n + \alpha_2 x^{n-1} + \cdots + \alpha_n x + \alpha_{n+1}$$

an obvious generalization is to find the optimum set of coefficients α_k in the *linear* combination of a more general set of functions, $f_1(x), f_2(x), \ldots, f_n(x)$,

$$y(x) = \alpha_1 f_1(x) + \alpha_2 f_2(x) + \cdots + \alpha_n f_n(x) \tag{14.10}$$

The generalization of χ^2 is then

$$\chi^2(\vec{\alpha}) = \sum_{i=1}^{N} \left(\frac{y_i - (\alpha_1 f_1(x) + \alpha_2 f_2(x) + \cdots + \alpha_n f_n(x))}{\Delta y_i} \right)^2$$

---

[6]See the discussion in Borse, *FORTRAN-77 and Numerical Methods*, 2nd ed., PWS Publishing, Boston, MA, 1991, p. 516.

Then, as long as Equation 14.10 remains *linear* in the coefficients α_k, regardless of how complicated the functions $f_k(x)$ might be, the equations for the least squares best fit can be written in strict analogy with the polynomial case. That is, form the generalized $N \times n$ Vandermonde matrix

$$\mathbf{M} = \begin{pmatrix} f_1(x_1) & f_2(x_1) & \cdots & f_{n-1}(x_1) & f_n(x_1) \\ f_1(x_2) & f_2(x_2) & \cdots & f_{n-1}(x_2) & f_n(x_2) \\ \vdots & \vdots & \ddots & \vdots & \vdots \\ f_1(x_N) & f_2(x_N) & \cdots & f_{n-1}(x_N) & f_n(x_N) \end{pmatrix} \qquad (14.11)$$

and solve the equation

$$\vec{\alpha} = M \setminus \vec{y}$$

for the best-fit coefficients, and if the uncertainties in y are known to be Δy, the uncertainties in the α's are given by Equation 14.9.

Fitting with Legendre Polynomials

Legendre polynomials are polynomials in x defined only over the range $-1 \leq x \leq +1$. They originate in the solution of Laplace's differential equation in three-dimensional spherical coordinates, $(\nabla^2 \Phi = 0)$. With the replacement $x = \cos\theta$, the Legendre polynomial $P_\ell(x)$ is a component of the angles part of the general solution.

Like many other special functions with a similar origin, these polynomials have found broad application in a variety of areas unrelated to Laplace's equation. First, let us define some of the basic properties of Legendre polynomials:

Symmetry Each Legendre polynomial, $P_\ell(x)$, is a polynomial of degree ℓ in x, and contains only even powers of x if ℓ is even, and only odd powers if ℓ is odd.

Normalization For any ℓ, $P_\ell(x = 1) \equiv 1$.

Orthogonality The Legendre polynomials satisfy the orthogonality condition

$$\int_{-1}^{1} P_{\ell'}(x) P_\ell(x)\, dx = \frac{2}{2\ell+1}\delta_{\ell'\ell} \qquad (14.12)$$

Recursion Relations The Legendre polynomials satisfy several recursion relations. One of the most important is

$$P_{\ell+1}(x) = \frac{2\ell+1}{\ell+1}x P_\ell(x) - \frac{\ell}{\ell+1}P_{\ell-1}(x) \qquad (14.13)$$

For example, using the symmetry condition and the fact that $P_\ell(1) = 1$, we can conclude that $P_0 = 1$ and $P_1(x) = x$. Also, $P_2(x) = (x^2 + a)/(1 + a)$, with a determined in the equation that follows.

The orthogonality condition applied to these functions then requires that $\int_{-1}^{1} P_0 \, dx = 1$, $\int_{-1}^{1} P_1 \, dx = 0$, and

$$0 = \int_{-1}^{1} P_2(x) \, P_0(x) \, dx$$

$$= \frac{1}{(1+a)} \int_{-1}^{+1} (x^2 + a) \, dx$$

$$= 2\frac{(a + \frac{1}{3})}{a + 1}$$

$$\Rightarrow a = -\frac{1}{3}$$

Using the recursion relation along with the explicit expressions for P_0, P_1, and P_2, it is not difficult to write the MATLAB code for the general $P_\ell(x)$. You can verify your code by comparing it with the following explicit expressions for the first few Legendre polynomials:

$$P_0(x) = 1$$

$$P_1(x) = x$$

$$P_2(x) = \frac{1}{2}(3x^2 - 1)$$

$$P_3(x) = \frac{1}{2}(5x^3 - 3x)$$

$$P_4(x) = \frac{1}{8}(35x^4 - 30x^2 + 3)$$

$$P_5(x) = \frac{1}{8}(63x^5 - 70x^3 + 15x)$$

The main element in fitting an expansion in Legendre polynomials to data is the construction of the generalized Vandermonde matrix of Equation 14.11, where Legendre polynomials replace the arbitrary functions $f_i(x)$. The code to accomplish this is given in the next section.

MATLAB CODE FOR LEAST SQUARES FITTING WITH LEGENDRE POLYNOMIALS

The code will compute the $N + 1$ coefficients of a least squares fit to the n data points, (x_i, y_i), of the expansion

$$f(x) = \sum_{k=0}^{N} a_{N-k+1} P_k(x)$$

```
function a = Lgndrfit(N,x,y)
%Lgndrfit(N,x,y) will fit the n data points contained in
%  the vectors x and y to an expansion in Legendre poly-
%  nomials.  The expansion is through Legendre polynomials
%  of order N.  The Legendre polynomials are computed
%  using a recursion relation and the fit is obtained by
%  the generalized Vandermonde matrix.
%===========================================================
  n = length(x);
  if N==0
     a = mean(y);
     return
  elseif N == 1
     V = [x ones(n,1)];
     a = V\y;
     return
  elseif N>=2
     V = [.5*(3*x.^2 - 1) x ones(n,1)];

%     Use recursion relation for remaining polynomials
     for m = 3:N
        C = (2*x.*V(:,1)-V(:,2))-(x.*V(:,1)-V(:,2))/m;
        V = [C V];
     end
     a = V\y;
  end
```

$x = \cos\theta$	θ(degrees)	$R(\theta)$(km)
−1.000	180.000	6357.41
−0.875	151.045	6358.19
−0.750	138.590	6360.43
−0.625	128.682	6363.78
−0.500	120.000	6367.75
−0.375	112.024	6371.71
−0.250	104.478	6375.07
−0.125	97.181	6377.31
0.000	90.000	6378.07
0.125	82.819	6377.26
0.250	75.522	6375.03
0.375	67.976	6371.69
0.500	60.000	6367.75
0.625	51.318	6363.80
0.750	41.410	6360.44
0.875	28.955	6358.19
1.000	0.000	6357.39

Table 14.1   *Simulated earth radii*

EXAMPLE 14.2

Least Squares Fit of Legendre Polynomials to Earth Radii

The 17 data points in Table 14.1 approximate the radius of the earth in kilometers at various latitudes. The angle θ is measured in degrees with the North Pole at $\theta = 0$. The quantity x, the cosine of θ, is given in equally spaced values from -1 to $+1$. The values of $R(\theta)$ are approximate values of the earth's radius in kilometers at that angle.

The coefficients of a Legendre polynomial expansion through $P_5(x)$ computed with the code in this section are given in Table 14.2. (Be aware that the computation of the coefficients using the MATLAB code based on a generalized Vandermonde matrix gives the coefficients in the *reverse* of the normal order.)

i	a_i
0	6367.71506
1	−0.00064
2	−15.84563
3	0.01115
4	5.75435
5	−0.02442

Table 14.2 *Expansion coefficients from a least squares fit using Legendre polynomials*

This expansion is compared with the data in Figure 14.2.

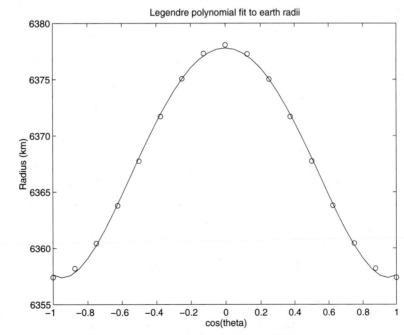

Figure 14.2 *Legendre polynomial fit to earth radii data.*

EVALUATING THE EXPANSION COEFFICIENTS USING THE ORTHOGONALITY CONDITION

There is another approach to expansions in terms of Legendre polynomials (and indeed all other forms of special polynomials and functions), and it is based on the orthogonality condition of Equation 14.12. If we wish to expand a *function* $f(x)$ as opposed to data values in terms of Legendre polynomials,

$$f(x) = \sum_{\ell=0}^{\infty} a_\ell P_\ell(x) \tag{14.14}$$

the coefficients a_ℓ can be "solved for" directly by first multiplying Equation 14.14 through by $P_{\ell'}(x)$, and then integrating both sides with respect to x from -1 to $+1$:

$$\int_{-1}^{1} P_\ell(x) f(x)\, dx = \sum_{\ell'=0}^{\infty} a_{\ell'} \int_{-1}^{1} P_{\ell'}(x) P_\ell(x)\, dx$$

$$= \sum_{\ell'=0}^{\infty} \frac{2\delta_{\ell'\ell}}{2\ell'+1} a_{\ell'}$$

$$= \frac{2}{2\ell+1} a_\ell$$

or

$$a_\ell = \frac{2\ell+1}{2} \int_{-1}^{1} P_\ell(x) f(x)\, dx \tag{14.15}$$

Because the function is presumed known for all $|x| \leq 1$, each of the coefficients can be computed, at least numerically, using this equation.[7]

In Example 14.2, the data were at 17 equally spaced points; thus, it seems they are ideally suited to a Romberg-type integration. (See Chapter 7 for a review of the Romberg algorithm.) The following MATLAB code evaluates a_ℓ, $\ell = 0, \ldots, 5$. In Table 14.3 these values are compared with coefficients obtained by the least squares expansion done earlier.

```
function [a,T] = Rmbrgdat(x,y,k)
%Rmbrgdat will integrate n equally spaced data points using
%      the Romberg algorithm.  The number of data points, n,
%      MUST be equal to 2^k +1.  The best value obtained is
%      returned as 'a', and the complete Romberg table is re-
%      turned in the k+1 by k+1 matrix T.
%=========================================================
   n = 2^k+1;    N = k + 1;    T = zeros(k+1,k+1);
   a = min(x); b = max(x); dx = (b-a);
   T(1,1) = .5*dx*(y(1)+y(n));
```

---

[7]Of course, we have yet to prove that the series converges uniformly to the function $f(x)$. We will ignore this subtlety and suggest that the reader consult any text on advanced calculus for details.

```
jn = 2^(k-1);
for i = 1:k
    jj = jn+1:2*jn:n;
    dx = dx/2;
    T(i+1,1) = .5*T(i,1) + dx*sum(y(jj));
    jn = jn/2;
end
for m = 1:k
  T(m+1:N,m+1) = T(m+1:N,m) + ...
                  (T(m+1:N,m) - T(m:N-1,m))/(4^(m)-1);
end
a = T(k+1,k+1);
```

	Using Least Squares	Using Orthogonality
i	a_i	a_i
0	6367.71506	6367.1335
1	−0.00064	−0.0017
2	−15.84563	−15.7078
3	0.01115	0.0072
4	5.75435	5.9878
5	−0.02442	−0.0312

Table 14.3 *Coefficients of an expansion in Legendre polynomials using the least squares and orthogonality methods*

The two sets of coefficients are not the same, which should not be surprising; they represent two quite different interpretations of the data. The least squares approach seeks to find the best agreement between the data within the limited model space of functions, in this case, P_0 through P_5. The solution for the coefficients via the orthogonalization integral returns *exact* values of the expansion coefficients in a complete expansion of the function $f(x)$ that passes through the data. For example, if the data happened to precisely match the values of the Legendre polynomial $P_6(x)$, the least squares method would, nonetheless, return nonzero values for at least the even terms, a_0, a_2, and a_4, and would represent the best we could do with these six functions. But the orthogonalization integral will return precisely zero for *all* a_i, except a_6. Which approach is better depends on what you are asking of the data. If you need a function that accurately summarizes the data, the least squares approach is suggested. However, frequently theoretical considerations predict the results of an experiment in the form of an expansion in terms of Legendre polynomials (or other such functions), and in that case, the actual coefficients in the expansion have physical significance, and their value *must* be computed using the orthogonalization method. That is, if it is important to determine how much "$P_4(x)$" is present in the data, we *cannot* use the least squares approach to answer the question.

In the current example, the data suggest that, not only does the earth bulge slightly at the equator, but it appears to have a very slight pear shape—it is somewhat larger in the Southern Hemisphere. (This is indicated by the nonzero values of the odd ℓ coefficients, a_1, a_3, and a_5, although the coefficient a_1 is likely equal to zero within error.) Theoretical models would then *predict* a shape specified by an expansion in Legendre polynomials. To compare the data with theory, we would use the calculation of the a_i's obtained with the orthogonality relation, not least squares.

Legendre polynomials are particularly useful when analyzing a function defined over a *limited* domain in the independent variable, x. If the range of x is $a \leq x \leq b$, then a simple change of variables, $x = \frac{1}{2}b(\xi + 1) - \frac{1}{2}a(\xi - 1)$ returns the problem to the -1 to $+1$ domain appropriate for Legendre polynomials.

CHEBYSHEV POLYNOMIALS

Another class of orthogonal polynomials defined over the domain $-1 \leq x \leq +1$ is the Chebyshev polynomials[8], $T_k(x)$. They have properties similar to those of Legendre polynomials:

Symmetry   Each Chebyshev polynomial, $T_k(x)$, is a polynomial of degree k in x, and contains only even powers of x if k is even, and only odd powers if k is odd.

Normalization   For any k, the *maxima* of $T_k(x)$ have a value of $+1$, and for $k \geq 1$, the *minima* have a value of -1. Thus $T_0 = 1$, $T_1 = x$, and $T_2(x) = 2x^2 - 1$. Also, the polynomial $T_k(x)$ has k roots in the interval $[-1, 1]$ located at the points

$$x_i = \cos\left(\frac{\pi(i - \frac{1}{2})}{k}\right) \qquad k \geq 1 \qquad (14.16)$$

Orthogonality   The Chebyshev polynomials satisfy two orthogonality conditions, an integral condition

$$\int_{-1}^{1} \frac{T_{k'}(x)T_k(x)}{\sqrt{1 - x^2}}\, dx = \left[\begin{array}{ll} 0 & k' \neq k \\ \frac{\pi}{2} & k' = k \neq 0 \\ \pi & k' = k = 0 \end{array} \right. \qquad (14.17)$$

and a discrete condition

$$\sum_{k=1}^{m} T_i(x_k)T_j(x_k) = \left[\begin{array}{ll} 0 & i \neq j \\ \frac{m}{2} & i = j \neq 0 \\ m & i = j = 0 \end{array} \right. \qquad (14.18)$$

where $[x_i]$ is the set of the m roots of the Chebyshev polynomial given by Equation 14.16.

---

[8]The reason that Chebyshev polynomials are universally characterized by the letter T has to do with a French spelling of the name, Tchebycheff.

Recursion   The Chebyshev polynomials satisfy the recursion relation
Relations
$$T_{k+1}(x) = 2xT_k(x) - T_{k-1}(x) \qquad (14.19)$$

The first few Chebyshev polynomials are

$$T_0(x) = 1$$
$$T_1(x) = x$$
$$T_2(x) = 2x^2 - 1$$
$$T_3(x) = 4x^3 - 3x$$
$$T_4(x) = 8x^4 - 8x^2 + 1$$
$$T_5(x) = 16x^5 - 20x^3 + 5x$$

You are invited to reformulate the previous MATLAB codes, both for least squares fitting, and for the exact calculation of expansion coefficients employing Legendre polynomials, into codes that use Chebyshev polynomials.

Unlike Legendre polynomials, Chebyshev polynomials rarely appear as a theoretical expansion of a model function; however, they are extensively used in curve fitting as an alternative to least squares analysis.

Problems

REINFORCEMENT EXERCISES

P14.1 The *Confidence Level* Function $Q(\chi^2, \nu)$. The confidence level function is defined in terms of MATLAB functions as $Q(\chi^2, \nu)$ =gamminc(chisqrd/2,nu/2). (See page 304 and Problem 13.8.) Plot this function versus χ^2 for $\nu = 2, 3$ over a range $0 \le \chi^2 \le 5$.

    a. What is the maximum value of χ^2 that corresponds to a 90% confidence level for a linear fit ($\nu = 2$)? However, the experimental errors were misread; they are actually half as large as assumed. For the same data that were assumed to give a 90% confidence level, what is now the confidence level? (Assume that all points have the same experimental error.)

    b. Repeat for a quadratic fit ($\nu = 3$).

P14.2 Minimizing the *Mean Absolute Difference*. If you suspect that your least squares fit of a line to data is being skewed by a few outliers, you may wish to try a procedure that is less sensitive to large deviations. Thus, in place of χ^2, you could define a measure of the goodness of fit to be

$$E(a, b) = \sum_{i=1}^{N} |y_i - (ax_i + b)|$$

and seek the values of a and b that yield a minimum of $E(a, b)$. The minimum of $\sum |y_i - \eta|$ is obviously obtained by $\eta = \text{median}(y)$, so we can formally solve for b in terms of a as

```
b = median(y - a*x)
```

thus E can be written as a function of the single variable a. The minimum then corresponds to a solution of $\partial E/\partial a = 0$, or

$$0 = \sum_{i=1}^{N} x_i \operatorname{sign}(y_i - ax_i - b)$$

(Recall $\partial|x|/\partial x = 1$ if $x > 0$ and -1 if $x < 0$). The zero of this function can be obtained by means of the MATLAB function **zero**. Thus, the code to find the slope and intercept of the best-fit line by this criterion is

```
%=====Code this as an M file=======
function f = MinAbs(a)
global Xdata Ydata
b = median(Ydata - a*Xdata);
f = sum(Xdata.*sign(Ydata-a*Xdata+b));
%
%=====And the MATLAB code is then===
global Xdata Ydata
a0 = ...   % Initial guess for a
a = fzero('MinAbs',a0);
b = median(Ydata - a*Xdata);
```

Use the data set
$$X data = \begin{bmatrix} 0 & 1 & 2 & 3 & 4 & 5 & 6 & 7 & 8 & 9 & 10 \end{bmatrix}$$
$$Y data = \begin{bmatrix} 0.0 & 0.9 & 8.2 & 2.5 & 3.4 & 4.7 & 5.4 & 6.5 & 8.0 & 8.8 & 10. \end{bmatrix}$$

to fit straight lines by the ordinary least squares method, and by the method of this problem, and compare the results. In particular, note the effect that the "outlier" at $x = 2$ has on each calculation.

P14.3 Ill-Conditioned Vandermonde Matrices.
The least squares problem is based on solving matrix equations of the form $\mathbf{M}\vec{a} = \vec{b}$, where the matrix $\mathbf{M}$ is a Vandermonde matrix or a generalization of the Vandermonde matrix. To demonstrate the instabilities of this method when the matrices are somewhat large, you could use the MATLAB function **rcond**, which returns the reciprocal of the *condition* of a matrix. (You will recall from Chapter 2 that the condition of a *singular* matrix is infinite, and a matrix is *ill conditioned* if its condition is large. Thus, if **rcond(A)** returns a number close to zero, the matrix is ill conditioned, and its inverse will be suspect.) For **n = 5:15**, let **x = 1:n** and **V = vander(x)**. Obtain the values of **rcond** in each case and, from these, determine the largest least squares problem that can be trusted on your computer. Use as a criterion, $c = \ln(\mathbf{rcond}/(10*\mathbf{eps})) > 1$.

P14.4 Least Squares Fit of a Line to Sorted Random Numbers. Duplicate the results of Example 14.1 on your computer.

P14.5 The Uncertainties in the Slope and the Intercept. Compute the slope and intercept of the least squares best-fit line to a sorted list of $N = 101$ random numbers between zero and one. Next, compute the uncertainties in the slope and the intercept using Equations 14.5 and plot the boundaries of the region that contains all lines based on the best-fit line plus or minus the uncertainties in this line. Finally, determine the number of points that fall outside this region. Increase the number of points. Does the fraction of points outside the region remain the same? Explain.

P14.6 Least Squares Fit of a Parabola. Determine the parameters and the uncertainties in the parameters of a least squares fit of a parabola to a sorted list of the *squares* of $N = 101$ random numbers in the range zero to one.

EXPLORATION PROBLEMS

P14.7 The MATLAB Functions `polyval` and `polyfit`. As in Example 14.1, produce a sorted list of $N = 101$ random numbers. Next, use the MATLAB function `polyfit` in the form

```
[a,S] = polyfit(x,y,1);
```

to determine the coefficients, a_1 and a_2, of the best-fit line. In this statement, the quantity S is a matrix that will be used by the MATLAB `polyval` to estimate the errors associated with each point. The corresponding statement

```
[Y,delta] = polyval(a,x,S)
```

will return the values of the best-fit line, Y, along with `delta`, a vector of equal length, containing estimates of the errors in each point, computed so that $Y_i \pm \delta_i$ contains at least 50% of the data.

Finally, compare the mean of `abs(delta)` with the estimate for Δy used in Example 14.1, namely, $\Delta y \approx 1/4\sqrt{N}$.

P14.8 Should We Add One More Parameter to Improve the Fit? As measured by their χ^2, a cubic equation with $\nu = 4$ parameters will always fit the same data better than a quadratic with $\nu = 3$. Likewise a quartic will fit better than a cubic, etc. How can we determine if a particular model fit is satisfactory and if it is a poor fit or an "overfit"? The confidence function, $Q(\chi^2, \nu)$, returns the probability that a given value of χ^2 could be that large for a particular fit; conversely, $1 - Q$ is the probability that the given χ^2 could be that small. Thus, $Q \approx 0$ is a poor fit; $Q \approx 1$ is likely an overfit.

For the data

$$x = [0 \quad 0.1 \quad 0.2 \quad 0.3 \quad 0.4 \quad 0.5 \quad 0.6 \quad 0.7 \quad 0.8 \quad 0.9 \quad 1.0]$$
$$y = [2.1 \quad 2.3 \quad 2.5 \quad 2.9 \quad 3.2 \quad 3.3 \quad 3.8 \quad 4.1 \quad 4.9 \quad 5.4 \quad 5.8]$$

with $\Delta y = 0.15$, compute χ^2 and Q for least squares fits of degree n = 2:7. (Use functions `polyfit` and `polyval`.) Acceptable fits will be those with $Q \approx 0.5$. Do your results agree with the qualitative statement that for a fit of N points, if $\chi^2 \gg N - \nu$, the fit is poor, whereas if $\chi^2 \ll N - \nu$, you probably have overfit?

Summary

LINEAR LEAST SQUARES

Best-Fit Line

The slope, α_1, and intercept, α_2, of the least squares best-fit line to a set of data, (x_i, y_i), are determined by solving the matrix equation

$$\begin{pmatrix} \overline{x^2} & \overline{x} \\ \overline{x} & 1 \end{pmatrix} \begin{pmatrix} \alpha_1 \\ \alpha_2 \end{pmatrix} = \begin{pmatrix} \overline{xy} \\ \overline{y} \end{pmatrix}$$

The approximate errors in the slope and intercept are given, respectively, by

$$\sigma^2(\alpha_1) = \frac{(\Delta y)^2}{N} \frac{1}{D_{xx}}$$

$$\sigma^2(\alpha_2) = \frac{(\Delta y)^2}{N} \frac{\overline{x^2}}{D_{xx}}$$

Best-Fit Polynomial

The results for fitting a line are readily generalized to fitting an nth degree polynomial. For example, the best-fit quadratic is obtained by solving

$$\begin{pmatrix} \overline{x^4} & \overline{x^3} & \overline{x^2} \\ \overline{x^3} & \overline{x^2} & \overline{x} \\ \overline{x^2} & \overline{x} & 1 \end{pmatrix} \begin{pmatrix} \alpha_1 \\ \alpha_2 \\ \alpha_3 \end{pmatrix} = \begin{pmatrix} \overline{x^2 y} \\ \overline{xy} \\ \overline{y} \end{pmatrix}$$

Errors in Parameters

The errors in the computed polynomial coefficients are determined by

$$\sigma^2(a_k) = (\Delta y)^2 \mathbf{C}_{kk}^{-1}$$

where $\mathbf{C} = \mathbf{M}^T \mathbf{M}$, and $\mathbf{M}$ is the Vandermonde matrix,

$$M = \begin{pmatrix} x_1^n & x_1^{n-1} & \cdots & x_1 & 1 \\ x_2^n & x_2^{n-1} & \cdots & x_2 & 1 \\ x_3^n & x_3^{n-1} & \cdots & x_3 & 1 \\ \vdots & \vdots & \ddots & \vdots & \vdots \\ x_N^n & x_N^{n-1} & \cdots & x_N & 1 \end{pmatrix}$$

FITTING WITH LEGENDRE POLYNOMIALS

Expansion Equation

The expansion of $f(x)$ in terms of Legendre polynomials is

$$f(x) = \sum_{\ell=0}^{\infty} a_\ell P_\ell(x)$$

Expression for Coefficients

Using the orthogonality of the Legendre polynomials, the coefficients in the expansion are given by

$$a_\ell = \frac{2\ell + 1}{2} \int_{-1}^{1} P_\ell(x) f(x)\, dx$$

MATLAB Functions Used

polyfit
: If x and y are two vectors of the same length, then `polyfit` will determine the coefficients of the least squares best-fit polynomial of degree n to these data. The use is

```
c = polyfit(x,y,n)
```

where n is the degree of the fitting polynomial with coefficients c_i.

vander
: If x is a vector of length n, then `vander(x)` returns a matrix composed of the powers, k, of x from $k = 0, \ldots, n - 1$, reading from the right column of the matrix. That is, $v_{ij} = x_i^{n-j}$.

rcond
: The *condition* number of a matrix is used to determine the reliability of a computed inverse. If the condition number is infinite, the matrix is singular, if it is very large, the matrix is *ill conditioned*. The reciprocal of the condition number of a matrix **A** is returned by `rcond(A)`.

15 / Nonlinear Least Squares

PREVIEW   As long as the model function depends linearly on the adjustable parameters, the minimization of χ^2 leads to a set of linear equations in the parameters which can be cast in the form of matrix equations, and we can use all the powerful procedures of linear analysis to obtain a solution. Yet, frequently, it is clear that the simplest form of the model function is not linear in the parameters, and alternative approaches are required. The progression in problem type from linear to nonlinear is always a major step. Most of the well-known algorithms used to solve linear problems no longer apply and cannot be adapted to the nonlinear case. Rarely can even the existence and uniqueness of a solution be established. Therefore, it is not uncommon to first attempt to *linearize* the problem; i.e., attempt to find a similar problem that can be treated by linear means, or at least segment the problem into linear and nonlinear parts. This avenue of approach is described first. Only if these methods fail or are for some reason unsatisfactory, is the full nonlinear problem attempted.

Because nonlinear problems are so often very difficult to solve, we proceed cautiously in describing their methods of solution. The fundamental step, as in the linear case, is to first compute the χ^2 that characterized the fit of a functional form to the data. In the nonlinear case, χ^2 will frequently be a highly complicated, nonlinear function of many parameters. We use a variety of means to obtain the optimum fit, corresponding to a minimum of this function. The first and easiest method to try is based on utilizing the toolbox function, fmins, which attempts to find the minima of multivariable, nonlinear functions. If a good set of starting values for the parameters is available, this procedure is often successful, rather efficient, and certainly easy to code in MATLAB. It is, most often, the preferred method.

However, in situations where the minimum of χ^2 is extremely elusive, and methods based on fmins require excessive computational time, we may have to resort to more sophisticated methods that are specially designed to find the minimum of χ^2 in a more efficient manner. We describe and illustrate one such method, due to Marquardt, at the end of this chapter.

Pseudolinear Least Squares

The easiest option for handling a nonlinear least squares problem is to find a different problem that should produce similar results, but that is *linear* in the parameters. This technique avoids all the complications associated with nonlinear analysis and very often provides a satisfactory result. We can most readily explain this concept by considering a few examples.

336

LINEARIZATION OF THE EXPONENTIAL MODEL FUNCTION

The simple exponential form, $y(x) = Ce^{\lambda x}$, is a common solution to many physical problems and is nonlinear in the parameter λ. We could form a function $\chi^2(C, \lambda)$ in the usual way and differentiate it with respect to C and λ to find the equations defining the minimum. However, the equations are nonlinear in λ and usually quite difficult to solve. Before attempting to find a solution to the nonlinear problem, it is worthwhile to try to find a change of variables that results in a *linear* dependence on the parameters.

Thus, if the model function is again, $y(x) = Ce^{\lambda x}$, we could take the natural logarithm of this equation to obtain,

$$y(x) = Ce^{\lambda x} \quad \Rightarrow \quad \ln y = \ln C + \lambda x$$

or, defining

$$Y \equiv \ln y$$
$$X \equiv x$$
$$\alpha_1 \equiv \lambda$$
$$\alpha_2 \equiv \ln C$$

the model equation written in the new variables becomes, $Y(X) = \alpha_1 X + \alpha_2$, which is precisely the same linear form as discussed previously in Chapter 14. The least squares analysis would then proceed by first replacing the data, (x_i, y_i), by $(X_i, Y_i) = (x_i, \ln y_i)$ and then by finding the parameters for the best-fit line, $Y(X)$, to the data set (X_i, Y_i). That is, we find the best fit to $\ln y_i$ rather than to y_i. Usually this is quite satisfactory, but in general, it will not produce precisely the same results as fitting directly to the nonlinear exponential. We examine the difference between the two procedures more carefully later in this chapter.

Once the parameters α_1 and α_2, are computed, the original parameters, C and λ, are obtained via the inverse of the equations for the α's; that is,

$$\lambda = \alpha_1$$
$$C = e^{\alpha_2}$$

An example involving three parameters is to transform the task of fitting a Gaussian distribution,

$$y(x) = Ce^{-(1/2\sigma^2)(x-x_0)^2}$$

to that of fitting a quadratic, $Y(X) = \alpha_1 X^2 + \alpha_2 X + \alpha_3$. This can be achieved by the change of variables,

$$Y \equiv \ln y \qquad \alpha_1 \equiv -\frac{1}{2\sigma^2}$$
$$X \equiv x \qquad \alpha_2 \equiv \frac{x_0}{\sigma^2}$$
$$\alpha_3 \equiv \ln C - \frac{x_0^2}{2\sigma^2}$$

LINEARIZATION OF OTHER NONLINEAR MODEL FUNCTIONS

Frequently the model function depends on only two parameters, and in such cases, it is often satisfactory to find a change of variables to recast the model in a form that is linear in the parameters. A few examples are given in Table 15.1.

Standard form of model	Rewritten as	Change of variables			
		Y	X	α_1	α_2
$y = \dfrac{ax}{1+bx}$	$\dfrac{1}{y} = \dfrac{1}{ax} + \dfrac{b}{a}$	$\dfrac{1}{y}$	$\dfrac{1}{x}$	$\dfrac{1}{a}$	$\dfrac{b}{a}$
$y = \dfrac{a}{x-b}$	$\dfrac{1}{y} = \dfrac{x}{a} - \dfrac{b}{a}$	$\dfrac{1}{y}$	x	$\dfrac{1}{a}$	$-\dfrac{b}{a}$
$y = \dfrac{ax}{b^2 - x^2}$	$\dfrac{x}{y} = \dfrac{b^2}{a} - \dfrac{x^2}{a}$	$\dfrac{x}{y}$	x^2	$-\dfrac{1}{a}$	$\dfrac{b^2}{a}$
$y = ax^b$	$\ln y = b\ln x + \ln a$	$\ln y$	$\ln x$	b	$\ln a$
$y = ae^{-x^2/b^2}$	$\ln y = -\dfrac{x^2}{b^2} + \ln a$	$\ln y$	x^2	$-\dfrac{1}{b^2}$	$\ln a$
$\dfrac{x^2}{a^2} + \dfrac{y^2}{b^2} = 1$	$y^2 = b^2 - \dfrac{b^2}{a^2}x^2$	y^2	x^2	$-\dfrac{b^2}{a^2}$	b^2

Table 15.1   *Examples of linearized model functions*

Elementary Nonlinear Methods

To optimize the fit by the method of least squares when the model function is *nonlinear*, we proceed, as before, by first constructing the chi-square function, which is now a function of the n parameters represented as the elements of the vector, $\vec{a}$. True, finding the minimum of χ^2 by the ordinary means of differentiating with respect to the a_i's will not lead to a set of linear, matrix equations. But we have previously encountered the problem of the nonlinear minimization of a function of several variables (see Chapter 10), and this problem seems no different from the ones successfully handled in Part III. There is, however, one major difference that may have a bearing on the method of solution to be selected. In the ordinary multivariable minimization problem, you have no control over the choice of the function in question and, as a consequence, the derivatives used in finding the minimum are often far too complicated to be helpful. In our present situation, *you* created the model function and it was chosen in part because of its simplicity. Thus, we can safely assume that derivatives are always available.

Now, a distinctive feature of chi-square functions is that they almost always have very shallow minima. And, because we are searching for a best fit to data that contain statistical noise, it is never prudent to compute the minimum to high precision. Two or three significant digits should be sufficient.

Depending on the complexity of the model function, there are several different approaches when using MATLAB to find the minima of chi-square in nonlinear cases. We describe them in increasing order of difficulty.

Use of the Toolbox Function `fmins` to Find the Optimum χ^2

If the number of parameters is not too large, for example, ≤ 5 or 6, *and* you have a fairly good guess for the starting values for the minimization procedure, by far the easiest approach is to let MATLAB handle the details. Simply code the function $\chi^2(\vec{a})$ as an M-file and call the toolbox function `fmins`. The data will have to be passed to the function as `global` variables (including the errors, dy). Recall that `fmins` uses a *simplex search* method and is effectively guaranteed to eventually find a minimum. However, this method, as described in Chapter 10, uses only information from the values of the function itself (no derivatives) and, as a result, sometimes takes a long time to find the minimum. Nonetheless, it is usually the first method attempted because it is so easy to understand and to implement.

EXAMPLE 15.1

Fitting to a Sum of Exponentials

The most common nonlinear model is constructed from exponential functions. For example, the *pseudodata* of Figure 15.1 (page 340) were constructed by adding random numbers of the form, `0.3*(rand(n,1)-.5)`, to the n values computed using the expression

$$y(x) = 3e^{-0.4x} + 12e^{-3.2x}$$

The MATLAB M-file function to compute χ^2 as a function of the four parameters is:

```
function E = TwoExps(a)
  global xData yData dy
  x = xData(:);   y = yData(:);
  Y = a(1)*exp(-a(3)*x) + a(2)*exp(-a(4)*x);
  E = sum(((y-Y).^2)./dy.^2);
```

Using `a0 = [1 1 1 1]';`, the following call to `fmins` will find the values of the parameters that minimize χ^2 and will display the minimum value computed for χ^2 along with the values of the best-fit parameters:

```
global xData yData dy
% compute pseudo-data-------
x = [0:.2:4]';
y = 3*exp(-0.4*x)+12*exp(-3.2*x);
yData = y + 0.3*(rand(size(x)-.5);
dy = 0.3;
% Find best fit of TwoExps.m ---
```

```
a = fmins('TwoExps',a0,1.e-2,1);
E = TwoExps(a);
disp([E a'])
```

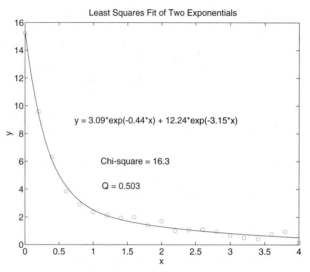

Figure 15.1 *A fit of two exponentials to data.*

Also, recall that the fourth parameter in the call to **fmins** produces a display of running results during the calculation. The number of calls to the function in this situation was 476, and the calculation took about 3 minutes on a PC. The value of dy was roughly estimated from the "scatter" of the data and estimated to be ≈ 0.27. The best-fit parameters were computed to be[1] $\vec{a} = [3.09 \quad 12.24 \quad 0.44 \quad 3.15]$, which should be compared with the values $[3 \quad 12 \quad 0.4 \quad 3.2]$. Using these values, we obtain $\chi^2 = 16.3$ and $Q = 0.50$. ●

EXAMPLE 15.2

Fitting to a Peak on Top of an Exponential Background

As a second, somewhat more complicated example, consider the results of an experiment represented again by pseudodata in Figure 15.2.

A common challenge in many experimental situations is to determine the parameters describing a particular phenomenon that is partially hidden by some form of background noise. Consider the problem of measuring a particular radioactive decay characterized by a decay energy, E_0, and described by a Lorentzian distribution (see Problem 13.10) in the face of considerable background radiation from cosmic rays and other radioactive decays. The model function should incorporate two elements: an exponential for the background and a Lorentzian-type function for the peak. We are particularly interested in determining the parameters of the Lorentzian.

---

[1]The reader should be aware that the solution of a nonlinear least squares problem is not always unique. There may be more than one set of parameters that will produce an equal fit.

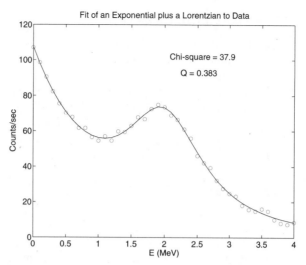

Figure 15.2 *A peak partially hidden by an exponential background.*

The M-file function for this problem is:

```
function E = HddnPeak(a)
  global x y dy
  Y = a(1)*exp(-a(2)*x) + a(3)*((x-a(4)).^(-2) + a(5)).^(-1);
  E = sum(((y-Y).^2)./dy.^2);
```

The pseudodata were computed using the values `a = [100 1 30 2 0.5]`, and `5*(rand(n,1)-.5)` was added to the values to simulate experimental error. For dy, we used `std(y-Y)`, which has a value of 1.35. Rather than start the search with all ones for a0, we first "eyeball" the data to estimate some of the parameters:

$$a_2 \approx \frac{\ln(y_1/y_n)}{x_n - x_1} \qquad \textit{Background decay constant}$$

$$= 0.629$$

$$a_1 \approx y_1 e^{a_2 x_1} \qquad \textit{Background amplitude}$$

$$= 107.0$$

$$a_3 \approx 20.00 \qquad \textit{Height of Lorentzian peak over background}$$

$$a_4 \approx 1.90 \qquad \textit{Location of Lorentzian peak}$$

$$a_5 \approx 0.75 \qquad \textit{Width of Lorentzian peak}$$

In this case the calculation required 440 function evaluations and a few minutes of PC time. The knowledgeable choice of starting parameters was critical in reducing the execution time to an acceptable value. ●

PARTIAL LINEARIZATION

Just about every model imaginable has a linear dependence on at least some of the parameters. The linear least squares methods of MATLAB are powerful and easy to use, thus it seems reasonable that employing these methods to attack the linear part of the problem should considerably reduce the execution time.

Recall that the backslash division operator will automatically solve an overdetermined set of equations in a least squares sense. That is, for n data points, (x_i, y_i), the n equations $\vec{y} = a\vec{x} + b$ can be solved for a and b in MATLAB by

```
M = [x ones(n,1)];
[a;b] = M\y;
```

Returning to the two-exponential model, we see that it is linear in the two parameters a_1 and a_2. Let us rewrite the model, labeling the linear parameters as `a` and the nonlinear ones as `b`; i.e.,

$$f(\vec{a}, \vec{b}, x_i) = a_1 e^{-b_1 x_i} + a_2 e^{-b_2 x_i}$$

Assume for the moment that the two nonlinear parameters, b_1 and b_2, are known. We can then solve for a_1 and a_2 by using the following code:

```
M = exp(-x*b');
a = M\y;
```

If this procedure can somehow be incorporated into the minimization method, χ^2 effectively becomes a function of only two parameters, not four, and the efficiency of the calculation should be enhanced. Furthermore, because the model is relatively simple, perhaps we should attempt to use derivative information to further speed the process. Thus, to find the minimum of χ^2, we need to solve the *two* simultaneous equations

$$0 = \frac{\partial(\chi^2)}{\partial b_1} = 2\sum_{i=1}^{n} \left[y_i - \left(a_1 e^{-b_1 x_i} + a_2 e^{-b_2 x_i} \right) \right] a_1 x_i e^{-b_1 x_i}$$

$$0 = \frac{\partial(\chi^2)}{\partial b_2} = 2\sum_{i=1}^{n} \left[y_i - \left(a_1 e^{-b_1 x_i} + a_2 e^{-b_2 x_i} \right) \right] a_2 x_i e^{-b_2 x_i}$$

(15.1)

plus use the previously described linear method to obtain a_1 and a_2. To find the roots of two simultaneous nonlinear equations, we could use the methods of Chapter 9, particularly the code for Newton's method on page 230. Adapting this code to find the root of the equations

$$\frac{\partial f(x, y)}{\partial x} = 0 \qquad \frac{\partial f(x, y)}{\partial y} = 0$$

we obtain the MATLAB code shown in Figure 15.3.

```
function r = Newton2d(dF,x0)
%Newton2d will find the root of the column vector
%  function whose name is contained in the string
%  constant, dF.  In the present context, this func-
%  tion corresponds to the two elements of the grad-
%  ient of a function F(x,y).
%
%  The root is found using Newton's method with dif-
%  ference expressions for derivatives starting with
%  a column vector initial guess, x0.  Convergence is
%  attained when |dx| < 1.e-6.  This function will
%  handle only 2 simultaneous equations, but can
%  easily be generalized.
%-------------------------
x   = x0;
dx = [0.01;0.02];
for i = 1:50
    f0   = feval(dF,x);
    dfdx = feval(dF,b+[db(1) 0]')-f0;
    dfdy = feval(dF,b+[0 db(2)]')-f0;
    M    = [dfdx/dx(1) dfdy/dx(2)];
    dx   = -M\f0;
    x    = x+ dx;
    if (norm(dx < 1.e-6));return; end
end
error('took more than 50 iterations')
```

Figure 15.3 *MATLAB code to find the root of two simultaneous nonlinear equations.*

Next, the MATLAB function dEda computes both expressions in Equation 15.1 and returns them as the elements of a 2×1 column vector $\vec{f}$. Notice, in particular, the use of backslash matrix division to obtain the least squares values of the linear parameters, followed by the use of ordinary matrix multiplication to duplicate the summations. The root of this function will be a column vector containing the values of the nonlinear parameters, b_1 and b_2. The linear parameters are then obtained once again by backslash division. The root of the two equations is obtained by using the function Newton2d of Figure 15.3.

```
%==============First the .m function ======
%    b contains the nonlinear parameters
%    a contains the linear parameters
function f = dEda(b)
global Xdata Ydata dY
  x = Xdata(:); y = Ydata(:);
  M = exp(-x*b');
  a = M\y;
  Yfit = M*a;
  f = ([x x].*M)'*(y-Yfit);
```

```
%=============Then the main MATLAB code ==
global Xdata Ydata dY
b0 = [1 1]';
b = Newton2d('dEda',b0);
a = [exp(-Xdata*b')\Ydata;b];
disp(a)
```

These latest values for the four parameters, a_i and b_i, should be compared with the values from the previous calculation. In this example, they turn out to be the same to four significant figures. However, the execution time for the latter calculation was approximately 10 times faster.

More Sophisticated Nonlinear Methods

If nothing you try reduces the execution time to an acceptable level, you have little choice but to resort to more sophisticated (thus more complicated) methods. Because we are attempting to find the minimum of χ^2, which is now a function of numerous parameters, this would seem to be a problem identical to those considered in Chapter 10, where several methods of finding the minimum of a multivariable function were discussed. This is indeed the case—you may wish to reread Chapter 10 at this time. The only change in the earlier methods will be to incorporate some special features of the chi-square function to produce a more efficient algorithm. The result is a highly efficient and popular method for solving nonlinear least squares problems, known as Marquardt's method.

THE METHOD OF MARQUARDT

The method of Marquardt adapts the most general of the methods of Chapter 10 for finding minima of multivariable functions specifically to the least squares problem. The method is based on the insights that (a) near the actual minimum, the function χ^2 should be very nearly parabolic in the parameters, and thus quadratic methods will be most efficient, and (b) far from the actual minimum, quadratic methods take too tiny a step, and *steepest descent* methods employing the gradient are more efficient.

The quadratic approximation is expected to be valid in the vicinity of the minimum, and is obtained by a truncation of the multidimensional Taylor series expansion of χ^2 after the quadratic terms. If χ^2 is expanded in *parameter* space about the point $\vec{a}_0$, we obtain the equation

$$\chi^2(\vec{a}) \approx \chi^2(\vec{a}_0) + \delta\vec{a}^T \cdot \nabla_a \chi^2(\vec{a}_0) + \frac{1}{2}\delta\vec{a}^T \cdot \mathbf{H} \cdot \delta\vec{a} \qquad (15.2)$$

where $\delta\vec{a} = \vec{a} - \vec{a}_0$, and the matrix, $\mathbf{H}$, is the *Hessian* matrix defined as, $\mathbf{H}_{ij} = \partial^2(\chi^2)/\partial a_i \partial a_j$. (See Equation 10.2, and the discussion on page 242.) Next, consider a similar expansion for the gradient of χ^2,

$$\nabla_a \chi^2(\vec{a}) \approx \nabla_a \chi^2(\vec{a}_0) + \mathbf{H} \cdot \delta\vec{a} + \cdots \qquad (15.3)$$

The gradient is zero at the point where χ^2 is a minimum, and thus, for small $\delta\vec{a}$,

$$\vec{a}_{\min} \approx \vec{a}_0 + \delta\vec{a}$$

where

$$\delta\vec{a} = -\mathbf{H}^{-1} \cdot \nabla_a \chi(\vec{a}_0)^2 \tag{15.4}$$

On the other hand, the steepest descent method simply takes a step $\delta\vec{a}$ along the direction of the negative gradient,

$$\delta\vec{a} = -\kappa \nabla_a \chi(\vec{a}_0)^2 \tag{15.5}$$

where κ is an undetermined proportionality constant and ∇_a is the gradient with respect to the parameters $\vec{a}$.

To incorporate features of both methods, Marquardt first computes the second derivatives of the Hessian matrix using the explicit form of chi-square:

$$\mathbf{H}_{ij} = \frac{\partial^2}{\partial a_i \partial a_j} \sum_{m=1}^{n} (y_m - f(\vec{a}, x_m))^2 \tag{15.6}$$

$$= 2 \sum_{m=1}^{n} \frac{1}{(\Delta y_m)^2} \left[\frac{\partial f(\vec{a}, x_m)}{\partial a_i} \frac{\partial f(\vec{a}, x_m)}{\partial a_j} - (y_m - f(\vec{a}, x_m)) \frac{\partial^2 f(\vec{a}, x_m)}{\partial a_i \partial a_j} \right]$$

The difference between the model and the data should contain as many positive as negative terms, so it is expected that the second term in the brackets, involving second derivatives of the model function, will average to approximately zero or at least be very small. In addition, this term tends to make the calculation unstable. At any rate, the convention is to drop the second derivative term and approximate the Hessian as

$$\mathbf{H}_{ij} \approx 2 \sum_{m=1}^{n} \frac{1}{\Delta y_m^2} \left[\frac{\partial f(\vec{a}, x_m)}{\partial a_i} \frac{\partial f(\vec{a}, x_m)}{\partial a_j} \right] \tag{15.7}$$

Next, to include steepest descent effects, the Hessian matrix is replaced by

$$\mathbf{H} \Rightarrow \mathbf{H} \cdot \mathbf{I}(1 + \kappa)$$

where $\mathbf{I}(1 + \kappa)$ is a diagonal matrix with $1 + \kappa$ on the diagonal.[2] That is, only the diagonal elements of $\mathbf{H}$ are altered by the adjustable factor $1 + \kappa$. This will, to some extent, incorporate the steepest descent approximation of Equation 15.5. Equation 15.4 is then solved for the improvement $\delta\vec{a}$.

The original choice for κ should be much less than one to avoid taking too large an initial step. The calculation proceeds as follows:

1. Compute $\chi^2(\vec{a_0})$, then $\delta\vec{a}$ using Equation 15.4, and finally $\chi^2(\vec{a_0} + \delta\vec{a})$.
2. If, in this step, χ^2 has *increased*, indicating that you are not near the minimum, increase the steepest descent aspect by increasing κ by a factor of 10 or so. But *do not* update the values of the parameters, $\vec{a}_0$.

---

[2]For a more complete description of the Marquardt method, including theoretical justification for the $(1 + \kappa)$ term, see Wong, *Computational Methods in Physics and Engineering*, Prentice-Hall, Englewood Cliffs, NJ, 1992, p. 373.

3. If, in this step, χ^2 has *decreased*, you may be near the quadratic region, so decrease κ by a factor of 10 or so and replace $\vec{a}_0 \Leftarrow \vec{a}_0 + \delta\vec{a}$. Proceed to the next iteration.

4. If the change in χ^2 is less than one, terminate the calculation.

EXAMPLE 15.3

Marquardt's Method Applied to Fitting Data with Multiple Exponentials

A model function particularly well suited to a MATLAB implementation of Marquardt's method is a linear combination of N exponentials.[3] The total number of parameters is $2N$. As discussed in the section on partial linearization, we separate the parameters into two classes: linear, a_m, $m = 1 : N$, and nonlinear, b_m, $m = 1 : N$. Thus,

$$f(\vec{a}, \vec{b}, x_i) = \sum_{m=1}^{N} a_m e^{-b_m x_i}$$

The data consist of the n points (x_i, y_i), and they are stored in column vectors.

The first element of Marquardt's method requires evaluating the derivatives of the model function with respect to each of the parameters:

$$(\nabla_a f)_m = \frac{\partial f}{\partial a_m} = e^{-b_m x_i} \qquad (\nabla_b f)_m = \frac{\partial f}{\partial b_m} = -a_m x_i e^{-b_m x_i}$$

In the MATLAB code that follows, each of these expressions will represent an $n \times N$ matrix. The Hessian matrix will be an $N \times N$ matrix computed using Equation 15.7. The MATLAB expressions to compute the matrices are:

```
dfa = exp(-x*b');
dfb = -(x*a).*dfa;
df  = [dfa dfb];
H   = df'*(df./dy.^2)
```

The gradient of χ^2 is

$$\frac{\partial(\chi^2)}{\partial a_k} = -2 \sum_{m=1}^{n} \frac{1}{\Delta y_m^2} [y_m - f(\vec{a}, \vec{b}, x_m)] \frac{\partial f(\vec{a}, \vec{b}, x_m)}{\partial a_k}$$

or in MATLAB code

```
gradChi = -2*df'*((y - exp(-x*b')*a)./dy.^2);
```

Figure 15.4 shows the remainder of the necessary MATLAB code assembled in the function MrqdExps. A MATLAB script that demonstrates fitting exponentials to data is included on the applications disk and is named TstMQ.m.

As with any least squares calculation, the approximate uncertainties in the computed best-fit parameters are obtained by using these values to first compute $\mathbf{H}^{-1}$, and then equating the parameter uncertainties to the diagonal elements of $\mathbf{H}^{-1}$.

---

[3] In most cases, attempts to fit a sum of more than three exponential functions to data will not yield satisfactory results. The parameter space is too large, and the fitting functions are too similar to obtain definitive results.

```
function [a,b] = MrqdExps(c,d,Xdata,Ydata,dy)
% MrqdExps will attempt to fit a sum of N separate exponential
% functions, where N is the length of the input vectors, c and d,
% which are each column vectors containing the initial extimates
% for the linear and exponential terms (i.e., exp(-x*d')*c). The
% best-fit values are returned as (a,b).  If the initial Chi^2
% is > 1e4*n, where n is the number of data points, the code is
% halted and better estimates are requested.  If the number of
% iterations exceeds 1000, the code terminates unsuccessfully.
% ===============================================================
  a = c;   b = d;   K  = 0.01;
  N  = length(a);   n = length(x)
  Chi= sum((Ydata-exp(-Xdata*b')*a).^2)/dy^2;
  if Chi > 10000*n
     disp('initial parameters yield too large a Chi^2 = ')
     disp(Chi)
     disp('improve guess and try again')
     return
  end
  dChi= 2;   COUNT = 0;
  while abs(dChi) > .05
     df= exp(-Xdata*b');
     Y = df*a;
     df = [df -(Xdata*a').*df];
     gradChi = -2*df'*(Ydata-Y)/dy^2;
     H  = 2*df'*df*diag((1+K)*ones(2*N,1))/dy^2;
     dc = -H\gradChi;
     da = dc(1:N,1);       db = dc(N+1:2*N,1);
     dChi=sum((Ydata-exp(-Xdata*(b+db)')*(a+da)).^2)/dy^2-Chi;
     if dChi >= 0
        K = 10*K;
     else
        K = K/10;
        a = a + da;   b = b + db;   Chi = Chi + dChi;
     end
     COUNT = COUNT + 1;
     if COUNT > 1000
        disp('failure due to excessive iterations')
        disp('latest values, Chi^2 = '); disp(Chi)
        disp('parameters a =    b = '); disp([a b])
        break
     end
  end
```

Figure 15.4 *MATLAB code using Marquardt's method to fit exponentials.*

An additional demonstration program that fits a function to a chart of wind chill factors can be found on the applications disk; it is named WindDemo.m. ●

Problems

REINFORCEMENT EXERCISES

P15.1 Two-Parameter Nonlinear Least Squares. Select one of the model functions in Table 15.1, assigning $a = 3$, $b = 2$. Use this function to compute a pseudodata set of 21 points, (x_i, y_i), for values of X = [1:.1:3]'; Y = f(a,b,x);. Simulate the effects of experimental noise in the data by replacing

$$y \Leftarrow Y + K * (\text{rand}(\text{length}(Y), 1) - .5);$$

where K = (max(Y)-min(Y))/20. Plot these data as discrete points. (Add the statement hold on after the plot.) Next, proceed to compute a number of least squares fits to these data as follows:

   a. Compute (X_i, Y_i) by the change of variables indicated in the table, and, from these, compute the least squares parameters (α_1, α_2). Invert the equations for (α_1, α_2) to obtain the model parameters a and b. Using these parameters, plot the model over the discrete plot of the data.

   b. Code an M-file function for $\chi^2(\vec{c})$, where $\vec{c} = \left(\begin{smallmatrix} a \\ b \end{smallmatrix}\right)$. Next, use the function **fmins** to directly find the best-fit values of the parameters. (Use dy = K and a = 3.0; b = 0.5 as the initial estimate of the parameters. Remember to declare x, y, and dy as global variables.) Again, plot the model function using these latest values for a and b. (Add the statement hold off after this plot.) Compare the value of χ^2 computed with this set with that computed using the values of part a.

   c. Analytically differentiate the expression for χ^2 with respect to a and with respect to b. Code an M-file function dModel(c) that returns a 2×1 column vector containing the values of the two derivatives, where c is a column vector containing the values of a and b. Finally, use the function Newton2d to find the roots of these equations, and compare the values with the results of parts a and b.

   d. Plot a three-dimensional perspective plot of $\chi^2(a, b)$ for a suitable range of a and b. Use **view** to orient the plot so that the minimum is clearly visible.

A MATLAB demonstration script for this problem is on the applications disk and is named Prb15p1.m.

P15.2 Three-Parameter Nonlinear Least Squares with Partial Linearization. Duplicate the pseudodata of Figure 15.2 using the same values for the parameters, a_i, $i = 1 : 5$. Reduce the problem from five to three parameters by using the ideas of partial linearization to compute a_1 and a_3 using backslash division. Recode the function HddnPeak using these ideas as a function of only three parameters. Use the function fmins to find the best-fit parameters and, from these, compute the remaining two. Compare both the values of the parameters and the number of function evaluations with the results of the direct method of Example 15.1. See also the MATLAB script file named Prb15p2.m on the applications disk.

P15.3 Singular Least Squares. The model function $y = ae^{bx+c}$ appears to have three parameters. However, the model could be written as $y = Ce^{bx}$ with $C = ae^c$, suggesting that the parameters a and c are not independent.

a. Show that two of the equations for the minimum of χ^2, namely $\partial\chi^2/\partial a = 0$ and $\partial\chi^2/\partial c = 0$, are actually the same equation.
b. How does this affect the calculation of the inverse of the Hessian matrix in Marquardt's method?
c. Devise a test for determining whether or not your set of parameters in the model function contains any superfluous terms; i.e., parameters that are not independent.

P15.4 Marquardt's Method and Two-Parameter Least Squares. Select one of the model functions in Table 15.1. Determine the expressions for

$$\frac{\partial f(a, b, x_m)}{\partial a}, \quad \frac{\partial f(a, b, x_m)}{\partial b}, \quad \frac{\partial(\chi^2)}{\partial a}, \quad \frac{\partial(\chi^2)}{\partial b}$$

and use them to write a MATLAB code to use Marquardt's method to find the best-fit parameters. Test your code by first creating pseudodata as indicated in Problem 15.1, and then least squares fitting to the data. First, use a *good* estimate for the parameters to verify that your code is correct. Second, experiment with *poor* choices for the initial guess. Does the program break down for any of your choices? Can you trace any of the failures to *pathologies* in the equation for χ^2? You may wish to execute a three-dimensional perspective plot of χ^2 over the range of the failure.

Summary

PSEUDOLINEAR LEAST SQUARES

Exponential Model Function

The fit of data to a *nonlinear* exponential model, $y(x) = Ce^{\lambda x}$, can be accomplished by introducing new variables, $Y \equiv \ln y$ and $X \equiv x$, so that the problem expressed in terms of the new variables,

$$Y = \alpha_1 X + \alpha_2$$

is *linear* in the parameters $\alpha_1 = \lambda$, $\alpha_2 = \ln C$. Similar variable changes can frequently be found for many other elementary nonlinear model functions.

USE OF THE TOOLBOX FUNCTION fmins
TO FIND THE OPTIMUM χ^2

Minimizing χ^2

The model function, f, is written in terms of parameters, $\vec{\alpha}$, and the expression for χ^2,

$$\chi^2(\vec{\alpha}) = \sum_{i=1}^{N} \left(\frac{y_i - f(\vec{\alpha}, x_i)}{\Delta y_i} \right)^2$$

is coded as a MATLAB M-file function. The minimum of this function is obtained by using the toolbox function fmins, and the result corresponds to the optimum fit. The confidence level associated with the fit is then estimated using Equation 13.18 on page 308.

Partial
Linearization

If the model function depends linearly on *some* of the parameters, the algorithm to minimize χ^2 is often segmented into two parts: (1) equations of the form, $\partial \chi^2 / \partial b_i = 0$, to determine the nonlinear parameters, and (2) equations of the form, $\mathtt{a} = \mathtt{M} \backslash \mathtt{y}$, employing backslash division to determine the linear equations.

THE METHOD OF MARQUARDT

Definition

Marquardt's method assumes that, as a function of the parameters, $\vec{a}$, χ^2 is: (1) *parabolic* near the minimum and adequately described by the first few terms of a Taylor series, and (2) *steeply banked* far from the minimum and handled best by *steepest descent* methods.

Basic Equations

The improvement to the estimate of the parameters is given by

$$\vec{a}_{\min} \approx \vec{a}_0 + \delta \vec{a}$$

with

$$\delta \vec{a} = -\mathbf{H}^{-1} \cdot \nabla_a(\chi^2)$$

where the Hessian matrix is approximated as

$$\mathbf{H}_{ij} = 2 \sum_{m=1}^{n} \frac{1}{\Delta y_m^2} \left[\frac{\partial f(\vec{a}, x_m)}{\partial a_i} \frac{\partial f(\vec{a}, x_m)}{\partial a_j} \right]$$

and in subsequent iterations, the Hessian is amended by the replacement

$$\mathbf{H} \Rightarrow \mathbf{H} \cdot \mathbf{I}(1 + \kappa)$$

where κ is chosen to be *large* if far from the minimum, or is reduced as the minimum is approached.

16 / *Interpolation and Spline Fits*

PREVIEW   Characterizing a large amount of data by a simple functional form with a few parameters determined by least squares fitting is the central element of data analysis. However, there are many situations when a different approach is required. If there are relatively few data points, and they are well separated, a common challenge is to use a few neighboring points to accurately estimate the value of a quantity *between* the individual data points, a process called *interpolation*. You can also use the last several points to predict the value of a quantity if the measurements were to be extended (*extrapolation*). Or perhaps, you need to draw a curve through the data to simply guide the eye. In this case, the driving criteria are that the curve be "smooth" and that it pass through each and every point. What we want to do is "connect the dots," or using the more technical name, *draw spline fits*.

In this chapter we describe a few relatively simple methods utilizing MATLAB functions to interpolate, extrapolate, and to draw spline fits through data.

Lagrange Interpolation

A line can be precisely drawn through any two points, a parabola through any three points, and an nth-degree polynomial through any $n + 1$ points. That is, an arbitrary nth-degree polynomial,

$$y(x) = a_1 x^n + a_2 x^{n-1} + \cdots + a_n x + a_{n+1}$$

has $n + 1$ undetermined coefficients that can be specified by $n + 1$ independent relations. Thus, if the polynomial is evaluated at $n + 1$ distinct points, we can write

$$
\begin{aligned}
y(x_1) &= a_1 x_1^n + a_2 x_1^{n-1} + \cdots + a_n x_1 + a_{n+1} = y_1 \\
y(x_2) &= a_1 x_2^n + a_2 x_2^{n-1} + \cdots + a_n x_2 + a_{n+1} = y_2 \\
&\ \vdots \\
y(x_n) &= a_1 x_n^n + a_2 x_n^{n-1} + \cdots + a_n x_n + a_{n+1} = y_n \\
y(x_{n+1}) &= a_1 x_{n+1}^n + a_2 x_{n+1}^{n-1} + \cdots + a_n x_{n+1} + a_{n+1} = y_{n+1}
\end{aligned}
$$

or in matrix notation

$$\begin{pmatrix} x_1^n & x_1^{n-1} & \cdots & x_1 & 1 \\ x_2^n & x_2^{n-1} & \cdots & x_2 & 1 \\ x_3^n & x_3^{n-1} & \cdots & x_3 & 1 \\ \vdots & \vdots & \ddots & \vdots & \vdots \\ x_n^n & x_n^{n-1} & \cdots & x_n & 1 \\ x_{n+1}^n & x_{n+1}^{n-1} & \cdots & x_{n+1} & 1 \end{pmatrix} \begin{pmatrix} a_1 \\ a_2 \\ a_3 \\ \vdots \\ a_n \\ a_{n+1} \end{pmatrix} = \begin{pmatrix} y_1 \\ y_2 \\ y_3 \\ \vdots \\ y_n \\ y_{n+1} \end{pmatrix}$$

or

$$\mathbf{M}\vec{a} = \vec{y}$$

The square matrix $\mathbf{M}$ is once again the *Vandermonde* matrix of the vector $\vec{x}$. It can be calculated directly using the toolbox function `vander(x)`. This equation can easily be solved for the coefficients of the polynomial that passes precisely through each and every data point. Given $n + 1$ data points (x_i, y_i), the MATLAB code to find the a_i's is simply:

```
M = vander(x);
a = M\y;
%                  then, as a check,
Y = polyval(a,x);
disp([Y y])
```

This polynomial is known as the *Lagrange* polynomial (prior to the existence of MATLAB, it was usually computed in a different manner). The same polynomial could be obtained using `polyfit(x,y,n)`, where n is the number of data points.

This doesn't look too hard. Instead of the laborious least squares procedures, why not fit a thousand points *exactly* with a degree 999 polynomial? Well, solving a 1000×1000 matrix equation is out of the question, with respect to both storage and computational time considerations. But, is it a useful alternative for smaller problems; for example 10 or 20 data points?

The key word is *exactly*. Every point that is a result of measurement has some experimental error; i.e., the complete measured value is $y_i \pm \Delta y_i$. There is no justification for insisting that the curve go precisely through y_i. Any point in the range $y_i - \Delta y_i$ to $y_i + \Delta y_i$ will do almost as well. Forcing a nonphysical requirement on the fitting function, that of fitting the points exactly, can be expected to have potentially disastrous effects. See, for example, the $n = 6$ degree polynomial that has been fit through the seven data points in Figure 16.1.

The data exhibit a mild fluctuation about a slightly increasing value from $x = 1$ to $x = 7$, but certainly they do not suggest any of the wild swings exhibited by the exact fit polynomial—those are clearly an artifact generated by the additional requirement of exact fit. In this case, an exact fit to the data would lead to serious

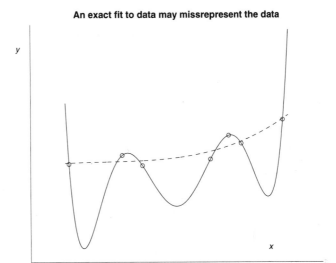

An exact fit to data may missrepresent the data

Figure 16.1 *An exact fit to seven data points.*

misconceptions about the nature of the data set. For comparison, a more well-behaved cubic least squares fit to the same points is also drawn on the figure.

If exact fits with Lagrange polynomials carry with them significant dangers of misinterpretation of the data, what useful purpose can they serve? Interpolation comes to mind. If we need to have an estimate of $y(x)$ at a point between data points, we could fit a polynomial through a *few* surrounding points and use the value of the polynomial at the desired point to estimate y.

MATLAB CODE FOR INTERPOLATION

The following code is an illustration of the intuitive ideas of interpolation. Given a value of $x = x_{\text{test}}$, the function scans the data to find the $m + 1$ nearest data points, fits an mth-degree polynomial through these points, checks to see if the polynomial at x seems to be far out of line with regard to the nearby data, and, if so, prints a warning. The function returns the value of the polynomial at x. There must be at least $m + 1$ data points in the vectors x and y.

```
function Y = PolyIntp(xtest,m,x,y)
%  PolyIntp(xtest,m,x,y) will fit the nearest m+1
%  data points to each element of the vector xtest
%  to a polynomial of degree m and return a vector
%  containing the values of the polynomial at xtest
%======================================================
   x = x(:); y = y(:); xtest = xtest(:)
   n = length(xtest);
```

```
   for j = 1:n
      [Z,imin] = sort(abs(x-xtest(j)));
         I = imin(1:m+1);    k = sort(I);
         a = polyfit(x(k),y(k),m);
      Y(j)= polyval(a,xtest(j));
         dy= max(y(k))-min(y(k));
         if abs(max(Y(j)-y(k))) > dy
           disp([xtest(j) Y(j)])
           disp('is a possibly a bad fit')
         end
   end
```

By using this code for several (*small*) values of m you can get some idea of the accuracy of the interpolation method.

You will not be surprised to learn that MATLAB has excellent interpolation functions in the toolbox. The most commonly used interpolation function is `interp1`, which is used in the following way:

```
Y = interp1(x,y,X,<option>);
```

The vectors (x,y) are the "data"; i.e., the points through which the interpolated curve must pass. The points specified in the vector X must be in sequence, and are the in-between points at which the data are to be estimated. The parameter `<option>` is a string variable that specifies the type of interpolation to be used. The default is a linear interpolation in which the points (x,y) are simply connected by straight lines. Other options are `'spline'`, where a cubic spline interpolation function is used (see the upcoming section, Cubic Spline Interpolation) or `cubic`, where a cubic equation is fitted through the points. (The `'cubic'` option requires that the values in X be equally spaced.)

Neville's Iterative Construction of the Lagrange Polynomial

A significant disadvantage associated with interpolation algorithms based on the Lagrange polynomial; i.e., the nth-degree polynomial that passes precisely through a given set of $n + 1$ points, is that they are susceptible to wild oscillations and, as a consequence, can lead to large associated errors when used to interpret data. An improvement is to successively increase the degree of the fitting polynomial, one degree at a time, and use the change in the computed values from one step to the next as an indication of the actual errors. This procedure, known as Neville's algorithm, also serves as the basis for a number of related extrapolation techniques, and it is found to be a particularly important element in recent techniques for the solution of differential equations. We return to this aspect of interpolation again in Chapter 19.

Neville's algorithm can easily be understood with the help of a simple example. Consider the construction of the Lagrange polynomial through five known points, (x_i, y_i), $i = 1, \ldots, 5$, by means of a *tableau* or pattern.

$$
\begin{array}{lll}
(x_1, y_1) & \bullet P_1 = y_1 \\
& & P_{12} \\
(x_2, y_2) & \bullet P_2 = y_2 & & P_{123} \\
& & P_{23} & & P_{1234} \\
(x_3, y_3) & \bullet P_3 = y_3 & & P_{234} & & P_{12345} \\
& & P_{34} & & P_{2345} \\
(x_4, y_4) & \bullet P_4 = y_4 & & P_{345} \\
& & P_{45} \\
(x_5, y_5) & \bullet P_5 = y_5
\end{array}
$$

The elements of each column of the triangular pattern represent polynomials that pass through the points indicated by the subscripts. That is, P_{234} is the Lagrange polynomial that passes through the three points indicated. Thus, the elements of the first column are simply constants equal to the corresponding y values. Neville's algorithm constructs the final polynomial through all five points, P_{12345}, by using the elements of one column of the tableau to construct the elements of the next column. For example, an element of column 3 is related to two elements of column 2 by

$$
P_{345}(x) = \frac{(x - x_3)P_{45}(x) - (x - x_5)P_{34}(x)}{x_5 - x_3}
$$

We can assume that $P_{34}(x)$ and $P_{45}(x)$ each pass through the two indicated points, so this construction guarantees that P_{345} will pass through the three indicated points. That is, $P_{345}(x_5) = P_{45}(x_5)$ and $P_{45}(x_5)$, passes through (x_5, y_5). Or more explicitly,

$$
P_{45}(x) = \frac{(x - x_4)P_5(x) - (x - x_5)P_4(x)}{x_5 - x_4}
$$

so that $P_{45}(x_5) = y_5$.

This idea can easily be generalized to

$$
P^{[m]}_{i \to (i+m)}(x) = \frac{(x - x_i)P^{[m-1]}_{(i+1) \to (i+m)}(x) - (x - x_{i+m})P^{[m-1]}_{i \to (i+m-1)}(x)}{x_{i+m} - x_i}
$$

where $P^{[m]}_{i \to (i+m)}(x)$ is the Lagrange polynomial of degree m that passes through all the points $y_i, y_{i+1}, \ldots, y_{i+m}$. Because the degree of the polynomial determines how many points the polynomial will pass through, we can shorten this notation to $P^{[m]}_i(x)$, or

$$
P^{[m]}_i(x) = \frac{(x - x_i)P^{[m-1]}_{i+1}(x) - (x - x_{i+m})P^{[m-1]}_i(x)}{x_{i+m} - x_i} \tag{16.1}
$$

Because we are primarily concerned with the changes in the polynomial from one step to the next, it is advantageous to rewrite Equation 16.1 in terms of the differences between the new value $P^{[m]}_i(x)$ and the two current elements, $P^{[m-1]}_{i+1}(x)$ and $P^{[m-1]}_i(x)$. Defining

$$
\begin{aligned}
D^{[m]}_i &\equiv P^{[m]}_i(x) - P^{[m-1]}_i(x) \\
U^{[m]}_i &\equiv P^{[m]}_i(x) - P^{[m-1]}_{i+1}(x)
\end{aligned} \tag{16.2}
$$

as the differences along the "down" and "up" directions along the diagonals of the tableau, the following recursion relations can be derived from Equation 16.1.

$$U_i^{[m+1]} = \left(\frac{x - x_{i+m+1}}{x_{i+m+1} - x_i} \right) (D_{i+1}^{[m]} - U_i^{[m]})$$

$$D_i^{[m+1]} = \left(\frac{x - x_i}{x_{i+m+1} - x_i} \right) (D_{i+1}^{[m]} - U_i^{[m]})$$

(16.3)

The relationships between the quantities $D_i^{[m]}$ and $U_i^{[m]}$ and the various order polynomials are illustrated in Figure 16.2.

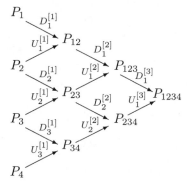

Figure 16.2   *A Neville interpolation table.*

The procedure is then as follows:

1. Start with a set of n known values of the function, (x_i, y_i), $i = 1, \ldots, n$ and a set of x values at which the function is to be interpolated. We will call this set X_j, $j = 1, \ldots, p$. Finally, decide on the maximum order at which to accomplish the interpolation. A typical value is $N \approx 9$. Of course, N must be less than n.

2. Initialize the quantities $D_i^{[0]}$ and $U_i^{[0]}$.

$$D_i^{[0]} = y_i$$
$$U_i^{[0]} = y_i$$

[*Note:* Because both sets of quantities will be updated in each step and then discarded, there is no need to retain the superscript in the MATLAB code.]

3. To minimize the error, the ordinary procedure is to next scan the x values to find the position of the one closest to the current X. Calling this position i_X, the interpolation starts at this position by assigning $y(X) \approx y_{i_X}$. In the remaining steps, the algorithm will "snake" through the tableau, ultimately reaching the rightmost vertex. This is relatively easy to code in MATLAB *if there is only one X value*. However, to interpolate several X values at once, *and* to accomplish this with a minimal use of `for` loops,

the code can become quite complicated. For this situation, we suggest instead, starting at the upper left vertex and following the upper diagonal down to the rightmost vertex. In either case, aside from differences in round-off errors, the answers are the same.

4. Use Equations 16.2 to compute the new values of $D_i^{[m]}$ and $U_i^{[m]}$ for $m = 1, \ldots, N-1$. The latest values of either $D_i^{[m]}$ or $U_i^{[m]}$ will be our error estimates for that step. The code to interpolate a single X value is illustrated in Figure 16.3. The code to interpolate a complete set of X values is given in Figure 16.4.

```
function [Y,dY] = Nevil1(x,f,N,X)
% Nevil1 interpolates the function f(x) represented by values
% contained in the column vector f evaluated at the points in
% the column vector x to the single point f(X).
%
% A total of N surrounding points are used in the interpola-
% tion.  The code first finds a band of data points of length
% N surrounding the point closest to X (labeled x(Ic)).  Next,
% the Nevile tableaux is computed, one column at a time. The
% best estimate is obtained by stepping through the table to-
% wards the rightmost vertex in the most direct manner.  The
% interpolated value and the error estimate are returned.
%=================================================================
 n = length(x);
 [a,Ic] = min(abs(x-X));
 dN= fix(N/2);
 if Ic > dN
    if Ic <= n-dN
       I = Ic-dN;
    else
       I = n-N+1:n;
    end
 else
       I = 1:N;
       Ic= n-dN;
 end
 F = f(I); D = f(I);  U = f(I);   xx = x(I);
 [a,Ic] = min(abs(xx-X));
 Y = F(Ic); Ic = Ic -1;
 for m = 1:N-1
    W = [D;0]-[0;U];
    W = W(2:N-m+1);
    xL= xx(1:N-m);  xR = xx(m+1:N);
    D = (xL-X)./(xL-xR).*W;   U = (xR-X)./(xL-xR).*W;
    if 2*Ic < N-m
        dY = D(Ic+1);
    else
```

Figure 16.3 *MATLAB code for Neville interpolation of a single point (continues).*

```
        dY = U(Ic); Ic = Ic - 1
    end
    Y = Y+dY;
end
```

Figure 16.3   *(Continued)*

```
function [Y,dY] = Nevil(x,f,N,X)
% Nevil interpolates the function f(x) represented by values
% contained in the column vector f corresponding to the x
% values contained in the column vector x.  The interpolation
% is then done at the x values specified in the ROW vector,
% X.  A total of N surrounding points are used for interpo-
% lation.  The code is basically the same as Nevil1, except
% no attempt is made to minimize round-off errors by finding
% an optimum path through the table.
%
% The interpolated values and the error estimates are returned
% as ROW vectors. The new portion of the code computes a mat-
% rix containing the indices of the data points surrounding
% each of the points to be interpolated.  This matrix, I, is
% then an N by p matrix, where p is the length of X.
%-----------------------------
dN= fix(N/2);
n = length(x);      p = length(X);
xx = x*ones(1,p);   X = ones(n,1)*X
%----------------------
%  Make a table of indices of points surrounding X
%
 [a,Ic] = min(abs(xx-X));
I1 = cumsum(ones(N, length(find(Ic < dN))));
I2 = find(Ic>dN & Ic<=n-dN);
I2 = cumsum([Ic(I2)-dN;ones(N-1,length(I2))]);
I3 = length(find(Ic> n-dN));
I3 = cumsum([(n-N+1)*ones(1,I3);ones(N-1,I3)]);
I  = [I1 I2 I3];
%----------------------
X  = X(1:N,:);
xx = reshape(xx(I),N,p);
F  = reshape(f(I),N,p);
D  = F;   U  = D;    Y  = F(1,:);
for m = 1:N-1
  W = D(1:N-m,:)-U(2:N-m+1,:);
  xL= xx(1:N-m,:);          xR= xx(m+1:N,:);
  X = X(1:N-m,:);
  D = (xL-X)./(xL-xR).*W;  U = (xR-X)./(xL-xR).*W;
  dY = D(1,:);             Y = Y + dY
end
```

Figure 16.4   *MATLAB code for Neville interpolation of a set of X values.*

In Table 16.1 we have used the values of $\sin x$ for x = 0:0.2:2 to interpolate the value at $x = 0.9$ and to extrapolate the value at $x = 2.3$.

| | | | Fractional errors | |
| | | | Approx. | Exact |
X	N	Y	dY/Y	$(Y - \sin X)/Y$
0.9	3	0.7830	−1.36e-2	−0.04e-2
	5	0.783324	−5.69e-5	−0.33e-5
	7	0.7833269	−3.97e-7	−0.27e-7
	9	0.783326909	−3.18e-9	−0.24e-9
2.3	3	0.740	−23.0e-2	−0.81e-2
	5	0.7458	19.2e-3	0.17e-3
	7	0.745708	−12.3e-4	0.03e-4
	9	0.7457048	66.8e-6	0.61e-6

Table 16.1 *Interpolation and extrapolation of* $\sin x$

Bulirsch-Stoer Interpolation and Extrapolation

The principal advantage of Neville's method of interpolation is that, in each step, by comparing successive interpolations, we are provided with an estimate of the actual error. Still, Neville's algorithm is based on polynomial interpolation and is, therefore, prone to all its potential problems. As we have seen in Chapter 4, the Padé approximation, using a ratio of polynomials, usually results in a dramatic improvement in both accuracy and efficiency, frequently yielding an approximation with an extended range of convergence. It would, therefore, seem highly desirable to develop an algorithm of the Neville type that could, step by step, construct a function that is a ratio of polynomials that will exactly fit through N points. Bulirsch and Stoer[1] have provided us with such a procedure. The idea is quite simple. The function $F_i^{[m]}(x)$ is a ratio of two polynomials with coefficients chosen so that the fraction passes through $m + 1$ points, starting at x_i; that is, through $x_i, x_{i+1}, \ldots, x_{i+m}$. In the Neville algorithm, the function $P_i^{[m]}(x)$ was a *polynomial* of degree m. In this case, the function $F_i^{[m]}(x)$ is given by

$$F_i^{[m]}(x) = \frac{p_1 x^\mu + p_2 x^{\mu-1} + \cdots + p_\mu x + p_{\mu+1}}{q_1 x^\nu + q_2 x^{\nu-1} + \cdots + q_\nu x + 1}$$

In the denominator polynomial, we have set $q_{\nu+1} = 1$ without loss of generality. The total number of parameters on the right is then $\mu + \nu + 1$, and if

---

[1]Stoer and Bulirsch, *Introduction to Numerical Analysis* (English translation by Bartels, Gautschi, and Witzgall), Springer-Verlag, Berlin and New York, 1980.

$F_i^{[m]}(x)$ is to pass through $m + 1$ points, we must have $m = \mu + \nu$. Further, if we are fitting through an *odd* number of points, we require that both polynomials be of the same degree ($\mu = \nu$); if fitting through an *even* number, we require that $\nu = \mu - 1$. For a given value of m, we can use the methods of Chapter 4 to compute the coefficients of the two polynomials. Alternatively, in the spirit of Neville's method, Bulirsch and Stoer have obtained an equation similar to Equation 16.1 that relates $F_i^{[m]}(x)$ to $F_i^{[m-1]}(x)$.

Rather than duplicating the rather extensive algebra involved in the derivation of the Bulirsch-Stoer relation, we refer the reader to the footnote 1, and simply quote the result:

$$F_i^{[m]}(x) = F_{i+1}^{[m-1]}(x) + \frac{F_{i+1}^{[m-1]}(x) - F_i^{[m-1]}(x)}{\left(\dfrac{x - x_i}{x - x_{i+m}}\right)\left[1 - \dfrac{F_{i+1}^{[m-1]}(x) - F_i^{[m-1]}(x)}{F_{i+1}^{[m-1]}(x) - F_{i+1}^{[m-2]}(x)}\right] - 1} \qquad (16.4)$$

From this point on, the development of the algorithm proceeds in exact analogy with Neville's algorithm, with the simple polynomials in each element of the tableau being replaced by a quotient of polynomials.

1. We begin by assigning $F_i^{[0]} = f(x_i)$ and defining $F_i^{[-1]}$ to be zero. That is, the first column of the triangular tableau will simply be constants or a polynomial of degree zero divided by a polynomial of degree zero.

2. The next column will be functions $F_i^{[1]}(x)$ that pass through two adjacent points (degree 1 divided by degree 0 polynomials). They can be labeled as $F_{12}(x), F_{23}(x), \ldots$ in direct analogy with the polynomials of Chapter 15, and they are computed using Equation 16.4.

3. The tableau is continued, column by column, until the rightmost vertex is reached. The value at the vertex represents the interpolated value of the function corresponding to the given value of the independent variable x.

4. To improve accuracy, as in the polynomial case, it is best to rewrite the iterative algorithm in terms of the *differences* in F values from column to column. Defining again "downgoing" and "upgoing" differences for this case as $D_i^{[m]}(x)$ and $U_i^{[m]}(x)$, where

$$D_i^{[m]}(x) = F_i^{[m]}(x) - F_i^{[m-1]}(x)$$
$$U_i^{[m]}(x) = F_i^{[m]}(x) - F_{i+1}^{[m-1]}(x)$$

and using Equation 16.4, the following relations can be derived:

$$U_i^{[m+1]}(x) = D_{i+1}^{[m]}(x)\left[\frac{D_{i+1}^{[m]}(x) - U_i^{[m]}(x)}{\left(\frac{x-x_i}{x-x_{i+m}}\right)U_i^{[m]}(x) - D_{i+1}^{[m]}(x)}\right] \qquad (16.5)$$

$$D_i^{[m+1]}(x) = \left(\frac{x - x_i}{x - x_{i+m}}\right)U_{i+1}^{[m]}(x)\left[\frac{D_{i+1}^{[m]}(x) - U_i^{[m]}(x)}{\left(\frac{x-x_i}{x-x_{i+m}}\right)U_i^{[m]}(x) - D_{i+1}^{[m]}(x)}\right] \qquad (16.6)$$

5. Finally, the code in Figures 16.3 and 16.4 is amended using Equations 16.5 and 16.6 to update $D_i^{[m+1]}(x)$ and $U_i^{[m+1]}(x)$, replacing the relations based on Equation 16.3.

The MATLAB code implementing Bulirsch-Stoer interpolation to either a single value or to a vector of values involves only minor changes to the Neville interpolation codes, and is therefore not included here.

There is one final modification to be addressed before we leave this topic. The function `BulStr` will interpolate or extrapolate a single function at a given set of X values. In the next chapter, we will need to extrapolate a *set of functions* to a *single value* of X, namely $X = 0$. Also, it will be more efficient if we can construct the code to augment an already existing tableau by adding a single outside diagonal. That is, suppose n values of a function are known for a sequence of x values converging to zero:

$$\begin{array}{ccccccc} x_1 & > & x_2 & > & x_3 & \cdots & x_n & \cdots & \Rightarrow 0 \\ f(x_1) & & f(x_2) & & f(x_3) & \cdots & f(x_n) & \cdots & \Rightarrow f(0) \end{array}$$

These values are then used to construct a Bulirsch-Stoer extrapolation tableau to obtain the value of $f(0)$. If a new point, $(x_{n+1}, f(x_{n+1}))$, is added to the sequence, we should be able to use the current table to obtain the new extrapolated value for $f(0)$. This is illustrated for $n = 3$ in Figure 16.5.

Augmenting a Neville table by adding one row

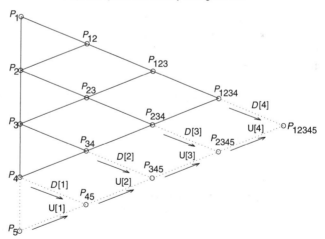

Figure 16.5 *Augmenting an existing tableau by adding an additional diagonal.*

You will notice that, to obtain the new value of $f(0)$, we merely need to add all of the new U's along the new diagonal to the value of f_{n+1}. Also, each element of the new pair (U, D) in a particular column depends only on the values of the pair (U, D) to the immediate left. That is, $U_2^{[2]}$ and $D_2^{[2]}$ can be written in terms of $D_3^{[1]}$ and $U_2^{[1]}$ and the corresponding values of the x's. Thus, the first new U on the new outside diagonal is obtained using Equation 16.4, as

$$B = \left[\frac{f_{n+1} - f_n}{\left(\dfrac{x_n}{x_{n+1}} \right) f(x_n) - f(x_{n+1})} \right]$$

$$d = \left(\frac{x_n}{x_{n+1}} \right) f(x_n) B$$

$$U_n^{[1]} = f(x_{n+1}) B$$

Thereafter, we can move up the new outside diagonal using Equations 16.5 and 16.6; in each step, the most recently calculated value for d is used along with the values of the U's from a previous calculation. The MATLAB code to accomplish this is contained in the function BulStr0, shown in Figure 16.6. The general-

```
function [Y,dY] = BulStr0(x,f,N)
% BulStr0 extrapolates the set of functions, f1(x), ..., fp(x),
% represented by values contained in the n by p matrix f to the
% single value, x = 0. It is assumed that the x values are mono-
% tonically decreasing and that the previous lower outside diag-
% onal has been saved in U.  (This is a global variable.) The
% temporary new outside diagonal is stored in V.  The extrapo-
% lated values and the error estimates are returned as ROW vec-
% tors, Y and dY.
%----------------------------
global U
n = length(x);
if n<N+1;
     error('Input for N is too large (> length(x)-1)');
end
for i = 1:N
   r = x(N-i+1)/x(N+1);
   if i == 1
       box = (f(N+1,:)-f(N,:))./(r*f(N,:)-f(N+1,:));
V(i,:)= f(N+1,:).*box;
d    = r*f(N,:).*box;
   else
       box = (d - U(i-1,:))./(r*U(i-1,:)-d);
       V(i,:)= d.*box;
       d    = r*U(i-1,:).*box;
   end
end
yf = f(N+1,:)+sum(V);
dy = abs(V(N,:));
U  = V;
```

Figure 16.6 *MATLAB code for simultaneous Bulirsch-Stoer rational fraction ex-trapolation of a set of p functions to $x = 0$.*

ization of the code to handle p simultaneous functions is trivial: simply replace $f(i) \Rightarrow f(i,:)$ and keep in mind that the extrapolated values consist of p vertical sequences contained in the $n \times p$ matrix of f values.

For example, to simultaneously extrapolate e^x and $x\cos x$ to $x = 0$, we could start with the values of the two functions at $x = 1, \frac{1}{4}, \frac{1}{9}, \ldots, \frac{1}{9^2}$ and use function BulStr0 to carry out the extrapolation. In MATLAB code we would have:

```
i = [1:9]';
x = i.^(-2);
for N = 1:6
    [f0,df] = BulStr0(x,f,N);
end
disp(df)
    1.0e-7 *
0.0001          0.1602
```

Cubic Spline Interpolation

Often it is best to abandon the attempt to use least squares analysis to fit a simple function to complex data and, instead, to merely "connect the dots" by drawing a smooth curve through the data using a French curve or a drafter's spline (a flexible elastic bar). This procedure is especially appropriate if the curve itself is to be used to interpolate between data points, because connecting the dots will not introduce the occasional wild swings associated with polynomial fits. Also, it is often the case that the purpose of the curve is not to approximate some experimental phenomenon, but rather, to guide the eye or to line-fill a drawing characterized by a set of points. The exact nature of the fitting function is unimportant; the main thing is that the fit be exact and that it be smooth.

CUBIC SPLINES

The most common of many mathematical algorithms to smooth-fit data points is called the *cubic spline*. The idea is to connect every adjoining pair of points by a cubic; a *different* cubic for each pair. The individual cubic functions are then joined together at the data points (the *knots*) into a smooth curve. The complete fitting function over the entire range of the independent variable, x, is called $F(x)$. If the data consist of n points, the function $F(x)$ will be represented by $n-1$ cubic segments of the form

$$f_i(x) = a_i(x - x_i)^3 + b_i(x - x_i)^2 + c_i(x - x_i) + d_i \qquad (16.7)$$

for $x_i \leq x \leq x_{i+1}$ and $i = 1, \ldots, n-1$.

Each of the segments contains four parameters that must be determined by imposing four conditions on that particular segment. Two of these are, obviously, that the function segment must pass through the points at either end. That is,

$$y_i = F(x_i) \quad = f_i(x_i) = d_i$$
$$y_{i+1} = F(x_{i+1}) = f_i(x_{i+1}) = a_i \Delta x_i^3 + b_i \Delta x_i^2 + c_i \Delta x_i + d_i$$

where Δx_i is the width of the ith interval, $\Delta x_i = (x_{i+1} - x_i)$.

Two additional conditions are required. These will be requirements on the *smoothness* of the overall function at the connections (knots) between the segments. Mathematically, the conditions will be that both the slopes and the second derivatives of the two adjoining segments match at the knots:

$$f'_{i+1}(x_{i+1}) = f'_i(x_{i+1}) \qquad \text{for } i = 1, \ldots, n-1$$
$$f''_{i+1}(x_{i+1}) = f''_i(x_{i+1})$$

These relations are illustrated graphically in Figure 16-7.

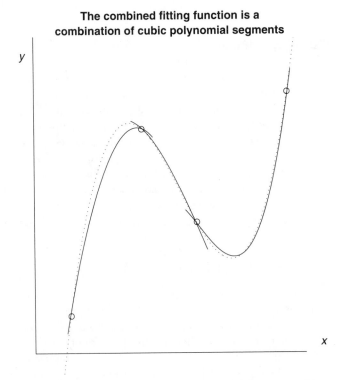

The combined fitting function is a combination of cubic polynomial segments

Figure 16.7   *Combined fitting function $F(x)$ is a combination of smoothly joined segments.*

Applying these four conditions to determine the coefficients of the cubic segments is straightforward, but it involves considerable algebra because the latter two conditions have introduced adjoining segments (along with their unknown coefficients) into the problem. The details of the solution for the coefficients is given in the following section.

SOLUTION FOR THE COEFFICIENTS OF THE CUBIC SEGMENTS

Because each of the function segments is a cubic, the second derivatives, which must match at the knots, can be expressed as a linear function,

$$f_i''(x) = f_i''(x_i) + \frac{(x - x_i)}{\Delta x_i}[f_{i+1}''(x_{i+1}) - f_i''(x_i)] \tag{16.8}$$

and using Equation 16.7, this can be written in terms of the coefficients b_i as

$$f_i''(x) = 2b_i + 2\frac{(x - x_i)}{\Delta x_i}(b_{i+1} - b_i) \tag{16.9}$$

Working backwards from $f_i''(x)$, we next obtain $f_i'(x)$ by integration,

$$
\begin{aligned}
f_i'(x) &= f_i'(x_i) + \int_{x_i}^x f_i''(x)\, dx \\
&= c_i + 2b_i(x - x_i) + \frac{(x - x_i)^2}{\Delta x_i}(b_{i+1} - b_i)
\end{aligned}
\tag{16.10}
$$

where we have used the fact that $f_i'(x_i) = c_i$. Imposing the condition that the slopes must match at the knots, we obtain

$$f_{i+1}'(x_{i+1}) = f_i'(x_{i+1}) = c_{i+1} = c_i + 2b_i \Delta x_i + \Delta x_i(b_{i+1} - b_i) \tag{16.11}$$

Integrating one more time yields the functions $f_i(x)$:

$$
\begin{aligned}
f_i(x) &= f_i(x_i) + \int_{x_i}^x f_i'(x)\, dx \\
&= y_i + c_i(x - x_i) + b_i(x - x_i)^2 + \frac{(x - x_i)^3}{3\Delta x_i}(b_{i+1} - b_i)
\end{aligned}
\tag{16.12}
$$

and using the last condition, $f_i(x_{i+1}) = y_{i+1}$, this can be written as

$$c_i = \frac{\Delta y_i}{\Delta x_i} - \frac{1}{3}\Delta x_i(b_{i+1} + 2b_i) \tag{16.13}$$

With this last equation, we can replace c_{i+1} and c_i in Equation 16.11 to obtain

$$\frac{1}{3}[b_i \Delta x_i + 2b_{i+1}(\Delta x_i + \Delta x_{i+1}) + b_{i+2}\Delta x_{i+1}] = \frac{\Delta y_{i+i}}{\Delta x_{i+1}} - \frac{\Delta y_i}{\Delta x_i} \tag{16.14}$$

for $i = 1, \dots, n - 2$

In this equation, the elements on the right-hand side are known quantities, and using the facts that $b_1 = b_n = 0$, the left-hand side can be written as a matrix times a column vector consisting of the nonzero elements of b_i. This equation is solved for the coefficients b_i. Once the b_i's are known, Equation 16.13 is used to compute the c_i's, and Equation 16.9 is combined with Equation 16.7 to obtain the expression for the a_i's; that is,

$$a_i = \frac{b_{i+1} - b_i}{3\Delta x_i} \qquad (16.15)$$

These equations simplify considerably if the x values are equally spaced. (See Problem 16-6.)

Constructing the code to implement the preceding algorithm to compute the set of cubic segments is straightforward but tedious. If you are secure in your understanding of the method, it is probably acceptable to forego the construction of your own spline code and make use of the very excellent cubic spline function in the MATLAB spline toolbox. This function is called `spline`, and its use is

```
Y = spline(x,y,X)
```

where (`x`,`y`) are vectors containing the discrete points through which the set of cubic segments is to be drawn. The vector `X` contains the values of the independent variable at which the spline interpolation function is to be evaluated, and the computed values are returned in the vector `Y`. This function is illustrated in Example 16.1.

EXAMPLE 16.1

Comparison of a Cubic Spline Fit Through Discrete Points with the Exact Function

The equation for damped oscillations is

$$y(x) = Ce^{-\alpha x}\cos\beta x$$

This function is first plotted over the range $0 \leq x \leq 5$ with $C = 10$, $\alpha = 0.5$, $\beta = 2\pi$. In order that the plot be "smooth," the function is evaluated at 251 points, `x = [0:.02:5]'`. The result is shown in Figure 16.8. Next, the same function is evaluated at 26 points, `x = [0:.2:5]'`, and plotted as a discrete plot with small circles. Finally, the values of x and y at the circles are used as the input to `spline`. The spline function is then interpolated at intervening points specified by `X = [.03:.033:4.97]'` and plotted on the same graph.

```
x = [0:.02:5]';
n = length(x);
C = 10;    alpha = 0.5;    beta = 2*pi;
y = C*exp(-alpha*x).*cos(beta*x);
plot(x,y)
hold on
xdata = x(1:16:n);  ydata = y(1:16:n);
plot(xdata,ydata,'o');
X = [0.03:0.033:4.94]';
Y = spline(xdata,ydata,X);
plot(X,Y)
hold off
```

Notice that the spline curve passes smoothly through the circle points and nicely follows the exact function. You may wish to least squares fit a damped oscillator

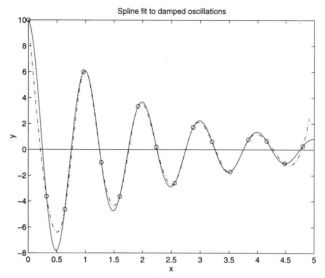

Figure 16.8 *A cubic spline is interpolated through the circle points, which were calculated from a damped oscillator function. The function is also plotted for comparison.*

function through the circle points to see if the original values are reproduced and thus determine if the fit is improved.   ●

Problems

REINFORCEMENT EXERCISES

P16.1 **Exact Polynomial Fits Using Vandermonde Matrices.** Compute the Vandermonde matrix for n equally spaced points over the interval $0 \leq x \leq 1$ for $n = 5, 6, \ldots, 10$. Obtaining a Lagrange polynomial that passes through n points at these values of x will depend on the value of the determinant of the Vandermonde matrix. Compute and collect, in the form of a table, the determinant of the Vandermonde matrix as a function of n.

P16.2 **Interpolation.** Make a table of x and $\ln(x)$ for $x = $ 1.0:0.2:3.0. Next, use the function `PolyIntp` on page 353 to interpolate the table values at $x = $ 1.5:0.2:2.5. Compute and display the difference between the interpolated values and the exact values as a function of the number of interpolating points m, for $m = $ 3:7.

P16.3 **Neville's Interpolation for Equally Spaced Points.**
   a. Reproduce Equations 16.1 through 16.3 under the assumption that all the x values are equally spaced.
   b. Evaluate the function $y = x^3 + 2x^2 + x + 1$ for values of **x** in the range $x = $ 0:0.5:2.0 and, *by hand*, interpolate the value of the function at $x = 0.75$. Compare with the exact result.

c. For an arbitrary point, how far must you carry out the procedure before the exact input polynomial is reproduced?

P16.4 Computer Application of Neville Interpolation. Use the functions `Nevill` and/or `Nevil` to duplicate the results of Table 16.1.

P16.5 Bulirsch-Stoer Interpolation. Make a table of the values of the function $y = (1 - x^2)/(1 + 2x)$ for $x = 11:-2:1$.

a. Use these values to extrapolate, *by hand*, the function to $x = 0$ in the following way:

(i) Fit each pair of adjacent points to the function $F_1(x) = a_1 x + a_2$ and evaluate this function at $x = 0$; i.e., $F_1(0) = a_2$.

(ii) Next, fit each set of three adjacent points to a function $F_2(x) = (a_1 x + a_2)/(b_1 x + 1)$ and evaluate this function at $x = 0$; i.e., $F_2(0) = a_2$.

(iii) Next, try a function $F_3(x) = (a_1 x^2 + a_2 x + a_3)/(b_1 x + 1)$. This time $F_3(x = 0)$ should be precisely one.

b. Compare these results with the values returned by `BulStr0`. Use the function `Bulstr0` to reproduce the results of Problem 16.4. Make a table of the results as a function of N, the number of points used in the extrapolation.

P16.6 Cubic Splines with Equally Spaced Points. Assume that the evaluation points for a cubic spline are equally spaced, with $x_{i+1} - x_i \equiv h$. Reproduce the analysis of Equations 16.7 through 16.14 under this assumption. Next use these equations to obtain, *by hand*, the cubic spline fit to the following data sets:

a.

x_i	y_i
0	0
1	1
2	4
3	8

b.

x_i	y_i
0	0
1	2
2	4
3	8
4	16

P16.7 Comparison of Cubic Splines and Polynomial Fits. Store the following data set in vectors x_i and y_i:

i	x_i	y_i	i	x_i	y_i
1	0.0	−2.20	9	0.8	0.40
2	0.1	−1.90	10	0.9	−0.10
3	0.2	0.00	11	1.0	−1.10
4	0.3	−0.10	12	1.1	−1.30
5	0.4	−0.01	13	1.2	−2.00
6	0.5	−0.30	14	1.3	−1.80
7	0.6	−0.20	15	1.4	−0.02
8	0.7	−0.09	16	1.5	1.00

Plot the data as discrete points labeled by "o"s. Use the function `spline` to compute the values of the cubic spline fit to the data for X = 0:.2:16, and (with hold on) graph the spline fit. Next, use `polyfit` to determine the coefficients of $n = $ 6th and $n = $ 9th degree polynomial fits to the data. Use `polyval` to evaluate these polynomials for the preceding X values and add these to the graph.

   a. Does either of the polynomials have large "swings" that are not representative of the data?

   b. If each of the y values has an associated error bar, approximately how small an error bar is required before the $n = 6$ polynomial can be rejected?

   c. Do the same for the $n = 9$ polynomial. Make sure that any possibly spurious swings of the polynomial are also covered by the error bars.

EXPLORATION PROBLEMS

P16.8 Extrapolation of Air Drag Data. A common experiment to measure the drag force of the air on a freely falling object involves taking a photograph of the falling object illuminated by a strobe light that flashes at a rapid rate. The sequence of images on the photograph then records the position of the object at regular time intervals. The data then correspond to the set of N values $y(t_0 + i\Delta t)$, $i = 0, \ldots, N - 1$, where Δt is the time between flashes of the light. The quantity, t_0, is the time at which the object is released. Generally, the release of the object and the flashing of the light are *not* synchronized, so that the value of t_0 is not known. To fit the data to a test function, the first step in the analysis must then be to determine t_0. The following data represent the distance of fall of an object in such an experiment. The time interval of the strobe light was 20 flashes/s, or $\Delta t = 0.05$ s, and the distances are in meters.

i	y_i
1	0.0078
2	0.0202
3	0.0570
4	0.1187
5	0.2043
6	0.3131
7	0.7914
8	0.9998
9	1.2330
10	1.4892

Use the function `Nevill` to extrapolate these data to the point $y = 0$ and thereby determine the corresponding value of t_0.

Summary

POLYNOMIAL FITS TO DATA

Lagrange Interpolation Polynomial

The $n + 1$ coefficients, $\vec{a}$, of the nth degree polynomial that is an *exact* fit to $n + 1$ data points, (x_i, y_i), $i = 1, \ldots, n + 1$, are obtained by solving the equation

$$\mathbf{M}\vec{a} = \vec{y}$$

where **M** is the Vandermonde matrix constructed from the values contained in the vector $\vec{x}$. This polynomial is known as the *Lagrange* polynomial.

Using polyfit *and* polyval

The Lagrange polynomial can also be evaluated using the toolbox function polyfit(x,y,n) to determine the coefficients of an nth-degree polynomial fit to the data. The fitting polynomial is then evaluated at points contained in the vector X using polyval(a,X).

NEVILLE INTERPOLATION

Definition

The Neville interpolation polynomial is computed by iteratively using adjacent points in the data to construct a new interpolated set. The process is continued, with one fewer point in each step, until the algorithm collapses to a single point, yielding the final interpolated value. The change in the last step can be used as an estimate of the error in the method.

Basic Equations

The method is then represented by

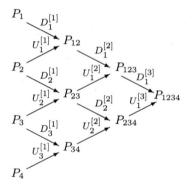

with the changes being given by the equations

$$U_i^{[m+1]} = \left(\frac{x - x_{i+m+1}}{x_{i+m+1} - x_i} \right) (D_{i+1}^{[m]} - U_i^{[m]})$$

$$D_i^{[m+1]} = \left(\frac{x - x_i}{x_{i+m+1} - x_i} \right) (D_{i+1}^{[m]} - U_i^{[m]})$$

BULIRSCH-STOER INTERPOLATION AND EXTRAPOLATION

Definition

The Neville polynomial interpolation scheme is generalized to apply to the interpolation or extrapolation of a *ratio* of polynomials; i.e., Padé approximations. The most common use is to *extrapolate* data to the point $x = 0$.

CUBIC SPLINE INTERPOLATION

Definitions

A *set* of cubic functions is fit to adjacent pairs of data points such that the combined fitting function, $F(x)$, passes through the data and "smoothly" connects each of the cubic subfunctions, $f_i(x)$, given by

$$f_i(x) = a_i(x - x_i)^3 + b_i(x - x_i)^2 + c_i(x - x_i) + d_i$$

The coefficients of each cubic are determined by matching the data *and* requiring that the first and second derivatives of the adjoining cubics be equal.

MATLAB Functions Used

interp1

Interpolation of data: If x and y are vectors of equal length containing the data, and if the values contained in x are monotonically increasing or decreasing, then interp1 will estimate values for Y at intervening points specified in the vector X. The use is

$$\text{Y} \; = \; \text{interp1}(\text{x}, \text{y}, \text{X}, 'method')$$

where the optional string constant *'method'* specifies the interpolation method to be used. The choices are *'linear'* for a linear interpolation, *'spline'* for a cubic spline fit, or *'cubic'*, which uses cubic interpolation. The latter requires that the x values be equally spaced. The default is *linear*.

spline

Cubic spline interpolation: Similar to interp1, the function spline will estimate values of Y at points specified in the vector X by fitting a cubic spline through the data contained in the equal-length vectors x and y. The use is Y = spline(x,y,X).

17 / Fourier Analysis

PREVIEW   A large class of phenomena can be described as *periodic* in nature. This includes any process involving any form of *waves*, such as sound, light, radio, or water waves. It is only natural then, to attempt to describe these phenomena by means of expansions in periodic functions. The *Fourier series* is an expansion of a function in terms of trigonometric sines and cosines. The story of the development of the Fourier series is a fascinating chapter in the history of mathematics that formed a much better understanding of continuity and convergence. However, for the purposes of data analysis, the most important step was the generalization of the expansion in terms of sines and cosines to expansions in terms of complex exponentials. This followed from the famous Euler identity

$$e^{i\theta} = \cos\theta + i\sin\theta$$

or the inverse equations

$$\sin\theta = \frac{1}{2i}(e^{i\theta} - e^{-i\theta}) \qquad \cos\theta = \frac{1}{2}(e^{i\theta} + e^{-i\theta})$$

In addition to opening up the study of functions of complex variables, this step permitted the Fourier analysis of functions to be expressed in terms of what are called *Fourier transforms*. A Fourier transform is a beautifully compact and symmetrical relationship between a function and its expansion coefficients in a complex Fourier series. Fourier transforms have found extensive application in the solution of almost every problem connected with wave phenomena. More relevant to our interests, they have played a dominant role in the modern analysis of data. There are two reasons for this. The first is the ability to use Fourier transforms to *filter* data; i.e., to remove large portions of the noise or other undesirable data segments that unavoidably accompany all measurements. The second is the result of discoveries made as early as the 1940s, but not extensively implemented until the 1960s, that enable coding of extremely fast, extremely efficient forms of the Fourier transform: the *fast Fourier transform* or FFT.

FFT's have very broad application in data analysis, not merely to treat periodic data, but in diverse data analysis applications that span photographic image enhancement to summarizing nuclear scattering experiments.

The Fourier Series

If a function, $f(x)$, is only defined over a finite range of the independent variable, $-L \leq x \leq +L$, a common practice is to extend the range of the function by adding the condition that $f(x)$ be *periodic* with period $2L$. That is, $f(x + 2L) =$

$f(x)$.[1] The trigonometric functions are periodic with period 2π, so it is natural to expand these functions in terms of trigonometric functions with an argument $[(x/2L)2\pi n]$, where n is any integer. That is,

$$f(x) = \frac{1}{2}a_0 + \sum_{m=1}^{\infty} a_m \cos\left(\frac{n\pi}{L}x\right) + \sum_{m=1}^{\infty} b_m \sin\left(\frac{n\pi}{L}x\right) \qquad (17.1)$$

where a_m and b_m are the Fourier expansion coefficients.

The sine and cosine functions satisfy the following integral orthogonality relations:

$$\frac{1}{L}\int_{-L}^{L} \cos\left(\frac{n'\pi}{L}x\right)\cos\left(\frac{n\pi}{L}x\right) dx = \delta_{n'n}$$

$$\frac{1}{L}\int_{-L}^{L} \sin\left(\frac{n'\pi}{L}x\right)\sin\left(\frac{n\pi}{L}x\right) dx = \delta_{n'n} \qquad (17.2)$$

$$\frac{1}{L}\int_{-L}^{L} \sin\left(\frac{n'\pi}{L}x\right)\cos\left(\frac{n\pi}{L}x\right) dx = 0$$

for all nonzero integers n and n', where $\delta_{n'n}$ is the Kronecker delta symbol and is defined to have a value of $+1$ if $n = n'$, and a value of zero otherwise. The values of the preceding integrals, if one or both of the integers n and n' are zero, follows from the facts that $\cos 0 = 1$, $\sin 0 = 0$. Next, if Equation 17.1 is multiplied by $\cos\left(\frac{n\pi}{L}x\right)$ and integrated from $x = -L$ to $x = L$, every term in the summation on the right is zero except for the term with $m = n$, and an expression for the coefficient a_n can be obtained. Likewise, multiplying by $\sin\left(\frac{n\pi}{L}x\right)$ and integrating yields an expression for the coefficients, b_n. Finally, the coefficient a_0 is obtained by simply integrating Equation 17.1, as is, from $-L$ to L, in which case all the terms in both sums yield zero. The expressions for the coefficients are then

$$a_n = \frac{1}{L}\int_{-L}^{L} f(x)\cos\left(\frac{n\pi}{L}x\right) dx$$

$$b_n = \frac{1}{L}\int_{-L}^{L} f(x)\sin\left(\frac{n\pi}{L}x\right) dx \qquad (17.3)$$

$$\tfrac{1}{2}a_0 = \overline{f}$$

where $\overline{f}$ is the average value of the function $f(x)$ over the interval.

Using these equations, it is usually a straightforward process to evaluate the expansion coefficients. For example, the sawtooth function, defined by $f(x) = |x|$ for $|x| \leq L$ and $f(x + 2L) = f(x)$, is plotted in Figure 17.1. To evaluate the coefficients in the expansion of this function, we first notice that $f(x)$ is *even* in x while $\sin x$ is *odd*. Thus, all the b_n's must be zero. The average value of the

---

[1]Functions defined over a different but finite range, such as $0 \leq x \leq b$ or $a \leq x \leq b$, can be rewritten in the preceding form by a simple change of variables.

function is clearly $\overline{f} = \frac{1}{2}$, and the remaining a's are obtained from

$$a_n = \frac{2}{L} \int_0^L x \cos\left(\frac{n\pi}{L}x\right) dx$$

$$= \frac{2L}{n^2\pi^2}[\cos(n\pi) - 1]$$

$$= -\frac{4L}{n^2\pi^2}, \qquad \text{if } n = \text{odd}$$

$$= 0, \qquad\qquad \text{if } n = \text{even}$$

so that the expansion of the sawtooth function is

$$f(x) = \frac{L}{2}\left(1 - \frac{8}{\pi^2}\sum_{n=\text{odd}}^{\infty} \frac{1}{n^2}\cos\left(\frac{n\pi}{L}x\right)\right)$$

The values from this series expansion including terms through $n = 1$, 3, and 5 are added to the plot of the sawtooth function in Figure 17.1.

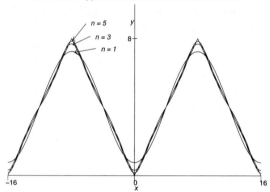

Successive approximations to a sawtooth function

Figure 17.1 *Fourier expansion approximations to the sawtooth function including terms through $n = 1, 3, 5$.*

The MATLAB script file `ForyrDmo.m` on the applications disk demonstrates the Fourier series expansion of a variety of functions.

A Complex Version of the Fourier Expansion

Replacing the sines and cosines by exponentials of the form $e^{\pm in\pi x/L}$ will yield the series expansion

$$f(x) = \sum_{n=-\infty}^{+\infty} c_n e^{(in\pi/L)x} \qquad\qquad (17.4)$$

where the coefficients, c_n, can easily be related algebraically to a_n and b_n. However, it is more instructive to solve for them anew by using the orthogonality conditions of the exponential function,

$$\int_{-\pi}^{\pi} e^{-in'x} e^{inx} \, dx = \int_{-\pi}^{\pi} e^{i(n-n')x} \, dx = 2\pi \delta_{n'n} \qquad (17.5)$$

where $\delta_{n'n}$ is the *Kronecker* delta symbol. Note that e^{-inx} is the *complex conjugate* of e^{inx}.

Again, multiplying Equation 17.4 by $e^{-i(n'\pi/L)x}$ and integrating from $-L$ to L, we obtain

$$c_n = \frac{1}{2L} \int_{-L}^{L} f(x) e^{-i(n\pi/L)x} \, dx \qquad (17.6)$$

Although one contains a summation over n and the other is an integration over x, there is a distinct similarity between Equations 17.4 and 17.6. This is the motivation for the next step—the introduction of the Fourier transform.

The Fourier Transform

Equations 17.4 and 17.6 can be combined to yield an important identity known as the *Fourier integral*:

$$f(x) = \frac{1}{2L} \sum_{n=-\infty}^{+\infty} e^{(in\pi/L)x} \int_{-L}^{L} f(\xi) e^{-i(n'\pi/L)\xi} \, d\xi \qquad (17.7)$$

If we next define a new variable, $\omega = n\pi/L$, and take the limit as $L \to \infty$, the difference between successive values of ω; i.e., $d\omega = \pi/L$, approaches an infinitesimal quantity and the summation in Equation 17.7 becomes an integral,

$$\frac{1}{L} \sum_{n=-\infty}^{\infty} \Rightarrow \frac{1}{\pi} \int_{-\infty}^{\infty} d\omega$$

Thus Equation 17.7 can be written as

$$f(x) = \frac{1}{\sqrt{2\pi}} \int_{-\infty}^{+\infty} d\omega \left(\frac{1}{\sqrt{2\pi}} \int_{-\infty}^{\infty} f(\xi) e^{-i\omega\xi} \, d\xi \right) e^{i\omega x} \qquad (17.8)$$

The expression in the parentheses is defined as the *Fourier transform* of $f(x)$, and it is designated as $\hat{f}(\omega)$.

Fourier Transform $\qquad \hat{f}(\omega) \equiv \frac{1}{\sqrt{2\pi}} \int_{-\infty}^{\infty} f(x) e^{-i\omega x} \, dx$

Replacing the term in the parentheses by this definition, Equation 17.8 then represents the inverse of this definition.

Inverse Fourier Transform $\qquad f(x) \equiv \frac{1}{\sqrt{2\pi}} \int_{-\infty}^{\infty} \hat{f}(\omega) e^{i\omega x} \, d\omega$

The function $\hat{f}(\omega)$ contains the expansion coefficients in a continuous expansion of $f(x)$ in terms of complex exponentials. The symmetry between the equations for $f(x)$ and $\hat{f}(\omega)$ is quite remarkable,[2] and it is often exploited by introducing a *Fourier transform operator*, $\mathcal{F}$, defined as

$$\hat{f}(\omega) = \mathcal{F}f(x) \qquad f(x) = \mathcal{F}^{-1}\hat{f}(\omega)$$

THE DIRAC DELTA FUNCTION

To easily convert from $f(x)$ to $\hat{f}(\omega)$, or from $\hat{f}(\omega)$ to $f(x)$, it is useful at this point to introduce the *Dirac delta function*, $\delta(x - a)$, which is defined by the properties:

$$\int_{x_1}^{x_2} f(x)\delta(x - a)\, dx = f(a) \qquad \text{if } x_1 \leq a \leq x_2$$

$$= 0 \qquad \text{otherwise}$$

(17.9)

Actually, the Dirac delta function is not a function at all; at least not in the sense of, "Give me an x and I'll give you a $y = \delta(x)$." It only has meaning within an integral. You can visualize $\delta(x - a)$ as a quantity that is zero for all $x \neq a$, and "infinity" at $x = a$, so that if included in an integral, it selects only the value of the integrand at $x = a$. Its importance in Fourier analysis stems from the relation:

$$\int_{-\infty}^{\infty} e^{i(\alpha - \beta)x}\, dx = 2\pi\delta(\alpha - \beta)$$

(17.10)

Again, this is not an "equation" in the normal sense, but you can verify that it satisfies the conditions of Equation 17.9 by considering

$$\int_{-\infty}^{\infty} f(x)\delta(x - a)\, dx = \frac{1}{2\pi} \int_{x_1}^{x_2} f(x) \int_{-\infty}^{\infty} e^{i(x-a)\omega}\, d\omega$$

$$= \frac{1}{2\pi} \int_{-\infty}^{\infty} e^{-ia\omega}\, d\omega \int_{-\infty}^{\infty} f(x)e^{ix\omega}\, dx$$

$$= \frac{1}{\sqrt{2\pi}} \int_{-\infty}^{\infty} \hat{f}(\omega)e^{-ia\omega}\, d\omega$$

$$= f(a)$$

where we have used the definition of the Fourier transform in the third line. Note also, that you can show directly from Equation 17.10 that $\delta(\alpha - \beta) = \delta(\beta - \alpha)$.

PROPERTIES OF THE FOURIER TRANSFORM

The symmetry between $f(x)$ and $\hat{f}(\omega)$ leads to a number of useful properties, some of which we quote here without proof. Each of these can be readily proved

---

[2]There is considerable variety among authors as to where to put the factors of 2π. You will encounter definitions of the Fourier transform without the prefactor of $1/\sqrt{2\pi}$, which then requires the factor in the inverse Fourier transform to be replaced by $1/2\pi$. Also, some authors replace the expansion variable $\omega \to f/2\pi$.

directly using the definitions of the Fourier transform and its inverse. Also, consult any good text on applied mathematics, e.g., Greenberg (1978).

Linearity:
$$\mathcal{F}(c_1 f_1(x) + c_2 f_2(x)) = c_1 \hat{f}_1(\omega) + c_2 \hat{f}_2(\omega)$$
$$\mathcal{F}^{-1}(c_1 \hat{f}_1(\omega) + c_2 \hat{f}_2(\omega)) = c_1 f_1(x) + c_2 f_2(x)$$

Transform of Derivatives:
$$\mathcal{F}\left(\frac{d^n}{dx^n} f(x)\right) = (i\omega)^n \hat{f}(\omega)$$
$$\mathcal{F}^{-1}\left(\frac{d^n}{d\omega^n} \hat{f}(\omega)\right) = (-ix)^n f(x)$$

Convolution:
Defining $\quad (f * g) \equiv \frac{1}{\sqrt{2\pi}} \int_{-\infty}^{\infty} f(x - \xi) g(\xi) d\xi$

then $\quad \mathcal{F}(f * g) = \hat{f}\hat{g}$

and $\quad \mathcal{F}^{-1}(\hat{f}\hat{g}) = f * g$

The *convolution* property is the item of greatest relevance to data analysis, as we illustrate in the next section.

INTERPRETATION OF THE FOURIER CONVOLUTION INTEGRAL

Consider a mechanical system consisting of a mass, m, attached to a spring with spring constant, k, and driven by a time-dependent and periodic force, $F(t)$, as shown in Figure 17.2.

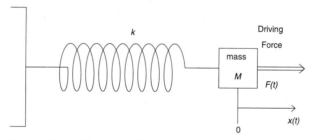

Figure 17.2 *Forced oscillations in a spring-mass system.*

Applying Newton's second law ($\mathbf{F} = m\mathbf{a}$) to this system yields
$$\frac{d^2 x(t)}{dt^2} + \omega_0^2 x(t) = \frac{F(t)}{m}$$
where
$$\omega_0^2 \equiv \frac{k}{m}$$

Next, by taking the Fourier transform of both sides of this equation and using the first two properties defined previously, we obtain
$$(\omega_0^2 - \omega^2)\hat{x}(\omega) = \frac{1}{m}\hat{F}(\omega)$$

or

$$\hat{x} = \hat{h}\hat{F}$$

where we have defined $\hat{h}(\omega) = (1/m)(1/(\omega_0^2 - \omega^2))$.

In this last equation, $F(t)$ and its transform, $\hat{F}(\omega)$, describe the *external* driving force applied to the spring-mass system. The system itself is characterized by the left-hand side of the differential equation or, in this case, by the transform function, $\hat{h}(\omega)$. The inverse transform of this function, $\mathcal{F}^{-1}\hat{h}(\omega) = h(t)$, presumably exists (see Problem 17.12) and describes the response of the system to the driving force. To reinforce these ideas, it is useful to consider a few special cases.

EXAMPLE 17.1

Response of a Mass-Spring System to a Variety of Driving Forces

1. **A δ-function driving force:** If $F(t) = F_0\delta(t - t_0)$, then, using the definition of Fourier convolution, we have for the solution for the motion of the mass:

$$y(t) = \mathcal{F}^{-1}\hat{y}(\omega)$$

$$= \mathcal{F}^{-1}[\hat{h}(\omega)\hat{F}(\omega)]$$

$$= h(t) * F(t)$$

$$= \frac{1}{\sqrt{2\pi}} \int_{-\infty}^{\infty} h(t - \tau)F(\tau)\,d\tau$$

$$= \frac{F_0}{\sqrt{2\pi}} \int_{-\infty}^{\infty} h(t - \tau)\delta(\tau - t_0)\,d\tau$$

$$= \frac{F_0}{\sqrt{2\pi}}h(t - t_0)$$

That is, the function $h(t - t_0)$ is proportional to the motion of the mass in response to a δ-function driving force at $t = t_0$. In other words, $h(t - t_0)$ is the motion we would expect of the mass if it were hit sharply with a hammer at time $t = t_0$.

Similar considerations apply to a wide variety of systems. For example, an electrical circuit consisting of a resistor of resistance R, a capacitor of capacitance C, and an inductor of inductance L is attached to a *driving* voltage source, $E_{in}(t)$ as shown in Figure 17.3. The differential equation for the current, $I(t)$, in the circuit is

$$L\frac{d^2I}{dt^2} + R\frac{dI}{dt} + \frac{1}{C}I = \frac{dE_{in}}{dt}$$

which, after taking the Fourier transform, becomes

$$\hat{I}(\omega) = \left(\frac{i\omega C}{1 + i\omega RC - \omega^2 LC}\right)\hat{E}_{in}(\omega) = \hat{h}(\omega)\hat{E}_{in}(\omega)$$

The solution for the voltage across the resistor is then

$$V_{out}(t) = RI(t) = R\,h(t) * E_{in}(t)$$

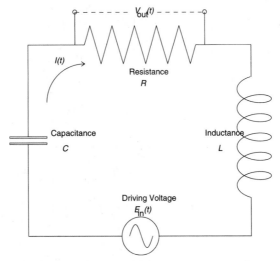

Figure 17.3 *A series LCR circuit connected to a driving voltage.*

where, as in the mass-spring system, $h(t - t_0)$ is the response of the system to a δ-function driving voltage applied at $t = t_0$.

2. **A constant driving force:** Returning to the mass-spring system, consider the effects of a constant force F_0. The solution is of course that the force simply stretches the spring by a constant amount, $\Delta x = F_0/k$. This result also follows from the Fourier transform equations,

$$\hat{F}(\omega) = \mathcal{F}[F_0]$$

$$= \frac{F_0}{\sqrt{2\pi}} \int_{-\infty}^{\infty} e^{i\omega t}\, dt$$

$$= \frac{F_0}{\sqrt{2\pi}} 2\pi \delta(\omega - 0)$$

so that

$$y(t) = \mathcal{F}^{-1}[\hat{h}(\omega)\hat{F}(\omega)]$$

$$= \frac{1}{\sqrt{2\pi}} \int_{-\infty}^{\infty} \frac{1}{m} \left(\frac{1}{\omega_0^2 - \omega^2} \right) \sqrt{2\pi} F_0 \delta(\omega - 0) d\omega$$

$$= \frac{F_0}{m\omega_0^2} = \frac{F_0}{k}$$

3. **A periodic driving force:** This time we choose $F(t) = F_0 \sin(at)$. It is not difficult to obtain the Fourier transform of $F(t)$,

$$\hat{F}(\omega) = \frac{F_0}{\sqrt{2\pi}} \int_{-\infty}^{\infty} e^{-i\omega t} \sin(at)\, dt$$

$$= \frac{F_0}{\sqrt{2\pi}} \frac{1}{2i} \int_{-\infty}^{\infty} \left(e^{-i(\omega - a)t} - e^{-i(\omega + a)t} \right) dt$$

$$= -iF_0 \sqrt{\frac{\pi}{2}} \left[\delta(\omega - a) - \delta(\omega + a) \right]$$

and the solution for $y(t)$ is then

$$y(t) = \mathcal{F}^{-1}\hat{y}(\omega) = \mathcal{F}^{-1}(\hat{h}\hat{F})$$

$$= \frac{F_0}{2im} \int_{-\infty}^{\infty} e^{i\omega t} \left(\frac{1}{\omega_0^2 - \omega^2} \right) \left[\delta(\omega - a) - \delta(\omega + a) \right] d\omega$$

$$= \frac{F_0}{2im} \left(\frac{e^{iat}}{\omega_0^2 - a^2} - \frac{e^{-iat}}{\omega_0^2 - a^2} \right)$$

$$= \frac{F_0}{m} \frac{\sin(at)}{\omega_0^2 - a^2}$$

which clearly suggests that the solution will diverge if the system is driven at a frequency that matches the system's natural frequency, ω_0.   ●

All of these examples illustrate that the solution to a particular problem is a *convolution* of two distinct parts. One part represents the external driving force, and the other component is the response of the system to a δ-function driving force.

Convolution and Data Filters   A slightly different interpretation of the result of the solution obtained by the convolution integral is that the *input* to the system is the driving force $F(t)$, whereas the *output* is the actual solution for the oscillations, $y(t)$. The connection between the two is contained in the relation, $\hat{y} = \hat{h}\hat{F}$. Thus, we can think of the response of the system as a *filter* that maps the input to the output. However, the mapping is not *one to one*, but is obtained by means of the Fourier convolution integral.

This interpretation is particularly important in many areas of data analysis. The goal could be to design a filter characterized by a function $\hat{h}(\omega)$ that accurately duplicates a portion of the input signal contained in $F(t)$, while suppressing all parts outside of a particular range. Or perhaps a filter can be designed to mask a portion of the random fluctuations or noise contained in the signal. Obviously, these and many similar applications of Fourier transforms are critically important to engineers and scientists. However, continuing this discussion will take us too far afield from our more modest goal of merely providing an introduction to the vast range of MATLAB capabilities in using Fourier transforms to analyze data. A more detailed treatment of this subject can be found in Strum and Kirk (1994).

EXAMPLE 17.2

Deconvolution of the Effects of a Filter

As a final example of the Fourier convolution integral, consider the untangling of the effects that measurements have on the data. Thus, if the x-ray intensity from a particular source is to be measured as a function of the wavelength, λ, you might use a spectrometer that is sensitive to a narrow band of radiation throughout some tunable range of wavelengths. You would then measure the intensity of the emitted x rays as you tuned the spectrometer through the range of wavelengths. However, the spectrometer has its own sensitivity window. If the spectrometer is set at λ_0, wavelengths in some range $\lambda_0 \pm \Delta\lambda$ will be accepted. A typical sensitivity curve for a spectrometer, $r(\lambda)$, is represented in Figure 17.4b. If the actual emission spectrum of the source is as shown in Figure 17.4a, we anticipate that the broadband nature of the measuring instrument will "wash out" those details of the source that are much narrower than the sensitivity curve of the spectrometer. In fact, the actual measurements, $I(\lambda)$, will be a convolution of the actual spectrum $I_0(\lambda)$, with the spectrometer sensitivity curve,

$$I(\lambda) = I_0 * r$$

$$= \frac{1}{\sqrt{2\pi}} \int_{-\infty}^{\infty} I_0(\lambda - \xi) r(\xi) \, d\xi$$

which is shown in Figure 17.5.

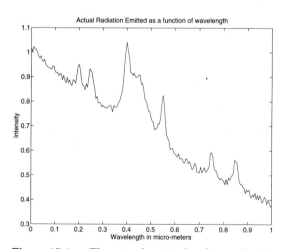

Figure 17.4a *The actual emitted radiation by the x-ray source as a function of wavelength.*

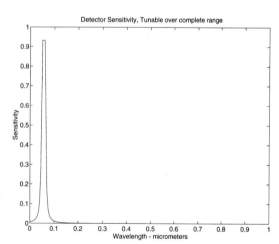

Figure 17.4b *The sensitivity curve for the spectrometer includes a range of wavelengths relative to the setting λ_0.*

Of these three curves, only the observed spectrum and the spectrometer sensitivity curve are known to the experimenter. However, we can use the knowledge of how the curves are related to untangle the effects of the spectrometer on the results and, in the absence of any other experimental noise or errors, obtain the curve of the actual spectrum of the source.

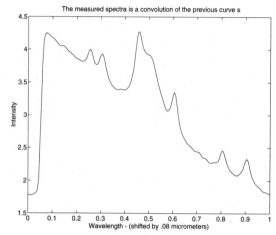

Figure 17.5   *The measured spectrum is the convolution of the sensitivity function and the actual spectrum.*

Thus, if $I(\lambda) = I_0 * r$, then $\hat{I}(\omega) = \hat{I}_0(\omega)\hat{r}(\omega)$, or

$$I_0(\lambda) = \mathcal{F}^{-1}[\hat{I}(\omega)/\hat{r}(\omega)]$$

which is how Figure 17.5 was obtained.   •

The Discrete Fourier Transform

Computer applications of the Fourier transform require that all of the definitions and properties of Fourier transforms be translated into analogous statements appropriate to functions represented by a discrete set of sampling points rather than by continuous functions. We will see that the extraordinary symmetry contained in the defining equation for $f(x)$ and its transform, $\hat{f}(\omega)$, carry over into their discrete cousins.

We begin with a set of N function values,

$$f_k = f(x_k), \qquad x_k = k\Delta x, \qquad k = 0, 1, \dots, N - 1$$

where Δx is the separation of the equidistant sampling points of the independent variable. We will also assume that the number of points, N, is *even*.

It is assumed that the set of values, f_k, to a good approximation, characterizes the complete function $f(x)$. That is, either the function is periodic with period $x_{k+N} = x_k$ or it is zero outside the range of the specified x's. Thus, the increment Δx is defined by the relation $2L = (x_{k+N} - x_k) = N\Delta x$ (*not* $(N-1)\Delta x$).

The discrete version of the Fourier transform then follows from

$$\hat{f}(\omega) = \int_{-\infty}^{\infty} f(x)e^{i\omega x}\,dx \approx C\sum_{k=0}^{N-1} f(x_k)e^{i\omega x_k}\Delta x$$

Because there are only N input values to the summation, we can only expect to compute N independent values of the transform, $\hat{f}_n$. These N values of the transform are chosen to be $\hat{f}(\omega_n)$, which are identified with the N equidistant values of the transform variable, ω_n. Furthermore, the constant of proportionality, $C\Delta x$, can be ignored, as it will be determined by requiring $\mathcal{F} \cdot \mathcal{F}^{-1} = 1$. Thus, the expression for the discrete Fourier transform is

$$\hat{f}_n = \sum_{k=0}^{N-1} f_k e^{i\omega_n x_k}$$

where

$$\begin{bmatrix} f_k = f(x_k) & x_k = k\Delta x \\ \hat{f}_n = \hat{f}(\omega_n) & \omega_n = n\Delta\omega \end{bmatrix}$$

and both n and k are positive integers in the range $[0, N-1]$.

It still remains to define the separation of the points in the transform variable, $\Delta\omega$. To this end, we note that the repeat distance in x is $2L = x_{k+N} - x_k = N\Delta x$, so that

$$e^{i\omega_n x_k} = e^{i(n\Delta\omega)(k\Delta x)}$$
$$= e^{i((2L/N)\Delta\omega)nk}$$

Noting that nk is an integer and that the exponential is periodic with a period of 2π, the value of $\Delta\omega$ is defined to be the smallest value that will preserve the periodicity $k \to k+N$. Thus,

$$\Delta\omega = \frac{\pi}{L}$$

$$\Delta x = \frac{2L}{N}$$

and so

$$e^{i\omega_n x_k} = \exp\left[\left(\frac{2\pi i}{N}\right)nk\right]$$

Further, defining the constant $W \equiv e^{2\pi i/N}$, we see that the discrete Fourier transform can be written as

$$\hat{f}_n = \sum_{k=0}^{N} W^{nk} f_k \tag{17.11}$$

which can also be written as a matrix equation

$$\vec{\hat{f}} = \mathcal{F}\vec{f}$$

where the element in row r, column c of matrix $\mathcal{F}$ is simply W^{rc}. The inverse discrete transform is then simply the inverse of this matrix:

$$\vec{f} = \mathcal{F}^{-1}\vec{\hat{f}}$$

THE DISCRETE INVERSE FOURIER TRANSFORM

We will next show that the matrix for $\mathcal{F}^{-1}$ is related to $\mathcal{F}$ in a remarkably simple manner—the row r, column c element of the matrix $\mathcal{F}^{-1}$ is proportional to W^{-rc}.

Consider the matrix product,

$$\left[\mathbf{W}^{+}\mathbf{W}^{-}\right]_{jk} \equiv \sum_{m=0}^{N} W^{jm} W^{-mk}$$

$$= \sum_{m=0}^{N} W^{m(j-k)}$$

The nondiagonal terms $(j \neq k)$ are easily seen to be zero by noting that $j - k$ is a fixed integer, say p, and the summation can be evaluated exactly by defining $\alpha \equiv W^{p}$ and writing

$$\sum_{m=0}^{N} \alpha^{m} = 1 + \alpha + \alpha^2 + \cdots + \alpha^{N}$$

$$= \frac{\alpha^{N+1} - 1}{\alpha - 1}$$

But $\alpha^{N+1} = e^{2\pi i p} = 1$. Further, each of the terms in the summation for the diagonal terms $(j = k)$ are equal to one, so that

$$[\mathbf{W}^{+}\mathbf{W}^{-}]_{jk} = (N+1)\delta_{jk}$$

Finally, requiring the matrix representing $\mathcal{F}^{-1}$ to satisfy the equation

$$[\mathcal{F}\mathcal{F}^{-1}]_{jk} = \delta_{jk}$$

we are led to the definition of the discrete Fourier transform as simply

$$f_k = \frac{1}{N+1} \sum_{n=0}^{N} W^{-nk} \hat{f}_n$$

In other words, any code we construct to compute the Fourier transform of a set of values, f_k, can, with a few minor modifications, be used to obtain the inverse transform as well.

EXECUTION TIME CONSIDERATIONS

In a typical realistic application of Fourier transforms the number of elements of f_k could easily be on the order of $N \approx 10^6$. The solution for the elements of $\hat{f}_n$ will require on the order of N^2 *complex* multiplications in addition to the evaluation of the elements of the matrix W^{km}. Thus, even though the discrete Fourier transform has an exceptionally simple form, the sheer number of terms to be processed appears to prohibit its use in these large data set applications. However, in the 1960s, a clever prescription for an extremely fast and efficient evaluation of discrete Fourier transforms gained rapid acceptance. The algorithm became known as the fast Fourier transform (FFT), and its popularity is due to the fact that, with it, we are able to reduce the number of basic operations from $\mathcal{O}(N^2)$ to $\mathcal{O}(N\log_2 N)$, or if $N = 10^6$, reduce the arithmetic by a factor of more than 50,000. The basic ideas of the FFT algorithm are outlined in the following section.

Fast Fourier Transforms (FFTs)

The fundamental idea of the fast Fourier transform derives from the fact that the discrete Fourier transform of an *even* number of points can be written as the sum of two distinct (but shorter) transforms. One consists of every other point, starting with x_0, and the other uses every other point, starting with x_1:

$$\hat{f}_n = \sum_{k=0}^{N-1} W^{kn} f_k$$

$$= \sum_{k=even}^{N-2} \left(e^{(2\pi i/N)} \right)^{nk} f_k + \sum_{k=odd}^{N-1} \left(e^{(2\pi i/N)} \right)^{nk} f_k$$

Let $k = 2\kappa$ in the first sum, and $k = 2\kappa + 1$ in the second, and $N' = N/2$ in both to obtain

$$\hat{f}_n = \sum_{\kappa=0}^{(N/2)-2} \left(e^{(2\pi i/N)} \right)^{n(2\kappa)} f_{2\kappa} + \sum_{\kappa=0}^{(N/2)-1} \left(e^{(2\pi i/N)} \right)^{n(2\kappa+1)} f_{2\kappa+1}$$

$$= \sum_{\kappa=0}^{N'-1} \left(e^{(2\pi i/N')} \right)^{n\kappa} f_{2\kappa} + W^n \sum_{\kappa=0}^{N'-1} \left(e^{(2\pi i/N')} \right)^{n\kappa} f_{2\kappa+1}$$

$$= \hat{f}_n^{[e]} + W^n \hat{f}_n^{[o]}$$

where $\hat{f}_n^{[e]}$ is clearly the Fourier transform of just the *even* elements of f_k, and $\hat{f}_n^{[o]}$ is the transform of just the odd elements. The number of terms in each of these is now $N/2$. If additionally, $N/2$ is also an even number, this subdivision can be repeated once more:

$$\hat{f}_n = \hat{f}_n^{[e]} + W^n \hat{f}_n^{[o]}$$

$$= \left(\hat{f}_n^{[ee]} + W^n \hat{f}_n^{[eo]} \right) + W^n \left(\hat{f}_n^{[oe]} + W^n \hat{f}_n^{[oo]} \right)$$

where $\hat{f}_n^{[ee]}$ is a Fourier transform of length $N/4$, consisting of every fourth term starting with f_0 and $\hat{f}_n^{[eo]}$ is a Fourier transform of length $N/4$, consisting of every fourth term starting with f_2, etc. If, in fact, N is a power of 2, this process can be continued until each Fourier transform consists of a single term.

To see the pattern that is emerging, consider the case where $N = 2^3$, and the set, f_k, consists of $[f_0, f_1, \ldots, f_7]$. Thus,

$$\hat{f}_n = \sum_{k=0}^{7} \left(e^{(i\pi/4)} \right)^{kn} f_k$$

$$= [f_0 + \alpha f_1 + \alpha^2 f_2 + \alpha^3 f_3 + \alpha^4 f_4 + \alpha^5 f_5 + \alpha^6 f_6 + \alpha^7 f_7]$$

where we have defined $\alpha = e^{(n\pi i/4)}$ to simplify the notation. Next, regroup the terms by successively taking every other term, duplicating the operation of successively shorter and shorter Fourier transforms:

$$\hat{f}_n = [f_0 + \alpha^2 f_2 + \alpha^4 f_4 + \alpha^6 f_6] + \alpha[f_1 + \alpha^2 f_3 + \alpha^4 f_5 + \alpha^6 f_7]$$

$$= \hat{f}_n^{[e]} + \alpha \hat{f}_n^{[o]}$$

$$= [(f_0 + \alpha^4 f_4) + \alpha^2(f_2 + \alpha^4 f_6)] + \alpha[(f_1 + \alpha^4 f_5) + \alpha^2(f_3 + \alpha^4 f_7)]$$

$$= [\hat{f}_n^{[ee]} + \alpha^2 \hat{f}_n^{[eo]}] + \alpha[\hat{f}_n^{[oe]} + \alpha^2 \hat{f}_n^{[oo]}]$$

$$= \left[(\hat{f}^{[eee]} + \alpha^4 \hat{f}^{[eeo]}) + \alpha^2(\hat{f}^{[eoe]} + \alpha^4 \hat{f}^{[eoo]})\right]$$

$$+ \alpha \left[(\hat{f}^{[oee]} + \alpha^4 \hat{f}^{[oeo]}) + \alpha^2(\hat{f}^{[ooe]} + \alpha^4 \hat{f}^{[ooo]})\right]$$

where, in the last line, each of the transforms originates from a single element f_k.

Thus, if we *begin* by writing the elements f_k in the order $f_0, f_4, f_2, f_6, f_1, f_5, f_3, f_7$, we can easily reconstruct the complete transform by simply repeatedly combining the elements two at a time as

$$\begin{array}{l} f_0 + \alpha^4 f_4 = \hat{f}^{[ee]} \\ f_2 + \alpha^4 f_6 = \hat{f}^{[eo]} \\ f_1 + \alpha^4 f_5 = \hat{f}^{[oe]} \\ f_3 + \alpha^4 f_7 = \hat{f}^{[oo]} \end{array} \right] \Rightarrow \begin{array}{l} \hat{f}^{[ee]} + \alpha^2 \hat{f}^{[eo]} = \hat{f}^{[e]} \\ \hat{f}^{[oe]} + \alpha^2 \hat{f}^{[oo]} = \hat{f}^{[o]} \end{array} \right] \Rightarrow \hat{f}^{[e]} + \alpha \hat{f}^{[o]} = \hat{f}_n$$

So, how do we determine the order of arranging the f_k's before they are to be combined by pairs? The rearrangement becomes very transparent if the original values for k are written in base-2.

k	0	1	2	3	4	5	6	7
$(k)_2$	000	001	010	011	100	101	110	111

In the first grouping, arranging by evens and odds, the even group consists of those numbers ending in a zero, and the odds end with a one, yielding

$$\begin{array}{cccccccc} 0 & 2 & 4 & 6 & 1 & 3 & 5 & 7 \\ 000 & 010 & 100 & 110 & 001 & 011 & 101 & 111 \end{array}$$

The next rearrangement regroups by looking at the next position in *from the right*, selecting zeros first; i.e.,

$$\begin{array}{cccccccc} 0 & 4 & 2 & 6 & 1 & 5 & 3 & 7 \\ 000 & 100 & 010 & 110 & 001 & 101 & 011 & 111 \end{array}$$

The values of k are arranged in a new sequence by evaluating their corresponding binary values in *reverse* order. This is the same as writing the numbers $0 \to 7$ in an order based on the value of their binary digits written in *reverse* order.

Thus, the complete algorithm is to first obtain the base 2 values of the numbers $k = 1, \ldots, N - 1$, then reverse the order of the digits in these numbers, and order the f_k's relative to the reversed binary values. The ordering of the f_k's is successively combined by pairs until a single number results.

MATLAB M-FILE FUNCTIONS TO COMPUTE FFTs

The Basic Discrete Fast Fourier Transform Function fft   The toolbox function fft(y) will compute the discrete Fourier transform of the function represented by the numbers contained in the vector y. If the length of y is a power of 2, the algorithm is based on reordering the numbers in *reversed binary sequence* as we have just described. If the length of y is not a power of 2, a slower, more complicated algorithm is used. Because speed is usually of great importance, it is suggested that you always use vectors of a length 2^k. If this is impossible, a preferred approach is to *pad* the vector y with sufficient zeros so that its length is a power of 2. This can be done automatically by adding a second argument, n, to fft, where n is a power of 2. Then, fft(y,n) will extend y to be of length n by adding zeros, or, if length(y) > n, the vector will be truncated. If y is a matrix, fft is applied to each column individually.

Keep in mind that the output from the fft function will be a set of *complex* amplitudes, c_n, of the individual exponentials, $e^{i\omega_n x} = e^{i(n\pi/L)x}$, which are themselves complex. Thus, when displaying the results of fft, the absolute values of the numbers are used.

For example, the expansion of the function $y(t) = e^{13(i\pi/L)x}$ in terms of these complex exponentials should produce a *single* term. To illustrate this, the following MATLAB code is used to graph $|z|$ as a function of n, where z is the Fourier transform of y:

```
n = 63;   L = 2;
t = -L:2*L/n:L;
y = exp(13*i*pi*t/L);
z = fft(y);
plot(0:63,abs(z))
```

The graph displays a narrow spike at $n = 13$ as it should. See Figure 17.6 on the next page.

If the function were $\sin(13\pi t/L)$ or $\cos(13\pi t/L)$ the result would be somewhat different. Figure 17.7 was obtained by replacing the line for y in the preceding code by y = cos(13*pi*t/L);. In this instance, there are two spikes: one corresponding to the frequency 13 of the cosine and the other its "negative." That is, because the cosine function is a combination of both a positive and a negative exponential, its expansion will contain these two corresponding terms. Notice, in particular, that although *negative* values of n are not permitted in the expansion, it is assumed that the Fourier transform is *periodic* with a period of $N = 64$ points. Thus, the spike at $n = -13$ is reproduced at $n = -13 + 64 = 51$.

The ability of the Fourier transform to identify periodic features is often a very useful tool in analyzing data. For example, the occurrence of sunspots seems to have cyclic behavior with a period of approximately 11 years. Data have been collected on sunspot activity for hundreds of years and provide a wealth of information for various forms of analysis. The monthly data on sunspot activity in the form of Wolf units are given in the supplied MATLAB function

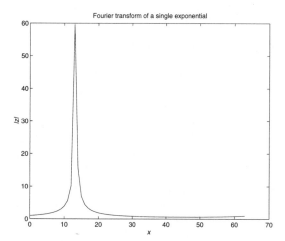

Figure 17.6   *The absolute magnitude of the Fourier transform of a single exponential returns nonzero coefficients for only the input exponential.*

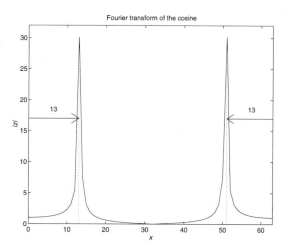

Figure 17.7   *The absolute magnitude of the Fourier transform of a single cosine function returns nonzero coefficients for the fundamental frequency (13) and for its negative.*

SunSpotD.m for a period from 1749 through 1960. A Wolf unit is defined as $W = k(10g + f)$, where g is the number of sunspot groups, f is the total number of sunspots in all groups, and k is a parameter to account for differences in observational conditions and techniques.[3] The data representing sunspot activity observed over a long period of time are plotted in Figure 17.8.

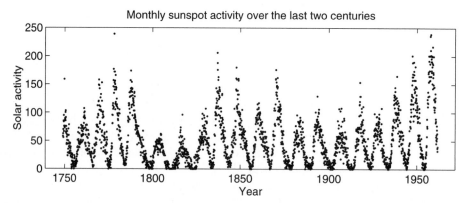

Figure 17.8   *Sunspot activity over the last two centuries. The points are monthly measurements.*

---

[3]The data are taken from Vitinskii, *Solar Activity Forecasting*, Academy of Sciences of the USSR, Leningrad, 1962; translated by Israel Program for Scientific Translations, available from the U.S. Dept. of Commerce.

From the graph of sunspot activity it is clear that the phenomenon is indeed periodic with a period of approximately 11 years. To obtain the dominant frequency more precisely, we first take the FFT of the data and plot its absolute magnitude versus a limited range of ω_n. The angular frequency is given by $\omega_n = 2\pi n/\Delta y$ with $\Delta y = (1960 - 1749) = 211$ years.

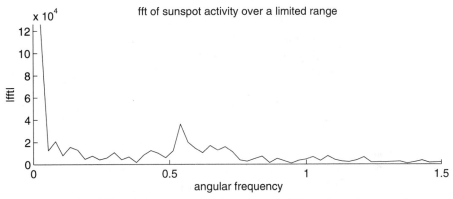

Figure 17.9 *The FFT of the sunspot activity is plotted for a limited range of ω_n.*

To confirm our interpretation that sunspot activity has a period of approximately 11 years, we next plot the same values for the norm of the FFT versus time rather than frequency, using $t_n = 2\pi/\omega_n$. This is shown in Figure 17.10, where a peak is clearly evident at $T = 11.64$ years.

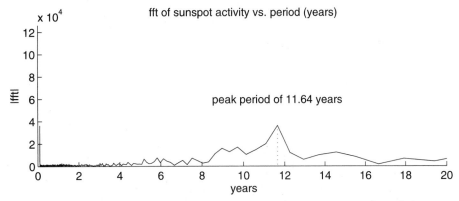

Figure 17.10 *The FFT of the sunspot activity plotted versus $T_n = 2\pi/\omega_n$. Only a portion of the complete range of T is shown.*

The Discrete Inverse Fast Fourier Transform Function `ifft`   As we have seen, the discrete inverse Fourier transform is a trivial variation on the ordinary discrete transform, and the MATLAB coding and use of this function is, in all respects, similar to that of `fft`. The name of the inverse function is `ifft`. As a simple test of this function, you should compute the `fft` of some simple functions and then use the function `ifft` to attempt to recover the original function. Again, the results generated by `ifft` will be complex numbers. If the sequence of operations $y = y(x)$, $\hat{y}(\omega) = \mathcal{F}y$, and $Y(x) = \mathcal{F}^{-1}\hat{y}(\omega)$ is duplicated in terms of the functions `fft` and `ifft`, the absolute magnitude of the difference between y and Y; i.e., $|y - Y|$, should be approximately equal to the machine accuracy.

THE PROBLEM OF WRAPAROUND

A fundamental assumption in the derivation of the discrete Fourier transform and its inverse is that *both* the function $f(x)$ and its transform, $\hat{f}(\omega)$, are *periodic*. This can cause substantial problems when the transform is applied to nonperiodic functions. For example, the Gaussian function

$$f(x) = \frac{1}{\sqrt{2\pi}\sigma}e^{-(1/2)(x^2/\sigma^2)}$$

can be inserted into the definition for the Fourier transform on page 375, and $\hat{f}(\omega)$ computed by completing the square in the integral. The result is

$$\hat{f}(\omega) = \frac{1}{\sqrt{2\pi}}e^{-(1/2)\sigma^2\omega^2}$$

which is also a Gaussian distribution. The function $f(x)$, with $\sigma = 1$, is graphed over the interval $-6 \le x \le 6$ in Figure 17.11. It is clearly not periodic, but as far as the Fourier transform and inverse transform are concerned, the function extends from $-\infty$ to $+\infty$, and the peak at $x = 0$ reappears at $x = \ldots, -24, -12, 0, 12, 24, \ldots$.

A Gaussian distribution function

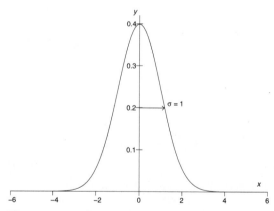

Figure 17.11   *A Gaussian function $f(x)$ with standard deviation $\sigma = 1$.*

This situation has serious consequences with respect to the inverse transform. The actual transform is, of course, the preceding Gaussian distribution in ω. However, computing the `fft` of $f(x)$ and plotting its absolute magnitude yield the result depicted in Figure 17.12.

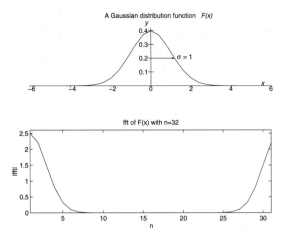

Figure 17.12 *The FFT of a Gaussian function yields a non-Gaussian as a result of wraparound.*

Thus, the transform function at $n = 31$ is the image of the point $n = -1$, etc. The anticipated Gaussian distribution of the transform has an unexpected feature at the end of the frequency spectrum that is a result of the assumption of periodicity. This feature is called "wraparound" for obvious reasons and can complicate the analysis. Basically, if $f(x)$ is zero outside some bounds, then $\hat{f}(\omega)$ cannot be, and vice versa. Nevertheless, the `ifft` of this function yields the original Gaussian $f(x)$.

One result of wraparound is that the solution of a differential equation via `fft`'s is not as trivial as it first appears. For example, the general solution of the mass-spring problem was expressed as $y(x) = \mathcal{F}^{-1}(\hat{h}\hat{F})$, where $\hat{F}$ is the Fourier transform of the driving force, $F(t)$, and $\hat{h}(\omega)$ is a Fourier transform of a *response* function determined exclusively by the mass and the spring. We generally know $\hat{h}(\omega)$ and $F(t)$. Thus, for an arbitrary driving force, it might seem that the solution can easily be obtained by multiplying `fft(F)` by $\hat{h}(\omega)$ and then taking the inverse transform. Unfortunately, this will not work because $\hat{h}(\omega)$ will lack the wraparound contributions from negative frequencies. Clearly, these could be included without great difficulty, but this would lead our discussion too far astray.

There are numerous other problems with interpreting FFTs as a consequence of wraparound, but they are beyond the scope of this text. The interested reader should consult Press et. al (1986).

Problems

REINFORCEMENT EXERCISES

P17.1 **Fourier Series.** Write a MATLAB M-file function `FourierX(fnc,L,n)` that uses the toolbox function **quad** to evaluate the first n expansion coefficients, a_i and b_i, $i = 1, \ldots, n$, by means of Equation 17.3. The function name is to be passed to the function in the string variable `fnc`. Test your function on each of the following functions. For each function, simultaneously graph the function along with the series representation of the function, including $n = 3$, 5, and 10 terms.

a. $f(x) = \begin{cases} -1 & \text{for } -1 \leq x < 0 \\ 1 & \text{for } 0 < x \leq 1 \\ 0 & \text{if } x = 0 \end{cases}$

b. $f(x) = x^2$, for $-2 \leq x \leq 2$.

c. $f(x) = x^3 e^{-x^2}$, for $-2 \leq x \leq 2$.

d. `humps(x)` for $-1 \leq x \leq 1$. `humps(x)` is a strongly peaked toolbox function.

P17.2 Orthogonality Conditions. Use the trigonometric identities

$$\cos \alpha \cos \beta = \frac{1}{2}[\cos(\alpha + \beta) + \cos(\alpha - \beta)]$$

$$\sin \alpha \sin \beta = -\frac{1}{2}[\cos(\alpha + \beta) - \cos(\alpha - \beta)]$$

$$\sin \alpha \cos \beta = \frac{1}{2}[\sin(\alpha + \beta) + \sin(\alpha - \beta)]$$

to prove the orthogonality conditions in Equation 17.2.

P17.3 Complex Fourier Series. Using Equation 17.6, compute the complex expansion coefficients, c_n, for the following functions:

a. $f(x) = \begin{cases} -1 & \text{for } -1 \leq x < 0 \\ 1 & \text{for } 0 < x \leq 1 \\ 0 & \text{if } x = 0 \end{cases}$

b. $f(x) = e^{-2x}$, for $-2 \leq x \leq 2$.

c. $f(x) = 1 + \sin 2x$, for $-\pi \leq x \leq \pi$.

P17.4 Fourier Series for the Range $0 \leq x \leq a$. By a change of variables in Equations 17.1 and 17.2, obtain the Fourier series expansion coefficients a_n, b_n, and c_n for a function that is defined over the interval $0 \leq x \leq a$ rather than $-L \leq x \leq L$.

P17.5 Dirac Delta Functions. The indefinite integral in Equation 17.9 is actually defined in terms of the integral

$$\int_{-t}^{t} e^{ikx} \, dx = 2\frac{\sin kt}{k}$$

The limit of this equation as $t \to \infty$ should represent the Dirac delta function, $\delta(t - 0)$. For $t = 10$ and 100, graph the absolute value of this integral as a function of k for $-1 \leq k \leq 1$.

P17.6 Fourier Transform. Determine the Fourier transform of

$$f(x) = \begin{cases} F_0 & \text{for } -a \leq x < a \\ 0 & \text{elsewhere} \end{cases}$$

The *width* of $f(x)$ is $\Delta x = 2a$. If the width, $\Delta \omega$, of $\hat{f}(\omega)$ is defined as the distance between the zeros of $\hat{f}(\omega)$, show that $\Delta x \Delta \omega = 2\pi$. This illustrates the general property that if $f(x)$ is narrowly peaked, $\Delta x \ll 1$, then the width of $\hat{f}(\omega)$ will be very broad, $\Delta \omega \gg 1$. This reflects the fact that it will require a large number of Fourier components to duplicate the narrowly peaked $f(x)$, and the converse is also true.

P17.7 Fourier Transform of $\delta(x)$. Determine the Fourier transform of the Dirac delta function, $\delta(x - a)$.

P17.8 Proof of the Convolution Theorem. Starting from the definitions of the Fourier transform and inverse Fourier transform, use the fact that

$$\int_{-\infty}^{\infty} e^{(x-x')\omega} \, d\omega = 2\pi \delta(x - x')$$

to prove that

$$\mathcal{F}^{-1}[\hat{h}(\omega)\hat{g}(\omega)] = \frac{1}{\sqrt{2\pi}} \int_{-\infty}^{\infty} h(x')g(x - x')dx' \equiv h * g$$

P17.9 Mass-Spring Problem with an External Impulse. The general solution of the mass-spring problem was given in terms of the convolution integral by the equation, $y(t) = h * F$. Thus, the solution for an arbitrary driving force can be obtained once the inverse Fourier transform of $\hat{h}(\omega) = m^{-1}(\omega^2 - \omega_0^2)^{-1}$ is determined. The determination of $\mathcal{F}^{-1}\hat{h}(\omega)$ is obtained using *contour integration*, and this topic is beyond the scope of this text. However, the result is

$$h(t) = \frac{1}{m}\sqrt{\frac{\pi}{2}}\frac{\sin \omega_0 t}{\omega_0}$$

where $\omega_0^2 = k/m$. Use this expression for $h(t)$ and the convolution integral to obtain the motion of the mass if the external force is an impulse; i.e.,

$$F(t) = \begin{bmatrix} F_0 & \text{for } 0 \leq t < t_0 \\ 0 & \text{elsewhere} \end{bmatrix}$$

P17.10 Spectrum Analysis and Convolution. Interpreting the solution of a differential equation expressed in terms of the convolution integral,

$$y(t) = \mathcal{F}^{-1}[\hat{h}(\omega)\hat{F}(\omega)] \equiv h * F$$

as an *input* $F(t)$ mapped into an *output*, $y(t)$, by means of a *filter*, $h(t)$. In particular, take the input and filter functions to be $h(t)$ and $F(t)$ of Problem 17.9 (*Note:* The reversal of roles of the two functions $h(t)$ and $F(t)$). If the filter function is very narrow, the solution should resemble $h(t)$. Show that this is true. What is the approximate result if the filter function is very broad?

P17.11 Data Filters. The function

$$f(x) = 4\sin^2(t - 2)e^{-50(t-2)^2} + \frac{1}{2}\left[(t - 5)^2 + \frac{1}{4}\right]^{-1/4}$$

has a broad peak centered at $x = 5$ and two small features near $x = 2$. This is the input to an instrument with a filter described by

$$r(x) = \begin{bmatrix} 0.50 & \text{for } x \leq 0.15 \\ 0.25 & \text{for } 0.15 < x \leq 0.30 \\ 0.10 & \text{for } 0.30 < x \leq 0.60 \end{bmatrix}$$

Compute the inverse Fourier transform, `ifft`, of the product of the Fourier transforms of these functions and compare with the actual function $f(x)$. Will the two small features be discernible?

EXPLORATION PROBLEMS

P17.12 **Fourier Transform of a Differential Equation.** A horizontal flexible beam is supported on an elastic foundation, and a distributed *load* is placed on the beam. Assuming that the beam is infinitely long, that the foundation can be represented by "springs" characterized by spring constants k per unit length along the beam, and that the load is given by the function $w(x)$, the deflection of the beam in the y direction, as a result of the load, is determined by the equation

$$YI_2 \frac{d^4y}{dx^4} + ky = w(x)$$

where the constant YI_2 is the *rigidity* of the beam. (Y is Young's modulus and I_2 is the second moment of the beam.) By taking the Fourier transform of this equation, show that the solution for the transform function, $\hat{y}(\omega)$, is given by

$$\hat{y}(\omega) = \hat{h}(\omega)\hat{w}(\omega)$$

where $\hat{w}(\omega)$ is the Fourier transform of the load. Determine the function $\hat{h}(\omega)$. The inverse transform of $\hat{h}(\omega)$ yields $h(x)$, which is the response of the beam to a delta function load. Show that if the delta function load is placed at x_0, then

$$h(x) = \frac{1}{\pi YI_2} \int_0^\infty \frac{\cos(x - x_0)\omega}{\omega^4 + (k/YI_2)} \, d\omega$$

Let $k = YI_2 = 1$, $x_0 = 1.5$, and evaluate this integral approximately for values of x in the range $[0, 4]$. Graph the result.

Summary

THE FOURIER SERIES

Definition

The Fourier series expansion of a function in terms of sines and cosines is

$$f(x) = \frac{1}{2}a_0 + \sum_{m=1}^{\infty} a_m \cos\left(\frac{n\pi}{L}x\right) + \sum_{m=1}^{\infty} b_m \sin\left(\frac{n\pi}{L}x\right)$$

where a_m and b_m are the Fourier expansion coefficients.

Expansion Coefficients

The expansion coefficients are determined by the equations,

$$a_n = \frac{1}{L} \int_{-L}^{L} f(x) \cos\left(\frac{n\pi}{L}x\right) dx$$

$$b_n = \frac{1}{L} \int_{-L}^{L} f(x) \sin\left(\frac{n\pi}{L}x\right) dx$$

$$\frac{1}{2}a_0 = \overline{f}$$

Complex Fourier
Series

The Fourier series can also be written as an expansion in terms of complex exponentials as

$$f(x) = \sum_{n=-\infty}^{+\infty} c_n e^{(in\pi/L)x}$$

with the expansion coefficients, c_n, given by

$$c_n = \frac{1}{2L} \int_{-L}^{L} f(x) e^{-i(n\pi/L)x} \, dx$$

FOURIER TRANSFORMS

The Fourier
Transform

By replacing the summation index, n, by a continuous variable, the summations in the expression for the Fourier series are replaced by integrals, yielding the equation for the Fourier transform $\hat{f}$ of the function $f(x)$:

$$\hat{f}(\omega) \equiv \frac{1}{\sqrt{2\pi}} \int_{-\infty}^{\infty} f(x) e^{-i\omega x} \, dx$$

The Inverse
Fourier
Transform

The inverse equation relating $\hat{f}$ to f is

$$f(x) \equiv \frac{1}{\sqrt{2\pi}} \int_{-\infty}^{\infty} \hat{f}(\omega) e^{i\omega x} \, d\omega$$

The Dirac Delta
Function

The definition of the Dirac delta function is

$$\int_{x_1}^{x_2} f(x)\delta(x - a) \, dx = f(a) \qquad \text{if } x_1 \leq a \leq x_2$$

$$= 0 \qquad \text{otherwise}$$

The Dirac delta function is only defined if it appears within an integrand. An important special case is contained in the relation

$$\int_{-\infty}^{\infty} e^{i(\alpha-\beta)x} \, dx = 2\pi\delta(\alpha - \beta)$$

Convolution

The convolution integral of two functions, $f(x)$ and $g(x)$, is defined as

$$f * g \equiv \frac{1}{\sqrt{2\pi}} \int_{-\infty}^{\infty} f(x - \xi)g(\xi) \, d\xi$$

which is related to their Fourier transforms by the relations

$$\mathcal{F}(f * g) = \hat{f}\hat{g}$$
$$\mathcal{F}^{-1}(\hat{f}\hat{g}) = f * g$$

DISCRETE FOURIER TRANSFORMS

Definition

The approximate Fourier transform of a discrete set of N points is given by

$$\hat{f}_n = \sum_{k=0}^{N} W^{nk} f_k$$

where $W \equiv e^{2\pi i/N}$. The inverse discrete transform is then

$$f_k = \frac{1}{N+1} \sum_{n=0}^{N} W^{-nk} \hat{f}_n$$

MATLAB Functions Used

fft   The fast Fourier transform of data contained in the vector x is computed by the function fft(x). Although this function will compute Fourier transforms of vectors of arbitrary length, the speed is greatly increased if the length is a power of 2.

ifft   The fast inverse Fourier transform of data. The algorithm is essentially the same used by fft and the same restrictions apply.

Bibliography for Part IV

1. P. R. Bevington, *Data Reduction and Error Analysis for the Physical Sciences*, McGraw-Hill, New York, 1969.

2. P. Bloomfield, *Fourier Analysis of Time Series—An Introduction*, John Wiley, New York, 1976.

3. G. J. Borse, *FORTRAN-77 and Numerical Methods*, 2nd ed., PWS Publishing, Boston, MA, 1991.

4. E. O. Brigham, *The Fast Fourier Transform*, Prentice-Hall, Englewood Cliffs, NJ, 1974.

5. C. C. Champeney, *Fourier Transforms and Their Physical Applications*, Academic Press, New York, 1973.

6. C. DeBoor, *A Practical Guide to Splines*, Springer-Verlag, Berlin and New York, 1978.

7. W. Feller, *An Introduction to Probability Theory and its Applications*, vols. I and II, John Wiley, New York, 1966.

8. M. D. Greenberg, *Foundations of Applied Mathematics*, Prentice-Hall, Englewood Cliffs, NJ, 1978.

9. P. G. Guest, *Numerical Methods of Curve Fitting*, Cambridge University Press, New York, 1961.

10. R. W. Hamming, *Numerical Methods for Scientists and Engineers*, McGraw-Hill, New York, 1962.

11. D. E. Knuth, "Seminumerical Algorithms," in *The Art of Computer Programming*, 2nd ed., vol. 2, Addison-Wesley, Reading, MA, 1981.

12. M. J. Maron, *Numerical Analysis: A Practical Approach*, Macmillan, New York, 1982.

13. B. R. Martin, *Statistics for Physicists*, Academic Press, New York, 1971.

14. H. J. Nussbaumer, *Fast Fourier Transform and Convolution Algorithms*, Springer-Verlag, Berlin and New York, 1982.

15. S. Pizer, *Numerical Computing and Mathematical Analysis*, Science Research Associates, Chicago, 1975.

16. W. H. Press, B. P. Flannery, S. A. Teukolsky, and W. T. Vetterling, *Numerical Recipes, The Art of Scientific Computing*, Cambridge University Press, Cambridge, England, 1986.

17. ———, *Numerical Recipes in FORTRAN*, 2nd ed. Cambridge University Press, Cambridge, England, 1992.

18. E. M. Pugh and G. H. Winslow, *Analysis of Physical Measurements*, Addison-Wesley, Reading, MA, 1966.

19. T. J. Rivlin, *The Chebyshev Polynomials*, John Wiley, New York, 1974.

20. L. L. Schumaker, *Spline Functions: Basic Theory*, John Wiley, New York, 1981.

21. J. Stoer and R. Bulirsch, *Introduction to Numerical Analysis* (English translation by R. Bartels, W. Gautschi, and C. Witzgall), Springer-Verlag, Berlin and New York, 1980.

22. Robert D. Strum and Donald E. Kirk, *Contemporary Linear Systems Using MATLAB*, PWS Publishing, Boston, MA, 1994.

23. Y. I. Vitinskii, *Solar Activity Forecasting*, Academy of Sciences of the USSR, Leningrad, 1962; translated by Israel Program for Scientific Translations, available from U.S. Dept. of Commerce.

24. Samuel S. M. Wong, *Computational Methods in Physics and Engineering*, Prentice-Hall, Englewood Cliffs, NJ, 1992.

PART V /
Differential Equations

PREVIEW   The subject of differential equations is certainly one of the most important in all of applied mathematics and one in which every practicing engineer or scientist must be well versed. The common relationships between physical quantities almost always contain terms that are proportional to *rates of change* of the quantities as well—rates of change with respect to time, position, strain, temperature, or a variety of other physical parameters. Differential equations are thus the most popular means we have for describing phenomena.

The primary distinction between the concepts involved in solving differential equations and those described earlier is this: Unlike an algebraic equation, whose solution yields a single number or a single vector of numbers, the solution of each differential equation is an entire *function*. Obviously, a question whose answer is $e^{-2x^2} \sin \pi x$ is likely to be more difficult than one whose answer is simply 2.7. Numerically, this will mean that the solution must always consist of a vector of values representing the solution function over some range.

Ordinary differential equations (ODEs) are those with only *one* independent variable, say x. The differential equation itself is then an equation relating the unknown dependent variable, for example, $y(x)$, and any number of its derivatives to functions of x. We can solve for the elements of $y(x)$, one at a time, in an incremental procedure; i.e., starting with y_0 to compute y_1; then knowing y_1, move on to compute y_2, and so forth. Or, we could attempt to solve for all the elements of $y(x)$ at once, in a manner analogous to solving matrix equations. Both ideas appear to be well suited to the *vectorized* mathematics of MATLAB. However, as we shall see, particularly for the stepping methods, this is not the case. We need to add to the existing MATLAB capabilities available in the form of toolbox functions.

In addition, solution procedures for differential equations are much more varied than are the methods for solving algebraic or matrix equations. One method may be the ideal procedure for most of the equations you will encounter, but you will never be able to apply it blindly. Some equations are inherently unstable; i.e., small changes in certain parameters yield large changes in the solution; these equations require special care. Other equations may have no solution whatsoever. Singularities in the solution function will clearly cause problems, as will infinite ranges of the independent variable. It will be necessary to introduce numerous classifications for

399

differential equations, each dictating that a particular avenue be followed to obtain a solution. Because the goal is to compute a sufficiently large number of values of the solution so that the complete function over some range can be inferred, most problems will require a substantial amount of computing, perhaps more than can be tolerated by your computing budget or your patience. Thus, the delicate balance between accuracy and speed becomes a central criterion when deciding on the appropriate algorithm to use on a particular differential equation. For example, in the stepping procedure, it may be necessary to write the algorithm so that it will always take the largest step possible, consistent with some accuracy constraints.

However, insofar as numerical methods are concerned, the most important classification of differential equation problems is whether the problem is an *initial value problem* or a *boundary value problem*. In an initial value problem we can combine sufficient information about the solution at a specific starting point with the information contained in the given differential equation to predict the solution one step further along the axis of the independent variable. Such problems are ideally suited for stepping procedures. The prototype for all stepping techniques is the method of Euler, described in Chapter 19. We next discuss the refinements of this algorithm, which include the midpoint method and the methods of Runge-Kutta. The MATLAB toolbox routines for initial value problems, ode23 and ode45, are based upon these ideas. Most of the time, the algorithm of choice for solving initial value problems will be one of these two MATLAB toolbox functions. However, differential equations can often be stubbornly unyielding to ordinary methods, so we introduce an alternative procedure to be used in those rare instances when the toolbox functions are not up to the task. This procedure, called the Bulirsch-Stoer extrapolation method, attempts at each stage to take the largest step possible, and is based on extrapolation techniques similar to Romberg integration or Aitken extrapolation. We compare and evaluate all these methods for suitability for particular problems, and also evaluate the available MATLAB toolbox stepping procedures.

We investigate boundary value problems in Chapters 20 and 21. In a boundary value problem the information available may be sufficient to guarantee a unique solution, but be insufficient to determine the starting point for a stepping procedure. There are no explicit MATLAB toolbox functions for solving boundary value problems. Therefore, we describe and illustrate a variety of techniques in detail in Chapter 20. In Chapter 21, we investigate special techniques applicable to the most common boundary value problem, a two-point boundary condition applied to linear, second-order equations.

Finally, in Chapter 22, we provide an introduction to the vast literature on the numerical solution of partial differential equations (PDEs). We describe the classification of PDEs and illustrate elementary techniques for each class of partial differential equation.

In this part, we hope to provide at least an introduction to many of the state-of-the-art procedures for solving differential equations. In addition, it is hoped that along the way the practitioner of these methods will acquire a *feel* for what is required to solve a differential equation, for this is the main ingredient in the art of numerical analysis.

We begin in Chapter 18 by reviewing some of the basic ideas central to ordinary analytic solutions to differential equations.

18 / *Elementary Concepts*

PREVIEW   If, in the equation relating functions of the independent variable x, the dependent variable $y(x)$, and derivatives of y with respect to x, the highest derivative is $d^n y/dx^n$, the equation is of *degree* n, or order n. Most of the commonly encountered differential equations of science and engineering are of second order. However, as we shall see, second-order (or even higher-order) equations can always be recast in terms of coupled first-order equations. For this reason it is essential that you be both familiar and comfortable with the common properties and solutions of first-order equations. In addition, existence and uniqueness criteria that apply to first-order equations can be generalized to higher-order equations when they are written as coupled first-order equations.

The analytic solution of second-order differential equations is a major component in the training of any engineer or scientist. However, the numerical solution of higher-order equations is usually much easier to understand and begins with simple *stepping method* solutions to first-order equations, which are then generalized to coupled sets of first-order equations.

In this chapter then, we review properties of first-order differential equations and describe how higher-order equations can be cast in terms of coupled first-order equations.

First-Order Differential Equations

Any first-order differential equation; i.e., an equation containing only the first derivative of $y(x)$, can be written formally as

$$y' = f(x, y)$$

If the function f is independent of y, the solution for $y(x)$ then merely amounts to integration,

$$y(x) = y_0 + \int_{x_0}^{x} f(x)\, dx$$

where y_0 is the value of the solution at x_0. Thus, to obtain a solution the *initial*[1] value of y must be specified as part of the problem.

---

[1] Of course, the *initial* value could just as well be at the *end* of the interval with the integration proceeding in a negative direction.

If the function f depends on y, a formal solution of the first-order equation can be obtained by iteration,

$$y^{[1]}(x) = y_0 + \int_{x_0}^{x} f(x, y_0) dx$$

$$y^{[2]}(x) = y_0 + \int_{x_0}^{x} f(x, y^{[1]}(x)) dx \tag{18.1}$$

$$y^{[3]}(x) = y_0 + \int_{x_0}^{x} f(x, y^{[2]}(x)) dx$$

$$\vdots \qquad\qquad \vdots$$

The difference between successive approximations to $y(x)$, namely, $\Delta_n(x) = y^{[n]}(x) - y^{[n-1]}(x)$, must approach zero *uniformly*[2] over the interval $x_0 \longleftrightarrow x$. If this is the case, then a *unique* solution to the equation *exists*. Using this criterion, it can be shown (see Kreyszig (1983), p. 59) that the conditions for the existence of a unique solution of the equation $y' = f(x, y)$, with $y_0 = y(x_0)$ given, are as follows:

Uniqueness/ Existence Conditions for First-Order Equations

Define a region of the plane around the point (x_0, y_0); i.e., $|x - x_0| < \Delta x$, $|y - y_0| < \Delta y$. Then:

- If $f(x, y)$ is bounded in the region; i.e., $|f(x, y)| < M$, and
- If $|\partial f / \partial y|$ is also bounded in the same region,

a *unique* solution exists in the region bounded by

$$|x - x_0| < \Delta x$$
$$|y - y_0| < M \Delta x$$

where M is the bound of $|f(x, y)|$ in the region.

Although this formulation of the solution to the differential equation is most useful in proving the uniqueness and existence of a solution, it can be extremely cumbersome to use in practice. (See Problem 18.3.) More tractable algorithms that avoid explicit integration will be described shortly.

There are a variety of techniques for obtaining analytic solutions to first-order equations, and it is expected that the reader is familiar with several of them. A few are illustrated in Example 18.1. See Kreyszig (1983) for a more complete summary of standard techniques for first-order equations.

EXAMPLE 18.1

Examples of Analytic Solutions to First-Order Equations

SEPARATION OF VARIABLES

Frequently the function $f(x, y)$ can be separated into two distinct functions, $f(x)$ and $g(y)$, in which case writing y' as dy/dx, the differential equation can

---

[2]Basically, this means that the area under $\Delta_n(x)$ must approach zero as $n \to \infty$. Or, put another way, $\lim_{n \to \infty} \Delta_n(x) = 0$ for all but a countable, i.e., discrete, set of x's where the discrepancies must remain finite.

be recast as

$$\frac{dy}{g(y)} = f(x)\,dx$$

and a solution is obtained by integrating the two sides of the equation independently. Again, note that what is obtained as a solution is an entire function $y(x)$, not merely a number.

EXACT DIFFERENTIALS

By comparing the differential equation written in the form

$$M(x,y)\,dy + N(x,y)\,dx = 0 \tag{18.2}$$

with the expression for the *exact* differential of a function of two variables, $F(x,y)$; i.e.,

$$dF = \frac{\partial F}{\partial y}\,dy + \frac{\partial F}{\partial x}\,dx = 0$$

and noting that $\partial^2 F/\partial x \partial y = \partial^2 F/\partial y \partial x$, we see that a condition that Equation 18.2 represent the exact differential of some unknown function F is that $\partial M/\partial x = \partial N/\partial y$. If this is the case, the solution of Equation 18.2 is simply $F = \text{const}$.

Thus, the equation

$$y' = \frac{-e^y}{xe^y + 2}$$

can be written as

$$e^y\,dx + (xe^y + 2)\,dy$$

which satisfies the condition of an exact differential. Thus, the solution is $F = C$, where $\partial F/\partial x = e^y$, so that $F(x,y) = xe^y + g(y)$, where $g(y)$ is an *arbitrary* function of y. However, the remaining condition, $\partial F/\partial y = xe^y + 2$, is then equivalent to $g'(y) = 2$. The complete solution to the differential equation is then

$$C = xe^y + 2y$$

The integration constant C is to be determined by using any initial condition of the form $y_0 = y(x_0)$.

A SPECIAL FIRST-ORDER EQUATION

An equation of the form

$$y' + M(x)y = N(x)$$

does not in general satisfy the conditions of an exact differential. However, by explicit differentiation, you can show that the form of the solution is

$$y(x) = e^{-\int M\,dx}\left(\int Ne^{\int M\,dx}\,dx + C\right) \tag{18.3}$$

where C is the constant of integration. Thus, in the equation

$$y' + \frac{y}{x} = \sin x$$

we can make the identifications $M(x) = 1/x$, $N(x) = \sin x$, so that the solution becomes

$$y(x) = e^{-\ln x}\left(\int e^{\ln x} \sin x \, dx + C\right)$$

$$= x\left(\int x \sin x \, dx + C\right)$$

$$= x(\sin x - x\cos x + C) \qquad \bullet$$

Naturally, a closed-form, analytic solution is preferable to a numerical solution. Before attempting a numerical solution, it is a good idea to spend a few minutes attempting to obtain an analytic solution. In any event, before embarking on either type of solution, you must first convince yourself that a solution *exists* and that, when obtained, it will be unique.

Second-Order Differential Equations

The most common type of differential equation encountered is the second-order differential equation, whose general form is

$$F(x, y, y', y'') = 0$$

The equation is termed *linear* if y, y', and y'' occur to the first power only,

Linear, $$p(x)y'' + q(x)y' + r(x)y = h(x) \qquad (18.4)$$
second-order

where p, q, r, and h are arbitrary functions of x. If the function on the right, $h(x)$, is zero, the equation is called *homogeneous*, otherwise it is *inhomogeneous*. Because the solution of a second-order equation is equivalent to *two* integrations, each introducing a constant of integration, it is clear that unique specification of a solution to a second-order equation will require that *two* independent extra conditions on the solution be specified. If *both* conditions are specified at the *same* value of x; e.g., $a = y(x_0)$ and $b = y'(x_0)$, the equation is known as an *initial* value problem. In all other forms for the original conditions the equation is known as a *boundary* value problem. As we shall see, the specification of the extra conditions as either *initial* or *boundary* values will greatly affect the method used to obtain a numerical solution to the equation.

The analytical solution of equations of the form of Equation 18.4 is a very important part of applied mathematics. Techniques such as series solutions, transform methods, and Green's functions are commonly used to obtain solutions to linear second-order differential equations. And all of these should be explored before resorting to a numerical approach.

Numerical methods of solution of differential equations are much less specific, often applying to nonlinear as well as linear equations, and to equations of arbitrary, not just second, order. Much more significant is the nature of the extra conditions specified in the problem. In this chapter we will deal exclusively with *initial* value problems. That is, it is assumed that if the equation is of nth order, there are n independent extra conditions on $y(x)$ or any of the derivatives through the $n - 1$st, *and* that all conditions apply at the same value of x_0.

Replacing an nth-Order Equation by n Coupled First-Order Equations

It is always possible to rewrite a problem characterized by an nth-order differential equation by a system of coupled first-order differential equations. For example, the second-order differential equation for $\mathbf{F} = m\mathbf{a}$ used to describe the motion of an object of mass m attached to a spring is

$$m\frac{d^2x(t)}{dt^2} = -kx, \qquad x(t_0) = x_0$$

$$\left.\frac{dx}{dt}\right|_{t=0} = 0$$

where k is the spring constant.

By introducing a *superfluous* variable, we can replace the single second-order equation by two coupled first-order equations. The natural choice for the new variable in this case is the velocity, $v(t)$. Thus, the two equations

$$\frac{dx}{dt} = v(t) \qquad x(t_0) = x_0$$

$$\frac{dv}{dt} = -kx \qquad v(t_0) = 0$$

represent an equivalent statement of the problem. These equations can be written somewhat more compactly by denoting the set of dependent variables to be components of a *vector*, $\vec{x}$,

$$\frac{dx_1(t)}{dt} = x_2(t), \qquad x_1(t_0) \equiv c_1 = x_0$$

$$\frac{dx_2(t)}{dt} = -kx_1(t), \qquad x_2(t_0) \equiv c_2 = 0$$

or

$$\frac{d\vec{x}(t)}{dt} = \vec{f}(\vec{x}, t) \qquad \vec{x}(t_0) = \vec{c}$$

Notice that these equations are *coupled*; that is, they must be solved in tandem. In each step of the solution, to compute the next value of $x_1(t)$ or $x_2(t)$, you must know the current values of *both* x_1 and x_2. The generalization to nth-order equations is clear.

There is considerable freedom in the selection of the $n - 1$ new variables that are introduced into the problem. The most common choice is $x_1 = x$, $x_2 = dx/dt$, $x_3 = d^2x/dt^2, \ldots$; however, any independent set of definitions is acceptable. In some problems you may find that suitably chosen combinations of derivatives perhaps multiplied by a variety of functions will be more convenient, particularly when the initial conditions involve such combinations of derivatives. These choices can affect the execution time of a numerical procedure considerably.

The reduction of an nth-order equation to a set of coupled first-order equations means that we can concentrate on the development of numerical algorithms suitable to first-order equations only. We will next review the most common of these as applied to a *single* first-order equation. The generalization to a set of n coupled equations will follow.

Problems

REINFORCEMENT EXERCISES

P18.1 Solutions to First-Order Equations. Solve the following first-order equations by the methods of this chapter:

a. $y' = \dfrac{y^2 - 1}{2xy}$

b. $y' = (y^2 - 1) \ln x$

c. $y' = \dfrac{y}{x} - \left(\dfrac{y}{x}\right)^2$

d. $y' = \sin x - 2\dfrac{y}{x}$

e. $y' = 1 + 2xy$

P18.2 Second-Order Equations as Coupled First-Order. Rewrite each of the following equations in terms of coupled first-order differential equations:

a. Bessel's equation,

$$xy'' + y' + kxy = 0; \qquad y(0) = 1, \ y'(0) = 0$$

b. Damped oscillator,

$$m\ddot{x} + c\dot{x} + kx = F_0 \sin \omega t; \qquad x(0) = x_0, \ \dot{x}(0) = 0$$

c. Duffing's equation,

$$\ddot{x} + c\dot{x} + \alpha x + \beta x^3 = F_0 \cos \omega t; \qquad (0) = x_0, \ \dot{x}(0) = 0$$

P18.3 Iterated Solutions. Using Equations 18.1, carry out three iterations towards a solution of $y' = 1 - y$, $y(0) = 0$, and compare your result with the exact solution.

Summary

FIRST-ORDER DIFFERENTIAL EQUATIONS

Definitions

The first-order differential equation, $y' = f(x, y)$ with $y(x_0) = y_0$, is formally given by

$$y(x) = y_0 + \int_{x_0}^{x} f(x, y(x)) \, dx$$

Iterative Algorithm

The formal solution can be solved iteratively as

$$y^{[n+1]}(x) = y_0 + \int_{x_0}^{x} f(x, y^{[n]}(x)) \, dx$$

starting with $y^{[0]}(x) \equiv y_0$.

A Special Linear ODE

The equation

$$y' + M(x)y = N(x)$$

has a solution given by

$$y(x) = e^{-\int M dx} \left(\int N e^{\int M dx} \, dx + C \right)$$

where C is the constant of integration.

SECOND-ORDER LINEAR DIFFERENTIAL EQUATIONS

Replacing by Coupled Equations

The second-order differential equation,

$$p(x)y'' + q(x)y' + r(x)y = h(x)$$

with extra conditions, $y(x_0) = y_0$, $y'(x_0) = v_0$, can be replaced by two coupled first-order differential equations by introducing two dependent variables, $y_1(x), y_2(x)$, defined as

$$y_1 \equiv y \qquad y_1(x_0) = y_0 \qquad y_1' = y_2$$

$$y_2 \equiv y' \qquad y_2(x_0) = v_0 \qquad y_2' = \frac{h(x)}{p(x)} - \frac{q(x)}{p(x)} y_2 - \frac{r(x)}{p(x)} y_1$$

This procedure is generalized in an obvious fashion for higher-order equations.

19 / *Initial Value Differential Equations*

PREVIEW If a differential equation is of nth-order, *and* the function and its first $n - 1$ derivatives are all specified at the same point, the differential equation is called an *initial value problem*. The differential equation itself can then be used to determine the nth and *all* higher derivatives. Thus, from the properties of the Taylor series, the function is completely determined in the neighborhood of the point; i.e., within the radius of convergence of the Taylor series. The natural procedure is to use the information about the function at the initial point and, upon expanding the function about this point, compute the value of the function and its derivatives at a nearby neighboring point. Once the set of information is complete at the new point, the process can be repeated across a complete interval. This *stepping* process is the universal basis for numerical solutions to initial value problems.

The most elementary implementation of a stepping procedure is called *Euler's method*, and it merely corresponds to a *linear* approximation to the function over a small step size. This method is far too inaccurate to ever be used in a realistic solution of a differential equation, but it does provide the basis for all of the more sophisticated stepping procedures. Improvements to the basic Euler algorithm are of two types: improvements in *accuracy*, and improvements in *efficiency*. However, because the accumulation of round-off error usually prohibits simply letting a coarse algorithm run longer to achieve more accuracy, efficiency and accuracy are interrelated. Our goal is to balance simplicity and ease of coding against the accuracy and efficiency needs of most problems. We want to find a handful of procedures that can be relied upon to solve the majority of initial value problems, and to be familiar with the inner workings of these procedures to know the direction to take when they fail to perform as desired.

The first criterion involves building into each method a prescription for determining the current accuracy of the calculation. This is done by repeating the calculation for a different step size and comparing the two results for consistency. A bonus is that, if the current result is more accurate than needed, the step size can be increased, thereby improving the efficiency. In this chapter, we describe this dynamic variation of the step size, and it is a part of all professionally written *black-box* algorithms.

The more sophisticated improvements to Euler's method all seek to increase the *order* of the computational algorithm. An algorithm is of order p if its accuracy is proportional to Δx^p, where Δx is the current step size. Basically, this is done by retaining more terms in the Taylor series expansion of the function. We describe, in detail, how Euler's method (first order in p) is improved to a second-order procedure. This leads to the second-order *Runge-Kutta* algorithms. We then quote and illustrate the results for the very popular fourth-order Runge-Kutta procedure. The MATLAB toolbox functions for solving initial value problems are based on Runge-Kutta algorithms.

Finally, to provide an alternative to normal stepping methods, we describe a more modern technique, known as Bulirsch-Stoer extrapolation. This method combines the ideas that were so successful in the design of the highly accurate Romberg integration methods with those of approximating a function by extrapolation. The result is a highly sophisticated algorithm that is a competitive alternative to the toolbox routines.

Elementary Stepping Procedures for First-Order Equations

EULER'S METHOD

The conceptual basis for all first-order stepping procedures is Euler's method, which is based on the definition of the derivative:

$$\frac{dy}{dx} \approx \frac{y(x + \Delta x) - y(x)}{\Delta x}$$

In the rest of this chapter, $y(x_i)$ will denote the exact value of the solution at the position x_i, and y_i will be the computed value. Because Δx is finite, the preceding equation is equivalent to the first two terms of a Taylor series expansion:

$$y(x_i + \Delta x) \approx y(x_i) + \frac{dy}{dx}\Delta x$$

And because the first-order differential equation reads $dy/dx = f(x, y)$, we arrive at the computational algorithm

$$y_{i+1} = y_i + f_i \Delta x \tag{19.1}$$

where f_i is the function $f(x, y)$ evaluated at the point (x_i, y_i). Euler's method is illustrated in Figure 19.1 (page 410).

Euler's method simply replaces the function $y(x)$ at the point x by a straight line with the same slope as the tangent line at that point. If this approximation is not particularly good [e.g., if Δx is not small enough, or if $y(x)$ is rapidly changing], the value calculated for y_{i+1} will not be very accurate. The subsequent point, y_{i+2}, which depends on the values calculated for y_{i+1}, will be even worse, and the error will accumulate for subsequent points. Despite the obvious inadequacies of Euler's method, its compelling simplicity makes it useful as a starting point for the development of more sophisticated procedures for solving differential equations. Even though Euler's method is never used in a serious solution of a problem, it is the basis for understanding almost all methods, and thus, we begin by investigating its implementation.

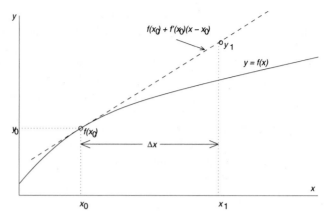

Figure 19.1   *Euler's method for computing the step from y_i to y_{i+1}.*

Assuming that $f(x, y)$ has been coded as an M-file function, the MATLAB code to implement N steps of Euler's method to compute y_i at $x_i = x_0 + i\Delta x$, with $\Delta x = (b - a)/N$, and a and b the limits of the independent variable x, is as follows:

```
function Y = Euler(a,b,y0,N,fnc)
%Euler solves the first order differential equation
%   y' = fnc(x,y) by a simple stepping procedure using
%   the first two terms of a Taylor series expansion of
%   the function y(x).  The function named in the string
%   variable fnc must expect SCALAR input for both x and
%   y, and returns a SCALAR.  The program is not suitable
%   for realistic solutions of a differential equation.
%   The input values are the range of x's [a<=x<=b], the
%   initial value of y [y0], and the number of steps [N].
%   A row vector of computed values is returned in Y.
%   The use is Euler(a,b,y0,N,fnc)
%=========================================================
   dx = (b-a)/N;
   x = a; Y(1) = y0;
   for i = 1:N
       x = x + dx;
       Y(i+1) = Y(i) + dx*feval(fnc,x,Y(i));
   end
   Y = Y(:);
```

The error in one step of Euler's method is proportional to the first term dropped in the Taylor expansion, so the error in each step is proportional to Δx^2. Thus, the accumulated error is approximately $N(\Delta x)^2 = (b - a)\Delta x$, meaning that Euler's method is a *first-order* method. This can be verified by using Euler's method to solve a simple test equation, first with N steps, and then with, for example, $10N$ steps. If the two sets of results are then compared with the known solution to the problem, the ratio of the actual errors in the two sets should be very nearly 10 at each computed point. (See Problem 19.1.) In Table 19.1 we have collected some of the results of solving the equation $y' = y$, $y(0) = 1$, from $x = 0$ to $x = 1$ in N steps by using Euler's method. The exact solution is $y(x) = e^x$.

N	y_N	% error
10	2.59...	4.6
10^2	2.705...	0.50
10^3	2.7169...	0.049
10^4	2.71814...	0.0104

Table 19.1 *Comparison of the solution of $y' = y$ in N steps with the exact result at $x = 1$; $y_{\text{exact}} = e = 2.718243\ldots$*

One of the most important characteristics of the methods that will be developed will be the *order of accuracy*, p, which tells us that the accumulated error is proportional to $\mathcal{O}(\Delta x)^p$. The ultimate goal is to obtain a method that has the largest value of p and requires the minimum amount of computation.

Mathematically, in the limit $\Delta x \to 0$, Euler's method, or any method based on Euler's method, will yield the exact result. In practice, round-off error will limit the attainable accuracy in any procedure. That is, in any stepping procedure, there is a point at which further reduction of the step size will result in an *increase* in the accumulated error. Notice in Table 19.1 that the increase in accuracy follows the increase in the number of steps very nicely through $N = 10^3$. However, the $N = 10^4$ calculation is only about five times more accurate than the $N = 10^3$ value, not ten times as expected. The minimum Δx has not yet been reached, but clearly round-off error is beginning to have a measurable effect.

You will notice that Euler's method requires on the order of 10^4 steps to attain reasonable accuracy, even for this very simple test equation. Further, solving an equation via a stepping procedure in MATLAB requires using `for` loops, or their equivalent. Although there are exceptions, most solution procedures cannot be "vectorized," and, as a result, can take an undesirably long time to execute unless they are "tuned" to fit the equation. It is, therefore, imperative to find improvements to Euler's method that can both increase the accuracy and the speed of computation.

In addition, in a real problem we will not have the exact solution available to monitor the accumulated error. Therefore, all algorithms must incorporate techniques to keep the solution on track and return an estimate of the accuracy of the result. Before describing the modern improvements to Euler's method, we will outline how the step size in any method can be dynamically adjusted to maximize efficiency while maintaining accuracy.

ADJUSTING THE STEP SIZE DURING THE CALCULATION

Regardless of how small the step size, without the exact solution you will have very little knowledge of the accuracy of your calculation. With Euler's method, you know the error is proportional to Δx, but the proportionality constant could conceivably be quite large. To monitor the estimated error is, however, quite easy. Simply redo the calculation with a step size half as large. The difference between the two results should be reflective of the actual error. Taking this idea a step further leads to the following important accuracy monitoring device, which should be part of every solution to a differential equation:

1. Compare the results of *each step* of width Δx with a calculation of two steps of width $\Delta x/2$. Call the absolute difference between the two computed y values Δy. If the basic algorithm is accurate to order p, then the error in an individual step is proportional to $(\Delta x)^{p+1}$. We then have

$$\Delta y = \left| y(x_i + \Delta x) - y\left(x_i + 2\frac{\Delta x}{2}\right) \right|$$

$$\approx k(\Delta x)^{p+1}\left(1 - \frac{1}{2^p}\right)$$

$$\approx k'(\Delta x)^{p+1}$$

2. If the accuracy you wish to maintain in each step is ϵ; i.e., $|\Delta y/y| < \epsilon$, there is a corresponding ideal step size, δx. For this ideal step size, $|\Delta y/y| = \epsilon = k(\delta x)^{p+1}$, whereas for the actual step size, $|\Delta y/y| = k(\Delta x)^{p+1}$. Thus, the ratio of this ideal step size to the actual step size is then

$$R = \frac{\delta x}{\Delta x} = \left(\frac{\epsilon y}{\Delta y}\right)^{1/(p+1)} \qquad (19.2)$$

If $R > 1$, the actual step size is too small, and the next step should be increased. If $R < 1$, the current step is not sufficiently accurate, and the step must be repeated with a smaller step size. To be somewhat conservative, it is common to reduce R by a factor slightly less than one before the replacement is made. Finally, if there is a possibility that the computed value of y may be approximately zero, you should replace $|y| \to |y| + |\Delta y|$. If you find that your calculation is repeatedly alternating between too small and too large a step size, you should reduce R by a slightly greater factor.

3. The step size adjustment algorithm is then:

- If $R = [\epsilon(1 + |y/\Delta y|)]^{1/(p+1)} > 1$, let $\Delta x \Leftarrow 0.9R\Delta x$ for the next step. Thus the *next* step will likely be larger, allowing the calculation to proceed quickly through a region where the function is quite smooth and has been accurately determined.

- If $R \leq 1$, the *current* step is not sufficiently accurate. Again, replace Δx by $0.9R\Delta x$ and *repeat* the step.

Improvements to Euler's Method

The simplest and most direct improvements to Euler's method are based on the following observation:

The next value of y (i.e., y_{i+1}) is obtained in Euler's method by approximating the function by a straight line with the same slope as the tangent to the curve at the *left* end of the interval. However, much better results are usually obtained if, instead, the slope of the tangent drawn at the *middle* of the interval is used for the approximating straight line.

This idea, which is illustrated in Figure 19.2, forms the basis for two modifications to Euler's method: the *midpoint method* and *Heun's method*. Both of these new methods use an algorithm of the form

$$y_{i+1} = y_i + f_{i+(1/2)}\Delta x \tag{19.3}$$

where $f_{i+(1/2)}$ represents the slope of the function $y(x)$ at the midpoint of the interval x_i to x_{i+1}. The two methods differ in their way of estimating this slope.

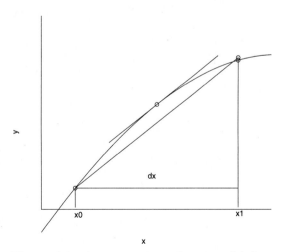

Figure 19.2 *Approximating a function $y(x)$ by a straight line using the slope at the midpoint of the interval.*

THE MIDPOINT METHOD

To simplify the notation, let the fundamental step size be $2\Delta x$, taking us from (x_i, y_i) to (x_{i+2}, y_{i+2}), with the midpoint at (x_{i+1}, y_{i+1}). Equation 19.3 can be implemented by first estimating y at the midpoint using Euler's method,

$$y_1 \approx y_0 + f(x_0, y_0)\Delta x$$

The slope at the midpoint of the interval $x_0 \to x_2$ is approximately $f(x_1, y_1)$. The procedure works across the entire interval $x_0 = a \leq x \leq b$, using the relations

$$x_{i+2} = x_i + 2\Delta x$$
$$y_{i+2} = y_i + 2f(x_{i+1}, y_{i+1})\Delta x$$

It can be shown that the accumulated error (not including round-off error) in this method is proportional to $(\Delta x)^2$; that is, the midpoint method is a second-order method. The procedure goes by a variety of names; we call it here the *improved polygon method* or the *second-order Runge-Kutta method*.

The method is based on the Euler-like equation

$$y_{i+2} \approx y_i + 2f(x_{i+1}, y_{i+1})\Delta x$$

Equations of this type are called *predictor* equations. They use currently known information—in this case, the value of y at earlier points (x_i and x_{i+1}) and the slope at x_{i+1}—to predict the next value of y.[1]

We could also estimate the next value of y by using the slope at the *right* end of the interval as

$$y_{i+1} \approx y_i + f(x_{i+1}, y_{i+1})\Delta x$$

In this equation, however, to compute a value for y_{i+1}, we must already have a value for y_{i+1}. This situation suggests a solution by iteration; that is, by guessing y_{i+1}, inserting it into the preceding equation, and iterating the equation several times. This type of equation is called a *corrector* equation.[2] This idea suggests an alternate way of estimating the slope at the midpoint; namely, we compute f at both the left and the right ends of the interval and use the average of the two. This is the basis of Heun's method.

HEUN'S METHOD

Heun's method proceeds by first estimating the change in y using the slope at the left end of the interval. Calling this Δy_L, we have $y_{i+1} \approx y_i + \Delta y_L$, which can be used to estimate the slope on the right and the change in y using this slope; that is, $\Delta y_R \approx f(x_{i+1}, y_i + \Delta y_L)$. Consequently,

$$y_{i+1} = y_i + \frac{1}{2}(\Delta y_L + \Delta y_R)$$

This procedure is illustrated in Figure 19.3.

Notice that this method is equivalent to:

- One application of a *predictor* equation

$$y_{i+1} = y_i + f(x_i, y_i)\Delta x$$

---

[1] Predictor-like equations are also commonly referred to as *open* equations.
[2] Corrector-like equations are also known as *closed* equations.

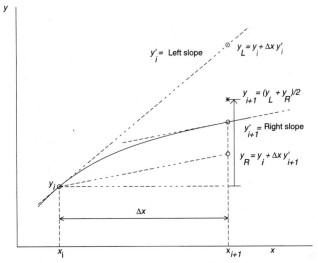

Figure 19.3 *Heun's method approximates the slope at the middle of the interval by averaging the slope on the left and right ends.*

• One application of a *corrector* equation

$$y_{i+1} = y_i + \frac{1}{2}[f(x_i, y_i) + f(x_{i+1}, y_{i+1})]\Delta x$$

Heun's method is also a second-order method.

THE MODIFIED MIDPOINT METHOD

Both the midpoint method and Heun's method can be used as the basis for solving a set of first-order differential equations. However, as we will see, the next level of improvements will be based on these two second-order procedures, and they will result in a dramatic increase in accuracy and efficiency.

The *modified midpoint method* is precisely the same as the ordinary midpoint method, except for the last step. The code is:

```
dx    = (b-a)/N;   x0 = a;
x(1) = x0 + dx;   x(2) = x(1) + dx;
y(1) = y0 + dx*f(x0,y0);
y(2) = y0 + 2*dx*f(x1,y1);
for i = 3:N
      x(i) = x(i-2) + 2*dx
      y(i) = y(i-2) + 2*dx*f(x(i-1),y(i-1));
end
```

The last value in the calculation, y_N, is then "refined" by using a corrector-type equation. The particular form of this corrector equation is a combination of all three methods, Euler, midpoint, and Heun's:

$$y_N \Rightarrow \frac{1}{2}\left[y_N^{\text{midpt}} - y_N^{\text{Euler}}\right] + y_N^{\text{Heun}}$$

$$= \frac{1}{2}\left[(y_{N-2} + 2\Delta x f_{N-1}) - (y_{N-1} + \Delta x f_{N-1})\right]$$

$$+ \left[y_{N-1} + \frac{1}{2}\Delta x \left(f_{N-1} + f_N\right)\right]$$

$$= \frac{1}{2}\left[y_N^{\text{midpt}} + y_{N-1} + \Delta x f_N\right]$$

With this particular combination of terms, and the corrector equation applied to the last step, it can be shown [see Wong (1992) p. 518] that the overall error in the method is once again second order, *but* all odd powers of Δx have canceled out. That is,

$$y(x_N) - y_N = c_2(\Delta x)^2 + c_4(\Delta x)^4 + c_6(\Delta x)^6 + \cdots$$

This means that if we calculate the value of y at the end of the interval, $y(x = b)$, first using a step size of Δx, and then calculate it with steps half as large, $\frac{1}{2}\Delta x$, we have

$$y(b) - y_N = c_2(\Delta x)^2 + c_4(\Delta x)^4 + \cdots$$

$$y(b) - y_{2N} = c_2\left(\frac{\Delta x}{2}\right)^2 + c_4\left(\frac{\Delta x}{2}\right)^4 + \cdots$$

By eliminating c_2, we have

$$y(b) = y_{2N} + \frac{y_{2N} - y_N}{3} - \frac{3}{4}c_4(\Delta x)^4 - \frac{15}{16}c_6(\Delta x)^6 + \cdots$$

This estimate for $y(b)$ is of *fourth* order. Thus, we are able to combine results from interval halving to increase the accuracy order in steps of two. This is entirely analogous to the procedure used to evaluate integrals by the Romberg method; indeed the equations are identical. By next considering steps of size $\Delta x/4$, you can show that

$$y(b) = y_{4N} + \frac{y_{4N} - y_{2N}}{15} + \mathcal{O}(\Delta x^6)$$

the same as the next level of Romberg integration. (See Chapter 7.)

To implement the ideas contained in this very powerful method, we could proceed to consider each step Δx as the end of an interval and then successively refine the computed value by constructing a Romberg-like table for steps $\Delta x/2, \Delta x/4, \ldots$. In practice, this procedure is found to be rather cumbersome and is rarely used except near very troublesome points in a solution. However, the modified midpoint method does form the basis for the very important and popular Bulirsch-Stoer extrapolation method, which we will describe later in this chapter.

We close this section by listing a MATLAB code for implementing the modified midpoint method. (See Figure 19.4.) We will make use of this code in the Bulirsch-Stoer method.

```
function y = Mdpt(dydx,N,xf,x0,y0)
% Mdpt integrates a set of first order differential equations
%    of the form, y' = dydx(x,y), with initial conditions,
%    y(x0)=y0.  The equations are integrated in N steps from
%    x0 to xf. Each of the quantities dx, x0, xf, y0, y', and
%    Y are COLUMN vectors.  dydx is a string containing the
%    name of the derivative function, which expects two column
%    vectors (x,y) as input, and returns a single column vec-
%    tor as output.
%
%    This code is intended to be part of a larger code and no
%    adjustment of the step size is included here.  NOTE: Only
%    the value of the solution at the END of the interval is
%    returned.
%-------------------------------
   dx = (xf-x0)/N;
   x = x0;
   Y = [y0 y0+dx.*feval(dydx,x0+dx,y0)];
   for i = 2:N
      x = x + dx;
      f =  feval(dydx,x,Y(:,2));
      Y = [Y(:,2) Y(:,1)+2*dx*f];
   end
   y = .5*(Y(:,2)+Y(:,1)+dx*f);
```

Figure 19.4 *MATLAB code for the midpoint method.*

The Method of Runge-Kutta

SECOND-ORDER RUNGE-KUTTA ALGORITHMS

The elementary refinements of the basic Euler method were based on the idea that using the average slope of the tangent over an interval to extrapolate a function to the next point should result in greater accuracy. The method of Runge-Kutta carries this a bit further. The slope that is used in the linear extrapolation is taken to be a *weighted* average of the slope at the left end of the interval plus some intermediate point. Thus the algorithm will read

$$y_{i+1} = y_i + f_{\text{avg}}\Delta x$$

where f_{avg} is determined by the equation

$$f_{\text{avg}} = af_i + bf_{i'}$$

where a, b are the "weights," f_i is the function evaluated at the point (x_i, y_i), and $f_{i'}$ is the function evaluated at some intermediate point defined as

$$(x_{i'}, y_{i'}) = (x_i + \alpha\Delta x, y_i + \beta f_i\Delta x)$$
$$f_{i'} = f(x_i + \alpha\Delta x, y_i + \beta f_i\Delta x)$$

where the two parameters α, β specify the position of the intermediate point.

Using this form of the algorithm, the four parameters are next chosen to optimize the accuracy of the computed result. The parameters a, b, α, and β are not totally free parameters. By expanding the function $f(x, y)$ in a two-dimensional Taylor series, it is possible to obtain the following constraint equations on the parameters. (See also Problem 19.6.)

$$a + b = 1$$
$$\alpha b = \beta b = \tfrac{1}{2}$$

These are three equations in the four unknown parameters, and thus there is still some freedom in how the parameters are chosen. The particular choice of parameters

$$a = 0 \qquad b = 1 \qquad \alpha = \beta = \tfrac{1}{2}$$

will result in an algorithm that is identical to the midpoint method discussed earlier. And a different choice, namely,

$$a = b = \tfrac{1}{2} \qquad \alpha = \beta = 1$$

yields Heun's method.

Both of these computational algorithms are known as second-order Runge-Kutta procedures, meaning that in both methods the accumulated error is proportional to $(\Delta x)^2$.

THE FOURTH-ORDER RUNGE-KUTTA ALGORITHM

By including more sampling points in the interval, the basic Runge-Kutta method can be improved to a procedure that has an accumulated truncation error proportional to $(\Delta x)^4$—that is, a fourth-order method. The determination of the many parameters is once again partially given by comparing with a two-variable Taylor series expansion and involves considerable algebra. The results are simply quoted here:

$$y_{i+1} = y_i + \tfrac{1}{6}[\Delta y_0 + 2\Delta y_1 + 2\Delta y_2 + \Delta y_3] \tag{19.4}$$

where

$$\begin{aligned}
\Delta y_0 &= f(x_i, y_i)\Delta x \\
\Delta y_1 &= f\left(x_i + \tfrac{1}{2}\Delta x, y_i + \tfrac{1}{2}\Delta y_0\right)\Delta x \\
\Delta y_2 &= f\left(x_i + \tfrac{1}{2}\Delta x, y_i + \tfrac{1}{2}\Delta y_1\right)\Delta x \\
\Delta y_3 &= f\left(x_{i+1}, y_i + \Delta y_2\right)\Delta x
\end{aligned} \tag{19.5}$$

You can easily show that for the special case where the function $f(x, y)$ is a function of x *only*, this procedure amounts to an integration of $\int f(x)\, dx$ by the Simpson's rule approximation; that is,

$$y_{i+1} = y_i + \frac{1}{3}\left(\frac{\Delta x}{2}\right)\left[f(x_i) + 4f\left(x_i + \frac{1}{2}\Delta x\right) + f(x_{i+1})\right]$$

The fourth-order Runge-Kutta algorithm is by far the most popular method for obtaining numerical solutions to differential equations. It is very easy to code and, except for especially perverse differential equations, is very stable and accurate. The method is *self-starting*; i.e., it depends on information to be found in the current step alone, and the step size can easily be adjusted during the calculation to accommodate a function that is rapidly varying. It is a relatively easy matter to write MATLAB code based on Equations 19.4 and 19.5 to implement the fourth-order Runge-Kutta method. However, you will not be surprised to learn that we have in the toolbox MATLAB functions that perform Runge-Kutta integration. These functions, which are described in the next section, are the suggested means for solving initial value differential equations.

Finally, when comparing two computational algorithms, keep in mind that the two prime considerations are efficiency and accuracy. The most common measure of efficiency is simply the number of "function calls" required by the two methods. Thus, in the fourth-order Runge-Kutta method, *four* function evaluations are required in each step, whereas only *two* per step are required in the modified midpoint method of the previous section. That is, the fourth-order Runge-Kutta method is preferred only if the same or better accuracy can be obtained using a step size twice that used in the modified midpoint method.

Interestingly, Runge-Kutta algorithms for orders p higher than 4 require either $p + 1$ or $p + 2$ function evaluations per step. As a result, the gain is less dramatic than that seen in going from 2 to 4. For this reason, most programmers rarely venture beyond fourth-order.

MATLAB Implementation of Runge-Kutta Algorithms

The two toolbox routines for solving one or a set of first-order differential equations are called ode23 and ode45. The first function uses Runge-Kutta equations of both second and third order to produce an algorithm that is third-order accurate. It also incorporates our suggested procedures for adjusting the step size during the calculation to improve efficiency and to monitor the accuracy. The second function operates in an identical fashion but is based on Runge-Kutta equations of fourth and fifth order, producing an algorithm that overall is of fifth order.

EXAMPLE 19.1

Earth-Moon Satellite Trajectories

The gravitational force on an object of mass m in orbit about the earth is

$$\mathbf{F} = -G\frac{mM_E}{r^3}\vec{r}$$

where $G = 6.672 \times 10^{-11}$ N $\cdot$ m^2/kg^2 is the gravitational constant, and $M_E = 5.97 \times 10^{24}$kg is the mass of the earth, and $\vec{r}$ is the vector distance from the center of the earth to the satellite. Using Newton's second law, $\mathbf{F} = m\mathbf{a}$, this becomes a second-order differential equation for the position of the satellite,

$$\mathbf{a} = \frac{d^2\vec{r}}{dt^2} = -G\frac{M_E}{r^3}\vec{r}$$

or in terms of x-y components,

$$a_x = -GM_E \frac{x}{r^3}$$

$$a_y = -GM_E \frac{y}{r^3}$$

$$r = \sqrt{x^2 + y^2}$$

See Figure 19.5.

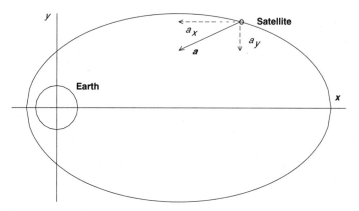

Figure 19.5 *The components of the acceleration of a satellite in orbit about the earth.*

To solve for the orbit of the satellite, the first task is to establish reasonable *initial* conditions for an orbit. To this end, consider the problem of placing a satellite in a *geosynchronous* orbit; i.e., an orbit that has a period of precisely 1 day so that it will remain in phase with the earth's rotation and will thus appear stationary over the equator. Communication satellites are placed in such orbits. The period, T, = 1 day = 86,400 s, so the velocity of the satellite is $v = 2\pi R/T$, where R is the to-be-determined orbit radius. This can be obtained by equating the centrifugal acceleration, v^2/R, to the gravitational acceleration,

$$\frac{v^2}{R} = \frac{4\pi^2 R}{T^2} = GM_E \frac{1}{R^2}$$

which yields

$$R = \left(\frac{GM_E}{4\pi^2} \right)^{1/3} T^{2/3} = 4.223 \times 10^7 \text{m} \qquad (= 26{,}240 \text{ miles or } 6.6R_E)$$

and the corresponding velocity is $v_0 = 3071\text{m/s}$. Thus, to place the satellite in a clockwise geosynchronous orbit, we use $(x, y)|_{t=0} = (-R, 0)$ and $(v_x, v_y)|_{t=0} = (0, v_0)$.

The next task is to rewrite the problem in terms of coupled *first*-order equations. This is easily accomplished by the definitions:

$$
\begin{array}{llll}
y_1(t) \equiv x(t) & y_1(0) = x_0 = -R & x' = v_x & y_1' = y_3 \\
y_2(t) \equiv y(t) & y_2(0) = y_0 = 0 & y' = v_y & y_2' = y_4 \\
y_3(t) \equiv v_x(t) & y_3(0) = v_{x0} = 0 & v_x' = a_x & y_3' = -GM_E y_1 / r^3 \\
y_4(t) \equiv v_y(t) & y_4(0) = v_{y0} = v_0 & v_y' = a_y & y_4' = -GM_E y_2 / r^3
\end{array}
$$

Finally, for this problem there is an additional check on the accuracy of the calculation. The total energy of the satellite should remain constant. The energy divided by the mass of the satellite is

$$
E = \frac{1}{2} v^2 - GM_E \frac{1}{r}
$$

The MATLAB code for the M-file function that returns the derivatives of the column vector y is given in Figure 19.6.

```
function YP = dydt(t,y)
%dydt returns the derivatives of the position and velocity
%    functions.  The total mechanical energy of the satellite,
%    E, is monitored as a check on the accuracy.  The original
%    energy E0 and the maximum fractional change, dE, are
%    global variables.  dE should be assigned a value of zero
%    before calling dydt.
%
%    The 4-element column vector y contains the initial values
%    of the orbit.  The (x,y) coordinates of the satellite cor-
%    respond to the first two elements of the vector y and the
%    output vector of derivatives YP.  The last two elements
%    in both y and YP refer to the velocity.
%===============================================================
global E0 dE
G = 6.67e-11;              % Gravitational Constant
ME= 5.97e24;              % Mass of the earth
y = y(:);
X = y(1:2); V = y(3:4);   r = sqrt(sum(X.^2));
DE = abs((0.5*sum(V.^2) - G*ME/r  - E0)/E0);
if DE>dE; dE = DE; end;
YP = [V; -G*ME*X/r^3];
```

Figure 19.6 *MATLAB code for satellite orbits about the earth.*

The MATLAB code to compute three orbits of the satellite and plot the result is then (see also the script file Exp19p1):

```
global E0 dE
dE = 0;
R = 4.223e7;      v0 = 3071;      T  = 60*60*24*3;
y0= [-R;0;0;v0];  t0 = 0;
```

```
tol = 1.e-4;      trace = 0;
[t,y] = ode23('dydt',t0,T,y0,tol,trace);
X = y(:,1);    Y = y(:,2);
plot(X,Y)
axis('image')
```

In the call to the toolbox function ode23, the arguments have the following meanings:

'dydt' A string containing the name of the M-file function that returns the derivatives as a column vector.

t0   The starting value of the scalar independent variable.

T   The final value of the independent variable.

y0   A column vector containing the initial value for the dependent variables.

tol   (*Optional*) The required accuracy of each step. The accuracy is based on the fractional size of y. If omitted, a value of 10^{-3} is used.

trace   (*Optional*) If other than zero, intermediate results will be printed.

The use of the more accurate routine, ode45, is precisely the same. You may wish to experiment with the preceding code using both ode23 and ode45 by varying the accuracy to determine execution time on your computer. You will notice that even though ode23 used a less accurate algorithm, the execution time will not be dramatically shorter. The reason for this is that ode45, although more complicated, will be able to take larger steps. The actual measure of efficiency is then a combination of both accuracy order and the total number of function calls. The MATLAB script file named Exp19p1.m on the applications disk duplicates the calculation of the satellite orbit.

We close this example with a similar, but more complicated, calculation. Suppose you wish to place a satellite in a *periodic* orbit about both the earth and the moon, and you wish to do this with the minimum expenditure of energy. The first job is to determine the differential equations that determine the orbit of the satellite. This can be done by using Figure 19.7.

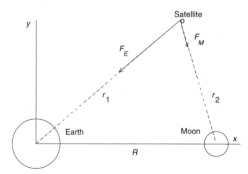

Figure 19.7   *The forces on a satellite in orbit about both the earth and the moon.*

Following the analysis for the earth orbit we easily obtain

$$a_x = -GM_E \frac{x}{r_1^3} - GM_M \frac{x - R}{r_2^3}$$

$$a_y = -GM_E \frac{y}{r_1^3} - GM_M \frac{y}{r_2^3}$$

$$r_1 = \sqrt{x^2 + y^2}$$

$$r_2 = \sqrt{(x - R)^2 + y^2}$$

Next, we establish the initial conditions. The minimum energy periodic orbit will correspond to a *figure-eight*-like orbit.[3] To get an idea of the energy of such an orbit, you could find the position, x, along the earth-moon axis where the gravitational forces cancel and compute the total potential energy at that point; i.e., $-G[M_E/x + M_M/(R - x)]$. Assuming that the kinetic energy is negligible at this point, we can use this value to estimate the kinetic energy ($\frac{1}{2}mv^2$) at the starting position. Using a starting point of $(x_0, y_0) = (-3 \times 10^7 \text{ m}, 0)$ this yields a starting velocity of $v_0 \approx 4800$ m/s.

The function `dydt` is then altered to include the force on the satellite from the moon as well, and the trajectory is computed and plotted. It will take some experimentation to obtain a stable orbit. You will have to increase the initial velocity somewhat to get the satellite to swing around the moon, and of course the period, T, will have to be increased to accommodate this much longer space trip. (Start with about 7–10 days expressed in seconds.) You will also find that the orbit is quite unstable; a small change in initial conditions will cause the satellite to fly off into space.

The result of such a calculation is shown in Figure 19.8.

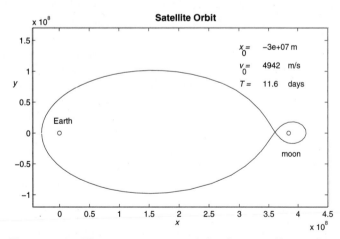

Figure 19.8   *The minimum energy orbit for a satellite circling both the earth and the moon.*

---

[3]Actually, the moon is a moving target. To see how to handle this complication, see Problem 19.9.

(§) The Bulirsch-Stoer Extrapolation Method

No one procedure is suitable for all problems. The Runge-Kutta methods on which the toolbox functions are based are powerful, especially when dynamical adjustment of the step size is included, so that 99% of the initial value ordinary differential equation problems that you face can be adequately integrated using these functions. There are, however, two competing classes of algorithms that you will encounter and with which you should be familiar. The first class includes a variety of *predictor-corrector* methods similar to, but rather more complicated than, the method mentioned on page 414. Predictor-corrector methods are difficult to code and can usually be bettered in efficiency and/or accuracy by either the Runge-Kutta methods or the Bulirsch-Stoer extrapolation methods that we will now discuss. As a result, we will not develop any further MATLAB codes based on predictor-corrector ideas. However, you may wish to at least review the concepts on page 414 so that you will understand what various researchers are talking about when they discuss their predictor-corrector codes.

The Bulirsch-Stoer extrapolation method has tremendous potential for beating all other methods when it comes to integrating a differential equation, particularly when the solution is quite "smooth." However, as we shall see, realizing this potential can be rather difficult, particularly when the code to beat is the toolbox function `ode45`.

The basic ideas behind the method are very simple:

1. To integrate the differential equation $\vec{y}' = \vec{f}(x, \vec{y})$, $\vec{y}(x_0) = \vec{y}_0$, over the interval $x_0 \leq x \leq x_f$, we begin by subdividing the interval into *large* steps Δx. This large step size could be $(x_f - x_0)/10$ or $(x_f - x_0)/50$.

2. The equation is then integrated over a single large step several times, each time using more and more subdivisions of the large step. Thus, we could first divide the large step into two subintervals, $\delta x = \frac{1}{2}\Delta x$, then into four, $\delta x = \frac{1}{4}\Delta x$, etc. We then have a set of values, $y_{\delta x}(x + \Delta x)$, representing the actual value of the integration over a single large step.

3. These values are then *extrapolated* down to $\delta x = 0$ for a more precise value of the function at the end of the large step. Using function `BulStr0` of Figure 16.6 to carry out the extrapolation, we are also provided with an estimate of the error in the computed value of $y(x + \Delta x)$. If the current accuracy is not sufficient, the function is integrated once more with a smaller substep size, δx, and the extrapolation repeated.

The ideas behind a single (large) step of the Bulirsch-Stoer method are illustrated in Figure 19.9.

To carry out the integration over one large step we will use the modified midpoint method on page 415 as implemented in the function `Mdpt` of Figure 19.4. Now for the details:

- The most obvious choice for the sequence of subinterval size is to simply repeatedly halve the intervals as was done in the similar problem of Romberg integration. However, this procedure causes the subinterval size

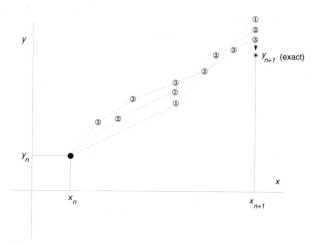

Figure 19.9 *The values of y at the end of the large step are estimated by integrating the function over finer and finer substeps and extrapolating to $\delta x \to 0$.*

δx to shrink too rapidly, resulting in excessive integration steps to obtain a small set of values to then extrapolate. A compromise is to use the sequence $\delta x = \Delta x \cdot [\frac{1}{2}, \frac{1}{4}, \frac{1}{6}, \frac{1}{8}, \ldots, \frac{1}{N}]$, where $N = [2\ 4\ 6\ 8\ 12\ 16\ 24\ 32\ 48\ 64\ 96]$, for $i > 1$, $N_{i+1} = 2N_{i-1}$.

- The integration is carried out using the modified midpoint method. After obtaining two values (from $N = 2$ and 4), the value at the end of the large step is obtained by extrapolation. If the error is still too large, the next value of N in the sequence is used and the extrapolation is repeated.

- To take the *largest* step possible with the least amount of intermediate integration, it is found[4] that six or seven intermediate integrations are optimum. Thus, if the current integration was successful after, say four intermediate integrations (i.e., an extrapolation using values obtained with $N = N_1$, N_2, N_3, and N_4), the current large step was likely too small and the next large step is increased. The converse applies if the number of intermediate integrations is greater than seven.

- In the rare event that the number of intermediate integrations goes all the way to 11 without success, we conclude that our choice for the large step, Δx, is at fault and we reduce Δx substantially (say by a factor of 8) and try again. If this happens repeatedly, the solution is likely approaching a singularity, and special techniques must be used to get past it. (I'd suggest that you use `ode45` to integrate over the limited range surrounding the offending point.)

---

[4]See Gear, *Numerical Initial Value Problems in Ordinary Differential Equations*, Prentice-Hall, Englewood Cliffs, NJ, 1971.

A MATLAB function that implements these ideas is the function `odeBS` shown in Figure 19.10. This function operates in exactly the same manner as ode23 and ode45.

```
function [t,y] = odeBS(dydt,t0,tf,y0,tol)
% odeBS operates similar to ode23 and ode45 using instead a
% Bulirsch-Stoer extrapolation technique.  The first large
% step is H = (tf-t0)/50.  The subsequent step size can be
% somewhat adjusted by changing Nfit (smaller Nfit yields
% smaller big steps).  This function calls Mdpt to carry out
% the intermediate integrations.  When Inside = 0, the
% integration has passed tf.
%---------------------------
 Inside = 1;
 Nfit = 6;
 Ns= [2 4 6 8 12 16 24 32 48 64 96];
 H = (tf-t0)/50;
 y = y0';  Y0 = y0; t = t0; T0 = t0;
 while Inside == 1
    if T0+H>tf; H = tf-T0;  end;
    X = H^2/4;
    Y =  feval('Mdpt',dydt,2,T0+H,T0,Y0);
    for i = 1:10
       N = Ns(i+1);
       F = feval('Mdpt',dydt,N,T0+H,T0,Y0);
       Y = [Y F];    X = [X (H/N)^2]; '
       [Yf,dY] = feval('BulStr0',X,Y',i);
       Err = max(abs(dY)./(abs(Yf)+abs(dY)));
       if Err/tol < 1
    T0 = T0+H;      Y0 = Yf';
          t = [t;T0];   y = [y;Yf];
         if T0>=tf; Inside=0; end;
         if  i ~= Nfit
             H = H*(Ns(Nfit)/Ns(i));
         end
            break
       end
 if i == 10; H = H/8;  end
    end
 end
%-----------------------------------------------------------
```

Figure 19.10 *MATLAB code for Bulirsch-Stoer extrapolation method.*

To compare this routine with the toolbox differential equation integrators, we have used it to again solve the figure-eight earth-moon satellite orbit. The results are shown in Figure 19.11.

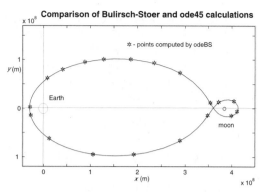

Figure 19.11 *The minimum energy satellite orbit about the earth and the moon is computed by both* ode45 *(solid curve) and by* odeBS *(circles).*

COMPARING THE EFFICIENCY OF THE BULIRSCH-STOER METHOD WITH ode45

In Figure 19.11 you will notice that odeBS has indeed succeeded in computing the orbit by taking rather large steps along the trajectory. Both calculations are of comparable accuracy. However, to compare the efficiency of the two codes, we could compare either the total number of function calls or the total execution time. And here we find a disappointment. For this particular calculation the ode45, Runge-Kutta fifth-order calculation is about twice as efficient as the Bulirsch-Stoer method using the function of Figure 19.10.[5]

Of course, with a method like Bulirsch-Stoer it is impossible to define an accuracy order because the algorithm combines extrapolation with a second-order midpoint method. However, the method seemed to have such great promise that you might have guessed that it would be more efficient than a simple marching method like Runge-Kutta. In fact, in most situations it is indeed more efficient. The problem here is that in the code of Figure 19.10 little thought was given to the overall efficiency of the calculation. Thus, every time the code takes a finer subdivision of the large step, it starts at the same place and needlessly recalculates the derivatives at the starting point. A simple improvement is then to calculate the derivatives at the starting point before any midpoint integration is attempted and to pass this value to each call to Mdpt. This will improve the efficiency by about 15%.

Next, we have paid a price in not using the same sequence used in Romberg integration; i.e., halving the intervals at each step. If that were the case, the *new* information in each substep would merely be the values of the functions at the midpoints of the current steps. Thus, repeat evaluations of the functions at the same points could easily be avoided. In this instance however, the sequence of big step divisions generates numerous duplication of effort while still not providing an easy remedy. In executing a midpoint integration over the big step for $N =$

---

[5]For differential equations with "smoother" solutions or when higher accuracy is asked for the two algorithms are closer in efficiency with odeBS occasionally beating out ode45. See Problem 19.7.

$2, 4, 6, 8, 12, 16, 24$ subdivisions, the derivative function is called a total of **79** times. However, only **33** of these are at distinct x values. Thus, there is the potential to improve the efficiency of the method by more than a factor of **2**. Also, each time the Bulirsch-Stoer method fails during a particular big step and is forced to redo the step after reducing it, **288** function calls are discarded. The recommendation is then as follows:

a. If your derivative function is very complicated and therefore expensive to call,

b. the solution is quite smooth, without wild oscillations or singular points, and

c. high accuracy is required,

then it is probably worth your while to spend some effort in optimizing the Bulirsch-Stoer code. Otherwise, the function `ode45` is hard to beat and will usually give satisfactory results.

A MATLAB script file that compares the results obtained using `ode23`, `ode45`, and `odeBS` to calculate the figure-8 orbit can be found on the applications disk; its name is `SatOrb.m`.

Problems

REINFORCEMENT EXERCISES

P19.1 Euler's Method. The differential equation, $y' = xy$, $y(0) = 1$, has the solution $y(x) = e^{x^2/2}$. Use the Euler method of page 409 to solve this equation over the interval $[0, 1]$ for $N = 10, 10^2, 10^3$ steps and compare with the exact result. Is the actual error proportional to Δx?

P19.2 Dynamic Step Size Adjustment. Adapt the code for Euler's method on page 409 to alter the step size according to Equation 19.2. The basic Euler algorithm requires on the order of 10^3 steps to achieve an accuracy of $\approx 10^{-3}$ at $x = 1$. Compare the number of floating-point operations in the amended method to that of the basic Euler method for the same accuracy.

P19.3 The Modified Midpoint Method. Use the modified midpoint method code in Figure 19.4 and the Euler code of page 410 to solve the test equation

$$xy' = (x^2 - y), \qquad y(1) = 0$$

over the interval $[1, 2]$ and compare each with the exact solution at $x = 2$.

P19.4 Initial Value Problems Solve the following nonlinear initial value problems using `ode23`.

	Equation	Initial Conditions	Range
a.	$y'' + 2\sin y = 0$	$y(0) = 1$, $y'(0) = 0$	$0 \le x \le \pi$
b.	$y'' + 2\sqrt{y} = 0$	$y(0) = 2$, $y'(0) = 1$	$0 \le x \le 4$
c.	$y'' + y \sinh x = 0$	$y(0) = 0$, $y'(0) = 2$	$0 \le x \le 4$

P19.5 Coupled Equations. The equations of motion for a projectile subject to air resistance are

$$\dot{v}_x = -\gamma |\vec{v}| v_x$$

$$\dot{v}_y = -\gamma |\vec{v}| v_y - g$$

where γ is a constant that characterizes the air drag, $g = 9.8$ m/s^2 is the acceleration of gravity, and $|\vec{v}| = \sqrt{v_x^2 + v_y^2}$ is the magnitude of the projectile velocity.

a. By defining

$$\vec{Y} = \begin{pmatrix} y_1 \\ y_2 \\ y_3 \\ y_4 \end{pmatrix} \equiv \begin{pmatrix} x \\ y \\ v_x \\ v_y \end{pmatrix}$$

reformulate the equations of motion as four coupled first-order differential equations. Code the four elements of $d\vec{Y}/dt$ as a single M-file function suitable for use by functions ode23 or ode45.

b. Use the initial conditions

$$x(0) = 0 \quad v_x(0) = v_0 \cos\theta \quad \text{with} \quad v_0 = 180 \text{ m/s}$$

$$y(0) = 0 \quad v_y(0) = v_0 \sin\theta \qquad\qquad \theta = 40°$$

to solve for $x(t)$ and $y(t)$. Obtain the trajectories for drag coefficients $\gamma = 0.1$ to 0.3 in steps of 0.05, and plot the results on a single plot.

P19.6 Runge-Kutta Constraints for 2nd Order. Starting from the equations,

$$\frac{dy}{dx} = f(x, y)$$

$$y_{i+1} = y_i + f_{\text{avg}} \Delta x$$

$$f_{\text{avg}} = a f_i + b f_{i'}$$

$$f_{i'} = f(x_i + \alpha \Delta x, y_i + \beta f_i \Delta x)$$

expand $f_{i'}$ about the point (x_i, y_i) through first order and compare the resulting expansion for y_{i+1} with the Taylor series,

$$y_{i+1} = y_i + f_i \Delta x + \frac{1}{2} \frac{df_i}{dx} (\Delta x)^2 + \cdots$$

Use the identity

$$\frac{df}{dx} = \frac{\partial f}{\partial x} + \frac{\partial f}{\partial y} f$$

to obtain the conditions on a, b, α, and β quoted on page 418.

P19.7 Comparison of ode23, ode45, and odeBS. The solution of the differential equation $y' = 1 - y$, $y(0) = 0$ is $y(x) = 1 - e^{-x}$. Solutions of this differential equation from $x = 0$ to $x = 2$, with a tolerance of 10^{-4} using the three MATLAB codes, ode23, ode45, odeBS, and the MATLAB function

```
function yprime = derv(t,y)
 yprime = 1-y
```

are compared in Table 19.2.

Function	Number of flops	Elapsed time (s)
ode23	66	0.070
ode45	102	0.103
odeBS	760	0.264

Table 19.2 *Comparison of MATLAB functions*

To next simulate a more complicated differential equation that has a very smooth solution, insert the following code into the function `derv`,

```
for k = 1:10
    s = bessel(3,k);
end
```

This addition has the effect of slowing down the calculation considerably without changing the results. Compare the three methods of solution for this problem and interpret the results.

EXPLORATION PROBLEMS

P19.8 Stiff Differential Equations.   The differential equation $y' = y - t^2$ has a general solution, $y(t) = Ae^t + t^2 + 2t + 2$. If the initial condition is $y(0) = 2$, the constant A is zero and the solution is a parabola. However, if $y(0) = 2.00001$, for example, or even as the result of round-off error, the coefficient of the exponential in the analytic solution will no longer be zero, and this term will ultimately dominate the numerical solution. Solve this equation for $0 \leq t \leq 10$, using first **ode23**, and then **ode45**. Plot the results.

P19.9 Nonlinear Differential Equations.   The equation for the oscillations of a pendulum is

$$\ddot{\theta} + \omega^2 \sin \theta(t) = 0$$

where $\theta(t)$ is the angular displacement of the pendulum from the vertical, and $\omega^2 = g/\ell$, $g = 9.8 \text{m/s}^2$, and ℓ is the length of the string. Use $\ell = 4.9$ m. We will take the boundary conditions for this problem to be

$$\theta(0) = \theta_0, \qquad \dot{\theta}(0) = 0$$

If θ is small ($\theta < 15° \approx 0.26$ radian), the sine can be approximated as $\sin \theta \approx \theta$, and the differential equation becomes the normal oscillator equation with solutions $\sin \omega t, \cos \omega t$. If θ is *not* small, the equation is *nonlinear* and quite difficult to solve. The exact solution is a complicated implicit function of elliptic integrals and can be found in Greenberg (1988). Instead, solve the nonlinear equation with $\theta_0 = 0.1$ over the range $0 \leq t \leq 2\pi/\omega$ using **ode23** and compare with the exact "linear" solution, $y(x) = \theta_0 \sin \omega t$. The match should be quite close. Next, if this step is successful, try larger values of $\theta_0 \approx 0.2, 0.3, \ldots$. Plot the difference between the numerical solution and the linear solution divided by θ_0 for the three values of θ_0.

P19.10 **Spaceship and a Moving Moon.** I hope you haven't already sent your astronauts off on their trip to the moon along the trajectory computed by using the code in Figure 19.6. There is a problem. Namely, the moon is itself a moving target and is orbiting the earth at an angular speed of $\omega_M = 2\pi/T_M$, where T_M is the period of the moon and is equal to 27.4188 days. This complicates the trajectory enormously, but the actual calculation is pretty much the same. First, consider a coordinate system that rotates with the moon, so that the moon remains at a fixed point. The calculation is then identical to that in function *dy dt*. However, because this coordinate system is rotating it *is not an inertial coordinate system!* To still use $\vec{F} = m\vec{a}$ we will have to add a *fictitious* force $\mathcal{F}$ to account for the rotations. This fictitious force is $\vec{\mathcal{F}} = +\omega_M^2 \vec{r}$. Thus, the expression for the acceleration on the satellite is

$$\vec{a} = -GM_E \frac{\vec{r}}{|\vec{r}|^3} - GM_M \frac{\vec{r} - \vec{d}}{|\vec{r} - \vec{d}|^3} + \omega_M^2 \vec{r}$$

where $\vec{r}$ is the vector from the earth to the satellite and $\vec{d}$ is the vector from the earth to the moon.

The orbit is calculated using this expression for the acceleration, and then the whole system is set into rotation with an angular speed ω_M to get the final trajectory. Perform the calculation using this expression for the acceleration. You will also have to write a new derivative function `DYDT2.m`. Use an initial velocity of $v_0 = 4849$ m/s, and a trip time of $T = 20$ days. The orbit will be very sensitive to these initial conditions. (*Note:* A figure-eight orbit is no longer possible.)

Summary

ELEMENTARY STEPPING ALGORITHMS

Euler's Method

Euler's method replaces the function over a small interval $[x, x + \Delta x]$ by a straight line whose slope is given by the derivative at the left end,

$$y(x_i + \Delta x) \approx y(x_i) + \frac{dy}{dx} \Delta x$$

and this is used repeatedly across the interval to compute the solution,

$$y_{i+1} = y_i + f_i \Delta x \tag{19.6}$$

with $f(x, y) = dy/dx$ and $x_i = x_0 + i\Delta x$ for $i = 1, \ldots, n$.

Adjusting the Step Size

To increase both the accuracy and the efficiency, the step size should be continually adjusted to accommodate the rate at which the solution is changing. This is done by repeating the calculation of each step with two half-steps and computing the quantity

$$R = \left[\epsilon \left(1 + \left| \frac{y}{\Delta y} \right| \right) \right]^{1/(p+1)}$$

where Δy is the difference between the two calculations, ϵ is the desired accuracy, and p is the order of the algorithm (Euler's is first order). The next increment, Δx, is then multiplied by 0.9R; i.e., $\Delta x \rightarrow 0.9R\Delta x$.

The Midpoint Method

The midpoint method uses an approximate expression for the slope at the *midpoint* of a double interval to yield the algorithm

$$x_{i+2} = x_i + 2\Delta x$$

$$y_{i+2} = y_i + 2f(x_{i+1}, y_{i+1})\Delta x$$

Heun's Method

Determining the slope of the line by averaging the slopes at the left and right ends of each interval results in Heun's method. This method requires that the slope at the right end be estimated in each step by first computing an approximate y_{i+1} using

$$y_{i+1} = y_i + f(x_i, y_i)\Delta x$$

then determining the right-end slope, $f(x_{i+1}, y_{i+1})$, from this value, and finally computing y_{i+1} once more using

$$y_{i+1} = y_i + \frac{1}{2}[f(x_i, y_i) + f(x_{i+1}, y_{i+1})]\Delta x$$

The Modified Midpoint Method

The modified midpoint method is a second-order method that plays an important role in many algorithms. The method is precisely the same as the midpoint method except for the last step, which becomes

$$y_N \Rightarrow \frac{1}{2}\left[y_N^{\text{midpt}} - y_N^{\text{Euler}}\right] + y_N^{\text{Heun}}$$

$$= \frac{1}{2}\left[y_N^{\text{midpt}} + y_{N-1} + \Delta x f_N\right]$$

THE METHOD OF RUNGE-KUTTA

Definition

The Runge-Kutta procedure uses the value of the solution and its derivatives at a variety of points within the current interval to estimate the value of the solution at the end of one step.

Fourth-Order Runge-Kutta

The most popular of the Runge-Kutta algorithms is given by the prescription

$$y_{i+1} = y_i + \frac{1}{6}[\Delta y_0 + 2\Delta y_1 + 2\Delta y_2 + \Delta y_3]$$

where

$$\Delta y_0 = f(x_i, y_i)\Delta x$$

$$\Delta y_1 = f(x_i + \frac{1}{2}\Delta x, y_i + \frac{1}{2}\Delta y_0)\Delta x$$

$$\Delta y_2 = f(x_i + \frac{1}{2}\Delta x, y_i + \frac{1}{2}\Delta y_1)\Delta x$$

$$\Delta y_3 = f(x_{i+1}, y_i + \Delta y_2)\Delta x$$

BULIRSCH-STOER EXTRAPOLATION METHOD

Description

This method is included as an alternative to the MATLAB codes **ode23** and **ode45**. It is based on the idea of first using one *large* step to compute one term in a solution, and then using values obtained from successively smaller steps over the same subinterval and *extrapolating* these numbers down to a step size of zero. The MATLAB code included is constructed so that it is interchangeable with **ode23** and **ode45**.

MATLAB Functions Used

ode23, ode45   Ordinary coupled initial value differential equations are solved in MATLAB using the functions **ode23** and **ode45**. Both functions use Runge-Kutta methods with automatic adjustment of the step size. Function **ode23** uses a combination of second- and third-order Runge-Kutta equations, whereas **ode45** is more accurate, employing a combination of fourth- and fifth-order Runge-Kutta equations. The use is

$$[\mathbf{x}, \mathbf{t}] = \mathbf{ode23}(\,'dxdt',\, \mathbf{t0}, \mathbf{tf}, \mathbf{x0}, \mathbf{tol}, \mathbf{trace})$$

where *'dxdt'* is a string constant containing the name of the M-file function that computes the derivatives of the n dependent variables, x_i, with respect to the *scalar* independent variable t. That is, the input to *'dxdt'* must be a vector $\mathbf{x}$ of length equal to the number of dependent variables or coupled equations, and a single scalar for t. The integration is over the range $t_0 \leq t \leq t_f$, the accuracy is **tol**, and if **trace** is nonzero, intermediate results of the calculation will be displayed. The default values for the optional **tol** are 10^{-3} for **ode23** and 10^{-6} for **ode45**. The default for optional **trace** is zero. The call to **ode45** is identical.

20 / Two-Point Boundary Value Problems

PREVIEW   The solution of an nth-order ordinary differential equation is an initial value problem if *all* of the n independent extra conditions needed to specify a solution are given at a *single* value of the independent variable. Any other combination of conditions classifies the problem as a *boundary value* problem.

The most common situation is where there are n_1 conditions at the start of the interval, x_0, and $n_2 = n - n_1$ conditions at the end of the interval, x_f. But there are other possibilities; for example, there may be a requirement that the function be finite at a particular interior singular point. We will restrict attention here to the case where the conditions are exclusively at the ends.

As a guiding example, consider the solution of the equation for a spring

$$my''(t) + ky(t) = 0$$

with boundary conditions that fix the position at both $t = 0$ and at $t = t_f$,

$$y(0) = 0 \qquad y(t_f) = 0$$

The general solution to the differential equation is

$$y(t) = C_1 \sin(\omega t) + C_2 \cos(\omega t)$$

where $\omega^2 = k/m$, so the condition at $t = 0$ requires that $C_2 = 0$; the one at t_f then forces $\omega t_f = n\pi$. Thus, the solution to the differential equation is

$$y(t) = C_1 \sin\left(n\pi \frac{t}{t_f}\right)$$

Notice that a solution to the differential equation exists *only* if the quantity $\sqrt{k/m}\,(t_f/\pi)$ is an *integer*. And, even if this is the case, the parameter C_1 is left undetermined, thus yielding an infinite number of solutions, one for each value of C_1. This is analogous to the solution of the matrix equation $\mathbf{A}\vec{x} = \vec{b}$. If the determinant of $\mathbf{A}$ is zero, either an infinite number of solutions exist, or no solution is possible. The reason that the constant C_1 is left undetermined is because the differential equation is linear *and* both the equation and the boundary conditions are *homogeneous* in y, and are thus unchanged by an arbitrary scaling of y.

This suggests that determining the existence and the uniqueness of a solution to a boundary value problem is going to be more problematical.[1] In an initial value problem, if the equation is linear, the solution can be started at the initial point, x_0, and proceed to a solution using the Taylor series. Thus, if the equation is second order, with $y_0, v_0 = dy/dt|_0$ as the initial conditions, *all* of the terms in the Taylor series are calculable:

$$f(x_0) = y_0$$
$$f'(x_0) = v_0$$
$$f''(x_0) = \text{obtained from differential equation}$$
$$f^{[3]}(x_0) = \text{from first derivative of differential equation}$$

$$\vdots \quad \vdots$$

This formulation provides the basis for all numerical stepping solutions, and, from it, follow essentially all statements concerning the existence and uniqueness of a solution.

However, for the boundary value problem, some of the necessary initial information is missing. In the preceding example, there is no initial value for dy/dt, and so the Taylor series cannot be used to provide either a stepping solution or information regarding the existence of a solution.

All of these considerations suggest that solving boundary value problems is going to be significantly more difficult than solving initial value problems.

The first step is to convince yourself that the boundary conditions supplied do indeed specify a unique solution. If the equation and the boundary conditions are *linear* and *homogeneous*, the equation can often be converted to a differential *eigenvalue* equation. For second-order equations the eigenvalue equations are called *Sturm-Liouville* equations; they are described and solved in any text on applied mathematics.[2] Essentially, all the well-known properties of solutions of ordinary matrix eigenvalue equations are transferable to this problem. Thus, the sample problem can be written as

$$-\frac{d^2 y}{dt^2} \equiv \mathcal{L}y = \lambda y; \qquad y(t_0) = 0, \qquad y(t_f) = 0$$

with the eigenvalue $\lambda = k/m$. The solutions, $u_n(t) = C\sin(\sqrt{\lambda_n}t)$ with $\lambda_n = (n\pi/t_f)^2$, as with the matrix equation, form a *complete* set of *orthogonal* functions, with the orthogonality condition

$$\int_0^{t_f} u_n(t)u_{n'}(t)\, dt = \delta_{nn'}$$

defining the value of C. A surprisingly large fraction of the classic differential equations of applied mathematics fall into this category and are the origin of many

---

[1]You can easily show that the only solution to a linear homogeneous initial value problem with homogeneous initial conditions is $y = 0$.

[2]See, for example, Birkhoff and Rota, *Ordinary Differential Equations*, 3rd ed., John Wiley, New York, 1978.

different sets of orthogonal functions; e.g., Legendre polynomials, Bessel functions, and Laguerre, Hermite, and Chebyshev polynomials. We always solve equations of this type in terms of these special functions, rather than attempting a numerical solution directly.

If the equation is non-Sturm-Liouville, then establishing the existence and uniqueness properties of the solution can be difficult.[3] Assuming a unique solution exists, there are basically three avenues of solution:

Superposition If the differential equation is of nth-order, and if it and the boundary conditions are *linear*, the solution can be obtained by solving n simultaneous *initial* value problems by any of the earlier stepping procedures; e.g., ode45, odeBS.

Shooting The problem can be converted to a search for the missing initial conditions. When they are obtained, the problem becomes a simpler initial value problem.

Relaxation A set of values at a sequence of intermediate *mesh* points that fit the boundary conditions is first guessed and then iteratively improved by using the differential equation.

For linear problems, the preferred method is usually superposition. Shooting is most often the method of choice for nonlinear problems. However, the method can be susceptible to round-off error, and for those situations where high accuracy is required, relaxation methods are employed.[4]

Of these methods, *shooting* is the easiest to understand, and we describe it in the first section. We then describe relaxation methods, which now include a vast literature. The methods of superposition require a short detour into some analytical properties of solutions; we postpone their discussion until the next chapter.

Shooting Methods for Boundary Value Problems

This method is best illustrated by way of an example. Consider the well-known problem of computing the trajectory of a cannonball. If the cannonball is fired from the initial position $(x_0 = 0, y_0 = 0)$ with an initial velocity v_0 at an angle θ, the equations for the trajectory are simply

$$ma_x = m\frac{d^2x}{dt^2} = 0; \qquad x(0) = 0; \qquad \frac{dx}{dt}\bigg|_0 = v_0 \cos\theta = v_{0x}$$

$$ma_y = m\frac{d^2y}{dt^2} = -mg; \qquad y(0) = 0; \qquad \frac{dy}{dt}\bigg|_0 = v_0 \cos\theta = v_{0y}$$

---

[3]Courant and Hilbert, *Methods of Mathematical Physics*, Interscience Publishers, New York, 1962.

[4]As with most iterative procedures, error of any type in one step is usually corrected in subsequent iterations.

which have the solution

$$x(t) = v_{0x}t$$

$$y(t) = v_{0y}t - \frac{1}{2}gt^2$$

You will note that the problem was stated as an initial value problem. If instead we wish to solve the same problem as a boundary value problem, the appropriate boundary conditions might be

$$x(0) = 0, \qquad x(t_f) = R$$
$$y(0) = 0, \qquad y(t_f) = 0$$

This then is the problem of firing the cannonball from the point $(0,0)$ in such a way that it strikes a target located at $(R,0)$ at a time $t = t_f$. The task is to find the correct firing velocity and angle.[5] In other words, we want to find the *missing initial conditions*, $\vec{v}_0 = (v_{0x}, v_{0y})$. The shooting approach is as follows:

1. *Guess* a complete set of values for the missing initial conditions. In this case, this would be $\vec{v}_0^g = \begin{pmatrix} c_1 \\ c_2 \end{pmatrix}$.

2. *Solve* the differential equation(s) over the entire range, $t_0 \le t \le t_f$, and determine the computed values of x, y at the end of the interval, $\vec{y}_f^c = \begin{pmatrix} x(t_f) \\ y(t_f) \end{pmatrix}$, based on this guess.

3. *Compare* the computed end conditions with the desired end conditions

$$\vec{f}(\vec{v}_0^g) = \vec{y}_f^c - \vec{y}_f = \begin{pmatrix} x(t_f) \\ y(t_f) \end{pmatrix} - \begin{pmatrix} R \\ 0 \end{pmatrix}$$

4. *Improve* the guess for $\vec{v}_0^g$ based on the value of $\vec{f}(\vec{v}_0^g)$ and iterate the process until $|\vec{f}| < tol$.

The statement of the method was framed in such a way that it is clear we are searching for the *root* of a particular function. Thus, the only remaining task is to design an M-file function that will have a value of zero when the end conditions are met. Of course, you should not fail to notice that each reference to this function will require a complete solution of the differential equation(s) by one of the earlier stepping methods. The root of the function will correspond to the missing initial conditions. The final solution to the boundary value problem will then be obtained by solving the equation(s), once again using these initial conditions. If there is only one missing condition, this could be accomplished by using the toolbox function **fzero** to find the root of f. If, as in this case, there are

---

[5]Note how the question of existence of a solution is much more subtle in this problem. If the firing velocity is too low or the required range is too far, there will never be a possibility of hitting the target, regardless of how large we make t_f or how much we adjust the firing angle.

two or more unknown initial conditions, a program employing a two-dimensional Newton's method could be used. (See Figure 9.1.) The code in Figure 20.1 is based on the secant method to find the root of a function of two variables and can be used in the larger problem of solving for the components of the initial firing velocity of the cannonball.

```
function dy = Fmissing(c)
% Fmissing uses estimates for the initial velocity contained
% in the column vector c to compute the complete trajectory
% of the cannonball.  The values of the (x,y) coordinates at
% the END of the interval are then compared with the actual
% end conditions and the difference is returned in the two
% element vector dy.
%
% If dy = 0, the values in c corresponding to the desired
% missing initial velocity have been found.
%---------------
  global y0 yf     % Column vectors containing the actual
                   % values of both (x,y) at each end; i.e.,
                   % y0 => [x0 y0]' and yf => [xf yf]'
  global t0 tf     % Interval of independent variable
  global dydt tol  % Name of the derivative function (as a
                   % string) and the tolerance.  These must
                   % also appear in calling code.  Each must
                   % be previously assigned a value.
%----------------------------------------------------------------
  c = c(:);
  Y0 = [y0;c];      % Y = [x y vx vy]', and c = [vx vy]' is
                    % is an estimate of the complete missing
                    % initial conditions.
  [t,y] = ode45(dydt,t0,tf,Y0,tol);
  dy = yf - y(length(t),1:2)';
%----------------------------------------------------------------
```

Figure 20.1 *MATLAB code for a function whose roots are the missing initial conditions for the cannonball trajectory.*

The two M-file functions shown in Figures 20.2 and 20.3 are then all that is required to solve for the missing values of the velocity components at $t = t_0$. For example, if the conditions of the trajectory are $(x_0, y_0) = (0, 0)$ and $(x_f, y_f) = (R, 0)$ over the time interval $0 \le t \le 25$ seconds, the preceding code computes the values of the initial velocity needed to hit the target at (x_f, y_f) when $t = t_f$. The computed values are $(v_{0x}, v_{0y}) = (R/t_f, 122.5)$; the *exact* values are $v_{0x} = R/t_f$ and $v_{0y} = \frac{1}{2}gt_f$.

```
function dy = Dervs(t,y)
% Dervs returns the values of the derivatives as a four ele-
% ment column vector using the given expressions for the
% acceleration.  That is,
%         y' = (x',y',vx',vy') = (vx,vy,ax,ay)
% A constant gravitational acceleration is assumed, with
% ax=0 and ay = -9.8.  The input t is a scalar and y is a
% four element column vector containing (x,y,vx,vy).
%---------------
 Y = y(:);
 dy = [Y(3) Y(4) 0 -9.8]';
%---------------
```

Figure 20.2 *MATLAB code for the derivative function.*

```
function r = Secnt2d(f,x0,dx,tol)
%Secnt2d uses a secant method generalized to 2-d to find a root
%  of the two equations f1(x)=0 and f2(x)=0, starting from the
%  two element initial guess x0.  Also needed is a two element
%  guess, dx, for an interval containing the root.  Both x and
%  dx are 2-element column vectors.  The name of the function
%  is contained in the string f and returns the values of both
%  equations.   This function must expect a two element input
%  and return a two element result. The maximum number of itera-
%  tions is 20.  The calculation terminates if |dx| < tol and a
%  root is returned in r.
%
% Secnt2df is a copy of the function SECANT2d of Figure 9-1
% adapted to this particular problem.
%============================================================
  for i = 1:20
      x0 = x;               f0 = feval(f,x0);
      x0 = x + [dx(1);0]; fx = feval(f,x0); fx = (fx-f0)/dx(1);
      x0 = x + [0;dx(2)]; fy = feval(f,x0); fy = (fy-f0)/dx(2);
      A  = [fx fy]'+eps;
      x  = x + dx; dx = -A\f0-dx;
      if (norm(dx) < tol);
          r = x;
          return;
      end
  end
  disp('still outside tolerance after 20 iterations, abs(dx) =')
  disp(abs(dx))
```

Figure 20.3 *MATLAB code to find the root of a function of two variables.*

Figure 20.4 shows the graphical representation of the cannonball trajectory.

Figure 20.4 *A cannonball trajectory that will hit a target at $(x_f, y_f) = (R, 0)$ is obtained by first solving for the appropriate firing velocity and angle.*

It is relatively easy to adapt the preceding code to the solution of any two-point boundary value problem. However, as with any problem requiring the solution for the roots of a multidimensional function, the progress of the calculation is usually extremely sensitive to the initial guess for the missing initial conditions. It is always advisable to spend considerable effort in obtaining a good first guess. Problems 20.3, 20.7, 20.9, 20.10, and 20.11 are more challenging.

DIFFERENTIAL EIGENVALUE EQUATIONS

A very common and important type of boundary value problem is the *eigenvalue* problem. If the differential equation *and* the boundary conditions are linear, the equation can usually be written as an eigenvalue problem. For example, the equation describing the oscillations of the spring given at the beginning of this chapter,

$$y'' + \omega^2 y = 0; \qquad y(0) = 0, \qquad y(t_f) = 0$$

is of the form $\mathcal{L}y = \lambda y$ with $\lambda = -\omega^2$ and $\mathcal{L} \equiv d^2/dt^2$ *plus* the conditions[6] $y(0) = 0$, $y(t_f) = 0$. It was found that this equation will have solutions *only* if the eigenvalue ω has a value determined by $\omega t_f = n\pi$. These types of differential boundary value equations are most often solved analytically, as was mentioned in the Preview. However, it is not difficult to adapt the shooting method to compute a value of one of the eigenvalues along with the calculation of the missing initial conditions. The procedure is simply to add an additional equation to the string of first-order equations when structuring the equation for a stepping solution.

---

[6]When writing a differential equation in terms of symbolic operators like $\mathcal{L}$, the boundary conditions must be included as part of the definition of the operator.

Thus, the preceding equation could be written as

$$y_1(t) \equiv y(t) \qquad y_1'(x) = y_2(x) \qquad y_1(0) = 0 \qquad y_1(t_f) = 0$$

$$y_2(t) \equiv \frac{dy}{dt} \qquad y_2'(x) = -(y_3)^2 y_1 \qquad y_2(0) = v_0$$

$$y_3(t) \equiv \omega \qquad y_3'(x) = 0 \qquad y_3(0) = \omega_0$$

(plus a normalization condition)

In this formulation the two quantities specifying the missing initial conditions; i.e., v_0, ω_0, must be guessed. Also, the second of the equations is now *nonlinear* and can be expected to extend the execution time and the potential for unforeseen problems.

You will note that there are, at the moment, only two extra conditions given to solve three first-order equations. The third condition should be a requirement for a solution. The original equation and the boundary conditions are homogeneous, so any solution obtained can be scaled by a constant and still remain a solution. This means that to specify a solution the third condition should be a requirement fixing the *normalization* of the solution. Almost any inhomogeneous condition on y or y' that is independent of the previous conditions will do; for example, $y'(t_f) = 1$. The shooting method then will solve for the *two* unknowns v_0, ω_0 that satisfy the boundary conditions.

NONLINEAR BOUNDARY VALUE PROBLEMS

Nonlinear differential equations are notoriously ill behaved and difficult to solve analytically. Although the conditions you use to establish the existence and uniqueness of a solution remain valid,[7] it is no longer possible to write the general solution as a *linear* combination of solutions to a homogeneous equation plus a particular solution. (See Chapter 21 for a discussion of this form of a solution for the linear case.) Thus, most of the techniques so successful for handling linear equations are useless. Each small change in a single parameter may yield a completely different analytic solution.

Still, if sufficient initial conditions are present, you can initiate a stepping procedure solution. Even if it fails, it can at least be followed out to a troublesome point, perhaps localizing the difficulties. Because stepping procedures are at the heart of the shooting method, this implies that if a reasonably smooth solution is expected, nonlinear boundary value problems are amenable to numerical solution as well. The basic shooting method remains exactly the same; only the potential for problems increases.

In fact, the shooting method is the *only* practical numerical method that can be applied to nonlinear boundary value problems. The relaxation methods

---

[7]Beware of the argument "reality demands that a solution exist." Your differential equation is merely a *model* of reality. Although the physical phenomena it is supposed to describe may indeed exist, the model may be flawed, and solutions to your equation may not exist.

of the next section can be used only in very special cases, and the method of superposition of Chapter 21 *requires* that the differential equation be linear. (An example of the solution of a nonlinear boundary value problem can be found in Problem 20.10.)

Relaxation Methods

The most common and simplest type of boundary value problem is one in which the equation is second order and the value of the function is specified at both ends of the interval. Procedures for finding a solution based on *relaxation* methods then subdivide the interval into a finite *mesh*, fix the function to the known values at each end, and guess values for all the intervening points. The differential equation is then used to refine the function at the mesh points. This process is shown graphically in Figure 20.5.

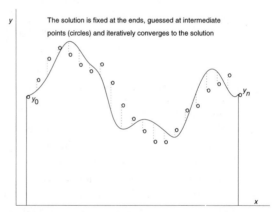

Figure 20.5 *A trial function is fixed at known values at each end of the interval and guessed at all intervening mesh points.*

THE FINITE DIFFERENCE METHOD

The most straightforward refinement method replaces the differential equation by a *finite difference equation*. It replaces all derivatives by approximate expressions based on the following approximations, which are in turn based upon central difference expressions.

$$\left.\frac{dy}{dx}\right|_{x_n} \approx \frac{y_{n+1} - y_{n-1}}{\Delta x_{n+1} + \Delta x_n}$$

$$\left.\frac{d^2y}{dx^2}\right|_{x_n} \approx 2\frac{\Delta x_n y_{n+1} - (\Delta x_n + \Delta x_{n+1})y_n + \Delta x_{n+1}y_{n-1}}{\Delta x_{n+1}\Delta x_n(\Delta x_{n+1} + \Delta x_n)}$$

(20.1)

with $\Delta x_n = x_n - x_{n-1}$. Next, if the mesh is chosen to be equally spaced, $\Delta x = (x_f - x_0)/(n+1) \equiv h$, with $x_k = x_0 + k\Delta x$, $k = 1, \ldots, n+1$, these equations can be written in a more familiar form as

$$\frac{dy}{dx}\bigg|_{x_n} \approx \frac{y_{n+1} - y_{n-1}}{2h}$$

$$\frac{d^2y}{dx^2}\bigg|_{x_n} \approx \frac{y_{n+1} - 2y_n + y_{n-1}}{h^2}$$

(20.2)

For example, consider the second-order equation

$$y'' + p(x)y' + q(x)y = g(x); \qquad y(x_0) = y_0, \qquad y(x_f) = y_f \qquad (20.3)$$

For reasons that will be clear in a moment, we first change variables to a new function $Y(x)$ that is *homogeneous* in the boundary conditions. The transformation is

$$y(x) = Y(x) + ax + b$$

with

$$\begin{pmatrix} a \\ b \end{pmatrix} = \frac{1}{x_f - x_0} \begin{pmatrix} 1 & -1 \\ -x_f & x_0 \end{pmatrix} \begin{pmatrix} y_0 \\ y_f \end{pmatrix}$$

so that the equation becomes

$$Y'' + p(x)Y' + q(x)Y = R(x); \qquad Y(x_0) = 0, \ Y(x_f) = 0 \qquad (20.4)$$

where $R(x) = r(x) - ap(x) - (ax+b)q(x)$. Then, replacing the derivative by the preceding approximate difference expressions, the differential equation can be written as

$$Y_{k+1} - 2Y_k + Y_{k-1} + \frac{h}{2}p_k(Y_{k+1} - Y_{k-1}) + h^2 q_k Y_k = h^2 R_k, \qquad k = 1, \ldots, n$$

with $Y_0 = Y_{n+1} = 0$. This set of equations can be written in matrix form as

$$\begin{pmatrix} -2+h^2q_1 & 1+\frac{h}{2}p_1 & 0 & \cdots & 0 \\ 1-\frac{h}{2}p_2 & -2+h^2q_2 & 1+\frac{h}{2}p_2 & \cdots & 0 \\ 0 & 1-\frac{h}{2}p_3 & -2+h^2q_3 & \cdots & 0 \\ \vdots & \vdots & \vdots & \ddots & \vdots \\ 0 & 0 & 0 & \cdots & -2+h^2q_n \end{pmatrix} \begin{pmatrix} Y_1 \\ Y_2 \\ Y_3 \\ \vdots \\ Y_n \end{pmatrix} = h^2 \begin{pmatrix} R_1 \\ R_2 \\ R_3 \\ \vdots \\ R_n \end{pmatrix}$$

(20.5)

If the number of mesh points is not too large, this matrix equation can be solved directly to obtain the values of Y_k. However, in most situations the number of mesh points is quite large, with values of $n > 10^6$ not uncommon, and storage of the coefficient matrix becomes unfeasible. In these situations an alternate procedure is to take advantage of the banded nature of the coefficient matrix and write the equations directly, solving the kth equation for Y_k. In the present case, where the coefficient matrix has two side bands, these equations read

$$Y_1 = \frac{1}{a_{11}}(h^2 R_1 - a_{12} Y_2)$$

$$Y_k = \frac{1}{a_{kk}}(h^2 R_k - a_{k,k+1} Y_{k+1} - a_{k-1,k} Y_{k-1}), \qquad k = 2, \ldots, n-1 \qquad (20.6)$$

$$Y_n = \frac{1}{a_{nn}}(h^2 R_n - a_{n,n-1} Y_{n-1})$$

where a_{ij} are the elements of the coefficient matrix.

Of course, to obtain a solution to these equations; i.e., values for the Y's on the left, we need to already know the values for the Y's on the right. This implies a solution by iteration.

JACOBI ITERATION

One iterative procedure for solving a set of linear (or nonlinear) equations is to guess a complete set of Y's, $\{Y_k^{[old]}, k = 1, \ldots, n\}$ and to use the equations to yield an improved set, $Y_k^{[new]}$. The new set is then substituted for the old and the procedure iterated until convergence is attained. This method is known as *Jacobi* or *Jacobi fixed-point* iteration. It can be shown[8] that, if the equations are linear and if the coefficient matrix is *diagonally dominant*, the method will converge for *any* initial guess. A diagonally dominant matrix is one in which the largest magnitude elements fall on the diagonal. More specifically, if

$$|a_{ii}| > \sum_{\substack{j=1 \\ j \neq i}}^{n} |a_{ij}| \qquad \text{for all } i$$

the matrix is diagonally dominant. (This requirement can also be shown to ensure that no zero pivot will be encountered when inverting the matrix.) Diagonal dominance is often too stringent a condition for convergence, and frequently matrices with simply the largest magnitude elements along the diagonal will be found to have a well-behaved solution. MATLAB coding of the Jacobi method makes use of the vectorized nature of MATLAB arithmetic, yielding an especially efficient implementation.

To illustrate the method we first rewrite Equation 20.5 formally as

$$\mathbf{A}\vec{Y} = \vec{R} \qquad (20.7)$$

and, as above, solve the ith equation for Y_i,

$$Y_i = \frac{1}{a_{ii}}\left(h^2 R_i - \sum_{k \neq i} a_{ik} Y_k\right)$$

---

[8]LaFara, *Computer Methods for Science and Engineering*, Hayden Book Co., Rochelle Park, N.J., 1973.

where the summation is over all k except for the term $k = i$. By adding and subtracting the term $a_{ii}Y_i$ to the sum, this can instead be written as

$$Y_i = Y_i + \frac{1}{a_{ii}}\left(h^2 R_i - \sum_{k=1}^{n} a_{ik}Y_k\right) \tag{20.8}$$

That is,

$$Y_i^{\text{new}} = Y_i^{\text{old}} + \Delta Y_i$$

The MATLAB version of this equation is then,

```
Y = Y + (h^2*R - A*Y)./diag(A)
```

where it is assumed that both Y and R are column vectors and that $\mathbf{A}$ is a square matrix. For later considerations, it is important to review how MATLAB executes this line of code. As in any computer language, the effect of the equal sign is to first evaluate the expression on the right and then replace the variable on the left by this value. Because Y is a column vector, this means that *all* of the elements of Y are first computed before the substitution is made, which is the essence of the Jacobi method. For a given coefficient matrix $\mathbf{A}$, right-hand-side vector $\vec{R}$, and an initial guess for the Y's, this equation is then iterated by placing it in a `for` loop until convergence is attained. The criteria used for convergence could be based on the size of the largest element of dY, or on its norm, $|dY|$, or on the fractional change in y; i.e., $|dY|/|Y|$. In all of the examples that follow we will use the latter criterion.

EXAMPLE 20.1

Solution of a Test Equation by Finite Differences Using the Jacobi Method

Consider the boundary value differential equation,

$$y'' + \frac{1}{4}y = x^2; \qquad y(0) = 0, \qquad y(\pi) = 0$$

To assess the accuracy of the numerical method, we can compare with the exact solution, which is

$$y(x) = 32\left(\sin\left(\frac{x}{2}\right) + \cos\left(\frac{x}{2}\right) - 1\right) + 4\left(x^2 - \pi^2\sin\left(\frac{x}{2}\right)\right) \tag{20.9}$$

By substitution, you can verify that this expression is a solution of the given differential equation.

To obtain a numerical solution, we next compare the preceding equation with the form in Equation 20.4 and obtain the following identifications:

$$p(x) = 0$$
$$q(x) = \tfrac{1}{4}$$
$$R(x) = x^2$$
$$x_0 = 0 \qquad\qquad x_f = \pi$$
$$h = \pi/(n+1)$$

so that the matrix equation, Equation 20.5, reads

$$
\begin{pmatrix}
-2+\frac{h^2}{4} & 1 & 0 & \cdots & 0 \\
1 & -2+\frac{h^2}{4} & 1 & \cdots & 0 \\
0 & 1 & -2+\frac{h^2}{4} & \cdots & 0 \\
\vdots & \vdots & \vdots & \ddots & \vdots \\
0 & 0 & 0 & \cdots & -2+\frac{h^2}{4}
\end{pmatrix}
\begin{pmatrix}
Y_1 \\ Y_2 \\ Y_3 \\ \vdots \\ Y_n
\end{pmatrix}
= h^4
\begin{pmatrix}
1 \\ 2^2 \\ 3^2 \\ \vdots \\ n^2
\end{pmatrix}
$$

$$(20.10)$$

The MATLAB code to solve this equation at 20 mesh points between 0 and π is then

```
n = 20;  h = pi/(n+1);   tol = 1.e-3;
A = (-2+h^2/4)*diag(ones(1,n));
A = A + (diag(ones(1,n-1),1)+diag(ones(1,n-1),-1));
R = h^4*([1:n]').^2;
Y = ones(n,1);   dY =  ones(n,1); iter = 0;
while norm(dY)/norm(Y) > tol
   dY = (R-A*Y)./diag(A);
     Y = Y + dY;
      iter = iter +1;
end
disp(iter)
disp('iterations required')
```

This code required ≈ 290 iterations to attain $|dY|/|Y| < 10^{-3}$. This does not mean that the solution is accurate to one part in 10^3, only that the method seems to be converging and that the final steps taken were quite small. The elapsed time[9] was ≈ 5 seconds.

For this example, the coefficient matrix is not large, so we can use MATLAB to obtain an "exact" solution to Equation 20.10 by means of backslash division. We next change the criterion for convergence to $|Y - y_{\text{exact}}|/|y_{\text{exact}}| < 10^{-3}$ and see how long it takes the Jacobi method to achieve a solution. The result is ≈ 845 iterations with an elapsed time of ≈ 10 seconds. Furthermore, even this "exact" solution is only an approximation to the true solution to the differential equation given by Equation 20.9. If the norm of the numerical solution is compared with this solution, the result is $|\vec{Y} - \vec{y}_{\text{true}}|/|\vec{y}_{\text{true}}| \approx 0.002$.

Thus, the Jacobi method has a very slow convergence rate and must be used with caution. Its redeeming feature is that the algorithm is *vectorized*, causing each individual iteration to be executed quite quickly. ●

You should keep in mind that the iterative procedure is merely an alternative to finding the inverse of the coefficient matrix. That is, the procedure is converging to the solution of Equation 20.10, which may or may not be a faith-

---

[9]To display the elapsed time, you can start a stopwatch with the command `tic` and stop the timer with `toc`.

ful representation of the solution of the original differential equation. Because Equation 20.10 is based on the finite difference equations, Equations 20.2, which are second-order accurate in $h = \Delta x$, the method can be expected to ultimately converge to a solution that is also accurate to second order.

CONVERGENCE RATE OF THE JACOBI METHOD

The convergence rate of the Jacobi iterative method is generally quite slow. Even more distressing, it can be shown[10] that the number of iterations required to attain a specified accuracy is at least proportional to the number of mesh points. In practice it is usually much worse. To see the reasons for this, we rewrite the matrix equation for the solution Y, Equation 20.10, by decomposing the coefficient matrix $\mathbf{A}$ into its diagonal part, $\mathbf{D}$, and its off-diagonal parts, $\mathbf{M}$,

$$(\mathbf{D} + \mathbf{M})\vec{Y} = \vec{R}$$

or

$$\mathbf{D}\vec{Y} = -\mathbf{M}\vec{Y} + \vec{R}$$

Because $\mathbf{D}$ is a diagonal matrix, its inverse is also diagonal with the reciprocals of the diagonal elements of $\mathbf{D}$ along its diagonal. Thus, this equation can be phrased in a manner appropriate to an iterative solution as

$$\vec{Y} = \mathbf{J}\vec{Y} + \vec{b}$$

with $\mathbf{J} = -\mathbf{D}^{-1}\mathbf{M}$, $\vec{b} = \mathbf{D}^{-1}\vec{R}$. If we start with a first guess for the solution labeled as $\vec{Y}^{(1)}$, the next approximation is obtained by

$$\vec{Y}^{(2)} = \mathbf{J}\vec{Y}^{(1)} + \vec{b}$$

The error in each $\vec{Y}^{(i)}$ is then $d\vec{Y}^{(i)} = \vec{Y}^{(i)} - \vec{Y}$. Because the vector $\vec{b}$ is the same in each iteration, we have the following expression for the successive errors:

$$d\vec{Y}^{(k+1)} = \mathbf{J}d\vec{Y}^{(k)} = (\mathbf{J})^k d\vec{Y}^{(1)} \qquad (20.11)$$

Then, the requirement that the procedure converge; i.e., $\lim_{k \to \infty} d\vec{Y}^{(k)} = 0$, means that $\lim_{k \to \infty} \mathbf{J}^k = 0$. For a matrix to satisfy this requirement, each and every one of its eigenvalues must have a magnitude less than one. Indeed, the convergence rate of the Jacobi method will be roughly determined by the largest magnitude eigenvalue of $\mathbf{J}$. If this largest magnitude eigenvalue is called $\lambda_{\max}$, the number of iterations, K, required to reduce the error by a factor 10^{-p} is determined approximately by

$$(\lambda_{\max})^K \approx 10^{-p} \qquad (20.12)$$

To illustrate this result we will use the matrix of the Example 20.1, given in Equation 20.10. The matrix $\mathbf{D}^{-1}$ is simply $(-2 + h^2/4)^{-1}$ times an identity matrix, and $\mathbf{M}$ is a matrix of two bands of ones parallel to the diagonal. The

---

[10]Press et al., *Numerical Recipes, The Art of Scientific Computing*, Cambridge University Press, New York and Cambridge, England, 1986, p. 654.

convergence matrix **J** is thus easily constructed and the MATLAB function `eig` can be used to determine its eigenvalues. In Table 20.1 this is done for matrices of various sizes and the maximum magnitude eigenvalue is determined. Finally, the approximate number of iterations required to obtain a solution accurate to 10^{-3} is computed by Equation 20.12.

Matrix size n	Maximum eigenvalue λ_{max}	Approx. no. of iterations K
5	0.89676	63
10	0.96938	222
20	0.99160	819
40	0.99780	3,134
80	0.99944	12,242
160	0.99986	48,375

Table 20.1   *The approximate number of iterations required to solve Equation 20.10 by Jacobi's method to an accuracy of 10^{-3}*

In this case the number of iterations required is roughly proportional to n^2.

GAUSS-SEIDEL ITERATION

A somewhat more efficient method is to use only one set of Y_k's, continually updating them with each subsequent equation. This variation on the Jacobi method is also guaranteed to converge if the coefficient matrix is diagonally dominant. However, because more recent information is then folded into each new equation, we expect this method to converge more quickly than the Jacobi method. Indeed, this is almost always the case. However, unlike the Jacobi method, the Gauss-Seidel method[11] is slightly more difficult to code using MATLAB. The trouble is that each element of the solution vector $Y(x_i)$ must be *successively* updated by using Equation 20.8, thereby excluding a *vectorized* MATLAB replacement statement. The MATLAB code for the Gauss-Seidel method is illustrated in Figure 20.6.

This code applied to Example 20.1, again using a convergence tolerance of $|dY|/|Y| < 10^{-3}$, yields a solution in ≈ 225 iterations, which is somewhat fewer than the number for the Jacobi method, indicating that the basic method is a bit more efficient. However, because the Gauss-Seidel method cannot make use of vectorized arithmetic, the elapsed time turns out to be much longer; ≈ 300 seconds or about 60 times longer for fewer iterations.

Both the Jacobi and the Gauss-Seidel methods are so slowly convergent that they are rarely used in realistic problems. Nevertheless, both methods play

---

[11] See also the discussion in Chapters 2 and 22.

```
function y = GAUSSD(A,r,yguess,tol)
% GAUSSD will iteratively solve for the solution to Ay=r
% by the Gauss-Seidel method.  This must be done using
% for-loops to replace the vectorized replacement state-
% ments of the Jacobi method.  The initial guess for the
% solution is the column vector yguess.  Convergence is
% attained when the fractional change in y is less than
% tol.  The coefficient matrix is assumed to be diagon-
% ally dominant and no row switching is employ. If one
% of the diagonal elements is less than 100*eps, the cal-
% culation is terminated.  The iterations limit is 1000.
%
% The solution is returned in Y and the number of itera-
% tions used in iter.
%========================================================
n = length(r); Y = yguess(:); dy = ones(n,1); iter = 0;
while norm(dY)/norm(Y) > tol
    for i = 1:n
        if abs(A(i,i))<100*eps;error('zero pivot found');end
        dY(i) = r(i)/A(i,i);
        for j = 1:n
            dY(i) = dY(i) - A(i,j)*Y(i)/A(i,i);
        end
        Y(i) = Y(i) + dY(i);
    end
    iter = iter + 1;
    if iter>1000; error('not converging in 1000 steps');end
end
```

Figure 20.6 *MATLAB code to solve the matrix equation* $\mathbf{A}\vec{y} = \vec{r}$ *using the Gauss-Seidel method.*

an important role in numerical analysis as the basis for modern, more rapidly convergent methods.

You should not interpret these comments as a dismissal of these two very simple procedures. If your coefficient matrix is diagonally dominant (or nearly so) and is relatively small (say $< 50 \times 50$), and you can afford to wait as the procedure traces through 500 to 1000 iterations, you will find the Gauss-Seidel method in particular to be extremely stable and reliable. Furthermore, keep in mind that these procedures, like all *iterative* procedures, are generally insensitive to a buildup of round-off error. Errors in one step are usually corrected in subsequent steps. A MATLAB script file that demonstrates the Gauss–Seidel method is on the applications disk and is named IterSol.m.

SUCCESSIVE OVERRELAXATION (SOR)

The basic iterative method at the heart of the Gauss-Seidel method is linear, thus we expect that it will take small but steady steps towards the final answer. It will only rarely overshoot and have to reverse direction. This suggests that it may be helpful to *artificially* accelerate the convergence by each time taking a slightly larger step than actually computed. Thus, the algorithm $Y_i^{\text{new}} = Y_i^{\text{old}} + \Delta Y_i$ can be replaced by

$$Y_i^{\text{new}} = Y_i^{\text{old}} + \omega \Delta Y_i$$

where ω is the *relaxation* factor and is expected to be somewhat larger than one. Relaxation factors cannot in general be used in the Jacobi method. The convergence is dictated by the maximum magnitude eigenvector of the convergence matrix, which is usually very close to one; therefore, multiplying by an $\omega > 1$ will cause the method to diverge. However, in the Gauss-Seidel method, errors in one step are, to some extent, corrected in the next line of the calculation, so the effects of a relaxation factor greater than one will be compensated.

The *optimal* value to use for ω is often difficult to predict and depends to some extent on the differential equation itself.[12] Most often the optimum value falls in the range $1.1 \leq \omega \leq 1.9$. Whenever using the SOR method it is prudent to first experiment with ω's in this range. The increase in the convergence rate can be quite dramatic. For example, Equation 20.8 was solved once again by the Gauss-Seidel method, but with a relaxation factor included. The number of iterations required to solve the problem to the same accuracy as the 200 iterations-no relaxation calculation is listed as a function of ω in Table 20.2.

Relaxation factor ω	Number of iterations n	Elapsed time (s)
1.1	190	195
1.2	159	175
1.3	133	137
1.4	110	121
1.5	89	98
1.6	69	70
1.7	50	49
1.8	29	26
1.9	56	58

Table 20.2 *Increased convergence of the Gauss-Seidel method as a function of relaxation factor ω is seen by comparing calculations of similar accuracy.*

---

[12]There is considerable literature available on the optimum value of ω for a variety of problems. See for example, Wong (1992).

SPARSE MATRICES

It has perhaps occurred to you that the preceding methods appear to be fundamentally flawed; not necessarily incorrect, but missing some key features of the basic structure of the problem. That is, the method is constructed to handle any diagonally dominant coefficient matrix. Yet, in every case, the coefficient matrices have a much more prominent characteristic than simple diagonal dominance. They are always *banded* and, if large, the vast majority of their elements are *zeros*. Thus, the algorithm will spend most of its time multiplying by zeros. It is therefore clear that a very important next step in improving the efficiency of these iterative algorithms is to design procedures to take advantage of this particular matrix structure. Matrices that contain a large number of zeros are called *sparse*, and it is not uncommon for Gauss-Seidel methods with relaxation to have coefficient matrices of order 10^3 or more. These matrices are usually sparse and banded. Even aside from arithmetic efficiency, storage considerations for such matrices usually dictate that special methods be used. Unfortunately, modern techniques for handling sparse matrices are well beyond the level of this text. Except for some elementary suggestions outlined in Problem 20.13, we refer you to the help files associated with the MATLAB `sparfun` directory references for a more detailed discussion.[13]

The Finite Element Method

An alternative to the finite difference method that is very popular for a wide class of boundary value problems is called the *finite element method*. The ultimate forms of the matrix equations that result are quite similar to those encountered in the finite difference method and are solved by similar techniques. However, the point of view is quite different. Whereas the finite difference method begins by using approximate expressions for derivatives, the finite element method attempts to fit the solution to a set of model functions by using the given differential equation to minimize the difference between the model solution and the actual answer. This difference is called a *residual*, and the derivation of this procedure is called the *method of weighted residuals*.

THE METHOD OF WEIGHTED RESIDUALS

We begin with an ordinary differential equation with homogeneous boundary conditions[14] written in symbolic form as

---

[13]It should be noted that MATLAB has a special toolbox for the solution of problems involving sparse matrices. The name of the toolbox is *SPARSE*, and it contains the usual help files to get you started if your problem forces you in this direction. We return to the topic of sparse matrices in Chapter 22 when discussing the solution of partial differential equations.

[14]If the boundary conditions are inhomogeneous, they can be "homogenized" using the method described on page 443.

$$\mathcal{L}y(x) = r(x); \qquad y(x_0) = 0, \; y(x_f) = 0$$

where the symbol $\mathcal{L}$ represents the combination of differential operators contained in the differential equation. For example,

$$y'' + 3e^x y' + 2y = 0 \quad \Rightarrow \quad \mathcal{L} \equiv \left(\frac{d^2}{dx^2} + 3e^x \frac{d}{dx} + 2 \right)$$

The next step in the procedure is to select a set of n independent *expansion* functions, $\{\phi_i(x)\}$, $i = 1, \ldots, n$. We will leave these functions unspecified for the moment. The solution to the differential equation is then approximated by an expansion in terms of these functions,

$$y(x) \approx Y(x) = \sum_{i=1}^{n} c_i \phi_i(x)$$

The goal of the method is then to determine the set of expansion coefficients c_i.

The criterion to be used to determine the c_i's is to first form a *residual* function,

$$\Delta(x) \equiv \mathcal{L}Y(x) - r(x) = \sum_{i=1}^{n} c_i \mathcal{L}\phi_i(x) - r(x)$$

and then to construct a mechanism for its minimization. Notice that, if the approximate solution, $Y(x)$, is *everywhere* equal to the exact solution, $y(x)$, then $\Delta(x)$ is precisely zero. Here we are attempting something less ambitious; namely, we seek an approximate solution that will be completely determined once the n unknown coefficients, c_i, are specified. To this end we must next establish n independent conditions on the c_i's. This is done by once again selecting a set of n independent functions $w_j(x)$, $j = 1, \ldots, n$, called *weight* functions. The weight functions are usually normalized so that

$$\int_{x_0}^{x_f} w_j(x)\, dx = 1 \quad \text{for all } j$$

The n conditions necessary to specify the solution are then to require "weighted" averages of the residual $\Delta(x)$ to be zero for each $j = 1, \ldots, n$. That is,

$$\int_{x_0}^{x_f} w_j(x)\Delta(x)\, dx = 0 \qquad j = 1, \ldots, n$$

In summary, to solve $\mathcal{L}y = r$,

1. Select a set of n *expansion* functions, $\phi_i(x)$, $i = 1, \ldots, n$.
2. Construct the *residual* function, $\Delta(x) = \sum c_i \mathcal{L}\phi_i - r$.
3. Select a set of n independent *weight* functions, w_j, $j = 1, \ldots, n$ and establish n conditions on the unknown expansion coefficients, c_i,

$$\int_{x_0}^{x_f} w_j(x)\Delta(x)\, dx = 0 \qquad j = 1, \ldots, n$$

or

$$\sum_{i=1}^{n} c_i \int_{x_0}^{x_f} \omega_j \mathcal{L}\phi_i(x)\, dx = \int_{x_0}^{x_f} \omega_j r(x)\, dx \qquad (20.13)$$

4. Assemble these relations into a matrix equation of the form $\mathbf{M}\vec{c} = \vec{b}$, where

$$M_{ji} = \int_{x_0}^{x_f} \omega_j \mathcal{L}\phi_i(x)\, dx \qquad b_j = \int_{x_0}^{x_f} \omega_j r(x)\, dx \qquad (20.14)$$

5. Solve this matrix equation for the c_i's by either the Jacobi or Gauss-Seidel method using successive overrelaxation.

The only remaining questions pertain to the choice of expansion and weight functions, $\phi_i(x)$ and $\omega_j(x)$. There is considerable freedom left in these choices, resulting in several well-known computational algorithms.

THE COLLOCATION METHOD

The most obvious choice for the weight functions is[15]

$$\omega_j(x) \equiv \delta(x - x_j)$$

Using the properties of the delta function, Equation 20.13 becomes

$$\sum_{i=1}^{n} c_i \mathcal{L}\phi_i(x_k) = r(x_k) \qquad (20.15)$$

Thus, this choice of weight functions is equivalent to requiring that the approximate solution $Y(x)$ satisfy the differential equation at each of the sampling points x_i, $i = 1, \ldots, n$. Once we have selected a set of expansion functions, this set of equations can be solved for the c_i's. Of course, we didn't need to go through such an elaborate procedure to merely obtain this relatively simple algorithm. The method of weighted residuals offers other possibilities. The most popular choice for the weight functions yields what is called the Galerkin method, which we describe in the next section.

THE GALERKIN METHOD

If we choose the set of weight functions to be identical to the set of sample functions, $\{\omega_i(x), i = 1, \ldots, n\} = \{\phi_j(x), j = 1, \ldots, n\}$, we obtain the method of

---

[15]You will recall that the property of the δ function (see page 376) is

$$\int_a^b f(x)\delta(x - x_k)\, dx = f(x_k)$$

if x_k is within the range of the integration variable.

Galerkin. In this instance, Equation 20.13 becomes

$$\sum_{i=1}^{n} c_i \left(\int_{x_0}^{x_f} \phi_i \mathcal{L}\phi_j \right) dx = \int_{x_0}^{x_f} \phi_j r(x) dx$$

which is a matrix equation, $\mathbf{M}\vec{c} = \vec{b}$, for the c's, where

$$M_{ji} = \int_{x_0}^{x_f} \phi_i \mathcal{L}\phi_j \, dx \qquad b_j = \int_{x_0}^{x_f} \phi_j r(x) \, dx$$

Specific implementations of the Galerkin method then differ in their choice for the set of expansion functions, $\{\phi_j(x), j = 1, \ldots, n\}$.

The "Hat" Functions, $\phi(x)$ The *finite element method* employs Galerkin's choice for the weight functions; i.e., $\{\omega_i(x), i = 1, \ldots, n\} = \{\phi_j(x), j = 1, \ldots, n\}$, and "pseudo" delta functions for the expansion functions. That is, the expansion functions are chosen to be *hat* functions, pictured in Figure 20.7.

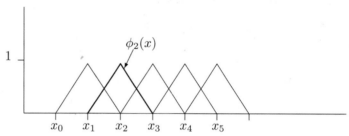

Figure 20.7   *The hat function, $\phi_i(x)$, is a triangle centered on x_i that extends from x_{i-1} to x_{i+1} and is zero elsewhere.*

If the sampling points, x_i, $i = 1, \ldots, n$ are chosen to be equally spaced with $\Delta x \equiv h$ and $x_0 = 0$, then the expression for these functions is

$$\phi_k(x) = \begin{cases} \frac{1}{h}(x - x_{k-1}) & = \frac{x}{h} - (k-1) & x_{k-1} \le x \le x_k \\ \frac{1}{h}(x_{k+1} - x) & = -\frac{x}{h} + (k+1) & x_k \le x \le x_{k+1} \\ 0 & \text{all other } x \end{cases} \qquad (20.16)$$

Each hat function has a value of one at different values of sampling points and is zero at *all* other sampling points. However, the functions are nonzero in the immediate neighborhood of the central point.

The finite element method then consists of the following steps:

1. Write the approximate solution as a linear combination of hat functions, $Y(x) = \sum c_i \phi_i(x)$.
2. Compute the elements of the matrix $M_{ji} = \int \phi_i \mathcal{L}\phi_j dx$ and the vector $b_j = \int \phi_j r(x) \, dx$.
3. Solve the matrix equation $\mathbf{M}\vec{c} = \vec{b}$ for the expansion coefficients c_i, noting that the solution at each sampling point, $Y(x_i)$, is equal to c_i. Either

the Jacobi method or the Gauss-Seidel method with SOR can be used to solve this equation.

Before illustrating the method, it is useful to summarize some properties of the hat functions. The integrals contained in the matrix $\mathbf{M}_{ij}$ will entail expressions like $\int \phi_i'' \phi_j \, dx, \int \phi_i' \phi_j \, dx$, and $\int \phi_i \phi_j dx$. Each of these can be evaluated using the expressions of Equation 20.16. The results are listed in Table 20.3.

$F(x)$	$\displaystyle\int_{x_0}^{x_f} \phi_i F(x) \, dx$	
ϕ_i	$2h/3$	
$\phi_{i\pm1}$	$h/6$	(zero for all other indices)
$\phi_{i\pm1}'$	$\pm1/2$	
ϕ_i''	$-2/h$	
$\phi_{i\pm1}''$	$1/h$	(zero for all other indices)
1	h	
x	ih^2	
x^2	$(i^2 + \frac{1}{6})h^3$	
x^3	$i(i^2 + \frac{1}{2})h^4$	

Table 20.3 *Representative integrals that may appear in the evaluation of $\mathbf{M}_{ij}$ and $\vec{b}_j$*

For example, if $\mathcal{L} = d^2/dx^2$, then $M_{ij} = \int \phi_j'' \phi_i \, dx$, or

$$\mathbf{M} = \frac{1}{h} \begin{pmatrix} -2 & 1 & 0 & \cdots & 0 \\ 1 & -2 & 1 & \cdots & 0 \\ 0 & 1 & -2 & \cdots & 0 \\ \vdots & \vdots & \vdots & \ddots & \vdots \\ 0 & 0 & 0 & \cdots & 2 \end{pmatrix}$$

EXAMPLE 20.2

Solution of the Previous Test Equation by the Finite Element Method

The boundary value problem solved in Example 20.1 by the finite difference method was

$$y'' + \frac{1}{4}y = x^2; \qquad y(0) = 0, \qquad y(\pi) = 0$$

The differential operator $\mathcal{L}$ is then

$$\mathcal{L} = \frac{d^2}{dx^2} + \frac{1}{4}$$

The computation of the matrix $\mathbf{M}$ using Table 20.3 yields

$$
\mathbf{M} = \frac{1}{h}
\begin{pmatrix}
-2 & 1 & 0 & \cdots & 0 \\
1 & -2 & 1 & \cdots & 0 \\
0 & 1 & -2 & \cdots & 0 \\
\vdots & \vdots & \vdots & \ddots & \vdots \\
0 & 0 & 0 & \cdots & 2
\end{pmatrix}
+ \frac{1}{4} \cdot \frac{h}{6}
\begin{pmatrix}
4 & 1 & 0 & \cdots & 0 \\
1 & 4 & 1 & \cdots & 0 \\
0 & 1 & 4 & \cdots & 0 \\
\vdots & \vdots & \vdots & \ddots & \vdots \\
0 & 0 & 0 & \cdots & 4
\end{pmatrix}
$$

and the vector $\vec{b}$ is

$$
\vec{b} = h^3
\begin{pmatrix}
1 \\ 2^2 \\ 3^2 \\ \vdots \\ n^2
\end{pmatrix}
+ \frac{1}{6} h^3
\begin{pmatrix}
1 \\ 1 \\ 1 \\ \vdots \\ 1
\end{pmatrix}
$$

The matrix equation for the c's is then,

$$
\begin{pmatrix}
-2 + \frac{h^2}{6} & 1 + \frac{h^2}{24} & 0 & \cdots & 0 \\
1 + \frac{h^2}{24} & -2 + \frac{h^2}{6} & 1 + \frac{h^2}{24} & \cdots & 0 \\
0 & 1 + \frac{h^2}{24} & -2 + \frac{h^2}{6} & \cdots & 0 \\
\vdots & \vdots & \vdots & \ddots & \vdots \\
0 & 0 & 0 & \cdots & -2 + \frac{h^2}{6}
\end{pmatrix}
\begin{pmatrix}
c_1 \\ c_2 \\ c_3 \\ \vdots \\ c_n
\end{pmatrix}
= h^4
\begin{pmatrix}
1 + \frac{1}{6} \\ 2^2 + \frac{1}{6} \\ 3^2 + \frac{1}{6} \\ \vdots \\ n^2 + \frac{1}{6}
\end{pmatrix}
$$

$$(20.17)$$

You will notice that this equation bears a strong similarity to the analogous equation obtained for the finite difference method, Equation 20.5. To understand further the similarities and differences of the two methods, we next solve the test equation once again for $n = 20$ intermediate points by both the finite difference method (Equation 20.10) and by the finite element method (Equation 20.17).[16] These solutions were then compared with the exact solution, Equation 20.9, and the average fractional error was computed. The average error and the average fractional error in the two calculations are compared in the following table.

	Finite difference method	Finite element method
Avg. error	0.0021	0.0022
Avg. frac. error	−0.00062	−0.00070

On the basis of these numbers there is little to distinguish the two methods. However, if the fractional error is plotted over the range of x values, a significant difference is revealed, as shown in Figure 20.8.

---

[16]In order to avoid being distracted by considerations relating to the iterative solution of these matrix equations, they were instead solved by finding the inverse of the coefficient matrices directly.

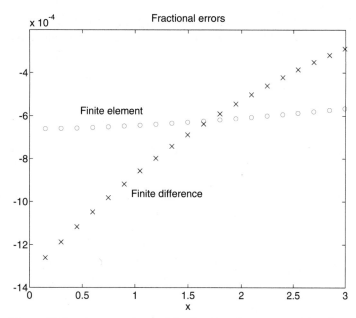

Figure 20.8   *A comparison of the fractional errors resulting from finite difference and finite element solutions of the same differential equation.*

This figure suggests an important distinction between the two methods. Although the accuracies of both methods are approximately the same and determined by the spacing of x values, the finite element error is more uniform across the complete range of the calculation.   ●

Problems

REINFORCEMENT EXERCISES

P20.1 Existence and Uniqueness.   Do the following boundary value problems have a unique solution? If not, are there then an infinite number of acceptable solutions or none?

     a. $y'' + 4y = 0$; $y(0) = 0,\ y(\pi) = 0$
     b. $y'' + 4y = 0$; $y(0) = 0,\ y(\frac{\pi}{4}) = 0$
     c. $y'' + 4y = 4$; $y(0) = 0,\ y(\pi) = 0$
     d. $y'' + 4y = 4$; $y(0) = 0,\ y(\pi) = 1$
     e. $y'' + 4y = 4$; $y(0) = 0,\ y(\frac{\pi}{3}) = 1$

P20.2 The Cannonball Problem.   In the cannonball problem, the missing initial conditions were the two components of the initial velocity, v_{x0}, v_{y0}, and the extra end conditions were $x(t_f) = R$, $y(t_f) = 0$. From the analytic solution to the problem, show that the condition for the existence of a solution is

$$v_{x0} v_{y0} = \frac{gR}{2}$$

P20.3 Cannonball Problem with Air Drag.   Solve the cannonball problem if there is an additional term in each of the differential equations that represents the effects of air drag; i.e.,

$$m\ddot{x} = -\gamma\dot{x}$$
$$m\ddot{y} = -\gamma\dot{y} - g$$

Use $R = 1000$, $t_f = 25$, $\gamma = 0.2$, and $m = 1$ and plot the final trajectory.

P20.4 Finite Difference.   Solve the boundary value problem,

$$y'' + \frac{y}{9} = 5\sin\frac{x}{2}; \qquad y(0) = 0, \; y(\pi) = 0$$

by the finite difference method, first with $\Delta x = \pi/10$, and then with $\Delta x = \pi/20$. Compare the actual error in the two calculations. [*Note:* The exact solution is $y_{\text{exact}}(x) = 24\sqrt{3}\sin(x/3) - 36\sin(x/2)$.]

P20.5 Finite Element Techniques.   The algebra involved in evaluating the integrals of the "hat" functions can be greatly reduced by first proving the following statement:

$$\int_a^b f(x)\phi_k(x) \approx -hP''(x)|_{x=kh} = hf(kh)$$

where $P(x) = \int xf(x)\,dx - x\int f(x)\,dx$. Use the explicit expressions for $\phi_k(x)$ to show that

$$\int_a^b f(x)\phi_k(x) = -\frac{1}{h}[P((k+1)h) - 2P(kh) + P((k-1)h)] \approx -hP''(kh)$$

with $P(kh) \equiv M(kh) - khN(kh)$ and $M(x) = \int xf\,dx$, $N(x) = \int f\,dx$.

P20.6 Finite Element.   Show that the finite element method applied to the differential equation in Problem 20.4 yields the following equation:

$$\left(\mathbf{A} + \frac{h^2}{54}\mathbf{B}\right)\vec{y} = \vec{b}$$

with

$$\mathbf{A} = \begin{pmatrix} -2 & 1 & 0 & \cdots & 0 & 0 \\ 1 & -2 & 1 & \cdots & 0 & 0 \\ \vdots & \vdots & \vdots & \ddots & \vdots & \vdots \\ 0 & 0 & 0 & \cdots & 1 & -2 \end{pmatrix} \qquad \mathbf{B} = \begin{pmatrix} 4 & 1 & 0 & \cdots & 0 & 0 \\ 1 & 4 & 1 & \cdots & 0 & 0 \\ \vdots & \vdots & \vdots & \ddots & \vdots & \vdots \\ 0 & 0 & 0 & \cdots & 1 & 4 \end{pmatrix}$$

and

$$\vec{b} = 5h^2 \sin\left(\frac{h}{2}\vec{n}\right)$$

Solve this equation for $n = 10$ and $n = 20$ and compare the accuracy with that of the exact solution.

P20.7 Shooting.   Rewrite the differential equation in Problem 20.4 as two coupled differential equations. From the expression for the exact solution, determine the value of the missing initial condition, $y_2(0)$. Solve this boundary value problem by the shooting method using an initial guess for the missing initial condition that is quite close to the exact value, say $y_2^g = 0.9y_2(0)$. Next, rerun the calculation, starting with a poor guess, say, $y_2^g = -y_2(0)$.

P20.8 Eigenvalue Boundary Value Problem. Bessel's equation of order zero,

$$y'' + \frac{1}{x}y' + k^2 y = 0; \qquad y(0) = 1, \qquad y(1) = 0$$

is a boundary value problem that is to be solved for both the solution $y(x)$ *and* for the appropriate eigenvalue k^2. Use the method outlined on page 437. For the normalization condition use $y'(1) = -1.248459$. [You will have to construct a function that finds the root of a function of *two* variables. Also, this problem is quite sensitive to the starting conditions, so start the integration from $x = 0.001$ instead.] Compare with the exact solution; i.e., $k = 2.4048\ldots$.

P20.9 Comparison of Methods. The boundary value problem,

$$y'' + y = 2e^{-x}; \qquad y(0) = 0,\ y(\pi/2) = 0$$

has an exact solution, $y_{\text{exact}}(x) = -e^{-\pi/2}\sin x - \cos x + e^{-x}$.
   a. Solve for the missing initial condition via the shooting method. Then compute the numerical solution to the equation and compare with the exact result. Use a tolerance of 10^{-4}.
   b. Solve the problem by the finite difference method for 9 interior points and compare with the exact result.
   c. Solve by using the finite element method for 9 interior points.
   d. Compare the accuracy of the finite element and the finite difference methods as a function of x across the range of the integration.

P20.10 Nonlinear Boundary Value Problem. The nonlinear boundary value problem,

$$(y-1)^2 y'' + 2(y-1)(y')^2 = 0; \qquad y(0) = 2,\ y(1) = 3$$

has an exact solution, $y_{\text{exact}}(x) = 1 + (7x+1)^{1/3}$. Solve this equation by the shooting method and compare with the exact solution.

P20.11 Duffing's Equation. The equation for an anharmonic oscillator under the influence of an external oscillatory driving force is known as *Duffing's* equation:

$$m\ddot{y} + \gamma\dot{y} + ky + \epsilon y^3 = F_0 \cos\omega_0 t$$

where ϵ is the coefficient of the anharmonic term and m, k, γ, F_0, and ω_0 are, respectively, the mass, the spring constant, the damping strength, the amplitude of the external force, and the angular frequency of the oscillations of the external force. Take $m = 1$, $k = 2$, $\gamma = 0.5$, $F_0 = 10$, $\omega_0 = \frac{\pi}{2}$, and $\epsilon = 0.1$. If the boundary conditions are $y(0) = 0$ and $y(1) = 1.5$, solve for the missing initial condition and then for the solution over the range $0 \leq t \leq 4$.

P20.12 Convergence Rate and Eigenvalues of the Jacobi Matrix. Duplicate some of the results summarized in Table 20.1. That is, for $n = 5, 10, 20, 40$ compute the maximum eigenvalue of the Jacobi matrix and use it to estimate the number of iterations required to attain an accuracy of 10^{-3}. Next, for $n = 20$, solve Equation 20.10 iteratively for the predicted number of iterations and compare the accuracy achieved with that predicted.

P20.13 Sparse Matrices.   Solve the boundary value problem of Example 20.1 for $n = 1000$ points. Do not use an iterative method. Instead, compute the inverse of the coefficient matrix directly using the toolbox routine to compute the inverse of a *sparse* matrix. Compare with the exact solution.

EXPLORATION PROBLEMS

P20.14 Infinite Boundary Conditions.   The exact solution of the boundary value problem,

$$y'' + 2xy' = 0; \qquad y(0) = 1, \ y(\infty) = 0$$

is

$$y(x) = 1 - \frac{2}{\pi} \int_0^x e^{-x^2} dx$$

$$= 1 - \mathrm{erf}(x)$$

To solve this problem by the shooting method, we must approximate the boundary condition $y(\infty) = 0$ by $y(\gamma) = 0$ with γ chosen large enough so that contributions to the integral for $x > \gamma$ are negligible, yet not so large as to force the procedure to spend most of its effort integrating zeros. Devise such a procedure and test it by comparing with the exact result. What is the effect of choosing a value for γ that is too small? (Try it.) What if γ is too large?

Summary

SHOOTING METHODS

Description

To solve a second-order *boundary* value problem by shooting, the problem is transformed into a search for the *missing* initial condition.

Shooting Algorithm

The problem is first recast as two coupled first-order equations, e.g.,

$$y_1' = f_1(x, y_1, y_2) \qquad y_1(x_0) = Y_0$$
$$y_2' = f_2(x, y_1, y_2) \qquad y_2(x_0) = \xi$$

plus the boundary condition, say $y_2(x_f) = Y_f$. The unknown initial condition is ξ. A function F is then constructed that will take a value of ξ, compute the solution by a stepping procedure across the entire range, determine $y_2(x_f)$, and assign a value to F as $F(\xi) = y_2(x_f) - Y_f$. Thus, the root of $F(\xi)$ will correspond to the missing initial condition.

RELAXATION METHODS

Finite Difference Method

The difference expression approximations for derivatives,

$$\frac{dy}{dx}\bigg|_{x_n} \approx \frac{y_{n+1} - y_{n-1}}{2h}$$

$$\frac{d^2 y}{dx^2}\bigg|_{x_n} \approx \frac{y_{n+1} - 2y_n + y_{n-1}}{h^2}$$

can be used to replace a differential equation by matrix equations that relate the solution at a point, y_i, to neighboring points. If the equation and the boundary conditions are linear, a change of variables can be found that will convert the boundary conditions to homogeneous; i.e., $y(x_0) = 0$, $y(x_f) = 0$. The matrix equations then uniquely define the solution at the chosen mesh points, within the validity of the approximations for the derivatives. Smaller values of Δx will improve the accuracy but will increase the size of the matrices.

Method of Weighted Residuals

The problem is again first transformed to one with homogeneous boundary conditions,

$$\mathcal{L}y(x) = r(x); \qquad y(x_0) = 0, \qquad y(x_f) = 0$$

where $\mathcal{L}$ represents the differential operators of the differential equation. Then the solution is expressed as an expansion in terms of n expansion functions $\phi_i(x)$, as

$$y(x) \approx Y(x) = \sum_{i=1}^{n} c_i \phi_i(x)$$

and *residual* functions that measure the accuracy of the solution are constructed as

$$\Delta(x) \equiv \mathcal{L}Y(x) - r(x) = \sum_{i=1}^{n} c_i \mathcal{L}\phi_i(x) - r(x)$$

Finally, a set of n *weight* functions is selected, and n conditions on the expansion coefficients are established by requiring that

$$\int_{x_0}^{x_f} \omega_j(x)\Delta(x)\,dx = 0 \qquad j = 1, \ldots, n$$

Finite Element Method

If *both* the expansion functions $\phi_i(x)$ and the weight functions $w_i(x)$ are chosen to be the same and equal to "hat" functions, defined as

$$\phi_k(x) = \begin{cases} \frac{1}{h}(x - x_{k-1}) = \frac{x}{h} - (k-1) & x_{k-1} \le x \le x_k \\ \frac{1}{h}(x_{k+1} - x) = -\frac{x}{h} + (k+1) & x_k \le x \le x_{k+1} \\ 0 & \text{all other } x \end{cases}$$

the method of weighted residuals is called the finite element method.

SOLUTION OF LARGE MATRIX EQUATION

Jacobi Iteration

A set of n linear equations can be iteratively solved by rewriting the ith equation for y_i, assuming a set of values for the quantities that appear on the right, and computing a new set of values for the y_i's on the left. An entire set of values y_i^{new} is computed from the starting set y_i^{old}. This process is repeated until convergence is attained.

Gauss-Seidel
Iteration

In this case the values of y_i^{new} replace those of y_i^{old} as they are computed during an iteration, thereby eliminating a need to store two sets of y values.

Successive
Overrelaxation
(SOR)

The convergence rate of the Gauss-Seidel method can be improved by slightly increasing the computed improvement to each y value by a relaxation factor ω, with ω most frequently in the range $1.1 \leq \omega \leq 1.9$.

Sparse Matrices

The matrices that result from either the finite difference method or the finite element method are always *banded*, *sparse*, and *large*. Such a matrix can then be mapped into a *sparse* matrix using the MATLAB function `sparse(A)` and the matrix equation $\mathbf{A}\vec{x} = \vec{b}$ solved in MATLAB as $x = \mathbf{A}\backslash\vec{b}$.

MATLAB Functions Used

`tic, toc`

It is possible to measure the elapsed time used by a statement or sequence of statements by preceding it with the statement `tic` to turn on the timer, and following with `toc` to turn the timer off and display the elapsed time.

`sparse`

The matrix `A` is converted to *sparse* form by `S = sparse(A)`. The ordinary arithmetic operators of MATLAB may then be used in conjunction with `S`.

21 / *The Method of Superposition*

PREVIEW Most numerical procedures are designed to have broad applicability. The stepping procedures of Chapter 19 can be used to solve almost any initial value problem of arbitrary order, linear or nonlinear. Similarly, the shooting method of Chapter 20 is usually the method of choice for general boundary value problems, and can often be used to successfully solve even nonlinear problems. Although the relaxation methods of Chapter 20 can only be applied to linear equations, they impose no limitation on the order of the equations.

The *method of superposition*, however, takes a different approach to solving differential equations. Rather than developing an algorithm that applies to a wide class of equations, we instead concentrate on procedures for a limited category, but one that includes the most commonly encountered equations. That is, the method of superposition is intended to be applied to a two-point boundary value problem that is *linear* and usually of second order. This type of equation is a standard feature of engineering and science and has been the subject of intense theoretical study for more than two centuries. As a result, there are a great many theoretical features peculiar to this type of equation that can be used to facilitate a numerical solution. We begin with a short review of some of these properties.

Theoretical Properties of Linear Boundary Value Problems

THE STANDARD FORM

Any nth-order linear differential equation can be written in *standard* form as

$$y^{[n]} + p_1(x)y^{[n-1]} + \cdots + p_n(x)y = r(x)$$

by dividing through by the coefficient of $d^n y/dx^n$. All the theoretical properties of solutions to differential equations will assume that the equation is written in standard form, *and* that each of the functions $p_i(x)$ is *continuous* and *bounded* over the range of x values in question.

THE HOMOGENEOUS SOLUTION

The most important property of linear differential equations is contained in the following theorem:[1]

---

[1] For a proof of this and other properties of linear differential equations, consult any textbook on differential equations, e.g., Greenberg, *Foundations of Applied Analysis*, Prentice-Hall, Englewood Cliffs, NJ, 1978.

A *linear, homogeneous* differential equation of order n has precisely n *linearly independent* solutions.

A homogeneous differential equation can be written symbolically as $\mathcal{L}y = 0$, where $\mathcal{L}$ represents all the combinations of differential operators and functions of the independent variable x acting on or multiplying the dependent variable y. Any terms not dependent on y are placed on the right-hand side of the equation. The n functions that are solutions of this equation are called *homogeneous* solutions. Thus, for an nth-degree equation, the set of homogeneous solutions is a set of functions, $\{y_1(x), y_2(x), \ldots, y_n(x)\}$.

For example, the equation $y'' + k^2 y = 0$ has homogeneous solutions

$$\{\sin kx, \cos kx\}.$$

An equivalent set is $\{e^{ikx}, e^{-ikx}\}$. That is, the first set can be expressed as a linear combination of the elements of the second. However, the members of either set are linearly independent of one another.

One particular set of linearly independent functions, called the *fundamental* set, is a collection of solutions corresponding to the following n different combinations of initial conditions, all at the same value of the independent variable x:

$$
\begin{array}{ccccc}
y_1(x_0) = 1, & y_1'(x_0) = 0, & y_1''(x_0) = 0, & \cdots & y_1^{[n-1]}(x_0) = 0 \\
y_2(x_0) = 0, & y_2'(x_0) = 1, & y_2''(x_0) = 0, & \cdots & y_2^{[n-1]}(x_0) = 0 \\
y_3(x_0) = 0, & y_3'(x_0) = 0, & y_3''(x_0) = 1, & \cdots & y_3^{[n-1]}(x_0) = 0 \\
\vdots & \vdots & \vdots & \ddots & \vdots \\
y_n(x_0) = 0, & y_n'(x_0) = 0, & y_n''(x_0) = 0, & \cdots & y_n^{[n-1]}(x_0) = 1
\end{array}
$$

Thus, the set $\{\cos kx, \sin kx\}$ is the fundamental set of homogeneous solutions to the equation $y'' + k^2 y = 0$ at $x_0 = 0$.

THE WRONSKIAN

The condition that the set of solutions be linearly independent means that the only set of constants $\{c_i, i = 1, \ldots, n\}$ that can be found such that

$$c_1 y_1(x) + c_2 y_2(x) + \cdots + c_n y_n(x) \equiv 0$$

where the expression equals zero for all values of x for which the functions y_i are defined, is the set $c_1 = c_2 = c_3 = \cdots c_n = 0$. Successive differentiation of this relation to obtain n conditions on the c's yields the matrix equation,

$$
\begin{pmatrix}
y_1(x) & y_2(x) & y_3(x) & \cdots & y_n(x) \\
y_1'(x) & y_2'(x) & y_3'(x) & \cdots & y_n'(x) \\
y_1''(x) & y_2''(x) & y_3''(x) & \cdots & y_n''(x) \\
\vdots & \vdots & \vdots & \ddots & \vdots \\
y_1^{[n-1]}(x) & y_2^{[n-1]}(x) & y_2^{[n-1]}(x) & \cdots & y_n^{[n-1]}(x)
\end{pmatrix}
\begin{pmatrix}
c_1 \\
c_2 \\
c_3 \\
\vdots \\
c_n
\end{pmatrix}
= 0
$$

The condition that the functions be linearly independent; i.e., that there *not* exist any nonzero solution vector for the c's, is that the determinant of the coefficient matrix be nonzero. This determinant, which is a function of x, is called the *Wronskian*, $W(x)$,

$$
W(x) = \begin{vmatrix}
y_1(x) & y_2(x) & y_3(x) & \cdots & y_n(x) \\
y_1'(x) & y_2'(x) & y_3'(x) & \cdots & y_n'(x) \\
y_1''(x) & y_2''(x) & y_3''(x) & \cdots & y_n''(x) \\
\vdots & \vdots & \vdots & \ddots & \vdots \\
y_1^{[n-1]}(x) & y_2^{[n-1]}(x) & y_3^{[n-1]}(x) & \cdots & y_n^{[n-1]}(x)
\end{vmatrix}
\tag{21.1}
$$

Again, the condition that n solutions of the nth-order, homogeneous differential equation, $\mathcal{L}y = 0$, be linearly independent, is that the Wronskian constructed from those functions be nonzero for all x. Thus, the functions $\{\sin kx, \cos kx\}$ are solutions of $y'' + k^2 y = 0$, and have a Wronskian

$$
W(x) = \begin{vmatrix}
\sin kx & \cos kx \\
\cos kx & -\sin kx
\end{vmatrix} = -1
$$

THE GENERAL HOMOGENEOUS SOLUTION

The general solution to the equation

$$
y^{[n]} + p_1(x)y^{[n-1]} + \cdots + p_n(x)y = 0
$$

exists wherever the functions $p_i(x)$ are continuous and bounded, for *any* linearly independent set of n boundary conditions, and can be *uniquely* expressed as a linear combination of the homogeneous solutions

$$
y(x) = \sum_{i=1}^{n} c_i y_i(x)
$$

The n coefficients in this expansion are determined from the n boundary conditions. Thus, any solution to $y'' + k^2 y = 0$ can be written $y(x) = A \sin kx + B \cos kx$.

With unspecified boundary conditions, the preceding equation is often called the *complementary solution*, and the coefficients c_i are undetermined. These constants may then be interpreted as the n constants of integration resulting from an integration of the differential equation.

THE PARTICULAR SOLUTION

The solution of the general *inhomogeneous* differential equation,

$$
y^{[n]} + p_1(x)y^{[n-1]} + \cdots + p_n(x)y = r(x)
$$

can be written as

$$
y(x) = \sum_{i=1}^{n} c_i y_i(x) + y_p(x)
\tag{21.2}
$$

$$
= y_c(x) + y_p(x)
$$

that is, as a sum of the *complementary* solution and a *particular* solution. Because there are already n undetermined coefficients in the complementary solution, the particular solution contains *no* undetermined constants.

The particular solution is *any* function that satisfies the inhomogeneous differential equation, whereas the complementary solution satisfies the equation $\mathcal{L}y = 0$.

Thus, the theoretical solution of an nth-order inhomogeneous differential equation will consist of first solving the homogeneous equation for the set of solutions $y_i(x)$ contained in the complementary solution and then attempting to find a particular function, $y_p(x)$, that satisfies the inhomogeneous equation.

The Superposition Method

The majority of initial value problems can be solved numerically using any of the several stepping procedures described at the beginning of this chapter. Our primary concern at the moment is to use some of the theoretical properties we have just outlined to facilitate numerical solutions of boundary value problems.

If the boundary conditions are *linear* conditions on the function $y(x)$ and its first $n - 1$ derivatives at only two points, x_0 and x_f, we can always find a change of variables such that the boundary conditions in the new coordinates are *homogeneous*.[2]

SOLUTION OF BOUNDARY VALUE PROBLEMS
USING THE FUNDAMENTAL SOLUTIONS

Each of the complete set of n independent solutions of the homogeneous equation, $\mathcal{L}y = 0$, belonging to the *fundamental* set of complementary solutions, is an *initial* value problem and can be determined by relatively simple stepping procedures. Furthermore, because the only requirement on $y_p(x)$ is that it be a solution of the inhomogeneous equation, we can likewise define it to be a solution of an initial value problem, namely,

$$\mathcal{L}y_p = y_p^{[n]} + a_1(x)y_p^{[n-1]} + a_2(x)y_p^{[n-2]} + \cdots + a_n(x)y_p = r(x)$$

with initial conditions

$$y_p(0) = y_p'(0) = y_p''(0) = \cdots = y_p^{[n-1]}(0) = 0$$

Thus, all the pertinent functions can be obtained by solving initial value problems. The solution to the *boundary* value problem at hand is a linear combination (or *superposition*) of this set of $n + 1$ functions, and it is given by Equation 21.2. Finally, the constants, c_i, in this equation are determined by imposing the boundary conditions.

The principal task is to formulate an algorithm that will simultaneously solve for all the fundamental solutions (or as many as may be required) plus the particular solution, in one pass of a stepping method like ode23 or ode45.

---

[2]See the discussion on page 443 of Chapter 20.

This algorithm will depend critically on the specific boundary conditions of the problem, and it is best illustrated with an example.

EXAMPLE 21.1

Solution of an Arbitrary Third-Order Inhomogeneous Boundary Value Problem as a Superposition of a Combination of Initial Value Solutions

Suppose the boundary value problem we wish to solve is third-order and inhomogeneous, e.g.,

$$\mathcal{L}y = y''' + a(x)y'' + b(x)y' + c(x)y = r(x) \tag{21.3}$$

with boundary conditions

$$y(0) = y''(0) = 0, \qquad y(L) = 0$$

Three independent functions that are solutions of $\mathcal{L}y = 0$ with the three independent sets of initial conditions,

$$
\begin{aligned}
y_1(0) &= 1, & y_1'(0) &= 0, & y_1''(0) &= 0 \\
y_2(0) &= 0, & y_2'(0) &= 1, & y_2''(0) &= 0 \\
y_3(0) &= 0, & y_3'(0) &= 0, & y_3''(0) &= 1
\end{aligned}
$$

can easily be obtained by any of the stepping procedures. Thus, the general solution is

$$y(x) = c_1 y_1(x) + c_2 y_2(x) + c_3 y_3(x) + y_p(x)$$

where y_p is the particular solution to the inhomogeneous differential equation. The particular solution, $y_p(x)$, can be computed by solving the initial value problem,

$$y_p''' + a(x)y_p'' + b(x)y_p' + c(x)y_p = r(x)$$

where

$$y_p(0) = y_p'(0) = y_p''(0) = 0$$

Thus, all four of the functions y_1, y_2, y_3, and y_p can be solved by solving an *initial* value problem using simple stepping procedures. The solution of the *boundary* value problem is a combination of these functions.

Notice that, using the boundary conditions associated with the original differential equation, we obtain

$$
\begin{aligned}
y(0) = 0 &\implies c_1 = 0 \\
y''(0) = 0 &\implies c_3 = 0 \\
y(L) = 0 &\implies c_1 y_1(L) + c_2 y_2(L) + c_3 y_3(L) + y_p(L) = 0 \\
& \qquad c_2 = -y_p(L)/y_2(L)
\end{aligned}
$$

so that

$$y(x) = \frac{[y_2(L)y_p(x) - y_2(x)y_p(L)]}{y_p(L)} \tag{21.4}$$

In retrospect, we see that we do not require the solution for either $y_1(x)$ or $y_3(x)$.

To be more specific, the preceding problem can be structured for one of the stepping codes like `ode23` in the following way. If we designate the only required element of the complementary solution as $y_c \equiv y_2(x)$, then

$$y_c'''(x) = -a(x)y_c'' - b(x)y_c' - c(x)y_c$$
$$y_p'''(x) = r(x) - a(x)y_p'' - b(x)y_p' - c(x)y_p$$

so that we could define the following relationships:

$$
\begin{array}{llll}
Y_1 = y_c & Y_1' = Y_2; & & Y_1(0) = 1 \\
Y_2 = y_c' & Y_2' = Y_3; & & Y_2(0) = 0 \\
Y_3 = y_c'' & Y_3' = -a(x)Y_3 - b(x)Y_2 - c(x)Y_1; & & Y_3(0) = 1 \\
Y_4 = y_p & Y_4' = Y_5; & & Y_4(0) = 0 \\
Y_5 = y_p' & Y_5' = Y_6; & & Y_5(0) = 0 \\
Y_6 = y_p'' & Y_6' = r(x) - a(x)Y_6 - b(x)Y_5 - c(x)Y_4; & & Y_6(0) = 0
\end{array}
$$

It is then an easy matter to code the function required for the derivative functions, $\vec{Y}'$, and to use `ode23` to solve for Y_i, $i = 1, \ldots, 6$, over the range $0 \le x \le L$. Finally, the solution to the original inhomogeneous equation, Equation 21.3, is a simple combination of Y_1 and Y_4.[3]   ●

A METHOD FOR SECOND-ORDER EQUATIONS

A very common task in science and engineering is solving a second-order, inhomogeneous, boundary value problem where the left-hand side of the equation; i.e., the differential form, corresponds to a "named" equation, such as Bessel's or Legendre's equation, and the right-hand side and the boundary conditions are specific to the particular problem at hand. To guide the discussion, we will use Bessel's equation written in standard form:

$$y'' + \frac{1}{x}y' + k^2 y = 0 \qquad \text{for } 0 < x \le b$$

Because this equation is second-order homogeneous, we know it must have two linearly independent solutions in the specified range of x values. We will call these two functions $J(x)$ and $Y(x)$.[4] The complete solution to the homogeneous differential equation is then

$$y(x) = AJ(x) + BY(x)$$

---

[3]Unfortunately, the MATLAB functions `ode23` and `ode45` and most other stepping-type differential equation solver programs use an adaptive step size to maintain accuracy. This means that each of the four independent solutions of the differential equation, $y_i(x)$, $i = 1, \ldots, 3$, and $y_r(x)$ will be determined at a different set of x values, and thus, cannot be combined as in Equation 21.4. Remedies for this situation can be found in Problem 21.1.

[4]These functions are technically Bessel functions (of zeroth order) of the first and second kind, and they are generally written as $J_0(kx)$ and $Y_0(kx)$. Bessel functions of higher integer order p ($J_p(kx)$ and $Y_p(kx)$) are solutions of the generalized Bessel equation,

$$y'' + \frac{1}{x}y' + \left(k^2 - \frac{p^2}{x^2}\right)y = 0$$

where, for problems of this type, the two functions, $J(x)$ and $Y(x)$, will be *known* functions.

The solution to the inhomogeneous problem, $\mathcal{L}y = r(x)$, is

$$y(x) = AJ(x) + BY(x) + y_p(x)$$

where $y_p(x)$ is the particular solution and A and B are unspecified constants. If the particular solution is known, we simply match this equation to the two given boundary conditions to obtain the unique solution.

Thus, the only problem left is to obtain the particular solution to equations of this type. The most common approach for computing $y_p(x)$ is by applying a method called *variation of parameters*.

Variation of Parameters

The complete solution to the inhomogeneous problem can be written as

$$y(x) = A(x)J(x) + B(x)Y(x) \tag{21.5}$$

where the expansion constants have been generalized to be completely unspecified functions. There is considerable flexibility in how these functions will be defined. Our only constraint is that $y(x)$ satisfy the inhomogeneous differential equation. The differential equation is second order, so we must next evaluate y' and y''. In the expression for y',

$$y'(x) = (AJ' + BY') + (A'J + B'Y)$$

we choose to set the terms in the second set of parentheses, $A'J + B'Y$, equal to zero. The second derivative is then

$$y''(x) = (AJ'' + BY'') + (A'J' + B'Y')$$

This time, the terms in the second set of parentheses are set equal to the inhomogeneous term, $r(x)$; i.e., $A'J' + B'Y' = r(x)$. Thus,

$$y = (AJ' + BY')$$
$$y' = (AJ' + BY')$$
$$y'' = (AJ'' + BY'') + r(x)$$

Except for the term $r(x)$, these are the same expressions for y, y', and y'' obtained when A and B were constants. Thus, this prescription for $y(x)$ is clearly a formal solution of the inhomogeneous equation $\mathcal{L}y = r$. The conditions that were imposed on the functions $A(x)$ and $B(x)$ can be summarized as

$$\begin{pmatrix} J & Y \\ J' & Y' \end{pmatrix} \begin{pmatrix} A' \\ B' \end{pmatrix} = \begin{pmatrix} 0 \\ r(x) \end{pmatrix} \tag{21.6}$$

which is a set of two, *coupled*, first-order differential equations. Expressions for A' and B' can be obtained by constructing the inverse of the 2×2 matrix, whose determinant is the Wronskian, $W(x)$. The result is

$$\begin{pmatrix} A' \\ B' \end{pmatrix} = \frac{1}{W(x)} \begin{pmatrix} -rY \\ +rJ \end{pmatrix}$$

These expressions can then be integrated from $x = x_0$ to $x = x$ to obtain $A(x)$ and $B(x)$,

$$\begin{pmatrix} A(x) \\ B(x) \end{pmatrix} = \begin{pmatrix} A_0 \\ B_0 \end{pmatrix} + \int_{x_0}^{x} \frac{r(\xi)}{W(\xi)} \begin{pmatrix} -Y(\xi) \\ +J(\xi) \end{pmatrix} d\xi$$

Finally, these expressions are substituted into Equation 21.5 to obtain

$$y(x) = A_0 J(x) + B_0 Y(x) + \int_{x_0}^{x} \frac{r(\xi)}{W(\xi)} [J(x)Y(\xi) - J(\xi)Y(x)] \, d\xi \qquad (21.7)$$

Although the analysis for deriving Equation 21.7 is somewhat complicated, you should, nevertheless, appreciate the significance of this result. Equation 21.7 solves a second-order inhomogeneous boundary value problem by means of a *single* integration. There is no searching for missing initial conditions as in the shooting method, nor any error terms associated with a finite step size as in all relaxation methods. The equation is, in principle, *exact*. As a result, whenever the differential equation has *known* homogeneous solutions; i.e., whenever the differential equation is one of a class of "named" equations, the method of superposition should be used to obtain a solution. A short list of some common "named" equations and the forms for their homogeneous solutions is given in Table 21.1.

Bessel's equation $\quad 0 \le x < \infty$	
$y'' + \dfrac{1}{x}y' + \left(k^2 - \dfrac{\nu^2}{x^2}\right)y = 0$	$\nu = \text{integer}$ $J_\nu(kx), \qquad J_\nu(0) = 0$ $Y_\nu(kx), \qquad Y_\nu(0) \to \infty$
	$\nu \ne \text{integer}$ $J_\nu(kx), \; J_{-\nu}(kx)$
Legendre's equation $\quad -1 \le x \le 1$	
$(1 - x^2)y'' - 2xy' + \ell(\ell + 1)y = 0$	$\ell = \text{positive integer}$ $P_\ell(x), \qquad P_\ell(1) = 1$ $Q_\ell(x), \qquad Q_\ell(1) \to \infty$
"Spherical" Bessel equation $\quad 0 \le x < \infty$	
$y'' + \dfrac{2}{x}y' + \left[1 - \dfrac{\ell(\ell + 1)}{x^2}\right]y = 0$	$\ell = \text{positive integer}$ $j_\ell(x), \qquad j_\ell(0) = 0$ $\eta_\ell(x), \qquad \eta_\ell(0) \to \infty$

Table 21.1   *Examples of standard second-order differential equations*

Problems

REINFORCEMENT EXERCISES

***P21.1* Second-Order Boundary Value Problem.**   The differential equation,

$$y'' + \frac{1}{4}y = x^2; \qquad y(0) = 0, \ y(\pi) = 0$$

has been solved by a variety of methods in Chapter 20.

a. Show that the solution can be written as

$$y(x) = y_p(x) - \frac{y_p(\pi)}{y_2(\pi)} y_2(x)$$

where y_p and y_2 are solutions of the two *initial* value problems

$$y_2'' + \frac{1}{4}y_2 = 0; \qquad y_2(0) = 0, \ y_2'(0) = 1$$
$$y_p'' + \frac{1}{4}y_p = x^2; \qquad y_p(0) = 0, \ y_p'(0) = 0$$

b. To numerically solve the problem, two independent differential equations must be solved. However, to combine these into the final solution for $y(x)$, both solutions must be evaluated at the same set of x values; something that **ode23** or **ode45** is unlikely to accomplish. Instead, write a code to solve both equations simultaneously; that is, a solution based upon the following definitions:

$$
\begin{array}{llll}
Y_1 = y_2 & Y_1' = Y_2 & Y_1(0) = 0 \\
Y_2 = y_2' & Y_2' = -\frac{1}{4}Y_1 & Y_2(0) = 1 \\
Y_3 = y_p & Y_3' = Y_4 & Y_3(0) = 0 \\
Y_4 = y_p' & Y_4' = -\frac{1}{4}Y_3 - x^2 & Y_4(0) = 0
\end{array}
$$

c. Compare the efficiency of this method with the shooting method.

***P21.2* Numerical Solution of Example 21.1.**   Code the solution to Example 21.1 in terms of a MATLAB M-file function $dY = DerivY(Y,x)$ that computes the six elements of the derivatives of Y. Use this function to solve the differential equation,

$$x^3 y''' - x^2 y'' - 2xy' + 6y = \sqrt{x}$$

subject to the boundary conditions, $y(1) = y''(1) = 0$ and $y(4) = 0$ over the interval $1 \leq x \leq 4$. Do not use a method that employs adaptive step size. Instead, use the modified midpoint method of Chapter 19 (see page 416) to construct a code that employs evenly spaced points. Next, use three independent homogeneous solutions to this equation, $y_1(x) = x^2$, $y_2(x) = x^3$, $y_3(x) = x^{-1}$, and a particular solution, $y_p(x) = (8/45)\sqrt{x}$, to obtain the exact solution. Compare the exact solution with your numerical result.

***P21.3* Second-Order Equations.**   Listed in the following table are several second-order equations along with their complementary solutions. Compute the Wronskian for each set of functions and, from an equation analogous to Equation 21.6, determine the expression for the complete solution.

Differential equation	Homogeneous solutions	Interval	Inhomogeneous term
$(x-1)y'' - xy' + y = r(x)$	x, e^x	$[0, 1]$	$(x-1)^2$
$x^2 y'' - 4xy' + 6y = r(x)$	x^2, x^3	$[0, 1]$	$x^2 + x \ln x$
$x^2 y'' + xy' + y = r(x)$	$\sin(\ln x), \cos(\ln x)$	$[1, 2]$	x^2

P21.4 Damped Oscillations. The differential equation for a damped oscillator driven by a periodic external force is

$$m\frac{d^2 y}{dt^2} + \gamma \frac{dy}{dt} + ky = F_0 \sin \omega_0 t$$

where m is the mass, k is the spring constant, γ is the damping constant, and F_0 is the strength of the external force, which has an angular frequency, ω_0. Defining

$$\alpha = \frac{\gamma}{2m}, \qquad \beta^2 = \frac{k}{m} - \alpha^2$$

the two homogeneous solutions are

$$y_1(t) = e^{-(\alpha + i\beta)t} \quad \text{and} \quad y_2(t) = e^{-(\alpha - i\beta)t}$$

If $\beta^2 < 0$, the system is called *overdamped*, and the homogeneous solution will simply execute exponential decay. If $\beta^2 > 0$, the homogeneous solution will oscillate with an angular frequency, β, with an amplitude that decreases exponentially. Use the method of variation of parameters to determine equations for the derivatives of the coefficients of y_1 and y_2 in the general solution, and then use the toolbox function ode23 to obtain a numerical solution for these coefficients. Use the values $m = k = 1$, $\gamma = \frac{1}{2}$, $F_0 = 10$, and $\omega_0 = \frac{\pi}{4}$, for the parameters, and $y(0) = 0$ and $y(\omega_0 t) = 0$ for the boundary conditions. Next, the particular solution can be shown to be

$$y_p(t) = \left[(\alpha^2 + \beta^2)^2 + 4\alpha^2 \omega_0^2 \right]^{-1} \frac{F_0}{m} \left[(\alpha^2 + \beta^2) \cos \omega_0 t + 2\alpha \omega_0 \sin \omega_0 t \right]$$

Use this expression to determine the exact solution and compare it with the numerical results.

P21.5 Resonance. If the driving frequency, ω_0, in Problem 21.4 is close to the *natural* frequency of the oscillator, the system will experience *resonance*. That is, for the *no damping* case, $\gamma = 0$, the particular solution will *diverge*, and if $\gamma \neq 0$, the solution will have a maximum amplitude of oscillation. This situation corresponds to the *resonant conditions* for the system. Plot the solution as a function of γ for the case of resonance.

Summary

THEORETICAL PROPERTIES

Existence

A *linear, homogeneous* differential equation of order n has precisely n *linearly independent* solutions.

Linear	If $y_i(x)$, $i = 1, \ldots, n$, are solutions of a linear homogeneous differential equation, they
Independence	form a linearly independent set if their Wronskian, defined as

$$W(x) = \begin{vmatrix} y_1(x) & y_2(x) & y_3(x) & \cdots & y_n(x) \\ y_1'(x) & y_2'(x) & y_3'(x) & \cdots & y_n'(x) \\ y_1''(x) & y_2''(x) & y_3''(x) & \cdots & y_n''(x) \\ \vdots & \vdots & \vdots & \ddots & \vdots \\ y_1^{[n-1]}(x) & y_2^{[n-1]}(x) & y_3^{[n-1]}(x) & \cdots & y_n^{[n-1]}(x) \end{vmatrix}$$

is nonzero.

The Fundamental	A set of n linearly independent boundary conditions that, in turn, defines a linearly
Set of LI	independent set of solutions is given by
Solutions	

$$\begin{aligned}
y_1(x_0) &= 1, & y_1'(x_0) &= 0, & y_1''(x_0) &= 0, & \cdots & & y_1^{[n-1]}(x_0) &= 0 \\
y_2(x_0) &= 0, & y_2'(x_0) &= 1, & y_2''(x_0) &= 0, & \cdots & & y_2^{[n-1]}(x_0) &= 0 \\
y_3(x_0) &= 0, & y_3'(x_0) &= 0, & y_3''(x_0) &= 1, & \cdots & & y_3^{[n-1]}(x_0) &= 0 \\
&\vdots & &\vdots & &\vdots & \ddots & & &\vdots \\
y_n(x_0) &= 0, & y_n'(x_0) &= 0, & y_n''(x_0) &= 0, & \cdots & & y_n^{[n-1]}(x_0) &= 1
\end{aligned}$$

The functions that result from these conditions are called the set of *fundamental* solutions.

General	The general solution to the homogeneous differential equation of order n is a linear
Homogeneous	combination of the n linearly independent solutions:
Solution	

$$y(x) = \sum_{i=1}^{n} c_i y_i(x)$$

The n expansion coefficients must be determined by imposing n initial or boundary conditions.

Particular	The solution to an inhomogeneous differential equation of order n is the general homo-
Solution	geneous solution plus the *particular* solution. The particular solution, $y_p(x)$, contains
	no undetermined coefficients and is any function that satisfies the inhomogeneous equa-
	tion.

METHODS OF SOLUTION FOR INHOMOGENEOUS EQUATIONS

The	If the extra boundary condition,
Superposition	
Method	

$$y_p(0) = y_p'(0) = y_p''(0) = \cdots = y_p^{[n-1]}(0) = 0$$

is added to the *fundamental* set, the $n + 1$ linearly independent functions y_i, $i = 1, \ldots, n$ plus y_p will be sufficient to determine both the homogeneous and the particular solutions, and thus the general solution to an inhomogeneous equation. *All* of these functions can be obtained simultaneously by a stepping procedure.

Variation of Parameters

If the expansion coefficients in the general homogeneous solution are generalized to be functions themselves,

$$y(x) = \sum_{i=1}^{n} c_i(x) y_i(x)$$

and this expression is further assumed to be a solution of the *inhomogeneous* equation, differential equations of order $n - 1$ result for each of the coefficients, $c_i(x)$. In particular, for a second-order inhomogeneous equation, the resulting first-order equations for $c_1(x)$ and $c_2(x)$ can be written as simple integrals, and the complete solution can be expressed in closed form.

22 / *Introduction to Partial Differential Equations*

PREVIEW   The solution of partial differential equations (PDEs), both analytical and numerical, is the single, most important topic in applied analysis. Virtually all areas of science and engineering phrase problems in the language of PDEs, and every practicing professional is expected to be adept at their solution. The literature is very extensive on this subject, and most of it is at a quite advanced level. However, the basic ideas and concepts related to solutions of PDEs are much the same as those used to solve ordinary differential equations (ODEs). Because PDEs are concerned with differential equations in two or more independent variables, it is more common to encounter computational problems.

Typically, for PDEs the memory and computation time requirements are increased by an order of magnitude, round-off error is more likely a concern, pathologies in the solution are more frequent, and, overall, the calculation is likely to be more complicated, bigger, and more expensive. Techniques that address each of these concerns have been developed, and for state-of-the-art methods, we refer the reader to several of the books in the Bibliography.

In this chapter, we attempt to understand the basic distinction between the various types of PDEs, to outline the computational approach appropriate to each type, and to illustrate the manner of obtaining a solution for representative equations using MATLAB. The numerical methods developed in this chapter will be suitable for the majority of problems encountered that relate to PDEs, and should provide the reader with sufficient background to pursue more advanced readings on the subject. It is hoped that the short introductory treatment contained in this chapter will entice the reader to experiment further in this very interesting subject.

Finally, two features of MATLAB are especially suited to facilitating the solution of PDEs: sparse matrix techniques, and the ease of obtaining three-dimensional graphs. Some numerical procedures for PDEs invariably require the solution of very large matrix equations. Frequently, these equations can now be solved *directly* in MATLAB using sparse matrices, thus considerably reducing the nature of the problem. Also, the interpretation of a solution of two independent variables is greatly aided by MATLAB's three-dimensional graphing capabilities.

Preliminary Concepts

We will restrict our attention in this chapter to the simplest partial differential equations, chosen to illustrate the principal computational techniques. In most cases, these techniques are readily generalized to more complicated equations. Our example equations will be second-order differential equations, linear in the dependent variable, and will be restricted to equations in only two independent variables. The first step in each of the procedures will be to approximate derivatives by a variety of finite difference equations; thus generalizing to higher order will usually entail using a higher-order difference expression. A nonlinear equation, frequently a very difficult analytical problem, quite often only requires a minor adjustment in the numerical procedure, although the potential for more complicated difficulties is legion. Problems involving more than two independent variables can be very difficult to solve numerically, especially in MATLAB, which restricts arrays to only two indices.

We begin with a few definitions and the classification of the types of partial differential equations. It is assumed that the reader has encountered our demonstration equations earlier and is familiar with the introductory techniques used to obtain analytical solutions, particularly the method of separation of variables.

THE STANDARD FORM

The standard form of a second-order, linear, partial differential equation in two independent variables is

$$A\frac{\partial^2 u}{\partial x^2} + 2B\frac{\partial^2 u}{\partial x \partial y} + C\frac{\partial^2 u}{\partial y^2} = F\left(x, y, u, \frac{\partial u}{\partial x}, \frac{\partial u}{\partial y}\right)$$

where $u(x, y)$ is the dependent variable, and $F(x, y, u, \partial u/\partial x, \partial u/\partial y)$ is an arbitrary function of x and y, and is linear in u and its first derivatives.

A few common examples are given in the following table:

		A	B	C	F
$\dfrac{\partial^2 u}{\partial x^2} = \dfrac{1}{c^2}\dfrac{\partial^2 u}{\partial t^2}$	One-dimensional wave equation	1	0	$-\dfrac{1}{c^2}$	0
$\dfrac{\partial^2 u}{\partial x^2} + \dfrac{\partial^2 u}{\partial y^2} = 0$	Two-dimensional Laplace's equation	1	0	1	0
$D\dfrac{\partial^2 T}{\partial x^2} = -\dfrac{\partial T}{\partial t}$	One-dimensional diffusion equation	D	0	0	$-\dfrac{\partial T}{\partial t}$

All of these equations are homogeneous, second-order PDEs. For numerical solutions, inhomogeneous forms of these equations are solved similarly.

BOUNDARY CONDITIONS

As with ordinary differential equations, the method of solution of a PDE is dictated by the types of boundary conditions specified in the problem. For ODEs, the choices are either initial value problems or boundary value problems. In either case, boundary values are required to guarantee that the equation has a unique and stable solution. In initial value problems, sufficient information regarding the function and its derivatives are given at a single point. Then, using finite difference expressions of the form

$$y_{i+1} = y_i + \left.\frac{dy}{dx}\right|_i \Delta x + \frac{1}{2} \left.\frac{d^2y}{dx^2}\right|_i (\Delta x)^2 + \cdots$$

the solution is constructed in the neighborhood of the point, in a stepping manner. (The initial conditions specify the function and its first derivatives, and the differential equation is used to determine higher-order derivatives in the series.) Thus, the solution is constructed emanating from the initial point, and the natural method of solution is a stepping procedure.

As you know, methods to obtain numerical solutions to boundary value problems differ substantially from those used for initial value problems. Procedures based on finite difference equations seek to iteratively refine an initial guess for the solution to the actual solution *simultaneously* at all points on the grid. This usually leads to solutions of a large set of simultaneous linear equations.

The distinction between stepping methods and matrix methods applies to PDEs as well, where, once again, the appropriate procedure is dictated by the type of boundary conditions given in the problem. However, a particular stepping method or a particular boundary value method is never suitable for all types of PDEs.[1] The form of the necessary boundary conditions to guarantee a unique and stable solution depends, to some extent, on the type of PDE. The appropriate boundary conditions will involve specifying the solution, its normal derivative, or both, on a boundary of the surface (or volume) containing the points where the solution is to be computed. The standard notation for these conditions is as follows:

Dirichlet   The values of only the *function*, u, are given on a boundary.

Neumann   The values of only the *normal derivatives* of the function are given on a boundary.

Cauchy   The values of *both* the function and its normal derivative are specified on the same boundary.

CLASSIFICATION OF PDEs

To specify the appropriate boundary conditions sufficient to guarantee a unique, stable solution to a particular PDE, the equation is first classified as being

---

[1]See Morse and Feshbach, *Methods of Theoretical Physics*, vol. 1, McGraw-Hill, New York, 1953.

one of three types: elliptic, parabolic, or hyperbolic, by evaluating the quantity $(B^2 - AC)$, where A, B, and C are the coefficients in the standard form of the PDE. The classification is then

$$B^2 - AC = \begin{array}{ll} < 0 & \textit{Elliptic} \\ 0 & \textit{Parabolic} \\ > 0 & \textit{Hyperbolic} \end{array}$$

Thus, of the examples given in the preceding table, the wave equation is hyperbolic, Laplace's equation is elliptic, and the diffusion equation is parabolic. These equations are the prototypical equations for the discussion of all second-order, linear PDEs. If a PDE is a function of more than two independent variables, it may be a combination of types; for example, elliptic in x and y, but hyperbolic in x and t. The determination of the appropriate boundary conditions for each type of equation can be found in any text on applied mathematics. The results are summarized in Table 22.1, which is adapted from a similar table in Morse and Feshbach (1953).

		TYPE OF PDE		
Type of boundary condition		**Elliptic** (*Poisson's equation*) $\frac{\partial^2 u}{\partial x^2} + \frac{\partial^2 u}{\partial y^2} = \rho(x,y)$ Equilibrium distribution	**Parabolic** (*Heat conduction*) $\frac{\partial T}{\partial t} = -D\frac{\partial^2 T}{\partial x^2}$ Diffusion	**Hyperbolic** *The wave equation* $\frac{\partial^2 u}{\partial t^2} = v^2\frac{\partial^2 u}{\partial x^2}$ Wave propagation
Dirichlet or *Neumann* on	OPEN Surface	*Not enough*	**UNIQUE, STABLE solution in one direction only**	*Not enough*
	CLOSED Surface	**UNIQUE, STABLE solution**	*Too much*	*Too much*
Cauchy on	OPEN Surface	*Unphysical results*	*Too much*	**UNIQUE, STABLE solution**
	CLOSED Surface	*Too much*	*Too much*	*Too much*

Table 22.1   *Boundary conditions appropriate to PDEs*

The contents of this table are easily remembered by considering the physical phenomena described by each differential equation.

Poisson's Equation—Elliptic   Poisson's equation can be used to determine the equilibrium values of the static electric potential. The natural conditions would be to specify the function or its derivative (the electric field); i.e., Dirichlet or Neumann conditions, on a surrounding boundary and then to determine the potential everywhere in the interior. If the source term, $\rho(x, y)$, is zero, the equation is known as Laplace's equation. Figure 22.1 shows the boundary conditions.

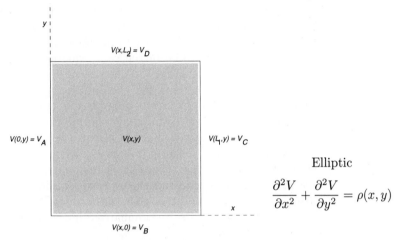

Elliptic

$$\frac{\partial^2 V}{\partial x^2} + \frac{\partial^2 V}{\partial y^2} = \rho(x, y)$$

Figure 22.1 *Appropriate boundary conditions for elliptic equations.*

The Diffusion Equation—Parabolic The diffusion or heat flow equation can be used to determine the temperature distribution in a metal bar as a function of position and time. The natural conditions would be to specify the temperature everywhere in the bar at time $t = 0$, and use the differential equation to determine the temperature at later times by stepping along the time axis. Figure 22.2 shows the boundary conditions.

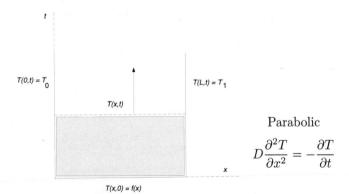

Parabolic

$$D\frac{\partial^2 T}{\partial x^2} = -\frac{\partial T}{\partial t}$$

Figure 22.2 *Appropriate boundary conditions for parabolic equations.*

The Wave Equation—Hyperbolic If a string is tightly connected between two horizontally separated points, ignoring gravity, the transverse or vertical displacement of the string is a function of position along the string, x, and time, t; i.e., $y(x, t)$. This function is then obtained by solving the wave equation. The appropriate boundary conditions for such a problem would be to fix the ends $(y(0, t) = 0, y(L, t) = 0)$ and to specify the initial displacement of the string. For example, the string might be set in motion by plucking it at the middle. That is, at $t = 0$, $y(x, 0) = f(x)$, where $f(x)$ is a function describing the shape of the initial displacement, *and* at $t = 0^-$, the string is stationary, $|\partial y / \partial t|_{t=0} = 0$. These conditions correspond to *Cauchy* conditions on an open surface, as is illustrated in Figure 22.3. [*Note:* It is implied that $\partial y / \partial x|_{x=0} = \partial y / \partial x|_{x=L} = 0$.]

The numerical solution of the wave equation uses known values for y and $\partial y / \partial t$ at a particular value of t to determine the value of y at a later time value. As with parabolic equations, like the diffusion equation, the natural method of solution is one of stepping in the time variable (the variable with open domain).

From the preceding examples, it appears that the methods to obtain a numerical solution of partial differential equations bear a strong resemblance to methods previously applied to ordinary differential equations. The methods are either of an *initial value* type, requiring stepping methods for a solution, or of a *boundary value* type, requiring iterative techniques. For numerical work, this distinction is more important than is the classification of elliptic, parabolic, or hyperbolic. These results are summarized in the following chart:

PDE classification	Example equations	Num. Method classification	Method of solution
Hyperbolic Parabolic	Wave equation Diffusion equation	Initial value boundary conditions	Stepping methods
Elliptic	Poisson's equation	Boundary value on closed surface	Iterative matrix methods

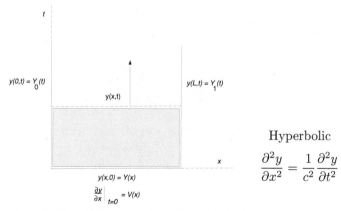

$$\text{Hyperbolic}$$

$$\frac{\partial^2 y}{\partial x^2} = \frac{1}{c^2} \frac{\partial^2 y}{\partial t^2}$$

Figure 22.3 *Appropriate boundary conditions for hyperbolic equations.*

Each of the two numerical method classes has its Achilles' heel. For stepping methods it is *stability*. Some seemingly sound stepping procedures for initial value problems will always diverge from the exact solution; i.e., they are unstable. For boundary value problems, stability is usually easy to obtain; the main concern is excessive computing time and memory size. Solving for the solution on even a modest grid will require the solution of matrix equations of extremely large order. The methods applied for each class are quite different, but both begin with the replacement of the partial derivative by *finite difference* equations.

FINITE DIFFERENCE EQUATIONS

Let the solution to the differential equation be designated as $u(x, y)$, where the two independent variables are x and y. If we then specify each of the independent variables on a grid as

$$x_i = x_0 + i\Delta x \qquad i = 0, \ldots, N_x$$
$$y_j = y_0 + j\Delta y \qquad j = 0, \ldots, N_y$$
$$u_{ij} \equiv u(x_i, y_j)$$

the partial derivatives of u can be approximated by finite difference expressions in the x and y directions.

Forward difference
$$\left.\frac{\partial u}{\partial x}\right|_{(x_i y_j)} \approx \frac{(u_{i+1,j} - u_{ij})}{\Delta x} + \mathcal{O}(\Delta x)$$

$$\left.\frac{\partial^2 u}{\partial x^2}\right|_{(x_i y_j)} \approx \frac{(u_{i+2,j} - 2u_{i+1,j} - u_{ij})}{(\Delta x)^2} + \mathcal{O}(\Delta x)$$

Central difference
$$\left.\frac{\partial u}{\partial x}\right|_{(x_i y_j)} \approx \frac{(u_{i+1,j} - u_{i-1,j})}{2\Delta x} + \mathcal{O}(\Delta x^2)$$

$$\left.\frac{\partial^2 u}{\partial x^2}\right|_{(x_i y_j)} \approx \frac{(u_{i+1,j} - 2u_{ij} - u_{i-1,j})}{(\Delta x)^2} + \mathcal{O}(\Delta x^2)$$

The forward difference equations are accurate to only first order in the step size Δx, whereas the central difference equations are accurate to second order. Finite difference equations accurate to higher orders can be derived[2] but are rarely used.

The first step in numerically solving a partial differential equation is to approximate the differential equation by a finite difference expression. But which finite difference expression should be used, forward, central, or reverse, and to what order? How do the results depend upon the step sizes chosen for the two variables? Does the step size in one variable affect the choice of step size in the second variable? To answer these questions, we begin with the simplest of procedures developed for the two classes of numerical solutions: *initial value type* (parabolic and hyperbolic) and *boundary value type* (elliptic).

---

[2]Hoffmann (1992), p. 175.

Initial Value Type Methods

STEPPING METHODS FOR PARABOLIC EQUATIONS

The prototypical parabolic equation is the diffusion or heat equation,

$$D\frac{\partial^2 T}{\partial x^2} = -\frac{\partial T}{\partial t} \qquad (22.1)$$

and boundary conditions sufficient to guarantee a unique solution would be

$$\begin{aligned} T(0,t) &= f_0(t) \\ T(L,t) &= f_L(t) \end{aligned} \qquad T(x,0) = g(x) \qquad (22.2)$$

A solution is then desired for the region $0 \le x \le L$, $0 \le t \le T$, where T could be infinity. A rectangular grid is next superimposed on this region, with equal increments, Δx, in the space coordinate, and Δt in the time coordinate, as illustrated in Figure 22.4.

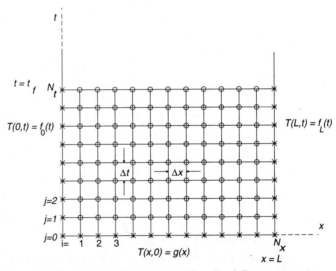

Figure 22.4   *Finite difference grid for the diffusion equation.*

At the beginning of the calculation, the points labeled with open circles are unknown, and those marked with an asterisk are known. The grid points are then

$$\begin{aligned} x_i &= i\Delta x & i &= 0,\dots,N_x & \Delta x &= L/N_x \\ t_j &= j\Delta t & j &= 0,\dots,N_t & \Delta t &= L/N_t \end{aligned} \qquad (22.3)$$

Explicit Methods—The FTCS Equations   To obtain a numerical solution, the differential equation is next approximated by a finite difference equation. We will first try a *forward time* and *centered space* (FTCS) set of difference expressions to obtain

$$D\frac{T_{i+1,j} - 2T_{i,j} + T_{i-1,j}}{(\Delta x)^2} = \frac{T_{i,j+1} - T_{i,j}}{\Delta t} \tag{22.4}$$

The calculation proceeds in the t direction. That is, to obtain the values of the solution along the $j+1$ line, the complete set of values along the earlier j line is used. If Equation 22.4 is rearranged to put the $j+1$ elements on the left, we have

$$T_{i,j+1} = \gamma T_{i-1,j} + (1 - 2\gamma)T_{i,j} + \gamma T_{i+1,j} \tag{22.5}$$

where $\gamma \equiv D\Delta t/(\Delta x)^2$.

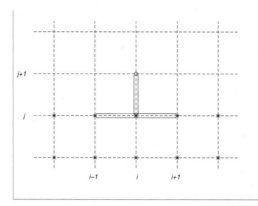

Figure 22.5   *Three points along the j line are used to determine the point at $(i, j+1)$.*

At each step in the procedure, all of the values along the current j line are known; thus, the entire set of values for $j+1$ can be determined directly. See Figure 22.5. This method, the FTCS method, is termed an *explicit* method.[3] The only remaining question concerns the accuracy and stability of the method, and the answer is not good. In a more comprehensive treatment of this method[4] it is shown that the method is *unstable* if

$$\gamma = \frac{D\Delta t}{(\Delta x)^2} > \frac{1}{2} \tag{22.6}$$

---

[3]Implicit methods are based on difference equations that employ more than one element on the unknown $j+1$ line; for example, $y_{j+1} = y_j + v_{j+1}\Delta t$, where $v_j = v(y_j, t_j) = dy/dt$. These equations cannot be solved directly for y_{j+1} starting from y_j, but are instead solved iteratively. That is, a set of values is guessed for y_{j+1}, and the difference equation is used to improve that guess. See also the discussion of predictor-corrector methods for ordinary differential equations in Chapter 19.

[4]Fox, *Numerical Solution of Ordinary and Partial Differential Equations*, Pergamon Press, New York, 1962.

For choices of Δt and Δx that yield $\gamma \leq 1/2$, the method is guaranteed stable and accurate. Although this sounds encouraging, in practice it is prohibitively restrictive. For example, consider solving the diffusion equation for the temperature, $T(x,t)$, in a bar of length $L = 1$m, with each end held at a fixed temperature. The constant, D, is called the thermal diffusivity and depends on the geometry and material of the bar. For a metal bar a typical value is $D = 10^{-4}$m^2/s. If we choose a value of Δx sufficiently small to obtain an accurate solution along a fixed t line, perhaps $\Delta x = 10^{-2}$m, in order for the method to be stable, the corresponding time step must be less than $(\Delta t)_{max} = (\Delta x)^2/2D = 5 \times 10^{-6}$s. Thus, obtaining a result for realistic values of time will require excessive computation time in even the simplest situations. In almost all cases, an alternative procedure must be found.

Implicit Methods—The Crank-Nicolson Equations   A slight variation in the previous procedure is to evaluate the second derivative along the $j+1$ line instead of using the known values along the j line. That is,

$$D\frac{T_{i+1,j+1} - 2T_{i,j+1} + T_{i-1,j+1}}{(\Delta x)^2} = \frac{T_{i,j+1} - T_{i,j}}{\Delta t} \tag{22.7}$$

Once again, the equation is rearranged by collecting the to-be-determined terms from the $j+1$ line on the left side of the equation as

$$\gamma T_{i+1,j+1} - (1+2\gamma)T_{i,j+1} + \gamma T_{i-1,j+1} = T_{i,j} \tag{22.8}$$

where, as earlier, $\gamma = D\Delta t/(\Delta x)^2$. This one equation contains three as yet unknown quantities, so it cannot be used directly in a stepping procedure. However, when all the equations for a specific value of j are collected together, they can be arranged into a single matrix equation,

$$\begin{pmatrix} 1+2\gamma & -\gamma & 0 & 0 & \cdots & 0 & 0 \\ -\gamma & 1+2\gamma & -\gamma & 0 & \cdots & 0 & 0 \\ 0 & -\gamma & 1+2\gamma & -\gamma & \cdots & 0 & 0 \\ \vdots & \vdots & \vdots & \vdots & \ddots & \vdots & \vdots \\ 0 & 0 & 0 & 0 & \cdots & 1+2\gamma & -\gamma \\ 0 & 0 & 0 & 0 & \cdots & -\gamma & 1+2\gamma \end{pmatrix} \begin{pmatrix} T_{1,j+1} \\ T_{2,j+1} \\ T_{3,j+1} \\ \vdots \\ T_{N_x-1,j+1} \\ T_{Nx,j+1} \end{pmatrix}$$

$$= \begin{pmatrix} T_{1,j} \\ T_{2,j} \\ T_{3,j} \\ \vdots \\ T_{N_x-1,j} \\ T_{Nx,j} \end{pmatrix} + \begin{pmatrix} f_0(t_j) \\ 0 \\ 0 \\ \vdots \\ 0 \\ f_L(t_j) \end{pmatrix} \tag{22.9}$$

or

$$\mathbf{A}\vec{T}_{j+1} = \vec{T}_j + \vec{b}$$

The coefficient matrix is, for sufficiently small γ, diagonally dominant, banded, and sparse. Thus, this equation is ideally suited for either iterative techniques or sparse matrix techniques. However, more importantly, this formulation of the difference equations can be shown[5] to be *stable* for all values of γ. The only restraints on the choice of step size are those dictated by accuracy and computation time considerations.

The Crank-Nicolson Method Equation 22.9 has the advantage of being unconditionally stable, and second-order accurate in Δx, but the difference expression for the time derivative is only first-order accurate in Δt. Crank and Nicolson[6] suggested amending the equation so that it is second order in both Δx and Δt, while still depending on only two values of the time variable. The idea is to use central difference expressions for both $(\partial^2 T/\partial x^2)$ and $\partial T \partial t$. The difference expression for $\partial T \partial t$ remains the same, but it is now interpreted as representing the derivative at the midpoint between j and $j+1$. The second derivative, $(\partial^2 T/\partial x^2)$, at the same point is then taken as the average of the difference expressions in x evaluated at j and $j+1$. That is,

$$D\frac{1}{2}\left(\frac{T_{i+1,j+1}-2T_{i,j+1}+T_{i-1,j+1}}{(\Delta x)^2}+\frac{T_{i+1,j}-2T_{i,j}+T_{i-1,j}}{(\Delta x)^2}\right)=\frac{T_{i,j+1}-T_{i,j}}{\Delta t}$$

(22.10)

This is illustrated in Figure 22.6.

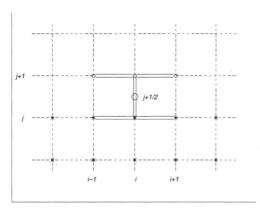

Figure 22.6 *Crank-Nicolson equations are a second-order approximation in both variables to the solution at the point $(x_i, t_{j+(1/2)})$.*

Rearranging terms in Equation 22.10 by collecting the $j+1$ terms on the left once again results in a matrix equation:

$$\mathbf{A}(\gamma)\vec{T}_{j+1} = \mathbf{A}(-\gamma)\vec{T}_j + \vec{b}$$

(22.11)

---

[5]Fox (1962).

[6]Crank and Nicolson, "A Practical Method for Numerical Evaluation of Solutions of Partial Differential Equations of the Heat-Conduction Type," *Proc. Cambridge Philosophical Soc.*, vol. 43, no. 50, p. 50, 1947.

where

$$\mathbf{A}(\gamma) = \begin{pmatrix} 1+\gamma & -\frac{\gamma}{2} & 0 & \cdots & 0 \\ -\frac{\gamma}{2} & 1+\gamma & -\frac{\gamma}{2} & \cdots & 0 \\ 0 & -\frac{\gamma}{2} & 1+\gamma & \cdots & 0 \\ \vdots & \vdots & \vdots & \ddots & \vdots \\ 0 & 0 & 0 & \cdots & 1+\gamma \end{pmatrix}$$

and

$$\vec{b} = \begin{pmatrix} \frac{1}{2}[f_0(t_0) + f_0(t_1)] \\ 0 \\ \vdots \\ 0 \\ \frac{1}{2}[f_L(t_{N_t-1}) + f_L(t_{N_t})] \end{pmatrix}$$

In essentially all problems involving the heat or diffusion-type partial differential equation, the Crank-Nicolson equations are found to converge more quickly than Equation 22.9, often much more quickly. Thus, the Crank-Nicolson method is usually the recommended approach. The reader should, however, be aware that there are very sophisticated codes that alternate between implicit and explicit equations, choosing the most appropriate and most accurate method at each step of the calculation. For our purposes, we will only illustrate the Crank-Nicolson method applied to a relatively simple problem for which an analytical solution is available.

NUMERICAL METHODS OF SOLUTION

The numerical solution of either Equation 22.7 or Equation 22.11 proceeds by either *direct* or *indirect* methods. Direct methods solve the matrix equations by Gaussian elimination. Because the matrices are often very large, banded, and sparse, the techniques of sparse matrices are almost always required. Indirect methods solve the equations at each j line iteratively using the method of Gauss-Seidel. Direct methods can suffer from round-off error and memory limitations. Accumulation of round-off error is not a problem in Gauss-Seidel methods; however, to optimize the convergence rate, the iterative procedure must be tuned by a procedure known as successive overrelaxation (SOR). If feasible, direct methods are preferred to indirect.

Direct Methods—Sparse Matrix Techniques   If the grid spacing along the x axis is not too small; e.g., $\Delta x > L/50$, it is easy to construct the matrices required in Equation 22.11 and then to solve the equations for each value of time by using backslash division. However, if more than 100 grid points are required along the x axis, memory limitations as well as excessive computation times will usually prohibit the execution of the calculation. The coefficient matrix is sparse; therefore, the techniques of sparse matrices are essential to obtaining a direct solution. The idea is simply to store *only* the nonzero elements along with their row-column location.

MATLAB has extremely easy-to-use sparse matrix functions. The most basic is `sparse(A)`. The use is `As = sparse(A)`, which will squeeze out all of the nonzero elements of `A` and store the remainder as a sparse matrix in `As`. Thus, if

$$A = \begin{pmatrix} 4 & -2 & 0 & 0 & 0 \\ -2 & 4 & 0 & 0 & 0 \\ 0 & 0 & 6 & 0 & 0 \\ 0 & 0 & 0 & 4 & -2 \\ 0 & 0 & 0 & -2 & 4 \end{pmatrix}$$

the MATLAB statement, `As = sparse(A)`, yields

```
As =
    (1,1)        4
    (2,1)       -2
    (1,2)       -2
    (2,2)        4
    (3,3)        6
    (4,4)        4
    (5,4)       -2
    (4,5)       -2
    (5,5)        4
```

Most importantly, the operations of matrix arithmetic: addition, multiplication, and backslash division, have been written in MATLAB to automatically convert to the more efficient sparse matrix algorithms whenever both matrices are sparse, or to automatically convert a sparse matrix to a full matrix if necessary.

For example, the statements

```
N = 450;
A = diag(ones(N,1)) + diag(2*ones(N-1,1),1);
tic;  C = A\A;  toc;
A = sparse(A);
tic;  C = A\A;  toc;
```

compare the execution times for backslash matrix division for a very large sparse matrix using ordinary arithmetic and sparse matrix arithmetic. If you have sufficient memory in your computer to run this code, you will observe more than a factor of 20 improvement in using the sparse matrix arithmetic. The comparison using similar, but smaller, matrices is less dramatic.

EXAMPLE 22.1

Numerical Solution of the Diffusion Equation

The conditions of a standard heat flow or diffusion equation problem are

$$T_0 = 100 \qquad T_L = 0 \quad \text{for } 0 \leq t \leq \infty$$
$$T(x)|_{t=0} = 0 \quad \text{for } 0 \leq x \leq L$$

and describe the temperature in a bar that is held at fixed temperatures, T_0 and T_L at each end, and starts at a uniform temperature of $0°$ at $t = 0$.

The quantity D, called the *thermal diffusivity*,[7] is tabulated here for a number of materials.

Material	Thermal diffusivity (m^2/s)
Copper	117×10^{-6}
Aluminum	85.5×10^{-6}
Iron	23.1×10^{-6}
Steel	11.7×10^{-6}
Water	1.14×10^{-6}
Glass	$(0.35\text{--}0.46) \times 10^{-6}$
Wood	0.11×10^{-6}

The code to solve for the temperature in the bar is then constructed by solving Equation 22.11. The computed temperature is shown graphically in Figure 22.7.

```
%Script file to solve the diffusion equation
%
  L   = 2;                % Length of bar
  dx = L/100              % step size
   x = [dx:dx:L-dx]';     % x-grid points, (interior of bar)
   n = length(x):         % size of matrices
  f0 = 100;  fL = 0;      % Fixed Temps at ends of bar.
%
   D = input('Please enter the diffusivity.  ');
%
  dt = 20                 % time step (in seconds)
   g = D*dt/dx^2;
% Construct the matrices A(+g) and A(-g)
%
  Ap = sparse(diag((1+g)*ones(n,1)) - ...
              diag(ones(n-1,1),1)*g/2 - ...
              diag(ones(n-1,1),-1)*g/2);
  Am = sparse(diag((1-g)*ones(n,1)) + ...
              diag(ones(n-1,1),1)*g/2 + ...
              diag(ones(n-1,1),-1)*g/2);
```

[7]The relation between the thermal diffusivity, D, and the thermal conductivity, k, is

$$D = \frac{k}{\rho C_p}$$

where ρ is the material density, and C_p is the material's specific heat at constant pressure (all in SI units).

```
T = zeros(n,1);          % Start with all zeros for T
% The right-hand-side vector b
%
  b = g*[f0 zeros(1,n-2) fL]';
  plot(x,T);hold on
  for t = dt:dt:30*60           % For thirty minutes
      T = Ap\(Am*T+b);
      if fix(t/60) == t/60; % plot every 60 seconds
          plot(x,T)
      end
  end
  hold off
```

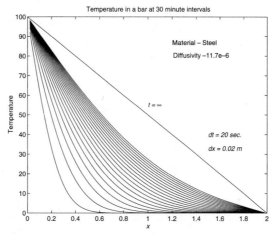

Figure 22.7 *The computed temperature in a bar as a function of time and position. The ends of the bar are fixed at 100 and 0 degrees, respectively.*

Indirect Methods—Gauss-Seidel Iteration The standard iterative method of solving the matrix equation $\mathbf{A}\vec{x} = \vec{b}$ is the Gauss-Seidel procedure, which was described in Chapter 20. Because direct methods employing sparse matrix techniques are so easy to use in MATLAB, iterative methods are competitive only in the following special situations: (a) If the coefficient matrix is *extremely large*; e.g., 1000×1000 or larger,[8] it is possible that even with sparse matrix techniques, memory limitations will prohibit a direct method. (b) If the accuracy of a direct calculation is less than satisfactory, the property of each step of an iterative method to correct errors in the previous step may tip the balance in favor of the Gauss-Seidel method. Thus, the only instances in which you will be using the Gauss-Seidel method are when solving relatively large matrix

---

[8] As we shall see, when solving elliptic equations, matrix equations of such enormous size are not uncommon.

equations. It is then essential that the rate of convergence be optimized by the method of *successive overrelaxation* (SOR). The Gauss-Seidel method converges monotonically, with each step edging closer to the solution, rarely overshooting. Thus, artificially increasing the correction in each step will likely improve the convergence rate. The correction factor is called the relaxation factor, ω, and is found to be between 1 and 2. (See Chapter 20.)

To employ the Gauss-Seidel method in the solution of a parabolic equation like the heat equation, the single line in the code to obtain the solution at the next time step on page 488; i.e.,

```
T = Ap\(Am*T+b)
```

must be replaced by an entire Gauss-Seidel iteration code.

STEPPING METHODS FOR HYPERBOLIC EQUATIONS

The prototypical hyperbolic partial differential equation is the wave equation,

$$v^2 \frac{\partial y}{\partial x^2} - \frac{\partial y}{\partial t^2} = 0$$

with boundary conditions

$$y(t,0) = f_0(t) \qquad y(t,L) = f_L(t) \quad \text{for } 0 \le t \le \infty$$

$$y(0,x) = Y_0(x) \qquad \left.\frac{\partial y}{\partial t}\right|_{t=0} = V_0(x) \quad \text{for } 0 \le x \le L$$

where v is the wave speed, Y_0 is the initial position, and V_0 is the initial transverse speed of the displacement $y(x,t)$.

Approximating the partial derivatives by finite difference expressions yields a stepping procedure similar to that used for parabolic equations. However, before doing that, it should be pointed out that the most effective and popular method of numerical solution of hyperbolic equations is via the *method of characteristics*. Although a detailed description of this method is beyond the scope of this text, a brief outline may be helpful in deciding whether or not to pursue this technique in obtaining a solution.

The Method of Characteristics   If the wave speed is a constant, the change of variables

$$\xi = x - vt$$
$$\eta = x + vt$$

transforms the wave equation into

$$\frac{\partial^2 y}{\partial \xi \, \partial \eta} = 0$$

which has the formal solution

$$y(\xi, \eta) = f(\xi) + g(\eta)$$

or

$$y(x,t) = f(x + vt) + g(x - vt)$$

where f and g are arbitrary functions. The initial conditions are then used to determine these functions,

$$f(x) + g(x) = Y_0(x)$$
$$f'(x) - g'(x) = V_0(x)$$

which yields the *d'Alembert* solution:

$$f(x) = \frac{1}{2}Y_0(x) + \frac{1}{2v}\int_0^x V_0(\xi)d\xi + \frac{1}{2}[f(0) - g(0)]$$

$$g(x) = \frac{1}{2}Y_0(x) - \frac{1}{2v}\int_0^x V_0(\xi)d\xi - \frac{1}{2}[f(0) - g(0)]$$

(22.12)

Notice that once the functions f and g are determined, the time evolution of the solution is easily visualized. The solution at $t = 0$ is, of course, $y(x,0) = f(x) + g(x)$. At a later time t_1, $y(x,t_1)$ is once again a sum of the exact same functions; however, the positions of the functions have been shifted along the x axis. The function f is shifted to the right by vt_1, and g is shifted to the left by the same amount. This is illustrated in Figure 22.8.

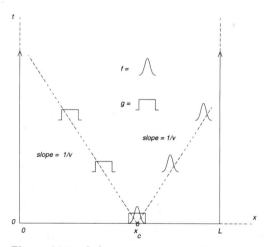

Figure 22.8 *Information contained in y and $\partial y/\partial t$ at $t = 0$ is propagated, undistorted, along the characteristic line.*

The characteristic curves at the point $t = 0$, $x = x_c$ are, in this case, straight lines with slope $\pm 1/v$. This strongly suggests that an efficient method for computing the solution to the wave equation is to integrate the solution out along each characteristic line for each value of x; that is, to replace stepping in t with stepping along a characteristic line. This is indeed the preferred method for solving hyperbolic equations, particularly when the wave speed, v, is a function of x and/or t, which clearly leads to situations wherein the characteristic lines are more complicated curves. For a detailed description of this method, see Abbott (1966).

Finite Difference Solution of the Wave Equation Approximating the second derivatives in the wave equation by central difference expressions yields

$$v^2 \left[\frac{y_{i+1,j} - y_{i,j} + y_{i-1,j}}{(\Delta x)^2} \right] - \left[\frac{y_{i,j+1} - y_{i,j} + y_{i,j-1}}{(\Delta t)^2} \right] = 0$$

where, as earlier, $y_{i,j} = y(x_i, t_j)$ and $x_i = x_0 + i\Delta x$, $t_j = t_0 + j\Delta t$. If we further define

$$\gamma^2 \equiv v^2 \frac{(\Delta t)^2}{(\Delta x)^2} \tag{22.13}$$

and rearrange the equation, putting the latest value of y on the left, we obtain

$$y_{i,j+1} = [\gamma y_{i+1,j}^y + 2(1 - \gamma^2)y_{i,j} + \gamma y_{i-1,j}^y] - y_{i,j-1} \tag{22.14}$$

It is helpful to display the structure of this equation by means of a graph connecting the points involved in the equation. See Figure 22.9.

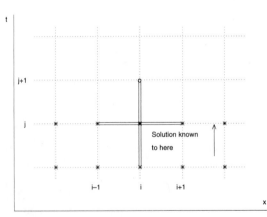

Figure 22.9 *Information needed to compute the value of $y_{i,j+1}$ is contained in the points $y_{i+1,j}$, $y_{i,j}$, $y_{i-1,j}$, and $y_{i,j-1}$.*

To construct a stepping procedure based on Equation 22.14, the equation is first written in matrix form, as

$$\vec{y}_{j+1} = \mathbf{A}\vec{y}_j - \vec{y}_{j-1} + \gamma^2 \vec{b}_j \tag{22.15}$$

with

$$\mathbf{A} = \begin{pmatrix} 2(1-\gamma^2) & \gamma^2 & 0 & \cdots & 0 \\ \gamma^2 & 2(1-\gamma^2) & \gamma^2 & \cdots & 0 \\ 0 & \gamma^2 & 2(1-\gamma^2) & \cdots & 0 \\ \vdots & \vdots & \vdots & \ddots & \vdots \\ 0 & 0 & 0 & \cdots & 2(1-\gamma^2) \end{pmatrix}$$

and

$$\vec{b}_j = \begin{pmatrix} f_0(t_j) \\ 0 \\ 0 \\ \vdots \\ 0 \\ f_L(t_j) \end{pmatrix}$$

Because each step of the algorithm requires two earlier values of y, we need to use the initial condition on the transverse velocity to determine the first two sets of y values. Thus,

$$\vec{y}_1 = \vec{y}_0 + \vec{v}_0\, dt$$

The stepping procedure based on Equation 22.14 can be shown to be stable if $\gamma^2 < 1$.

A code that illustrates the solution of the wave equation is given in Figure 22.10. In this code, the ends are fixed at $y(0, t) = y(100, t) = 0$; the initial displacement is zero, except for $25 \le x \le 33$, where there is a small parabolic pulse; and the initial transverse velocity is zero. The code then graphs the solution every 10 time steps, pausing for one-fifth step after each graph, and displays the result in the form of a movie. If you run this code, you will notice that the pulse initially splits in two, with each half propagating in opposite directions. When a pulse hits either end, the pulse flips over and reflects from the end. The two pulses pass through each other without distortion. After several reflections, you will notice the growth of round-off error in the solution in the form of small "wiggles" near the base of the pulse. This error can be reduced by reducing the step size (and increasing the computation time), or by going to a higher-order algorithm. Three frames of this movie are shown in Figure 22.11.

The reader should be warned that solutions to hyperbolic equations frequently contain features that make it extremely difficult to obtain a numerical solution. For example, the wave solution may develop *shock waves* representing discontinuities, the wave may oscillate wildly, making it impossible to accurately compute the solution, or small errors may grow in time and degrade the solution. As illustrated by the preceding example, if a solution is only required for a limited period of time, and if the resulting wave patterns make only a few passes through the region of interest during this time, the simple methods used here should suffice.

```
%Script file to solve the wave equation and
%display the results as a movie.
%
  g2 = .6;
  c  = 30;
  N  = 300;
  L  = 100;              dx = L/(N);
  dt = sqrt(g2)*dx/c;
  x  = [dx:dx:L-dx]';    n = length(x);
  nl = fix(n/4);         nr = fix(n/3);
  xl = x(nl);            xr = x(nr);
  Y0 = zeros(n,1);
   I = find(x >= xl & x <= xr);
   Y0(I) = (x(I)-xl).*(xr-x(I));
  Y0 = Y0/max(Y0);
  plot(x,Y0); hold on; axis([0 L -1 1]);
  plot([0 L],[0 0]);pause(1)
  A = sparse(diag(2*(1-g2)*ones(n,1)) + ...
      g2*diag(ones(n-1,1),1) + g2*diag(ones(n-1,1),-1));
  Y1 = Y0;        hold off
  for t = 2*dt:dt:1500*dt
      Y2 = A*Y1-Y0;
      Y0 = Y1;          Y1 =Y2;
      if(fix(t/10/dt) == t/10/dt)
          plot(x,Y1);    axis([0 L -1 1]);
          pause(.2)
      end
  end
  hold off
```

Figure 22.10 *MATLAB code to integrate the wave equation and display the results in the form of a movie. (See also file* wveqscpt.m *on the applications disk.)*

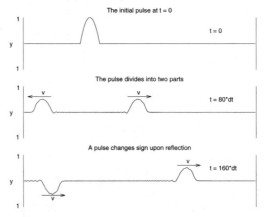

Figure 22.11 *Three frames of the movie representing a solution of the wave equation.*

Boundary Value Problems—Elliptic Equations

For elliptic partial differential equations to have a stable, unique solution, the solution must be specified on a *closed* boundary surrounding the region of interest. This requirement dramatically alters the mode of obtaining a numerical solution. For example, consider the prototypical elliptical equation, Poisson's equation,

$$\frac{\partial^2 V}{\partial x^2} + \frac{\partial^2 V}{\partial y^2} = \rho(x, y)$$

where $\rho(x, y)$ is a given *source* function, and $V(x, y)$ is the desired solution. This equation is next approximated by replacing the derivatives by central difference expressions to obtain

$$\left[\frac{V_{i+1,j} - 2V_{i,j} + V_{i-1,j}}{(\Delta x)^2}\right] + \left[\frac{V_{i,j+1} - 2V_{i,j} + V_{i,j-1}}{(\Delta y)^2}\right] = \rho_{i,j} \qquad (22.16)$$

where $V_{i,j} = V(x_i, y_j)$, and $x_i = x_0 + i\Delta x$, $y_i = y_0 + j\Delta y$. The five points that are related in Equation 22.16 are connected by a diagram analogous to that used for the wave equation, Figure 22.9. However, with the wave equation we begin with information about the solution on *two* j lines and are thus able to compute the solution on the third. With boundary conditions appropriate to Poisson's equation, the solution is initially known on only the single $j = 0$ line, and therefore there is insufficient information to start at $j = 0$ and step to the next j value. This is likewise true if we start on the $i = 0$ line and attempt to step in i values. Instead, a method of solution must be constructed that seeks to compute the solution at *all* interior points at once.

Consider a simple situation in which the solution is specified on a rectangular grid as illustrated in Figure 22.12.

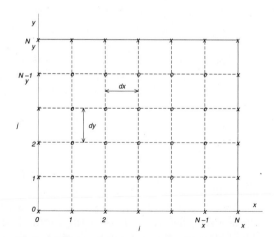

Figure 22.12 *A rectangular grid used to solve Poisson's equation. The solution is specified on all the exterior boundary points (∗) and is to be computed at the interior points (o).*

If the number of panels of width Δx and Δy are $N_x + 1$ and $N_y + 1$, respectively, the number of points at which the solution is as yet unknown is $N = N_x \cdot N_y$. Therefore, if Equation 22.16 is written for (i, j) values corresponding to each and every interior point, we would have N linear equations in N unknowns; clearly a matrix problem for which MATLAB is well suited. However, there is a problem. Even if the grid is relatively coarse; e.g., $N_x = N_y = 30$, the size of the coefficient matrix will be $30^2 \times 30^2$, or 900×900. Even using sparse matrix techniques, this matrix equation is probably too large to solve directly. For this reason, iterative techniques are more common when solving elliptic equations, and will be discussed first.

ITERATIVE TECHNIQUES FOR ELLIPTIC EQUATIONS

Assume, for simplicity, that the intervals in the x and y directions are equal, $\Delta x = \Delta y \equiv h$. Equation 22.16 can be written as

$$V_{i,j} = \frac{1}{4} \left[V_{i+1,j} + V_{i-1,j} + V_{i,j+1} + V_{i,j-1} \right] - h^2 \rho_{i,j} \qquad (22.17)$$

This equation states that the solution at any interior point is simply the average of the solution at the four nearest neighboring points less a contribution from the source term. There is an equation of this type for each of the interior points, so there are $(N_x - 1)(N_y - 1)$ equations in all. Of course, to determine the solution at a particular point, we must already have obtained the solution at the neighboring points, which is not the case. Nonetheless, Equation 22.17 does suggest an iterative approach. Values for the solution at all interior points are guessed and called $V_{i,j}^{(0)}$, and Equation 22.17 is used to compute a new set $V_{i,j}^{(1)}$. This method is the Jacobi method described in Chapter 20. The method will converge starting with any initial guess for the elements of $V_{i,j}$. The problem is that the formulation of the algorithm as a MATLAB code requires two nested **for** loops, and as you know, MATLAB is notoriously slow when executing loops. It is therefore essential that you take appropriate steps to accelerate the convergence rate.

    The first step in increasing the efficiency of the code is to replace the Jacobi method by the method of Gauss-Seidel. This involves a trivial change to the MATLAB code. Instead of using two separate arrays, $V_{i,j}^{(0)}$ and $V_{i,j}^{(1)}$, only one, $V_{i,j}$, is used, replacing elements in turn as they are calculated within each pass. Because more recent information concerning the solution is used in each step, the method can be expected to converge more quickly. In practice, the convergence rate is found to improve by 25 to 50%.

    Finally, as was pointed out in Chapter 20, the computed values in each pass converge monotonically to the solution, rarely overshooting the final result. This suggests that artificially increasing the improvement in each step may accelerate the convergence rate. This is the method of *successive overrelaxation* (SOR). The factor by which the improvement is increased is called ω. If $0 < \omega < 1$, the calculation is "underrelaxed" and convergence is usually slowed. If $1 < \omega < 2$, the convergence rate is accelerated, often dramatically. If $\omega > 2$, the method diverges.

The optimum value of ω can be determined analytically for a few simple equations, but more commonly it is found by trial and error. It is usually most efficient to repeatedly solve the problem with a relatively coarse grid and quite modest convergence criteria, varying ω to determine the approximate optimum value. You could also use the results of these calculations to determine the initial guess for the solution to be used in the final, more accurate, calculation.

EXAMPLE 22.2

Numerical Solution of Poisson's Equation for a Dipole

A dipole charge distribution consists of two equal but opposite charges, $\pm q$, separated by a small distance. The resulting electric potential from the dipole is a solution of Poisson's equation. The actual charges are point singularities.

To simulate this problem, we select a grid on the xy plane with $0 \leq x \leq 1$, $0 \leq y \leq 1$, and $\Delta x = \Delta y \equiv h = 0.04$. Thus, the grid is 25×25. Equation 22.17 is used to repeatedly improve each value of the solution on the grid. The convergence criterion requires that the maximum change in the solution at any point in the grid in a particular pass be less than one.

The code for this algorithm is given in Figure 22.13.

```
% Script file PoisIter solves Poisson's equation
% for a dipole distribution.
  h = 0.04;  L = 1; c = 100*(L/h)^2;
  Nx = L/h-1; Ny = Nx;
%
%    Simulate a dipole charge distribution
%
        Nh = fix(Nx/2);    % middle of grid
         q = 100*Nx^2;     % "magnitude" of the charge
       ixp = Nh:Nh+2;    ixm = Nh-1:-1:Nh-3;
       rho = zeros(Nx+1,Ny+1);
       rho(ixp,ixp) =  q*ones(k,k);
       rho(ixm,ixm) = -q*ones(k,k);
%
 w = 1.75;                 % The relaxation factor
%
 V = zeros(Nx+1,Nx+1);     % The initial guess for V
%
 Delta  = 10;
 ITER = 0;
%
 while Delta >= 1
    s = 0;
    for i = 2:Nx
       for j = 2:Ny
          dV = (V(i+1,j)+V(i-1,j)+V(i,j+1)+V(i,j-1))/4 + ...
                 -h^2*rho(i,j) - V(i,j);
          if abs(dV) > D; D = abs(dV); end
```

Figure 22.13 *MATLAB code to solve Poisson's equation (continues).*

```
        V(i,j) = V(i,j) + w*dV;
        end
   end
  Delta = D; disp(D)
   ITER = ITER + 1;
end
mesh(V)
```

Figure 22.13   *(Continued)*

The Jacobi method for this problem required 103 iterations and, on my computer, took 42 seconds of computing time. The Gauss-Seidel method required 70 iterations and 27 seconds. Next the relaxation factor is varied to find the optimum value. The results are illustrated in Figure 22.14, where the number of iterations is graphed as a function of ω. The electric potential computed by the MATLAB code is displayed in Figure 22.15.

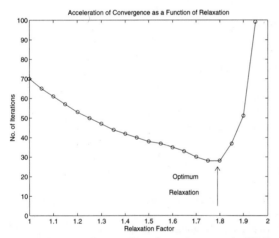

Figure 22.14   *The number of iterations in the SOR method is plotted as a function of the relaxation factor ω.*

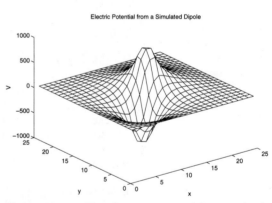

Figure 22.15   *The electric potential from a dipole computed by the SOR method.*

Example 22.2 illustrates the principal feature of elliptic equations—the iterative numerical techniques are almost always stable, and the accuracy of the result is most often strictly a function of computing time. However, with even the simplest of problems, computing time can easily be excessive. This is particularly true in MATLAB, where nested `for` loops can significantly slow a calculation. An alternative, somewhat more complicated, but direct method is therefore described in the next section.

DIRECT METHODS FOR ELLIPTIC EQUATIONS

The fundamental computational problem of Equation 22.17 is one of N linear equations in N unknowns; i.e., basically a matrix equation of the form $\mathbf{A}\vec{x} = \vec{b}$. The difficulties are: (1) writing Equation 22.17 in the standard form, $\mathbf{A}\vec{x} = \vec{b}$, and (2) dealing with the often enormous size of the coefficient matrix.

Rewriting the Finite Difference Equations in Matrix Form To recast Equation 22.17 into standard matrix form, the unknown quantities, $V_{i,j}$, with $i = 1, \ldots, N_x$, $j = 1, \ldots, N_y$, must first be written as a *vector*. To this end we consider, in turn, $V_{i,j}$ at each interior element of the grid, starting at the lower left-hand corner, counting first across and then up. That is,

$$
\begin{array}{ccccc}
k = & 1 & 2 & \cdots & N_y \\
(i,j) = & (1,1) & (1,2) & \cdots & (1, N_y) \\
 & N_y + 1 & N_y + 2 & \cdots & 2N_y \\
 & (2,1) & (2,2) & \cdots & (2, N_y)
\end{array}
$$

$$\vdots$$

or

$$k = j + N_y(i - 1) \quad \text{for } k = 1, \ldots, Nx \cdot N_y$$

Next, replacing (i,j) by k in Equation 22.17, we obtain

$$V_{k+1} - 4V_k + V_{k-1} + V_{k+N_y} + V_{k-N_y} = h^2 \rho_k \qquad (22.18)$$

In Equation 22.18, for simplicity, we have taken $\Delta x = \Delta y \equiv h$.

The boundary conditions on the four sides of the grid are:

$$
\begin{array}{llll}
V(x,y) = V_A(x) & 0 \le x \le L_x & y = 0 \\
 V_B(y) & 0 \le y \le L_y & x = 0 \\
 V_C(x) & 0 \le x \le L_x & y = L_y \\
 V_D(y) & 0 \le y \le L_y & x = L_x
\end{array}
$$

Equation 22.18 is next written for each grid point along a particular row. For example, for $i = 1$, we obtain

$$
\begin{array}{llllll}
j = 1 & & -4V_1 & +V_2 & +V_{N_y} & = -(V_A(x_1) + V_B(y_1)) \\
j = 2 & V_1 & -4V_2 & +V_3 & +V_{N_y+1} & = -V_A(x_2) \\
j = 3 & V_2 & -4V_3 & +V_4 & +V_{N_y+2} & = -V_A(x_3) \\
& \vdots \\
j = N_x & V_{N_y-1} & -4V_{N_y} & V_{2N_y} & & = -(V_A(x_{N_x}) + V_D(y_1))
\end{array}
$$

Repeating this for successive rows, we arrive at the matrix equivalent of Equation 22.18,

$$\begin{pmatrix} \mathbf{R} & \mathbf{I} & 0 & \cdots & 0 & 0 \\ \mathbf{I} & \mathbf{R} & \mathbf{I} & \cdots & 0 & 0 \\ 0 & \mathbf{I} & \mathbf{R} & \cdots & 0 & 0 \\ \vdots & \vdots & \vdots & \ddots & \vdots & \vdots \\ 0 & 0 & 0 & \cdots & \mathbf{R} & \mathbf{I} \\ 0 & 0 & 0 & \cdots & \mathbf{I} & \mathbf{R} \end{pmatrix} \begin{pmatrix} V_1 \\ V_2 \\ V_3 \\ \vdots \\ V_{N-1} \\ V_N \end{pmatrix} = \vec{b}_1 + \vec{b}_2 \qquad (22.19)$$

where $\mathbf{R}$ is an $N_x \times N_y$ matrix of the form

$$\mathbf{R} = \begin{pmatrix} -4 & 1 & 0 & \cdots & 0 \\ 1 & -4 & 1 & \cdots & 0 \\ 0 & 1 & -4 & \cdots & 0 \\ \vdots & \vdots & \vdots & \ddots & \vdots \\ 0 & 0 & 0 & \cdots & -4 \end{pmatrix}$$

$\mathbf{I}$ is the $N_x \times N_y$ identity, and $\vec{b}_1$ and $\vec{b}_2$ are N-element column vectors that depend upon the boundary conditions:

$$\vec{b}_1^T = [V_A(\vec{x})\ 0\ 0\ 0\ \cdots\ 0\ 0\ V_C(\vec{x})]$$
$$\vec{b}_2^T = [V_B(y_1)\ 0\ 0\ \cdots\ 0\ 0\ V_D(y_1)\ V_B(y_2)\ 0\ 0$$
$$\cdots\ 0\ 0\ V_D(y_2)\ V_B(y_3)\ \cdots\ V_D(y_{N_y})]$$

These vectors may appear to be awkward, but they are easily constructed in MATLAB. If the boundary functions have been coded in MATLAB, the MAT-LAB expressions for $\vec{b}_1$ and $\vec{b}_2$ would be

```
x = (1:Nx)*dx;    y = (1:Ny)*dy;    N = Nx*Ny;
b1 = zeros(Ny,Nx);
b1(:,1) = VA(x');   b1(:,Ny) = VC(x');
b1 = b1(:);
b2 = zeros(Nx,Ny);
b2(1,:) = VB(y);    b2(Nx,:) = VD(y);
b2 = b2(:);
```

Next the MATLAB code is needed to construct the matrices $\mathbf{R}$ and $\mathbf{I}$, and from these assemble the coefficent matrix. For clarity, we will assume that $N_x = N_y = N$.

```
A = zeros(N,N);
ix = i:Nx;    Iy = 1:Ny;
R = -4*eye(Ny,Nx) + diag(ones(Nx-1,1),1 ) + ...
                  + diag(ones(Nx-1,1),-1);
I = eye(Ny,Nx);
for k = 1:Ny
    A(Ny*(k-1)+ix,Ny*(k-1)+iy) = R;
    if k<Ny
```

```
       A(    Ny*k+ix,Ny*(k-1)+iy) = I;
       A(Ny*(k-1)+ix,    Ny*k+iy) = I;
   end
end
```

The matrix **A** is segmented into blocks of size $N_x \times N_y$, and each of these blocks is either zero, a banded sparse matrix **R**, or a sparse identity matrix. It is almost unavoidable that the size of the coefficient matrix will be quite large; therefore, the techniques of sparse matrices will be essential to obtain a solution.

MATLAB Implementation of the Direct Method Using Sparse Matrices Even for a coarse grid, the coefficient matrix in Equation 22.19 will be quite sparse. As an illustration of the use of sparse matrices in MATLAB, consider Example 22.3, which revisits the electric potential from a dipole.

EXAMPLE 22.3

Solution of Poisson's Equation for a Dipole by Direct Methods

The electric potential is a solution of Poisson's equation, where $-\rho$ represents the density of charge within the region. We will require the potential to be zero at all boundaries and will simulate a dipole charge distribution by equating ρ to a constant at a few points near the center of the grid, but displaced by x_d, y_d, and equal to the negative of this constant at points displaced by $-x_d, -y_d$. The charge density, ρ, is zero at all other grid points. The MATLAB code implementing Equation 22.19 is given in Figure 22.16 on page 502. The MATLAB script file for this example can be found on the applications disk and is named `PoisDir2.m`. The execution time for this code is approximately 15 times faster than for the earlier iterative method. ●

Beyond This Chapter

Although this chapter is intended as merely an introduction to the vast array of numerical procedures applied to partial differential equations, it is hoped that at this stage the reader has acquired sufficient understanding of the elementary approaches to appreciate the potential difficulties that can be encountered and to anticipate the more complicated remedies.

DIFFICULTIES ASSOCIATED WITH INITIAL VALUE PROBLEMS

Neumann Stability Analysis A central consideration in procedures for solving initial value problems is their stability. Stability criteria were merely quoted for the simplest, most common method, the Crank-Nicolson method. The serious practitioner of this and related methods should have an understanding of the origins of stability criteria. The idea is to attempt to find either exponential or power solutions in Δt, the step size, to the particular finite difference equation being considered. Positive exponentials or positive powers will lead to divergences and thus can be used to obtain stability criteria. This type of analysis is called Neumann stability analysis and is described in most of the more advanced texts devoted to either ordinary or partial differential equations.

```
% Script file PoisIter solves Poisson's equation
% for a dipole distribution.
  h = 0.04;  L = 1; c = 100*(L/h)^2;
  Nx = L/h-1; Ny = Nx
%
%    Simulate a dipole charge distribution
%
        Nh = fix(Nx/2);    % middle of grid
         q = 100*Nx^2;     % "magnitude" of the charge
       ixp = Nh:Nh+2;    ixm = Nh-1:-1:Nh-3;
       rho = zeros(Nx+1,Ny+1);
       rho(ixp,ixp) =  q*ones(k,k);
       rho(ixm,ixm) = -q*ones(k,k);
%
  V  = sparse(zeros(N,N));
rho = sparse(rho(:));   % Convert rho to a sparse
                        % vector.
A = zeros(N,N);
ix = i:Nx;   Iy = 1:Ny;
R = -4*eye(Ny,Nx)  + diag(ones(Nx-1,1),1 ) + ...
                   + diag(ones(Nx-1,1),-1);
I = eye(Ny,Nx);
for k = 1:Ny
   A(Ny*(k-1)+ix,Ny*(k-1)+iy) = R;
   if k<Ny
       A(    Ny*k+ix,Ny*(k-1)+iy) = I;
       A(Ny*(k-1)+ix,    Ny*k+iy) = I;
   end
end
A = sparse(A);
%
  V =h^2*A\rho;
%
%  To plot the results, the vector V must be converted
%  to an Nx by Ny full matrix.
%
V  = full(V);
VV = zeros(Ny,Nx);
for i = 1:Nx
    for j = 1:Ny
          VV(i,j) = V(j+Ny*(i-1));
    end
end
mesh(VV)
```

Figure 22.16 *MATLAB code to solve Poisson's equation by direct methods.*

Stiff Differential Equations *Stiff* differential equations, both ordinary and partial, have the property that small changes in boundary or initial conditions can cause huge changes in the solution away from the boundary. This is usually caused by positive exponential solutions to the equation that, although they may start out with zero amplitude, soon, because of round-off or other errors, acquire finite amplitude and thenceforth rapidly grow and swamp the desired solution. Even if the equations have no diverging solutions, the corresponding finite difference equations may have them, causing the solution to unexpectedly diverge. Problems involving stiff differential equations can be extremely difficult to correct, and there is no general prescription for their treatment. One avenue of attack is to use slower implicit (iterative) methods that are less susceptible to round-off errors.

Discontinuities or Singularities in the Solution As mentioned earlier, a common difficulty encountered in the solution of wave equations is the development of shock waves in the solution. Shock waves are a discontinuity in the solution, and naturally, the numerical solution will break down at such a point. About the only effective way to handle such a situation is to sharply decrease the increment, Δx, in the vicinity of the shock.

Adapting the step size to the changes in the solution was found to be an extremely efficient technique in the solution of ordinary differential equations. Not surprisingly, similar ideas have been and are being pursued in the solution of partial differential equations. In two or more dimensions, this often takes the form of a stretching or contraction of the grid. This topic is discussed in Hoffman, (1992).

Other Techniques for Initial Value Problems The method of tracing out the solution to hyperbolic equations along characteristic curves was mentioned in this chapter, and this is the most common and successful method in use today. The method of characteristics should be investigated if an accurate solution to a wave-type equation is desired over a range of time corresponding to several periods of oscillation.

DIFFICULTIES ASSOCIATED WITH BOUNDARY VALUE PROBLEMS

The primary concern when solving boundary value partial differential equations is the sheer size of the calculation required to obtain an accurate solution. Most modern techniques seek to address this problem.

Multiple Grids To speed the convergence of an iterative method on a grid, a rather obvious technique would be to solve the problem on a coarse grid, interpolate the results onto a finer grid, use these values as the initial guess in an iteration on the finer grid, and continue refining the solution by this means for several stages. This could, in turn, be coupled with a *warping* of the grid so that the mesh is finer at points where the function is most sensitive.

Fourier Reduction Techniques A popular technique for solving elliptic equations when the boundaries coincide with the coordinate axes, is to first express the solution and the source term at the grid points in terms of their discrete Fourier transforms. The finite difference equation approximation to Poisson's equation then becomes a simple relation between these two transform functions at each grid point. The solution is then obtained by the inverse fast Fourier transform. See Press (1986).

The Method of Images   A popular theoretical method of solving Poisson's equation is the method of images. If the solution due to a source term is to be zero on an infinite plane boundary, this can be accomplished by placing a like source term, but of opposite sign, on the other side of the boundary, much like a mirror. The solution is then computed in the combined region and is automatically zero on the surface separating the two sources. For several sources, or for curved or finite boundaries, this can lead to an infinite series of images.

Finite Element Methods   Finite element methods, as applied to ordinary differential equations, were described in Chapter 20. The chief advantage of finite element methods contrasted with the finite difference methods employed in this chapter is that with finite elements it is much easier to vary the grid size and in general to adapt the method to fit the equation at hand. This is often most important when the boundary is irregular.

Multidimensional Equations   A partial differential equation in three independent variables is typically more than an order of magnitude more difficult than a similar equation in only two independent variables. For example, the wave equation in two dimensions,

$$\frac{\partial^2 u}{\partial x^2} + \frac{\partial^2 u}{\partial x^2} = \frac{1}{v^2}\frac{\partial^2 u}{\partial t^2}$$

is elliptic in x, y, but is hyperbolic in x, t and y, t. The manner of solution is clearly to compute the solution in the xy plane for a particular value of t by methods appropriate to elliptic equations, and then to step in the t direction by Δt, using techniques designed for hyperbolic equations. Unfortunately, MATLAB makes this more difficult than it should be. The MATLAB restriction of matrices to only two subscripts makes the coding of three-dimensional problems more awkward.

Sparse Matrix Techniques   Sparse matrix techniques, particularly in MATLAB, are the easiest way to obtain a solution to an elliptic equation on a grid that is only moderately coarse. The principal difficulty is in establishing the basic structure of the coefficient matrix and the inhomogeneous right-hand-side vector that results from the boundary conditions. This is most evident when the boundaries are not rectangular. (See Problem 22.8.)

Irregular Boundaries   If a boundary is quite irregular, it may be very difficult to translate the resulting set of finite difference equations into a single matrix equation, and iterative techniques may be more suitable. In any event, irregular boundaries usually cause considerable complication in any method. For example, if the boundary does not pass through the nearest grid points, interpolated values are most frequently used at the grid points.

Nonlinear Equations Nonlinear equations were not discussed at all in this chapter. Needless to say, all problems associated with linear equations are present and greatly amplified in nonlinear equations. In addition, finite difference equations that result from nonlinear partial differential equations will likewise be nonlinear. If *direct* methods of solution are employed, this usually causes no problem. However, nonlinear equations are not so easy to solve by implicit or iterative methods. (For example, diagonal dominance no longer guarantees convergence of the Gauss-Seidel method if the equations are nonlinear.)

Finally, if you have access to the MATLAB partial differential equations toolbox on your system, you may wish to run some of the demonstration programs to get an appreciation of the level of sophistication that is possible in using MATLAB to solve PDEs. Type `help toolbox\pde` to get started.

Problems

REINFORCEMENT EXERCISES

P22.1 Unstable Solutions. Solve the FTCS equations, Equation 22.4 for the same situation as in Example 22.1. However, use a time step that is based upon $\gamma = 0.51$. Follow the solution for several time steps and determine how many steps are required to convince you that the solution is diverging.

P22.2 Short Times and the Diffusion Equation. Accurate solutions to the diffusion equation are frequently difficult to obtain for short times. For the problem of heat diffusion in an iron bar, solved numerically in Example 22.1, the exact solution is

$$T(x,t) = T_0 \left[1 - \frac{x}{L} - \frac{2}{\pi} \sum_{n=0}^{\infty} \frac{\sin(n\pi x/L)}{n\pi} e^{-(n\pi/L)^2 Dt} \right]$$

Plot the numerical solution and the exact solution for the first few time steps and compare the results.

P22.3 Standing Waves. Alter the code in Figure 22.10 so that the initial pulse on the string is a sine wave adjusted so that it is zero at both ends of the interval. For example, $y_0(x) = \sin(m\pi x/L)$, where m is an integer. The resulting motion is called a standing wave. Notice that the points along the x axis where the initial displacement is zero (called *nodes*) remain zero. Run the code for a variety of values of m.

P22.4 Reflections at an Interface. Consider the problem of a pulse traveling down a combination of two strings connected together, end to end. One string is thin, the other is thick. The speed of a wave in a thin string is greater than that in a thick string. Alter the code in Figure 22.10 so that the wave speed in the left half is $v_L = 10$, and the speed in the right half is $v_R = 30$. Run the code and pay particular attention to the added reflection that occurs at the interface between the two strings. Is there a phase change associated with this reflection?

P22.5 Driven Oscillations on a String. Alter the code in Figure 22.10 so that the left end oscillates as a function of time. For example, use $f_0(t) = 0.5\sin(2\pi ft)$, where f is the frequency of the driving force. (Use $f = 5/\Delta x, 10/\Delta x$.)

P22.6 Speed of Iterative Methods. The principal concerns in solving elliptic equations like Poisson's equation are computation time and memory restrictions.

    a. Adapt the code in Figure 22.13 to solve the dipole potential using the Jacobi method. (You will have to introduce two arrays, **Vold** and **Vnew**.)

    b. Determine the smallest **dx** that can be accommodated by the memory limits on your computer.

    c. Graph the number of iterations required to obtain a solution versus N_x, the number of points in one direction of the grid. Fit the results to kN_x^p, and determine p.

    d. Repeat part c for Gauss-Seidel iterations (without relaxation).

    e. Repeat the calculation in Figure 22.13 to determine the optimum relaxation factor, ω, for $N_x = 8, \ldots, 18$. Is the relaxation factor a function of N_x?

P22.7 Direct Method Matrices. Poisson's equation is to be solved on the square grid, $0 \le x \le 5, 0 \le y \le 5$, with $\Delta x = \Delta y = 1$. The boundary conditions are: $V(x,0) = V_A$, $V(0,y) = V_B$, $V(x,L) = V_C$, $V(y,L) = V_D$. Write Equation 22.18 for each of the 16 interior grid points, replace the (i,j) labels by an ordering of the points into a vector, and derive the form of the matrix equation, Equation 22.19.

P22.8 Different $\Delta x, \Delta y$ Generalize the code in Figure 22.13 to use different mesh sizes in the x and y directions. Test the code for $(N_x, N_y) = (8, 16)$ and $(16, 8)$. Is there any difference in the results?

EXPLORATION PROBLEMS

P22.9 Irregular Boundaries. Consider the solution of Poisson's equation on the irregular boundary drawn in the following diagram. (This is the boundary for the logo graph for MATLAB.)

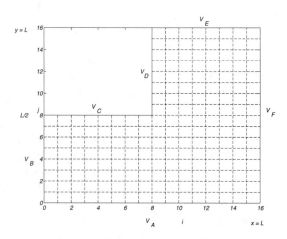

     a. Using a 16 × 16 grid, write Equation 22.18 for each of the interior grid points, replace the (i, j) labels by an ordering of the points into a vector, and devise a scheme for converting the solution $V_{i,j}$ on the irregular grid into a vector V_k. What is the form of the coefficient matrix?

     b. Solve the equation with $V_a = V_f = 10$, $V_b = V_e = -10$, $V_c = V_d = 0$, using sparse matrix techniques.

     c. Translate the vector solution V_k back into a full two-dimensional matrix. Plot the result.

P22.10 Nonlinear Differential Equations. If the diffusivity, D, in the diffusion equation is a function of temperature; e.g, $D(T) = D_0(1 + \alpha T)$, or if the velocity, v, in the wave equation is a function of x; e.g., $v(x) = v_0(1 + \beta \cos(10x/L))$, these equations are then nonlinear. For a variety of values of α or β, solve either equation, commenting on any difficulties that you encounter.

Summary

CLASSIFICATION OF PARTIAL DIFFERENTIAL EQUATIONS

Writing the second derivatives contained in a second-order partial differential equation as

$$A\frac{\partial^2 u}{\partial x^2} + 2B\frac{\partial^2 u}{\partial x \, \partial y} + C\frac{\partial^2 u}{\partial y^2}$$

the equation is classified by the value of $B^2 - AC$ as

Parabolic $B^2 - AC = 0$ Diffusion equation

Hyperbolic $B^2 - AC > 0$ Wave equation

Elliptic $B^2 - AC < 0$ Poisson's equation

Boundary Conditions   Parabolic and hyperbolic type equations require boundary conditions on an open surface. Most frequently these boundary conditions are typified by fixing the solution at the ends of an interval and specifying an initial spatial function (or functions) at $t = 0$. The domain is then open in the time variable. Elliptic functions require boundary conditions on a closed surface. The boundary conditions specify the solution or its derivative (but not both) on the surface containing the region of interest.

INITIAL VALUE PROBLEMS

Stepping methods are used to solve both parabolic and hyperbolic equations. A grid is established in the two independent variables, and the differential equation is approximated by finite difference expressions.

Crank-Nicolson Method   The Crank-Nicolson method uses central difference expressions for the second derivative of the function along the direction of the bounded variable (x), and relations that connect three grid points in this direction. The first derivative in the unbound variable, (t), involves only two grid points, and is interpreted as a central difference expression for the derivative at a point midway between two grid points. In this way, second-order expressions are used for both derivatives. These relations are rearranged to express the solution at advanced grid points in terms of the solution on grid points where it has been computed in an earlier step. The method is unconditionally stable and is based on Equation 22.10. The method is also applicable to hyperbolic equations, with the additional stipulation that two earlier grid points are required for each step.

BOUNDARY VALUE PROBLEMS

Elliptic equations cannot be solved in a stepping manner; instead, seek to obtain the solution at all interior grid points at once. This is done by establishing finite difference equations for each of the grid points, and solving the set of linear equations either by iterative techniques such as Gauss-Seidel or by direct matrix methods.

Implicit Methods   The set of $N = N_x N_y$ equations, where N_x and N_y are the number of grid points in the x and y directions, respectively, is solved iteratively, starting with an initial guess for the solution at each grid point. To obtain reasonably accurate solutions, the grid should be as fine as possible. However, the number of equations, N, increases dramatically as the grid is made finer, so it is essential that efficient means be used to solve the equation. Usually, the method of Gauss-Seidel is augmented with relaxation methods to improve the convergence rate. That is, anticipating that the method converges monotonically to the result, the improvement in each step is artificially increased by a relaxation factor, ω. The optimum value of ω is ordinarily not known beforehand, but it must be in the range $1 < \omega < 2$.

Direct Methods   By ordering the points of a two-dimensional grid into a single vector of length N, the N equations can be written as a single matrix expression and solved directly by the method of Gaussian elimination. Invariably, the size of the coefficient matrix is prohibitively large. The standard remedy is to employ the techniques of sparse matrices. In MATLAB, this is especially easy because sparse matrix arithmetic is, for the most part, transparent to the user. Direct methods are usually preferred, especially if the equation is nonlinear.

Bibliography for Part V

1. M. B. Abbott, *An Introduction to the Method of Characteristics*, American Elsevier, New York, 1966.

2. Forman S. Acton, *Numerical Methods That Work*, Harper & Row, New York, 1970.

3. G. Birkhoff and G.-C. Rota, *Ordinary Differential Equations*, 3rd ed., John Wiley, New York, 1978.

4. R. Courant and D. Hilbert, *Methods of Mathematical Physics*, Interscience Publishers, New York, 1962.

5. J. Crank and P. Nicolson, "A Practical Method for Numerical Evaluation of Solutions of Partial Differential Equations of the Heat-Conduction Type," *Proc. Cambridge Philosophical Soc.*, vol. 43, no. 50, 1947.

6. L. Fox, *Numerical Solution of Ordinary and Partial Differential Equations*, Pergamon Press, New York, 1962.

7. C. W. Gear, *Numerical Initial Value Problems in Ordinary Differential Equations*, Prentice-Hall, Englewood Cliffs, NJ, 1971.

8. Michael D. Greenberg, *Foundations of Applied Analysis*, Prentice-Hall, Englewood Cliffs, NJ, 1978.

9. ———, *Advanced Engineering Mathematics*, Prentice-Hall, Englewood Cliffs, NJ, 1988.

10. R. W. Hamming, *Numerical Methods for Scientists and Engineers*, McGraw-Hill, New York, 1962.

11. F. B. Hildebrand, *Methods of Applied Mathematics*, Prentice-Hall, Englewood Cliffs, NJ, 1952.

12. ———, *Introduction to Numerical Analysis*, McGraw-Hill, New York, 1974.

13. J. Hoffmann, *Numerical Methods for Engineers and Scientists*, McGraw-Hill, New York, 1992.

14. H. B. Keller, *Numerical Methods for Two-Point Boundary Value Problems*, Blaisdell, Waltham, MA, 1968.

15. Erwin Kreszig, *Advanced Engineering Mathematics*, 5th ed., John Wiley, New York, 1983.

16. Robert L. LaFara, *Computer Methods for Science and Engineering*, Hayden Book Co., Rochelle Park, NJ, 1973.

17. J. D. Lambert, *Computational Methods in Ordinary Differential Equations*, John Wiley, New York, 1973.

18. L. Lapidus and J. H. Seinfeld, *Numerical Solutions of Ordinary Differential Equations*, Academic Press, New York, 1971.

19. M. J. Maron, *Numerical Analysis: A Practical Approach*, Macmillan, New York, 1982.

20. P. Morse and H. Feshbach, *Methods of Theoretical Physics*, vol. 1, McGraw-Hill, New York, 1953.

21. M. Prenter, *Splines and Variational Methods*, John Wiley, New York, 1975.

22. W. H. Press, B. P. Flannery, S. A. Teukolsky, and W. T. Vetterling, *Numerical Recipes, The Art of Scientific Computing*, Cambridge University Press, New York and Cambridge, England, 1986.

23. A. Ralston and P. Rabinowitz, *A First Course in Numerical Analysis*, 2nd ed., McGraw-Hill, New York, 1983.

24. L. F. Shampine and M. K. Gordon, *Computer Solution of Ordinary Differential Equations*, Freeman, San Francisco, CA, 1975.

25. J. Stoer and R. Bulirsch, *Introduction to Numerical Analysis* (English translation by R. Bartels, W. Gautschi, and C. Witzgall), Springer-Verlag, Berlin and New York, 1980.

26. G. Strang and G. Fix, *An Analysis of the Finite Element Method*, Prentice-Hall, Englewood Cliffs, NJ, 1973.

27. Samuel S. M. Wong, *Computational Methods in Physics & Engineering*, Prentice-Hall, Englewood Cliffs, NJ, 1992.

28. O. C. Zienkiewicz, *The Finite Element Method in Engineering Science*, McGraw-Hill, New York, 1971.

Appendix / Programming in MATLAB

APPENDIX A.i /
Matrix Algebra
in MATLAB

PREVIEW MATLAB is much more than a language for manipulating matrices. It contains hundreds of application procedures that address a broad spectrum of problems, from the relatively simple to the state of the art. Moreover, it is accessible immediately, with very little instruction, to both the novice and the expert. However, the key to using MATLAB successfully and efficiently lies in the user's facility in constructing and manipulating square, rectangular, column, and row matrices. This is related to the central innovative ingredient of MATLAB. All previous languages are based on ordinary algebra, wherein a symbol or name is used to represent a single numerical quantity, and these symbols are then manipulated via the ordinary rules of arithmetic. But in MATLAB, *every* name is assumed to be a *matrix*, and the names are then manipulated via the rules of matrix arithmetic.[1]

The advantage of this system is that most simple operations can be automatically "vectorized" as matrix operations, and, as a result, 90% of the loops required in other computer languages can be avoided.

The purpose of this section is to introduce some of the elementary aids available in MATLAB for easily constructing matrices. We follow this with an illustration of some of their uses in computation. Like any skill, proficiency in this regard will require practice, and the reader should personally execute most of the examples and problems in MATLAB while following their description. A tutorial program illustrating the essentials of MATLAB programming that parallels the discussion in this appendix can be found on the applications disk.

Basic Arithmetic Operators

Mathematical expressions are constructed using the common arithmetic operators, which are evaluated according to the usual precedence rules. See Table A.i.1.

Arithmetic expressions are evaluated in successive scans, evaluating the highest precedence level expressions from left to right. This may be altered by using parentheses in the usual manner. The right and left divisions produce the same result when the operands are simple numbers, but represent quite different operations when the operands are entire matrices (more on this later). (See also the discussion of "backslash" division of matrices in Chapter 2 and the description

---

[1]The principal impediment to this concept is handling the perplexing question of matrix division. This is addressed in Chapter 2.

Precedence level	Operator	
0	$+$	addition
	$-$	subtraction
1	$*$	multiplication
	$/$	division
	$\backslash$	right division
2	$\wedge$	power

Table A.i.1   *Precedence rules for MATLAB operators*

of vectorized operators in Appendix Section A.ii.). That is, $3/2$ and $2\backslash3$ are the same value. The construction $2\backslash3$ is read as "2 divided into 3."

In addition, MATLAB has extended the ordinary arithmetic to include symbols to represent the results of arithmetic operations that give rise to either infinite or undefined values. Thus, the statements

$$a = \frac{1}{0}, \qquad b = \tan\left(\frac{\pi}{2}\right), \qquad c = \log(0)$$

will produce the results

$$\infty, \qquad 1.6325e+016, \qquad -\infty$$

respectively. Although $\tan(\pi/2)$ is mathematically infinity, round-off error in the procedures MATLAB uses to compute π and $\tan(\pi/2)$ result in a finite expression. (The actual value obtained is, of course, machine dependent.) Also, $d = 0/0$ and, using the value obtained for a above, $e = a/\log(0)$ will each produce the result NaN (*Not a Number*). Any further operation involving a NaN will yield NaNs. And, unlike FORTRAN and similar languages, the execution of the program *will continue* beyond the point where an indefinite or undefined value was obtained.

Entering Matrices and Vectors

Matrices are easily constructed in MATLAB either by explicitly entering all of the elements or by using some of the built-in MATLAB special functions. For example, to explicitly enter the 3×3 matrix

$$\mathbf{A} = \begin{bmatrix} 1 & 2 & 3 \\ 2 & 1 & 2 \\ 3 & 2 & 1 \end{bmatrix}$$

you would type

```
A = [1 2 3; 2 1 2; 3 2 1]
```

If this line is followed by striking the Enter key, the computer will display the result for **A**. MATLAB displays the latest result of computation after each line is entered, unless the display is suppressed by ending the line with a semicolon(;). Notice that the matrix begins and ends with square brackets, and the ends of successive rows are marked by semicolons (or by entering on successive lines). Each of the numbers in a row must be separated by a blank space. Unintentional blanks entered in a string of numbers can result in serious misinterpretations of the values intended. Decimal points, exponents, and mathematical expressions can be used to specify numbers. For example, the 1×3 matrix x (i.e., a row vector),

$$x = (\sqrt{\pi}, \cos(\pi/3), \ln(2.717))$$

is entered as

```
x = [sqrt(pi) cos(pi/3.0) log(2.717)];
```

Mathematical Functions Available in MATLAB

Ordinary arithmetic in MATLAB is quite similar to that of earlier computer languages and to ordinary arithmetic in general. In addition, unlike most computer languages, MATLAB has stored values for π (entered as `pi`), for $i = \sqrt{-1}$ (entered as `i` or `j`), and for ∞, entered as `Inf` or `inf`. Also, the variable name `ans` contains the last computed result, which may be an entire matrix, and `eps` contains a number reflecting the machine accuracy; i.e., the smallest distinction that can be made between two numbers of order unity.

Trigonometric functions		Miscellaneous functions			
MATLAB function name	Mathematical notation	MATLAB function name	Mathematical notation	MATLAB function name	Mathematical notation
`sin(c)`	$\sin c$	`abs(c)`	$\lvert c \rvert$	`real(c)`	real part
`cos(c)`	$\cos c$	`sqrt(c)`	$\sqrt{c}$		$\mathrm{Re}(a + ib) = a$
`tan(c)`	$\tan c$	`exp(c)`	e^c	`imag(c)`	imaginary part
`asin(c)`	$\sin^{-1} c$	`log(c)`	$\ln c$		$\mathrm{Im}(a + ib) = b$
`acos(c)`	$\cos^{-1} c$	`log10(c)`	$\log c$	`round(c)`	nearest integer
`atan(c)`	$\tan^{-1} c$	`rem(c/d)`	integer		$-7.4 \Rightarrow -7$
`sinh(c)`	$\sinh c$		remainder		$7.6 \Rightarrow 8$
`cosh(c)`	$\cosh c$		of	`ceil(c)`	rounds up
`tanh(c)`	$\tanh c$		c/d		$-7.6 \Rightarrow -7$
					$7.4 \Rightarrow 8$
				`floor(c)`	rounds down
					$-7.4 \Rightarrow -8$
					$7.6 \Rightarrow 7$

Table A.i.2 *Several of the built-in mathematical functions available in MATLAB*

Finally, keep in mind that MATLAB is case-sensitive. *All* toolbox functions are defined in lowercase letters.

The usual assortment of built-in mathematical functions is available in MATLAB. These are listed in Table A.i.2. (In the table, the input to the function is designated as c, which may be real, complex, or a matrix.) If the argument is a matrix, the function will return a like-size matrix of function values evaluated at each of the values of the matrix.

Other common mathematical functions are $\max(x)$ and $\min(x)$. If the input is simply a column or row vector, the MATLAB functions of the same name return the maximum or minimum elements, respectively. In addition, the functions will return the location of the maximum/minimum in the string of values. For example, if x = [1 5 3 8 2 7], then the MATLAB statement [y,i] = max(x) will return y = 8 as the maximum value and i = 4 as the index of the maximum value. If the input is an $m \times n$ matrix **A**, these functions will return a *row vector* containing the maximum/minimum values in each column of the matrix. Likewise, in the statement [y,i] = max(A), i will be returned as a row of integers containing the location of the maximum in each column.

Referencing Elements of a Matrix

Individual elements of a matrix may be referenced by specifying their indices within parentheses. Thus,

```
x    = [2 -1 1]
x(6) = acos(x(2))
```

results in

```
x =
      2.0000 -1.0000 1.0000   0      0    3.1416
```

Notice that the result of the statement is an extended vector x, now of length 6, and that the intervening elements (x(4) and x(5)) have been filled with zeros. Also, the result of a statement involving more than one element of x is the display of the entire matrix. The format for displaying the numbers can be adjusted, somewhat, by using the format statement in MATLAB.

The transpose of a matrix in MATLAB is effected by following the matrix with an apostrophe. For example, because the matrix **A** on page 514 is a symmetric matrix, the two matrices **A** and **A**′ are identical. Also, because x is a row-vector, x′ is a column vector.

AUGMENTING A MATRIX IN MATLAB

The statement

```
x = [x;x]
```

produces

```
x =
      1.7725   0.5   0.9995   0   0   3.1416
      1.7725   0.5   0.9995   0   0   3.1416
```

That is, the operation [x;x] places a copy of the matrix x *below* the current contents of x and produces a 2 × 6 matrix.

However, consider the result of

```
B  =  [A A]
```

where **A** is, once again, the 3 × 3 matrix given on page 514. The result for **B** is now

```
B  =
      1  2  3  1  2  3
      2  1  2  2  1  2
      3  2  1  3  2  1
```

Thus, the original matrix **A** has been augmented by inserting sufficient columns to accommodate a complete copy of **A** to be inserted next to the original **A**. Of course, the height of the two matrices joined in this fashion must match, or an error will result. (Try C = [x x'].) Next consider

```
C = [A  A;A  A]

C =
      1  2  3  1  2  3
      2  1  2  2  1  2
      3  2  1  3  2  1
      1  2  3  1  2  3
      2  1  2  2  1  2
      3  2  1  3  2  1
```

Two matrices can be joined, side by side, provided that they have the same number of rows. They can also be joined one on top of the other, provided they both have the same number of columns.

CONSTRUCTING EQUALLY SPACED VECTORS USING THE COLON

The colon operator (:) is a special feature in MATLAB for constructing row vectors of evenly spaced values. The statement

```
x = 1:5
```

will generate a row vector containing the integers from 1 to 5.

```
x =
      1   2   3   4   5
```

The general form of the statement is

```
x = a:dx:b
```

which, assuming a, b, and dx have been assigned values, yields

$$x = $$
$$a \quad a + dx \quad a + 2dx \cdots a + \text{fix} \left[\frac{b - a}{dx} \right] dx,$$

where fix(t) truncates the value of t to an integer. The step size, dx, can be negative. The vector x will be empty (which is not the same as zero) if b > a with dx negative, or if b < a with dx positive. If the increment is omitted, a value of +1 is assumed. Thus,

```
t = -5:0
```

yields

$$-5.0000 \quad -4.000 \quad -3.0000 \quad -2.0000 \quad -1.0000 \quad 0.0$$

and

```
t = 5:-pi:-5
```

results in

```
t =   5.0000   1.8584   -1.2832   -4.4248
```

Special Matrices in MATLAB

There are several built-in matrices available in MATLAB that can be easily accessed and used for experimentation. In particular, these matrices can be used as the input to the various matrix functions that we will describe shortly. The two simplest stored matrices are called magic(n) and hilb(n). Other, more exotic matrices are described in the MATLAB reference manual under the heading "Special Matrices."

magic(n)   Where n is an integer < 32. (The largest matrix that can be facilitated in the student edition of MATLAB is a square 32×32 matrix.) This function will return a square matrix whose elements constitute a magic square; i.e., the sum of elements along each row, column, or principal diagonal is the same value. By inspection, you should be able to determine the pattern for constructing magic squares for odd n. The algorithm for even n is a bit more complicated.

hilb(n)   Where n is an integer < 32. These Hilbert matrices are of the form

$$\begin{bmatrix} 1 & \frac{1}{2} & \frac{1}{3} & \cdots & \frac{1}{n} \\ \frac{1}{2} & \frac{1}{3} & \frac{1}{4} & \cdots & \frac{1}{n+1} \\ \vdots & \vdots & \vdots & \ddots & \vdots \\ \frac{1}{n} & \frac{1}{n+1} & \frac{1}{n+2} & \cdots & \frac{1}{2n} \end{bmatrix}$$

and are of note for being very ill conditioned. (See Chapter 2.)

Special MATLAB Functions for Constructing Large Matrices

Typing in all the elements of a large matrix can be tedious and is very prone to typing errors. MATLAB provides in the toolbox numerous functions for helping you construct large matrices. Several of the most common of these functions are described here.

`length(x)` — Returns a scalar (i.e., simply a number, a 1×1 matrix) representing the number of elements in the vector (either a row vector or a column vector). If x is a matrix, `length` returns the larger of the two dimensions of x.

`size(A)` — Returns two numbers (a 1×2 row vector) representing the row-column size of the matrix $[\mathbf{A}]$. The ordinary use is `[m,n]=size(A)`.

`zeros(n)` — Is an $n \times n$ matrix of all zeros.

`zeros(size(A))` — Produces an all-zero matrix of the same size as $\mathbf{A}$.

`ones(n)` — Is an $n \times n$ matrix of all ones.

`ones(m,n)` — Produces an $m \times n$ matrix of all ones.

`ones(size(A))` — Is a matrix of all ones that is the same size as $\mathbf{A}$.

`eye(n)` — Is the $n \times n$ identity matrix (ones on the diagonal, zeros elsewhere).

`eye(m,n)` — Is an $m \times n$ matrix with ones on the principal diagonal, zeros elsewhere.

`eye(size(A))` — Produces an identity matrix of the same size as $\mathbf{A}$.

`diag(X)` — If $\mathbf{X}$ is a *vector* of length n, `diag(X)` is an $n \times n$ matrix with the values of x along the diagonal and zeros elsewhere.

If $\mathbf{X}$ is a *matrix*, `diag(X)` is a column vector containing the elements of $\mathbf{X}$ along the principal diagonal.

`diag(x,k)` — Is a square matrix with the values of the vector x along the kth diagonal, zeros elsewhere. The main diagonal is $k = 0$ ($k > 0$ is above the main diagonal, and $k < 0$ is below it). The size of the matrix is $n + |k|$, where n is the length of x.

`diag(A,k)`	Is a column vector containing the elements of the kth diagonal of **A**, where the numbering of diagonals, k, is the same as described for `diag(x,k)`.

Thus,

```
x = [2 2 2 2 2];
y = [1 1 1 1];
B = diag(x) - diag(y,1) - diag(y,-1);
```

produces the banded matrix

$$\mathbf{B} = \begin{bmatrix} 2 & -1 & 0 & 0 & 0 \\ -1 & 2 & -1 & 0 & 0 \\ 0 & -1 & 2 & -1 & 0 \\ 0 & 0 & -1 & 2 & -1 \\ 0 & 0 & 0 & -1 & 2 \end{bmatrix}$$

Notice that the sum of two matrices, $\mathbf{A} + \mathbf{B}$, of the same size yields a matrix whose elements are the sums of the corresponding elements of **A** and **B**; i.e.,

$$\mathbf{C} = \mathbf{A} + \mathbf{B}; \qquad c_{ij} = a_{ij} + b_{ij}$$

The same matrix could also be built up as

```
B = diag(ones(5,1)) - diag(ones(4,1),1) - diag(ones(4,1),-1)
```

The Colon for Characterizing Parts of a Matrix

The colon operator is most often used to generate a vector of evenly spaced values, as illustrated earlier. However, it also plays the role of a *wild-card* marker when used to designate rows or columns of a matrix. For example, if **A** is an $m \times n$ matrix, then

`A(:,k)`	Is a submatrix of **A** containing *all* the rows of **A** but only column k. That is, it is simply equal to column k.
`A(k,:)`	Is equal to *row* k of **A**.
`A(4:6,:)`	Is a $3 \times n$ matrix containing rows 4 through 6.
`A(:,:)`	Is the same as **A**.
`A(:)`	Will *reshape* **A** into a single column matrix by stringing together all the columns of **A**.

If you are uncertain of the size of a matrix, it may be determined in MATLAB by using the function `size`. Likewise, the length of a vector is determined using `length`.

Some Cautions When Entering Matrices in MATLAB

- In a string of numbers, a *blank space* is interpreted to be the same as a comma; i.e., a separator of numbers.
- The number 6.67×10^{-11} is entered as **6.67e-11** (no spaces). Also, 10^{-11} is entered as **1.e-11**, not **e-11**. The latter would be interpreted as a variable "e" minus 11.
- Variable names must begin with a letter, contain only letters or digits, and although they may be of any length, MATLAB only retains the first 19 characters.
- Entering the command **who** or **whos** will display a list of all the variable names currently defined in a session. The command **clear** will remove all user-defined variable names from the list.
- You can change the format of displayed numbers by entering any of the following commands:

Normal Notation	*Scientific Notation*
format short (5 digits)	format short e (5 digits)
format long (15 digits)	format long e (15 digits)

  Rational Fractions
  format rat
  Numbers will be displayed as
  rational fractions (insofar as
  is possible)

- The contents of a matrix in an assignment statement must begin with a [and must end with a]. Successive rows are separated by a semicolon (;) or by entering them on separate lines.
- MATLAB is *case sensitive*; i.e., MATLAB and MatLab are not the same.
- If an arithmetic expression is entered alone (not part of an assignment statement), the value of the expression is returned in the variable **ans**.

Problems

PA.i.1 MATLAB Matrix Definitions   Predict and verify the output of the following MATLAB statements:

a. ```
x = 1:4;
y = 4:-1:1;
A = [x;y];
A = [A;A]
```

b. ```
A = [1:4;2:5;3:6;4:7]
```

c. ```
theta = 0:.1:pi;
    x = sin(theta);
    y = cos(theta);
    z = tan(theta);
table = [theta;x;y;z]'
```
```

```
d. A = 4*eye(2) - ones(2);
   B = ones(2);
   C = [A -B;-B A];
```

```
e. A = magic(4);
   A = [A zeros(4,1);[zeros(1,4) 1]];
```

```
f.    n = 8;
      x = ones(1,n);
      A = diag(x,3);
   [m,n]= size(A)
```

PA.i.2 MATLAB Matrix Multiplication   Enter the following vectors and matrices in MATLAB:

$$A = \begin{bmatrix} 1 & 2 & 3 & 4 \\ 2 & 4 & 6 & 3 \\ 3 & 6 & 4 & 2 \\ 4 & 3 & 2 & 1 \end{bmatrix} \qquad B = \begin{bmatrix} 2 & 3 & 3 & 2 \\ 3 & 5 & 5 & 3 \\ 3 & 5 & 5 & 3 \\ 2 & 3 & 3 & 2 \end{bmatrix}$$

and evaluate

a. A*B-B*A
b. C = .5*(B-3)∧2
c. A*C - C*A
d. A*B*C-C*B*A
e. C∧3

PA.i.3 MATLAB Matrix Multiplication   Enter the following vectors and matrices in MATLAB using diag, eye, and ones where possible:

$$Q = \begin{bmatrix} 2 & -1 & 0 & 0 \\ -1 & 2 & -1 & 0 \\ 0 & -1 & 2 & -1 \\ 0 & 0 & -1 & 2 \end{bmatrix} \qquad W = \begin{bmatrix} 4 & 3 & 2 & 1 \\ 3 & 6 & 4 & 2 \\ 2 & 4 & 6 & 3 \\ 1 & 2 & 3 & 4 \end{bmatrix}$$

$$L = \begin{bmatrix} 1 & 0 & 0 & 0 \\ -\frac{1}{2} & 1 & 0 & 0 \\ 0 & -\frac{2}{3} & 1 & 0 \\ 0 & 0 & -\frac{3}{4} & 1 \end{bmatrix} \qquad U = \begin{bmatrix} 2 & -1 & 0 & 0 \\ 0 & \frac{3}{2} & -1 & 0 \\ 0 & 0 & \frac{4}{3} & -1 \\ 0 & 0 & 0 & \frac{5}{4} \end{bmatrix}$$

and evaluate

a. Q*W
b. U*W*L
c. L*U*W
d. W*L*U
e. All of the above should produce the same result. From this can you conclude that any two of the four matrices *commute*?
f. If you additionally compute L*U, you should be able to determine which two of the matrices commute.

PA.i.4 Special Matrix Functions

a. For any matrix $\mathbf{A}$, what is the value of `length(size(A))`?

b. Use `diag` and `ones` to produce the matrices

$$
C = \begin{bmatrix} 8 & 7 & 6 & \cdots & 1 \\ 7 & 8 & 7 & \cdots & 2 \\ 6 & 7 & 8 & \cdots & 3 \\ \vdots & \vdots & \vdots & \ddots & \vdots \\ 1 & 2 & 3 & \cdots & 8 \end{bmatrix}
\qquad
D = \begin{bmatrix} 8 & 7 & 6 & \cdots & 1 \\ 7 & 7 & 6 & \cdots & 1 \\ 6 & 6 & 6 & \cdots & 1 \\ \vdots & \vdots & \vdots & \ddots & \vdots \\ 1 & 1 & 1 & \cdots & 1 \end{bmatrix}
$$

c. The sums of the elements of each of the columns of a magic square add to the same number, S. Use `sum` to determine this value for magic squares of odd order from 3 to 11. From these numbers, attempt to determine the equation relating S and the order p.

d. By considering several values of p, demonstrate that the sum of all the elements of the Hilbert matrix `hilb(p)` is approximately proportional to p.

PA.i.5 MATLAB Matrix Functions The ordinary MATLAB mathematical functions, such as `sin()`, `cos()`, and `exp()`, will accept a square matrix as input. However, the result is perhaps not what you might expect. For example, if $\mathbf{A}$ is a 2×2 matrix, then `exp(A)` will return a 2×2 matrix with elements computed as

$$
\begin{bmatrix} e^{a_{11}} & e^{a_{12}} \\ e^{a_{21}} & e^{a_{22}} \end{bmatrix}
$$

which is not the same as the definition of $e^{\mathbf{A}}$ in terms of an infinite series,

$$
e^{\mathbf{A}} \equiv \mathbf{I}_2 + \mathbf{A} + \frac{1}{2!}\mathbf{A}^2 + \frac{1}{3!}\mathbf{A}^3 + \cdots
$$

However, MATLAB has a subclass of toolbox functions that will actually use the series definitions for the functions by computing $\mathbf{A}^n$ for each term in the series. These functions (`expm()`, `logm()`, and `sqrtm()`) are accessed by simply adding an m onto the name of the function.

a. Use the MATLAB function `expm()`, along with the ordinary `sin()` and `cos()`, to numerically verify the results of Problem 1.22 in Chapter 1.

b. Use the MATLAB function `expm()` to evaluate $e^{\mathbf{A}}$, where $\mathbf{A}$ is *nilpotent* (see Problem 1.9 in Chapter 1) and is given by

$$
\mathbf{A} = \begin{bmatrix} 0 & 1 & 1 & 1 \\ 0 & 0 & 1 & 1 \\ 0 & 0 & 0 & 1 \\ 0 & 0 & 0 & 0 \end{bmatrix}
$$

Compare with a calculation using the series expansion for $e^{\mathbf{A}}$.

Summary

MATLAB CONSTRUCTION OF MATRICES AND VECTORS

	`x = [3 4 -1 0 5]`	*Row vector*
Vectors $\vec{x}$:	`x = [3 4 -1 0 5]'`	*Column vector*
	`x = [3:2:19]`	*Evenly spaced row vector, length 9*

$A = [1\ 4;3\ 2;0\ 1]$	A 3×2 *matrix,*	$\begin{bmatrix} 1 & 4 \\ 3 & 2 \\ 0 & 1 \end{bmatrix}$

Matrices   **M** :

$B = [1:3]'*[3:-1:1]$	A 3×3 *matrix,*	$\begin{bmatrix} 3 & 2 & 1 \\ 6 & 4 & 2 \\ 9 & 6 & 3 \end{bmatrix}$

$C = [B\ B;A'\ A']$ A 5×6 *matrix*

Special matrices

ones():	For constructing matrices of various sizes consisting of all *ones*.
zeros():	For constructing matrices of various sizes consisting of all *zeros*.
eye():	For constructing matrices of various sizes consisting of ones on the main diagonal and zeros elsewhere.
diag():	For constructing *diagonal* matrices of various sizes *or* for extracting a diagonal from a given matrix.

Formats

format short	Five digits
format short e	Five digits in scientific (e) notation
format long	Fifteen digits
format long e	Fifteen digits in scientific notation
format rat	Numbers displayed as rational fractions

MATLAB Functions Used

Inf, NaN, eps   Three constants added to the set of numbers normally available in computer arithmetic. **Inf** represents $+\infty$, and it is the result of overflow operations like dividing by zero. **NaN** represents an "undefined" numerical quantity, and it results from operations like $0/0$ or $\infty - \infty$. Any operation containing a **NaN** will yield the result **NaN**. The quantity **eps** is the machine accuracy; i.e., the smallest positive floating-point number that satisfies the relation $1 + eps \neq 1$.

ans   When a MATLAB operation is not assigned to a variable by the replacement operator, "=", the result is automatically stored in the variable named **ans**.

max, min   The maximum, minimum of a variable. If **x** is a vector, a single value is returned. If a matrix, a row vector containing the maximum, minimum of each column is returned. **max(max(A))** is the overall maximum of a matrix. Also, **[mx,i] = max(x)** will return the location of the maximum element(s) of **x** in the index **i**.

m:n, m:d:n   The colon operator can be used to generate a vector of equally spaced values from **m** to **n**. If $m > n$ the vector is empty. The quantity **d** is the spacing between elements; if omitted, a value of one is assumed.

[x y], [x;y]   If **x** and **y** are the same size and shape, then **[x y]** is an augmented matrix constructed by adding the elements of **y** to the right of those of **x**. **[x;y]** augments **x** by placing the elements of **y** below those of **x**.

`magic(n)`, `hilb(n)`	Two of several MATLAB toolbox functions that can be used to automatically construct matrices of size $n \times n$. `magic` returns a "magic" square, and `hilb` returns a matrix whose i,jth element is $(i+j-1)^{-1}$. They are useful for experimentation with matrices. See also the description of Specialized Matrices in the MATLAB reference manual.
`length, size`	If `x` is a vector, `length(x)` returns the number of elements; if a matrix, it returns the maximum dimension. The function `size(A)` returns the dimensions of a vector or a matrix.
`ones, zeros`	`ones(n)`, `ones(m,n)` return a matrix of all ones of size $n \times n$ or $m \times n$, respectively. `zeros` operates analogously and returns matrices of all zeros. `ones(size(A))` will return a matrix of all ones that is the same size as matrix `A`; `zeros` operates similarly.
`eye`	`eye(n)` will return the $n \times n$ identity matrix; i.e, a matrix with ones on the main diagonal and zeros elsewhere. `eye(m,n)` is an $m \times n$ matrix with ones on the main diagonal and zeros elsewhere.
`diag(x)`	This function is used to create or extract diagonals of a matrix. The position of diagonals is characterized by an integer k, with $k = 0$ referring to the main diagonal, and $k = +1$ (-1) designating the diagonal above (below) the main diagonal, etc. If $\vec{x}$ is a vector of length n, then `diag(x,k)` is a $p \times p$ matrix, with $p = n + |k|$ and the kth diagonal equal to $\vec{x}$ and zeros elsewhere. Simply using `diag(x)` returns a matrix with $\vec{x}$ along the main diagonal. `diag` can also be used to extract a diagonal from an existing matrix `M`. `diag(M,k)` returns a column vector equal to the kth diagonal of `M`.
The colon :	The colon operator can also be used as a *wild card* in designating the indices of a matrix. Thus, `x(:,3)` is a column vector constructed from column 3 and *all* the rows of the matrix `x`. `x(:,:)` is equal to `x`. Also, `x(:)` will reshape the matrix `x` into a single column vector composed of the individual columns of `x`, one after the other. This is particularly useful when the vectors transferred to a function are column vectors. That is, if `x` is either a row or a column vector, `x(:)` is a column vector. When on the left of an assignment statement, `x(:) = ...`, the elements of `x` are assigned using the previous shape for `x`.
`format`	A MATLAB command that controls the display of numerical results. See the description in the Summary.

APPENDIX A.ii /
Polynomials and
Other Functions

PREVIEW   The reason for our interest in polynomials is that effectively every function defined for use on a computer is in the form of a series expansion or a ratio of series expansions. And because the evaluation of a series expansion will always require that the infinite number of terms in the expansion be truncated to a polynomial, computer calculations are primarily in terms of the evaluation and manipulation of polynomials. Of course, most of these operations are invisible to the user. Nonetheless, the reader will find it useful to be comfortable with the numerical evaluation and manipulation of polynomials.

There are several toolbox functions that will efficiently evaluate a polynomial at point x or at a collection of points, given the vector containing the coefficients, or that will perform a variety of manipulations on one or more polynomials. In this chapter, we describe several of these MATLAB functions in detail.

MATLAB was designed to efficiently implement the elements of matrix arithmetic and, as described in Part I, can greatly simplify the solution of problems that are primarily "matrix problems." However, MATLAB is much more than simply a matrix manipulator. In addition to the arithmetic operators, [+, -, \, /, *, ^, :], MATLAB contains structures to perform summations, loops, comparisons, and input-output operations; in short, all the elements of a computer language. The construction of MATLAB code to evaluate a polynomial provides an excellent introduction to the basic elements of these MATLAB computing structures.

In addition, as in any computer language, MATLAB provides the capability of user-written *modules* or functions, here called M-file functions, which can be created and subsequently used in a fashion exactly like the already existing toolbox functions. In this chapter we introduce the subject of M–file creation, and illustrate it using various features of polynomials.

Evaluation of Polynomials

We restrict our definition of polynomials to those containing only positive integer powers, n, of the independent variable:

$$y(x) = a_n x^n + a_{n-1} x^{n-1} + \cdots + a_1 x + a_0$$

A polynomial is completely determined once the set of coefficients, a_i, has been specified. Note that the degree of the preceding polynomial; i.e., the highest power of x with nonzero coefficient, is n, and that there are $n+1$ coefficients. To bring the standard form of a polynomial in line with the usage in MATLAB, we will redefine the labeling of the coefficients and write the standard polynomial as

$$y(x) = c_1 x^n + c_2 x^{n-1} + \cdots + c_n x + c_{n+1}$$

That is, the $1 \times (n+1)$ row vector of coefficients, c, will contain all the coefficients of the nth-degree polynomial. Remember, in MATLAB, the coefficients, c_i, of the terms of a polynomial are stored in *reverse* order; the first element is the coefficient of the highest-order term.

The Toolbox Function `polyval`   The evaluation of the polynomial for a particular value of x is effected using the toolbox function `polyval()`. The usage is

$$\texttt{y = polyval(c,x),}$$

where c is the row vector containing the coefficients of the polynomial. Thus if c = [1 2 3 4], then `polyval(c,2)` returns a value of 26 (i.e., $2^3 + 2(2^2) + 3(2) + 4$).

However, as with other toolbox functions, the most useful feature is that, if the input value, x, is a vector or even a matrix, `polyval(c,x)` will compute the value of the polynomial at each value contained in x and will return a vector or matrix of the same size and shape as x containing those values. Thus, if $x = \begin{pmatrix} 1 & 2 \\ 3 & 4 \end{pmatrix}$, `polyval(c,x)` returns a *matrix* of values `polyval(c, x)` $\Rightarrow$ $\begin{pmatrix} 10 & 26 \\ 58 & 112 \end{pmatrix}$.

Next, we consider how the code to evaluate a polynomial is constructed.

Loop Structures in MATLAB

The algorithm to evaluate a polynomial will require a repetitive *loop* structure of the form:

$$x = (\text{specify some value for } x)$$

step no.:

1	$y = c_{n+1}$
2	$y = y + c_n x$ (so far, $y = c_n x + c_{n+1}$)
3	$y = y + c_{n-1} x^2$ (i.e., $y = c_{n-1} x^2 + c_n x + c_{n+1}$)
$\vdots$	$\vdots$
$n+1$	$y = y + c_1 x^n$

The structure in MATLAB to implement this loop is the `for` loop, which has the following form:

MATLAB `for` loop

`for i = i_1:Δi:i_2`

    The body of the loop contains
    a set of MATLAB statements to
    be executed for each successive
    value of `i`.

`end`

Here i_1 and i_2 are the initial and final values for `i` and Δi is the step size (if omitted, a value of $+1$ is assumed).[1] The loop structure is terminated by the MATLAB statement `end`.

    MATLAB will first determine how many steps are required and then execute the body of the loop that many times for the successive values of `i`. Of course, this means that if, for example, $i_2 < i_1$ with $\Delta i > 0$, the loop is ignored by MATLAB.

    For example, consider a loop structure that could be used to compute the sum of the integers from 1 to 100.

`sum = 0`	The variable `sum` starts out as zero.
`for i = 1:100`	
`sum = sum + i^2;`	`sum` is successively replaced by the previous value plus the new term, i^2.
`end`	
`sum = 338350`	

It is always a good idea to indent the body of the loop statements with respect to the `for` and the `end` statements.

    The MATLAB program to evaluate a 5th-degree polynomial, $y(x) = 2x^5 + 5x^4 - x^3 + 2x + 1$, with coefficients contained in the row vector `c`, at, say $x = 2$, would then be

---

```
c = [2 5 -1 0 2 1];
x = 2;
y = 0;
for i = 1:5
    y = y + c(i)*x^(n+1-i);
end
```

---

[1] Of course, $i_1 : \Delta i : i_2$ is simply a row vector. The more general form of the `for` loop is then `for i = `*matrix expression*, which proceeds by replacing `i` successively by the *columns* of the matrix expression. Thus, in addition to the usual situation where `i` is simply a numerical counter, more elaborate loops are possible wherein the loop counter, `i`, could be a sequence of column vectors.

However, evaluating a polynomial by using the exponentiation operator is extremely inefficient and, especially if the polynomial is to be repeatedly computed for various values of x, can severely increase the computing time of a program.

Horner's Method for Polynomial Evaluation   A much more efficient procedure to use when evaluating a polynomial like $y(x) = 2x^5 + 5x^4 - x^3 + 2x + 1$, would be to regroup the terms and write the polynomial as

$$y = ((((2x + 5)x - 1)x + 0)x + 2)x + 1$$

so that only multiplication and addition/subtraction are involved. This is known as Horner's method for polynomial evaluation. The MATLAB version of this method would then be:

```
c = [2 5 -1 0 2 1];
x = 2;
y = c(1);
for i = 2:5
    y = y*x + c(i)
end
```

This is the prescription that is used by the MATLAB toolbox function `polyval(c,x)` to evaluate the polynomial with coefficients `c` at the value of the independent variable specified by `x`.

NESTED for LOOPS

Once we have the prescription for evaluating a particular function, $y(x)$, coded in MATLAB, it would be useful to be able to evaluate the function for a complete range of x values, thereby producing a table of $x_i, y(x_i)$ values. These could then be scanned, if looking for a root, or plotted to display the general features of the function. You could do this by *nesting* two `for` loops, one completely inside the other, as:

```
Nesting of for loops

c = [2 5 -1 0 2 1];
for x = 1:0.1:3;
    y = c(1);
    for i = 2:5
        y = y*x + c(i)
    end
    disp(y)
end
```

Nesting of loops in this manner is permitted to any depth, provided that no two loops *overlap*; i.e., each *inner* loop must be completed before an encompassing *outer* loop can be terminated. For each of the values of x in the range $x = 1$ to $x = 3$, the preceding code will compute the values of the polynomial and will display the result. Note, however, that the values were not stored from one x to the next. This could be corrected by arranging for the code to successively store values for $x_i, y(x_i)$, in column or row vectors for a range of i values; for example,

```
n = 20;   dx = 1/n;
x = 1:dx:n*dx;
y = zeros(1,n+1);
for i = 1:n+1;
   y(i) =  c(1);
   for k = 2:5
       y(i) = y(i)*x(i) + c(k)
   end
end
disp([x;y])
```

The final line of the code will now display the entire set of **y** values, which have been stored, along with the **x**'s, in row vectors. These number pairs, $x_i, y(x_i)$, can easily be displayed in the form of a table by combining the x and y values into a $2 \times (n + 1)$ matrix and calling for the display of the transpose, such as `disp([x;y]')`.

The third line of the code, **y = zeros(1,n+1);**, deserves some special consideration. If we had omitted this line, the code would still compute all the values of y_i correctly but would be found to execute measurably more slowly. The reason for this is that as each new element of y is computed, MATLAB must "resize" the array y to include it among the others. To avoid this, we size y to its final size and shape before the loop by assigning it all zeros. If your code really requires **for** loops, avoid resizing matrices within the body of the loop to optimize efficiency.

Vectorized Arithmetic Operations

In place of the preceding nested **for** loop structure, it would be most convenient if, as with other mathematical functions like **sin, cos, exp,** etc., the evaluation of the polynomial $y(x)$ would return a simple number if x is just a scalar, but would return a complete matrix if x is a matrix. Recall, that if **x = [1.1 1.2 1.3;2.1 2.2 2.3]**, then **sin(x)** returns a 2×3 matrix of values corresponding to

$$\begin{matrix} \sin(1.1) & \sin(1.2) & \sin(1.3) \\ \sin(2.1) & \sin(2.2) & \sin(2.3) \end{matrix}$$

MATLAB has special arithmetic operators that can facilitate this type of calculation. They are called *element-by-element* operators, or *vectorized* operators. For example, the ordinary arithmetic operators of multiplication and division will perform the expected result between ordinary numbers and have been extended to execute matrix operations if the operands are matrices:

```
x = [1 0 -1;0 2 1];
y = [1 -1;2 2;0 -3];
```

$$A = x*y \Rightarrow \begin{pmatrix} 1 & 0 & -1 \\ 0 & 2 & 1 \end{pmatrix} \begin{pmatrix} 1 & -1 \\ 2 & 2 \\ 0 & -3 \end{pmatrix} \Rightarrow \begin{bmatrix} 1.0000 & 2.0000 \\ 4.0000 & 1.0000 \end{bmatrix}$$

$$y/A = yA^{-1} \Rightarrow \begin{pmatrix} 1 & -1 \\ 2 & 2 \\ 0 & -3 \end{pmatrix} \cdot \begin{pmatrix} -0.1429 & 0.2857 \\ 0.5714 & -0.1429 \end{pmatrix}$$

$$\Rightarrow \begin{bmatrix} -0.7143 & 0.4286 \\ 0.8571 & 0.2857 \\ -1.7143 & 0.4286 \end{bmatrix}$$

THE FORM OF ELEMENT-BY-ELEMENT OPERATORS

The operator that implements *element-by-element* multiplication is designated as .*; i.e., the ordinary operator for multiplication preceded by a period.[2] The rules for applying this operator are simply that the operands, if matrices, *must be of the same size and shape.* Then A.*B generates the complete set of products $a_{ij}b_{ij}$ for the full range of i and j.

```
x = [1 0 -1;0 2 1];
y = [1 -1;2 2;0 -3];
x.*y =
```
ERROR. The matrices x and y are not of the same shape. However,

$$x.*y' = \begin{bmatrix} 1 & 0 & 0 \\ 0 & 4 & -3 \end{bmatrix}$$

Element-by-element division and exponentiation work the same way:

$$B = x./y' \Rightarrow \begin{pmatrix} 1 & 0 & -1 \\ 0 & 2 & 1 \end{pmatrix} ./ \begin{pmatrix} 1 & 2 & 0 \\ -1 & 2 & -3 \end{pmatrix}$$

$$\Rightarrow \begin{pmatrix} \frac{1}{1} & \frac{0}{2} & -\frac{1}{0} \\ -\frac{0}{1} & \frac{2}{2} & -\frac{1}{3} \end{pmatrix}$$

$$= \begin{bmatrix} 1.0000 & 0.0000 & -\infty \\ 0.0000 & 1.0000 & -0.3333 \end{bmatrix}$$

---

[2]Be cautious when typing this operator to never introduce a space between the period and the asterisk.

$$C = x.\hat{}(y') \Rightarrow \begin{pmatrix} 1 & 0 & -1 \\ 0 & 2 & 1 \end{pmatrix} . \wedge \begin{pmatrix} 1 & 2 & 0 \\ -1 & 2 & -3 \end{pmatrix}$$

$$\Rightarrow \begin{pmatrix} 1^1 & 0^2 & (-1)^0 \\ 0^{-1} & 2^2 & 1^{-3} \end{pmatrix}$$

$$= \begin{bmatrix} 1 & 0 & 1 \\ \infty & 4 & 1 \end{bmatrix}$$

Returning to the earlier MATLAB code to evaluate a polynomial, we can now adapt the program to automatically return a matrix of results if the input, x, is itself a matrix. The actual code in the toolbox function `polyval`[3] that is used to evaluate a polynomial with coefficients contained in the row vector c for a collection of input values contained in the matrix x is:

```
[m,n] = size(x)
nc = max(size(c))
y   = zeros(m,n)
for i = 1:nc
    y = x.*y + c(i)*ones(m,n)
end
```

Basically, all the MATLAB toolbox functions are written so as to return a matrix of values corresponding to a matrix of input values by making use of the *element-by-element* arithmetic operators. You should always put a little extra effort in the writing of the functions you create and design them in a similar fashion to accept either scalars, vectors, or $m \times n$ matrices as input.

Multiplication and Division of Polynomials

To multiply a polynomial of degree n, specified by coefficients, $c_i, i = 1, \ldots, n+1$, by a monomial factor, $(x - a)$, we begin by formally writing out the product as

$$(x - a)(c_1 x^n + c_2 x^{n-1} + \cdots + c_n x + c_{n+1})$$

$$= c_1 x^{n+1} + (c_2 - ac_1)x^n + (c_3 - ac_2)x^{n-1} + \cdots + (c_{n+1} - c_n)x - ac_{n+1}$$

---

[3]There is a second version of this function in the toolbox called `polyvalm`, which, like the functions `expm`, `sinm`, `logm`, carries out the arithmetic in a *matrix* sense. That is, if **A** is a *square* matrix and the polynomial has coefficients, say c = [4 3 2 1], then `polyvalm(c,A)` will carry out the operations to evaluate 4*A*A*A + 3*A*A + 2*A + 1, where the multiplication is matrix multiplication, not element by element.

The coefficients, d_i, of the product polynomial are then given by

```
d(1) = c(1);
n    = length(c);
for i = 2:n
    d(i) = c(i) - a*c(i-1);
end
d(n+1) = -a*c(n);
```

Similarly, by writing out the long division of a polynomial of degree n, specified by coefficients, d_i, $i = 1, \ldots, n+1$, by a monomial factor, $(x - a)$, you can easily verify that the result is

$$(d_1 x^n + d_2 x^{n-1} + \cdots + d_n x + d_{n+1})/(x - a)$$
$$= d_1 x^{n-1} + (d_2 + ad_1)x^n + (d_3 + a(d_2 + ad1))x^{n-1} + \cdots + \frac{r}{x - a}$$

where r is the *remainder*. The coefficients of the resulting polynomial are then given by

```
c(1) = d(1);
n    = length(c);
for i = 2:n-1
   c(i) = d(i) + a*c(i-1);
end
r = d(n) + a*c(n-2);
```

where r is the remainder term. Note the similarity with the previous relationship between coefficients when multiplying a monomial times a polynomial.

More complicated, but equally straightforward, is the multiplication and division of two polynomials of degrees n_1 and n_2, respectively, with coefficients contained in the row vectors, c_i, $i = 1, \ldots, n_1 + 1$; d_j, $j = 1, \ldots, n_2 + 1$. Rather than deriving the result for the coefficients of the product and quotient, we simply make note of the fact that there are toolbox functions for their computation.

The Toolbox Function for Multiplying Polynomials, conv   The function conv(c,d) will determine the coefficients of the product of two polynomials with coefficients contained in the row vectors c,d, respectively. It returns a single row vector, p, containing the coefficients of the product polynomial. For example,

$$c = [1\ 3\ 3\ 1]; \qquad \text{i.e., } (x+1)^3$$
$$d = [1\ 4\ 6\ 4\ 1]: \qquad \text{i.e., } (x+1)^4$$
$$p = conv(c,d)$$
$$p =$$
$$1\ 7\ 21\ 35\ 35\ 21\ 7\ 1 \quad \Rightarrow (x+1)^7$$

The Toolbox Function for Division of Polynomials, deconv   The function deconv(d,c) returns *two* row vectors: q, containing the coefficients of the degree $n_2 - n_1$ quotient, and r, containing coefficients of the degree $n_1 - 1$ remainder. The usage is:

$$[q,r] = \texttt{deconv(p,c)} \quad \text{i.e., } (x+1)^7/(x+1)^3$$

```
q =
      1 4 6 4 1          ⇒ (x + 1)⁴
r =
      0 0 0
```

$q =$ 1 4 6 4 1 $\Rightarrow (x+1)^4$

$r =$ 0 0 0

Elementary MATLAB M-Files and Polynomials

Frequently, even the simplest calculation in MATLAB requires several lines of MATLAB commands and perhaps a dozen or more numerical values to be typed at the keyboard. And every time you wish to duplicate the calculation after a minor change or two in the code or input, you must retype the whole set of instructions all over again.[4] Clearly, what is needed is a procedure for saving a complete set of instructions and a mechanism for easily reusing that set at a later time. In a computer language like FORTRAN, this is accomplished with the use of subroutines or functions. The remedy in MATLAB is quite similar using *script* or *function* files. These are separate files labeled with the name of the script or function and always bearing the suffix .m.

A self-contained collection of MATLAB commands can be included in what is called a *script* file and given a name. When referenced in MATLAB by that name, the sequence of commands is then automatically executed, much like the operation of a FORTRAN subroutine. A script file will ordinarily not be expected to return a particular value but simply to duplicate a sequence of MATLAB instructions.

The second type of M-file is the *function* file, which enables the user to extend the basic library functions by adding his or her own favorite computational procedures. Function M-files are expected to return one or more results. Both function and script files may include references to other MATLAB toolbox routines, and, unlike functions in many computer languages, may even reference themselves. We will be primarily concerned here with function files.

MATLAB M-FILE FUNCTIONS

Each MATLAB function file must begin with a header statement of the form:

$$\texttt{function} \quad \begin{pmatrix} \text{name of} \\ \text{result or} \\ \text{results} \end{pmatrix} = \texttt{name} \begin{pmatrix} \text{argument} \\ \text{list} \end{pmatrix}$$

---

[4]Perhaps you have already discovered that while in MATLAB you can "back up" in the set of most recently entered lines by use of the up-arrow or the Ctrl-P operations. (These operations are system dependent and may not work the same on all systems.)

The body of the function subprogram consists of MATLAB statements that compute a value that is assigned to the output variable(s) named in the header statement. The values of the variables contained in the argument list of the function are available for use within the function. Variable names in the argument list and multiple names on the result side of the relation must be separated by commas. The function does not have a specific *end* statement.

For example, a partial listing of the toolbox function `polyval.m` contains the following MATLAB code:

```
function y = polyval(c,x)
%POLYVAL Polynomial evaluation.
%   If V is a vector whose elements are the coefficients of a
%   polynomial, then POLYVAL(V,S) is the value of the polyno-
%   mial evaluated at S.  If S is a matrix or vector, the
%   polynomial is evaluated at all points in S.  See POLYVALM
%   for evaluation in a matrix sense.
%
%   Polynomial evaluation using Horner's method
%
 [m,n] = size(x);
 nc = max(size(c));
 y = zeros(m,n);
 for i = 1:nc
    y = x.*y + c(i)*ones(m,n);
 end
```

Thus, for this function, the input quantities are the array of coefficients, `c(i)`, and the dependent variable, `x`, which may be a scalar or a matrix of values. The computed values are returned to the quantity, `y`, which is a matrix whose size and shape are determined by x. If the function is to return values for more than one variable, the independent variable names are enclosed in square brackets and separated by commas, as in

`[m,n] = size(A)`

The only connection the outside has with the internal workings of a function is through its argument list. Variables not included in the argument list that are defined within a function remain undefined outside that function and *vice versa*. For example, to compute a set of y values using the equation

$$y(x) = e^{-\alpha x} \sin \beta x$$

the definition line of the MATLAB function would resemble

`function y = damped(x,alpha,beta)`

Occasionally, however, the requirement that all variables used by the function be passed to it through the argument list is in conflict with other restrictions on the function. For example, the MATLAB function `fzero('fname',x0)` will attempt to find a root of the function `fname` starting with an initial guess, x0, for the root. The name of the function that is to be used by `fzero` is enclosed in single quotes. However, `fzero` expects a function *of a single variable*, and so, the form of the function above, `damped`, would result in an error. The alternative to passing variables or parameters directly to a function through the argument list is to pass them indirectly by means of the *global* statement.

GLOBAL VARIABLES

The procedure in MATLAB to permit the *sharing* of a variable between two functions, or between a function and the main MATLAB program, is performed via the `global` statement.[5] The idea is simply to define certain variable names in *both* of the locations in which the name is to be used by inserting a `global` statement of the form

```
global   name1   name2   name3
```

where the variable names are separated by blank spaces (*not commas*).

The `global` declaration must appear in the main MATLAB program or the "calling" *script* file or M-function file *and* within any function or script file where the variables are used. Also, a `global` statement cannot be placed within `for` or `if` loops. Global variables retain their character until the entire workspace is cleared or the computer session is terminated. Global variables should not appear in any function argument lists.

You should be very careful and conservative in your use of this statement. Declaring simple variable names to be `global` can easily cause conflicts with names used and forgotten in various function files. A good suggestion is to include a special character, such as an underscore or a period, or to use all capitals, in the names of global variables.

Thus, to find the roots of the function defined on the previous page, we could use a sequence of MATLAB statements like:

```
global  alpha_1  beta_1
alpha_1 = 0.25
beta_1 = 1.5
root = fzero('damped',3.);
root
    =  2.0944
```

Statements Related to `global`   The following statements are often found to be useful when your MATLAB code employs global variables:

---

[5]`global` statements in MATLAB serve the same purpose as `COMMON` statements in FORTRAN.

clear global   Will clear all global variables from the workspace. For example, if your choice of global variable names could cause a conflict with various other *dummy* variable names, you may wish to clear the workspace of all global variables and start over with a unique set of names for global variables.

isglobal   Is used to determine whether or not a particular variable name has earlier been declared global. The use is, isglobal(*name*). A value of +1 is returned if the name is currently global, and zero if not.

Writing M-files with an Editor

To use MATLAB for more than the simplest of operations, it is important that you become familiar with the procedures for writing script and function M-files. This cannot be done within MATLAB. You will need to have access to an ASCII-type text editor to write the files, and you will also need some knowledge of the computer system under which you are operating to store and retrieve the files you write. An ASCII-type editor is one that does not automatically add any strange characters to your file—characters that are not printed, but are used for arranging or sizing the output. If you are unsure about your current editor, try creating a file containing a sentence or two with your editor; then leave the editor and use the system command for listing the file on the screen.[6] If your editor does not satisfy these requirements, consult your computing center for advice about a suitable editor.

Next, assuming you know how to access the text editor, write and save a file. You will need to know something about files, directories, subdirectories, and the PATH statements that are used by your system when the computer searches through the contents of your files for a specific file name, particularly if you wish to create subdirectories to hold all the new M-files you will be creating. For the moment, we will use a simpler alternative.

INTERRUPTION OF MATLAB WITH THE ! COMMAND

The exclamation point, !, entered from within MATLAB and after the MATLAB prompt (>>), will cause the computer to temporarily interrupt MATLAB and enable you to enter system commands. For example, if your editor is initiated by the command edit *filename*, you could then use the editor to write a function M-file, and save it in the current MATLAB directory, or in another directory, by specifying the path. When the system command has completed execution, in this case when you leave the editor, the MATLAB prompt again appears, indicating a return to MATLAB. Other uses of this command could be to change directories, to search directories for a particular file, or to execute a FORTRAN or other

---

[6]The commands type *filename* on MS-DOS or VAX systems and cat *filename* or more *filename* on UNIX systems will list the contents of a file.

language program that may then, in turn, create data files that can be read by MATLAB programs. All of these applications require a familiarity with machine-dependent protocols, making it impossible to be very specific at this point in the discussion. As a minimum, you should determine the system-specific commands needed to do the following:

1. Access an ASCII text editor and use it to write and save files. If you are executing a Microsoft-Windows version of MATLAB, clicking the mouse on the *File/New* command will open the *Notepad* editor, which may then be used to write, edit, or print an M-file.
2. Create directories and subdirectories.
3. Change directories and the search path (PATH) that the computer uses to find files. If using the Microsoft-Windows version of MATLAB, the easiest way to add a new path to the collection of standard MATLAB paths is to edit (using *Notepad*) the file *MATLABRC.M*, and to simply add a new line to the collection of paths that you will find there. (Make sure that the path you add ends with a semicolon and is enclosed in single quotes.)
4. Execute FORTRAN, C, BASIC, or Pascal programs.
5. Obtain a printed copy of a file.

EXAMPLE A.ii.1

Creating and Using an M-file Function

The function

$$f_n(x) = x^2 \left(1 + \frac{1}{1 + x^2} + \frac{1}{(1 + x^2)^2} + \cdots + \frac{1}{(1 + x^2)^n} \right) \qquad \text{(A.ii.1)}$$

has interesting properties as $x \to 0$ and $n \to \infty$. Clearly, at $x = 0$, $f_n(0) = 0$ for all n. However, the terms within the parentheses can be explicitly summed by using the identity

$$1 + t + t^2 + \cdots + t^n = \frac{1 - t^{n+1}}{1 - t}$$

and thus, with a little algebra, the function can also be written as

$$f_n(x) = (1 + x^2) \left[1 - \frac{1}{(1 + x^2)^{n+1}} \right] \qquad \text{(A.ii.2)}$$

This expression suggests that, because for small, but nonzero values of x,

$$\lim_{n \to \infty} (1 + x^2)^{-n} = 0$$

the function should approach $(1 + x^2)$ for large n,

$$\lim_{n \to \infty} f_n(x) = 1 + x^2 \qquad \text{(for } x \neq 0) \qquad \text{(A.ii.3)}$$

We could write a MATLAB M-file function duplicating Equation A.ii.1 and numerically check the behavior of the function in the limits as $x \to 0$, and $n \to \infty$.

We now illustrate a possible session in MATLAB that creates an M-file function for Equation A.ii.2 and then uses this M-file function to investigate $f_n(x)$ near $x = 0$ and in the limit $n \to \infty$. The first step is to interrupt MATLAB via the ! operation and access an editor. It is assumed that you have determined the relevant editor commands to write and save a file.

```
>>!ed DISCONTF.m
```

On my computer, this command calls up an editor program that will write a file called "DISCONTF.m". The analogous command on your computer will differ. Note that the file name must end with a ".m" suffix. If you are operating under the Microsoft-Windows version of MATLAB, click the mouse on the *File* command and then on option *New*. A separate window will open as part of the *Notepad* program. Once in the editor program the lines will not begin with the MATLAB prompt, >>.

```
function f = DISCONTF(x,n)
  y = (1+x.^2).^(-1);
  c = ones(1,n+1);
  f = x.^2*polyval(c,y);
```

Note the use of *element-by-element* exponentiation to compute the entire set of values of $(1 + x^2)^{-1}$ in one statement.

The coefficients of the polynomial in Equation A.ii.1.

The complete set of function values is now computed. Save the file and exit from the editor program. The MATLAB prompt will again appear.

```
>>x = 0.001:0.005:0.2;
>>for p = 1:4
>>   n = 10^p;
>>   y = DISCONTF(x,n);
>> plot(x,y)
```

A set of 40 x values near zero.

This function will graph y versus x. This topic will be discussed in more detail in Appendix Section A.iv.

```
>> hold on
```

This is needed to "freeze" the plot figure so that subsequent plots will appear superimposed. Again, the details will be discussed in Appendix Section A.iv.

```
>>end
```

The calculation for $n = 10^4$ will take considerable time, and you may wish to substitute Equation A.ii.3 for this case. The plots of the function for the four values of n are drawn in Figure A.ii.1.

```
>>hold off
```

To "unfreeze" the plot.

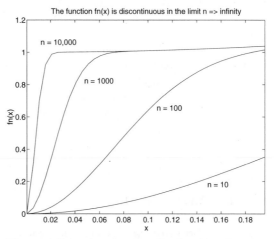

Figure A.ii.1 *As $n \to \infty$, the function becomes discontinuous at $x = 0$.*

Problems

REINFORCEMENT EXERCISES

PA.ii.1 Comparison of Polynomial Algorithms. Let the coefficients of an nth-degree polynomial be the diagonal elements of the Hilbert matrix of order $(n + 1)$; i.e., `c = diag(hilb(n+1))`. Next, for a set of x values, `x = -1:0.02:0`,

    a. Compute the value of the polynomial at the specified x values using the code on page 528. Use $n = 10$ and have the computer display the elapsed execution time. This can be done by preceding the `for` loop by the MATLAB stopwatch start command, `tic`, and closing the loop with an `end`; followed by the stopwatch stop statement, `toc`.

    b. Next repeat the calculation using the vectorized arithmetic operators as in the code on page 529. Again display the execution time.

    c. Finally, repeat the calculation using the toolbox function, `polyval`, display the execution times, and compare the times for the three calculations.

PA.ii.2 Element-by-Element Arithmetic. Starting with the vector `k = [1:3]`, predict the results of the following and verify using MATLAB:

    a. `k.*k - k*k'`

    b. `k./k - k.*(k.^(-1))`

    c. `(k*k')^(k*k')`

    d. `(k'*k).^(k'*k)`

    e. `k'*k.^k`

PA.ii.3 Nested Loop Structures in MATLAB.

    a. The product of an $m \times n$ matrix, $\mathbf{A}$, and an n-component vector, $\vec{x}$, yields an m-component vector, $\vec{b} = \mathbf{A}\vec{x}$. Using nested `for` loops, write the MATLAB code to compute the m components of the vector $\vec{b}$.

    b. If the matrices $\mathbf{A}$ and $\mathbf{B}$ are, respectively, $n_1 \times n_2$ and $n_2 \times n_3$, use nested `for` loops to compute the $n_1 \times n_3$ elements of the product $\mathbf{C} = \mathbf{AB}$.

    c. If the matrices $\mathbf{A}, \mathbf{B}$, and $\mathbf{C}$ are, respectively, $n_1 \times n_2$, $n_2 \times n_3$, and $n_3 \times n_4$, use nested `for` loops to compute the $n_1 \times n_4$ elements of the product, $\mathbf{D} = \mathbf{ABC}$.

PA.ii.4 Polynomial Functions. If $y(x)$ is a polynomial in x with coefficients contained in the vector c, and if $f(y)$ is a polynomial in y with coefficients contained in the vector d, write an M-file function to obtain the polynomial $f(y(x))$.

PA.ii.5 Polynomial Multiplication and Division. Simplify the following using the polynomial function, `deconv`:

    a. $\dfrac{2x^4 - 17x^3 + 34x^2 - 17x + 2}{(x - 2)(2x - 1)}$
    b. $\dfrac{x(x^6 + 4x^5 + 8x^4 + 10x^3 + 8x^2 + 4x + 1)}{(x^2 + x + 1)^2}$

EXPLORATION PROBLEMS

PA.ii.6 Vectorized Operators and the Toolbox Function sum. If x is a row or column vector, the toolbox function `sum(x)` simply returns the sum of the elements of x. If X

is an $n \times m$ matrix, sum(X) sums each of the m columns of X separately and returns these values in a *row* vector of m elements. Assuming that x is a column vector of length m, and k is a row vector containing the integers from 1 to n, for each of the following, write MATLAB code employing for loops to evaluate the expression, and then write *single* line MATLAB statements employing the function sum that duplicates the calculation.

a. $\displaystyle\sum_{i=1}^{m} \frac{1}{x_i}$ b. $\displaystyle\sum_{k=1}^{n} \frac{1}{1+k}$ c. $\displaystyle\sum_{i=1}^{m} \frac{x_i}{1+\sin(x_i)}$ d. $\displaystyle\sum_{i=1}^{m} x_i e^{-x_i^2}$

PA.ii.7 Vectorized Sums Involving Matrices. The row vector x and the column vector k have the same size as in Problem A.ii.6.

   a. Show that the matrices formed from these vectors as,

$$X = \texttt{ones(length(k), 1)} * x,$$
$$K = k' \texttt{ones(1, length(x))}$$

     are the same size and shape.

   b. What is the (i, j)th element of the element-by-element product, X.*K?

   c. Use the matrices X and K and the function sum to write single line MATLAB statements to evaluate the following:

     (i) $\displaystyle y_i = \sum_{k=1}^{n} k x_i^k$ (ii) $\displaystyle N_k = \sum_{i=1}^{n} x_i^{1/k}$

PA.ii.8 Continued Products. If x is a row or column vector, the toolbox function prod(x) simply returns the *product* of the elements of x. If X is an $n \times m$ matrix, prod(X) computes the product of all the elements in *each column* of X separately and returns these values in a *row* vector of m elements. Assuming that x is a column vector of length m, and k is a row vector containing the integers from 1 to n, write MATLAB codes employing for loops for each of the following expressions, and then write *single* line MATLAB statements employing the function prod to duplicate the calculation.

   a. $\displaystyle k! = \prod_{i=1}^{k} i = 1 \cdot 2 \cdot 3 \cdots (k-1) \cdot k$

   b. A famous product expression for π is

$$\frac{\pi}{2} = \left(\frac{2 \cdot 2}{1 \cdot 3}\right) \left(\frac{4 \cdot 4}{3 \cdot 5}\right) \cdots \left(\frac{(2n) \cdot (2n)}{(2n-1)(2n+1)}\right) \cdots$$

     Write a MATLAB expression using prod to evaluate the first k terms of this product.

PA.ii.9 Cumulative Sums and Products. The MATLAB function cumsum is used to compute the "cumulative" sums of a vector x. Thus, if x = [1 1 1 1], cumsum(x) returns [1 2 3 4]; i.e., the sums of the first $1, 2, \ldots$ terms of x, and thus, returns a vector of the same size as x that contains these cumulative sums. Similarly, if X is an $n \times m$ matrix, cumsum(X) returns a matrix of the same size as X with the cumulative sums of the columns of X as its elements.

The MATLAB function cumprod operates in a similar fashion. If k = [1 2 3 4 5], then cumprod(k) returns [1 2 6 24 125], the cumulative products of the elements of the vector k. And if K is an $n \times m$ matrix, `cumprod(K)` returns a matrix of the same size as K with the cumulative products of the columns of K as its elements. Use these functions (plus other MATLAB functions) to write single line MATLAB statements to evaluate the following:

a. A vector containing the values of $k!$, where `k = 1:10`

b. $\displaystyle\sum_{k=1}^{10} \frac{1}{k!}$ c. $\displaystyle\sum_{k=1}^{10} \frac{2^k}{k!}$

Summary

POLYNOMIALS IN MATLAB

Definition

$$y(x) = c_1 x^{n-1} + c_2 x^{n-2} + \cdots + c_{n-1}x + c_n$$

Evaluation

`y = polyval(c,x)`
   Horner's Method: $y = ((((c_1 x + c_2)x + c_3)x + c_4)x + c_5)x + c_6$

LOOP STRUCTURES IN MATLAB

The `for` *loop*

```
for i = i1:di:i2
    a set of MATLAB
    statements to be
    executed for each
    value of i
end
```

ELEMENT-BY-ELEMENT ARITHMETIC

Vectorized operators, `(.*)` `(./)` `(.^)`

If matrices **A** and **B** are the same size, `A.*B` is a matrix of the same size as **A** or **B** that contains the elements of **A** times the corresponding elements of **B**. The same is true for `(./)` and `(.^)`. Many structures containing less efficient `for` loops can be replaced by statements using vectorized element-by-element operators.

MULTIPLICATION AND DIVISION OF POLYNOMIALS

Multiplication

If a_i and b_j are the coefficients of two polynomials, the coefficients of the polynomial that is their product are given by `c = conv(a,b)`.

Division

The division of the two polynomials results in a polynomial with coefficients, c_i, and a remainder polynomial with coefficients, r_j. These are returned by the function, `[c,r] = deconv(a,b)`.

MATLAB M-FILE FUNCTIONS

M-file functions

New functions created by the user are accessible by MATLAB if structured as a separate file within the MATLAB "PATH" and with a name ending with the suffix ".m". The first line must be of the form

$$\text{function } [\text{a,b,c}] = name(argument\ list)$$

The variables a,b,c will be computed within the function and returned to MATLAB upon completion of the statements in the function. These functions are created with a separate text editor.

Global variables   In addition to the variables contained in the argument list of an M-file function, other variables may be passed to the function by declaring them to be global, both in the calling function (usually the basic MATLAB workspace itself) and in the referenced function. Do not separate the variable names in a global statement by commas.

Interruption of   The MATLAB program may be temporarily interrupted by typing an exclamation
MATLAB   point, !. Control is then transferred to the operating system. This is a common way to access a text editor to write M-files. (This feature is presently not available on Macintosh computers.)

MATLAB Functions Used

polyval   If $\vec{c}$ is a vector of length n containing the coefficients of a polynomial of order $n-1$, then for a scalar x, polyval(c,x) will return a single value corresponding to $c_1 x^{n-1} + c_2 x^{n-2} + \cdots + c_{n-1} x + c_n$. If x is a matrix or a vector, the value of the polynomial at each element of x is returned in a matrix of the same size as x.

polyvalm   If $\mathbf{A}$ is a square matrix, polyvalm(c,A) works much the same as polyval, except that the result is a single matrix obtained by using matrix exponentiation and multiplication in the expression for the polynomial.

for *loop*   A block of MATLAB statements preceded by the statement for   *index* = *matrix*, and followed by a simple end statement, that will be repeatedly executed for each of the successive values of *index* in the for assignment statement. Usually, *index* is a simple scalar and *matrix* is a vector. For example,

```
for k = 1:.1:4
   x(k) = k*pi/6;
   y(k) = sin(x(k));
end
```

Note that there are frequently more efficient alternatives to for loops that employ element-by-element arithmetic. Also, this code should be preceded by statements like

```
x = zeros(1,31);
y = zeros(1,31);
```

to avoid the time-consuming reshaping of the vectors in each cycle of the loop.

end   Used as a terminator of for loops and while loops and as a terminator of MATLAB if blocks.

disp(A)   Displays the contents of the matrix **A** without being preceded by A = .

A.*B   If **A** and **B** are two matrices of the same size, the operations of multiplication (*), division, (/), and exponentiation, (^), can be *vectorized* by immediately preceding the operator by a period. Thus, A.*B is a matrix of the same size as **A** (and **B**) whose elements are the products of the corresponding elements of **A** and **B**.

conv, deconv   If two polynomials are specified by coefficients a_i and b_j in the order specified by polyval, conv(a,b) returns the coefficients of the polynomial that is their product. Similarly, the function **deconv** will return the coefficients of the polynomial that is the quotient of two polynomials, plus the coefficients of a remainder polynomial. The use is [q,r] = deconv(b,a). For example, if a = [1 2 1] and b = [1 4 6 4 1], corresponding to $(x+1)^2, (x+1)^4$, respectively, then [q,r] = deconv(b,a) returns q = [1 2 1], r = [0 0 0 0 0], and [q,r] = deconv(a,b) returns q = 0, r = [1 2 1].

sum, cumsum   If x is a vector, sum(x) is a scalar equal to the sum of the elements of x, and cumsum(x) is a vector of the same length as x, with elements equal to the cumulative sum of the elements. Thus, cumsum([1 2 3])⇒[1 3 6]. If X is a matrix, sum(X) is a vector containing the sums of each of the columns of X, and cumsum(X) is a matrix of the same size as X, containing the cumulative sums of the elements of each of the columns.

prod, cumprod   If x is a vector, prod(x) is a scalar equal to the product of the elements of x, and cumprod(x) is a vector of the same length as x with elements equal to the cumulative product of the elements. Thus, cumprod([1 2 3])⇒[1 2 6]. If X is a matrix, prod(X) is a vector containing the products of the elements of each column of X, and cumprod(X) is a matrix of the same size as X containing the cumulative products of the elements of each of the columns.

APPENDIX A.iii /
Logical Structures
in MATLAB

MATLAB Logical Structures

Computer arithmetic consists of a variety of operations that take two numerical values and uniquely return a third whose value depends on the type of operation (sum, difference, product, quotient, or power), as well as the values of the original two numbers. However, every computer also has the ability to evaluate a different class of operations related, not to arithmetic, but rather to comparisons. Once again, two quantities can be used to uniquely determine a result where, in this instance, the operation will compare the two and determine things like "same," "greater than," "less than or equal to," etc. These comparison operators in MATLAB are called *logical relational operators*. We describe them in the next section.

LOGICAL OPERATORS AND EXPRESSIONS

Logical expressions in MATLAB are constructed from comparisons between various quantities. The symbols that effect the comparisons are called *Relational operators*; they are listed in the following table:

Relational operator	Meaning
$<$	less than
$<=$	less than or equal to
$>$	greater than
$>=$	greater than or equal to
$==$	equal
$\sim=$	not equal

Logical comparisons are evaluated as either *true* or *false*. In MATLAB, *true* has a numerical value of $+1$, and *false* has a value of zero. Thus

$$12.2 > 3.1 \qquad \text{This expression has a value of } +1$$
$$1 \sim= 1 \qquad \text{This expression has a value of zero.}$$

Of course, as with ordinary arithmetic operations, MATLAB logical relational operators can be applied to complete matrices, as well as simple scalar

545

quantities. Thus, if A and B are two matrices of the *same size and shape*[1], then A *rel-op* B, where *rel-op* is one of the operators in the preceding table, will produce a matrix of like size composed of 1's and 0's by comparing the two matrices A and B, element by element, producing a $+1$ where the comparison is true, 0 where false.

EXAMPLE A.iii.1

Monitoring the Tension on a String

If a rock of mass $m = 2$ kg is tied to a string of length $L = 1.5$ m, and swung in a vertical circle so that the speed at the top of the circle is $v_{\text{top}} = 6$ m/s, the tension, T, in the string when it makes an angle of θ with the horizontal is

$$T = \frac{mv_{\text{top}}^2}{L} + mg(2 - 3\sin\theta)$$

See Figure A.iii.1. Also, it is known that the string is likely to break if $T > 85$N. The angles θ for which the tension exceeds the breaking strength of the string can be tagged by the following MATLAB code:

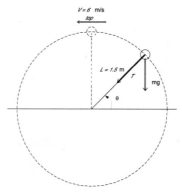

Figure A.iii.1   *Force diagram for a rock tied to a string and twirled in a vertical circle.*

```
m     = 2           % mass of rock
L     = 1.5;        % length of string
Tmax  = 145;        % breaking strength of string
g     = 9.8;        % gravitational acceleration
vtop  = 6;          % rock speed at top of circle
theta = 0:10:360;   % angle in degrees in steps
                    % of 10 degrees.
thrad = 2*pi*theta/360; % convert angles to radians
T     = m*(vtop^2/L + g*(2-3*sin(thrad)));
table = [theta; T>Tmax]
```

[1] As with ordinary arithmetic operators, an exception to the *same size and shape* rule is when one of the two operands is a simple scalar or number, as in A + 2 or A < pi. In these cases, the scalar is first expanded into a matrix of the size of A before the operation is executed. Thus, A < pi will test whether each element of A is less than π.

```
Columns 1 through 12
   0   10   20   30   40   50   60   70   80   90  100  110
   0    0    0    0    0    0    0    0    0    0    0    0
Columns 13 through 24
 120  130  140  150  160  170  180  190  200  210  220  230
   0    0    0    0    0    0    0    0    0    0    0    0
Columns 25 through 36
 240  250  260  270  280  290  300  310  320  330  340  350
   0    0    1    1    1    0    0    0    0    0    0    0
Column 37
 360
   0
```

Thus, it appears that the string is likely to break within about 20 degrees of the bottom of the circle. ●

More complicated expressions can be built up by combining two or more relational expressions by means of the following logical *combinational* operators.

Combinational operator	Meaning
&	AND
|	OR
~	NOT

Combinational operators are most often used to combine logical expressions that are simple scalars and have values *true* $= +1$ or *false* $= 0$. However, when employed with non-true/false variables,

Combinatorial operators evaluate *all nonzero* operands as *true*; i.e., $+1$. They return, however, only 1's and 0s.

In normal situations, the evaluation of logical expressions is fairly transparent, as can be seen from the following examples.

1. All arithmetic operations are processed before the logical expression is evaluated. Thus

$$\texttt{Size} \; < \; 10 * \texttt{sqrt}(100)$$

has the same value as `Size < 100`.

2. Parentheses may be added for clarity or to alter the value of the expressions:

$$(\texttt{Size} \; - 6) \; < \; (10 * \texttt{sqrt}(100) \; - \; 50)$$

3. Logical expressions may be combined by using the operators & and |, and they are then evaluated according to the following rules:

$$(true) \; \& \; (true) = (true)$$
$$(true) \; \& \; (false) = (false)$$

That is, the entire expression is *(false)* if either side of the & is *(false)*.

$$(true) \mid (true) = (true)$$
$$(true) \mid (false) = (true)$$

The entire expression is *(true)* if either side of the | is *(true)*. For example, if the variables a and b have values 2 and 8, respectively, then

$$a > 6 \ \& \ b < 20$$

is evaluated as *false* AND *true* = *false*, whereas

$$a*b \ == \ 0 \ \mid \ a < 0$$

is evaluated as *false* OR *false* = *false*.

However, complicated expressions containing both ANDs and ORs can have an ambiguous meaning. For that reason, the following hierarchy rule is followed when MATLAB evaluates a logical expression:

A logical expression is evaluated by first processing all arithmetic expressions. Then the logic operators are processed scanning left to right. The subexpressions are combined (& and | operators processed) from left to right, with the &'s processed *before* the |'s.

Consider the meaning of the following logical statement:

$$(a \ == \ b) \ \mid \ (x < 1) \ \& \ (z >= 0)$$

which has the same meaning as

$$(a \ == \ b) \ \mid \ ((x<1) \ \& \ (z>= 0)),$$

i.e., the & is processed *before* the |. If the values associated with the variables are $a = 3$, $b = 3$, $x = 0$, and $z = -1$, the expression reads

$$(true) \mid ((true) \ \& \ (false))$$

and since *(true)* & *(false)* = *(false)*, the expression is equivalent to

$$(true) \mid (false)$$

which has a meaning of *(true)*.

However, the expression forcing | before & has a different value:

$$((a \ == \ b) \ \mid \ (x<1)) \ \& \ (z>+0),$$

Using the same values for the variables, this is equivalent to

$$[(true) \mid (true)] \ \& \ (false)$$
$$(true) \ \& \ (false)$$

which has a value of *(false)*.

As with the relational operators, if combinational operators are used with *matrices* of the same size and shape, they act by combining the matrices element by element, returning 1's where true, and 0's where false.

The *NOT* operator, ~, although grouped with the combinational operators, is really a unary operator that simply interchanges *true* and *false*. Thus B = ~A returns a matrix that is zero, wherever A is not, and is +1 wherever A is precisely zero.

LOGICAL RELATIONS INVOLVING MATRICES

Not surprisingly, all the logical relational and combinational operations in MAT-LAB have been structured so as to apply to comparisons between matrices as well as between simple numbers. In this regard, logical operations involving matrices operate in an *element-by-element* sense. Thus, a relational operation like

```
A < B
```

will only be valid if the two matrices are of the *same size and shape*. The result of the comparison will be a matrix of the same size with +1 in the positions where $a_{ij} < b_{ij}$, and zeros elsewhere.

An exception is in the comparison of a *scalar*; i.e., a simple number, with a matrix. Scalars can always be compared with either a matrix or another scalar. In a comparison like `A == 3`, MATLAB will first build up a matrix of the same size as **A** using `3*ones(size(A))` before executing the comparison. Thus `eye(n) < 2` $\Rightarrow$ `ones(n,n)`, whereas `eye(n) ~= 1` $\Rightarrow$ `ones(n,n) - eye(n)`.

Thus, a comparison between two matrices of the form `A == B` does not determine whether the two matrices are identical, but rather returns a matrix of comparisons of their individual elements. The MATLAB function `all` can be used to test for mathematical equality of matrices. This function will return a single +1 if *all* of the elements of a matrix are *nonzero*, and a zero otherwise. Thus, `all(A==B)` will return a +1 if the two matrices are identical. A companion function is `any`; `any(A)` will return a +1 if *any* of the elements of **A** are nonzero, and zero otherwise.

THE MATLAB FUNCTION find

Once you have a little experience in writing MATLAB programs, you will discover that one of the most useful of all the basic toolbox functions in the entire lexicon of MATLAB names is the function `find`. This function will find the locations of nonzero entries in either a vector or a matrix. Thus, if x is a vector, the statement

```
i = find(x)
```

will return the locations of each of the nonzero elements of x. For example,

```
x = [0 1 2 3 4];
y = [4 3 2 1 0];
i = find(x==y);
```

will return a value of `i = 3` to the location of the only two elements of the vectors x and y that match. The operation of `find` for a matrix is similar. If A is a matrix, then

```
[i,j] = find(A)
```

will return the row-column locations (i, j), of each of the nonzero elements of the matrix A. Obviously, the most frequent use of this function is to determine the indices of the matrix result of a logical comparison. Thus, `find(A==B)` will determine *which* of the elements of A and B are equal.

By combining relational comparisons, quite elaborate searches can be effected. Thus, `i = find(abs(polyval(c,x)) < .01 & (x<1 | x>3))` will determine for which of the x values, when $x < 1$ or $x > 3$, the polynomial has an absolute value less than 0.01.

MATLAB `if` Structures

The simplest form of the MATLAB `if`-structure is:

`if` *logical expression*

> The body of the `if` block is a set of statements that will be executed only if the value of the expression is *true*; i.e., +1. Otherwise the block is skipped and the program proceeds with the statements after the `end` statement. To improve readability, the block is usually indented with respect to the `if` and the `end` statements.

`end`

In addition to the basic single option version, there are `if` structures available with multiple options resulting from one or more comparisons between logical expressions. These structures add either an `else` and/or an `elseif` statement to the `if` block, between the `if` and the `end` statements, along with their associated execution blocks.

`if` *logical expression* 1

> A block of statements executed only if *expression* 1 is *true*. After execution, control branches to the statement following the `end` statement.

`elseif` *logical expression* 2

> This set is executed only if *expression* 1 is *false*, and *expression* 2 is *true*. (Note: There should not be a space between the `else` and the `if`.)

`else`

> If both *expression* 1 and *expression* 2 are *false*, this block of statements is executed.

`end`

Frequently an `if` test is used along with the `for` loop to monitor the progress of a calculation, and, when a particular quantity becomes small or when there is evidence that the procedure is diverging, the execution of the loop is halted and the computed result or an appropriate error condition is displayed. Thus, in addition to the normal termination of a loop by means of the `end` statement, we will need to make use of additional MATLAB commands to interrupt the calculation. These commands are listed as follows:

break	Terminates the execution of MATLAB `for` and `while` loops (see the next section). In nested loops, `break` will terminate only the innermost loop in which it is placed.
return	Primarily used in MATLAB functions, `return` will cause a normal return from a function from the point at which the `return` statement is executed.
error(*'text'*)	Terminates execution and displays the message contained in *text* on the screen. Note, the text must be enclosed in single quotes.

In addition, the function `disp` can be used to display results on the screen. If **A** is a matrix, `disp(A)` will display the contents of **A**, and it differs from the normal `A =` statement only in the absence of the line `'A ='` on the screen. This function can also be used to display *text* on the screen by enclosing the text in single quotes, as `disp(`*'text'*`)`. Unfortunately, you can only supply one argument at a time to `disp`. Thus, `disp('latest value of x = ',x)` will cause an error.

MATLAB `while` STRUCTURES

Finally, there is a structure in MATLAB that combines the `for` loop with the features of the `if` block. This is the `while` loop, and it has the form:

`while` *logical expression*

> This set of statements is executed *repeatedly* as long as the logical expression remains true (i.e., equals +1), or if the expression is a matrix rather than a simple scalar variable, as long as *all* the elements of the matrix remain nonzero.

`end`

For example, the following two MATLAB functions sum the series for e^x by continuing to add terms until the magnitude of a term is less than some small quantity, `small`. The first function uses a standard `for` loop, and discontinues the summation when the absolute size of a term is less than `small`.

```
function y = EtoXa(x,small)
%EtoXa sums the series for e^x using for-loops.
%   The summation terminates when |term| < small,
%   where small is a convergence tolerance.  The
%   input x is assumed to be a column vector.
%===============================================
    y = ones(length(x),1)       % a column vector of ones
    term = x;
    for i = 1:100               % Hopefully, 100 terms
                                % will  suffice.
        y = y + term;           % Summation
        term = x.*term/(i+1);   % redefine term
        if abs(max(term))<small % test for smallness applied
                                % to largest element of term
```

```
            return              % if small enough, quit
          end
    end
    error('did not converge in 100 steps')
```

The second function accomplishes the same result using a `while` loop.

```
function y = EtoXb(x,small)
%EtoXb sums the series for e^x using while-loops.
%   The summation terminates when |term| < small.
%   and x is assumed to be a column vector.
%=====================================================
  y = ones(length(x),1);        % a column vector of ones
  term = x;    i = 1;
  while abs(max(term)) > small % convergence test
      y = y + term;             % Summation
      term = x.*term/(i+1);     % redefine term
      i = i + 1;                % increment counter
  end
```

Notice that

1. The functions have been structured to expect a *column vector* as input. Thus, the defining statement for the accumulating sum, y, is `y = ones(length(x),1)`, rather than simply, `y = 1`. If you are unsure as to whether the input vector to your function is a column or a row vector, the statement `y = y(:)` will cause any vector (or any matrix for that matter) to be replaced by a column vector of the elements in sequence. Also, the product in the line redefining `term` uses element-by-element multiplication.

2. The quantity used to test for convergence is named `small` rather than `eps`. This is because the variable name `eps` is predefined in MATLAB to be equal to the numerical *resolving power* of your computer. It is the smallest distinction that can be made between numerical values; more specifically, it is the smallest numerical value such that $1 + \epsilon \neq 1$. (The numerical value of `eps` is approximately 10^{-16}.)

Use of the MATLAB Function `feval`

The program for the function `padecoef` on page 144 to compute the Padé approximation for $\ln(1 + x)$ was written specifically for that function. It would be much more convenient if it could be adapted to calculate the Padé coefficients for an *arbitrary* function. That is, we would like to be able to transfer the *name* of a function through the argument list of `padecoef`, as well as the numerical limits of the calculation. The way this is accomplished in MATLAB is via the function `feval`. First, a generic *dummy* name of a function, say `fname`, is added to the argument list of `padecoef` in its definition line.

```
function [a,b] = padecoef(x0,x1,n,fname)
```

The variable, `fname`, is not intended to be the name of an actual function, but simply a variable that *contains* the name of an existing function. Thus, prior to the execution of `padecoef`, we must assign a value to `fname` in the form of a *string* of symbols enclosed in single quotes representing the name of the function being approximated. Thus,

```
fname = 'exp'
[a,b] = padecoef(x0,x1,n,fname)
```

or

```
[a,b] = padecoef(x0,x1,n,'exp')
```

will cause the name of the function `exp` to be transferred to the function `padecoef` by the variable `fname`. The function so referenced may be any of the toolbox or built-in functions, or it may be a user-written M-file.

Next, *within* the function `padecoef`, at points where the transferred function `fname` is to be evaluated, `fname` must be referenced by using the MATLAB function `feval` as follows:

$$\texttt{feval(fname,} \left. \begin{array}{l} \text{A list of the variables required} \\ \text{in the argument list of } \texttt{fname} \end{array} \right)$$

Thus,

```
y1 = feval('exp',3.0);
fname = 'asin';
x = 3.0;
y2 = feval(fname,x);
```

are the same as

```
y1 = exp(3.0);
y2 = asin(3.0);
```

Of course, if the function within the function requires a number of parameters, you must take care that either all of the parameters are passed through the argument list as well, or that they are declared to be `global` in the calling routine.

The changes that need to be made within the preceding code for function `padecoef` are to replace the two lines making reference to the specific function `log` by,

```
f0 = feval(fname,x0);
f = feval(fname,x);
```

This technique permits us to construct functions that will perform a variety of tasks on an arbitrary function. Thus, in Chapter 5, `feval` is used to design code for functions that will find the roots or will integrate the generic function $f(x)$ by using the MATLAB function `feval`. To apply the code to specific functions, either those in the toolbox or your own, simply insert the name of the function, enclosed in quotes, in the calling statement for root-finding or for integration.

Problems

REINFORCEMENT EXERCISES

***PA.iii.1* MATLAB Logical Structures.**   Write the MATLAB expressions that are true when

    a. $a > b > c$
    b. $|x - y| > 0$ and $x + y \leq 1$
    c. $c < d$ or $a < b$, but not both
    d. $x = y \neq z$
    e. $\mathbf{A}$ and $\mathbf{B}$ are matrices, and $\mathbf{B} = \mathbf{A}^{-1}$
    f. The sizes of $\mathbf{A}$ and $\mathbf{B}$ are appropriate for multiplication, $\mathbf{AB}$.
    g. Three numbers, a, b, and c can form the sides of a triangle; i.e., $|a - b| \leq c \leq a + b$

***PA.iii.2* Using the Functions `find`, `any`, and `all`.**   Write MATLAB statements that accomplish the following:

    a. Determine whether or not a matrix is symmetric.
    b. For a range of x values contained in a vector, determine which values x are in the range $-1 \leq x \leq 1.5$ and satisfy $\sin x - \cos x > \frac{1}{2}$.
    c. Replace all negative elements of a matrix by $+1$

***PA.iii.3* MATLAB `if` and `while` Structures.**   Often the roots of a nonlinear equation can be solved by an iterative procedure known as the *method of successive substitutions*, provided (a) the equation can be written in a form

$$x = f(x)$$

and (b), $|f'(x_r)| < 1$, where x_r is the root. Write a MATLAB M-file function `Subst('name',x0,tol)` that will iterate the preceding equation starting with an initial guess, x_0, where 'name' is a string of characters designating the name of a second M-file function containing the function $f(x)$. Use these two functions to search for the roots of the following equations. Use `if` or `while` statements to continue the calculation until either subsequent changes in x are less than `tol = 1.e-6`, or the procedure appears to be =.

    a. $F(x) = \frac{x}{3} - e^{-x^2} = 0$. First write the equation as $x = \sqrt{\ln(3/x)}$ and then as $x = 3e^{-x^2}$. Start with $x_0 = 1$.
    b. $F(x) = x^{10} + 5x^3 - 7 = 0$. First write the equation as $x = (7 - 5x^3)^{1/10}$, and then as $x = (7 - x^{10})^{1/3}/5^{1/3}$. Start with $x_0 = 1$.

PA.iii.4 Logical Structures and Matrices, all any. Write an M-file function that will determine, within a tolerance **tol**, whether or not two matrices satisfy $AB = BA = I$. [Do not use **for** or **while** loops.]

PA.iii.5 Use of the Function feval. Write the MATLAB code for an M-file function,

$$cr = \text{STEP}(a, b, dx, \text{funct})$$

that will step from $x = a$ to $x = b$ in steps of size dx and look for the point where the function, **funct**, changes sign. The most recent interval is then returned in **c**. This function should accept an *arbitrary* function specified in the name **funct**, and it should stop the search if $x > L$.

Summary

MATLAB LOGICAL STRUCTURES

Logical expressions

Two matrices of the same size can be compared, element by element, using the six relational operators, (**<**, **<=**, **>**, **>=**, **==**, **=**), resulting in a matrix of the same size with 1's in the positions where the relation is true, and zeros elsewhere. Two independent logical expressions can be combined using the two combinatorial operators, **&** (AND) and **|** (OR). The result of a logical expression can be used as input to MATLAB **if** and **while** decision structures to produce multipath programs.

if *structures*

The MATLAB **if** structure begins with the statement

if *variable*

and is terminated with an **end** statement. All statements in between are executed only if the *variable* is "true" (i.e., is nonzero). Alternate possibilities are conceivable by inserting separate blocks of statements, each preceded by **elseif** *variable-n*. If a block of statements is to be executed only if the variables in the **if** and **elseif** tests are false (*zero*), the statements are inserted after the **if** and **elseif** blocks and preceded by an **else** statement.

while *structures*

Similar to the **if** loop, the form of a MATLAB **while** loop is a block of statements preceded by a statement of the form **while** *variable*, and it is terminated by an **end** statement. The block of statements will be repeatedly executed as long as the expression contained in *variable* remains nonzero. Because there is a potential for endless loops in a **while** structure, there should always be a mechanism for automatically terminating the calculation if the test variable never assumes a value of zero.

MATLAB Functions Used

Relational operators

A MATLAB *logical relation* is a comparison between two variables **x** and **y** of the same size effected by one of the six operators, **<**, **<=**, **>**, **>=**, **==**, **~=**. The comparison involves corresponding elements of **x** and **y**, and yields a matrix or scalar of the same size with values of "true" or "false" for each of its elements. In MATLAB, the value of "false" is zero, and "true" has a value of one. Any nonzero quantity is interpreted as "true."

Combinatorial
operators
all,any

The operators & (AND) and | (OR) may be used to combine two logical expressions.

If x is a vector, all(x) will return a value of one if *all* of the elements of x are nonzero, and a value of zero otherwise. If X is a matrix, all(X) returns a row vector of ones or zeros obtained by applying all to each of the columns of X. The function any operates similarly if *any* of the elements of x are nonzero.

find

If x is a vector, i = find(x) will return the indices of those elements of x that are nonzero (i.e., true). Thus, replacing all the negative elements of x by zero could be accomplished by

$$i = \text{find}(x < 0);$$
$$x(i) = \text{zeros}(\text{size}(i));$$

If X is a matrix, [i,j] = find(X) operates similarly and returns the row-column indices of nonzero elements.

if, else, elseif

The several forms of MATLAB if blocks are

```
if variable                 if variable 1               if variable 1
    block of state-             block of state-             block of state-
    ments executed              ments executed              ments executed
    if variable                 if variable 1               if variable 1
    is "true,"                  is "true,"                  is "true,"
    i.e., nonzero               i.e., nonzero           elseif variable 2
end                         else                            block of state-
                                block of state-             ments executed
                                ments executed              if variable 2
                                if variable 1               is "true,"
                                is "false,"             else
                                i.e.,zero                   block of state-
                            end                             ments executed
                                                            if neither
                                                            variable is
                                                            "true"
                                                        end
```

break

Will terminate the execution of a for or while loop. Only the innermost loop in which break is encountered will be terminated.

return

Causes the function to return at that point to the calling routine. MATLAB M-file functions will return normally without this statement. It is mostly used for abnormal returns.

error('text')

Within a loop or function, if the statement error(*'text'*) is encountered, the loop or function is terminated, and the *text* is displayed.

while

The form of the MATLAB while loop is

`while` *variable*

block of statements executed as
long as the value of *variable*
is "true"; i.e., nonzero

`end`

`feval` Used when a function F itself calls a second "dummy" function "f". For example, the function F might find the root of an arbitrary function identified as a generic $f(x)$. Then, the name of the actual M-file function, say `fname`, is passed as a *character string* to the function F either through its argument list or as a global variable, and the function is evaluated within F by means of `feval`. The use is `feval(name,x1,x2,...,xn)`, where `fname` is a variable containing the name of the function as a character string; i.e., enclosed in single quotes, and `x1,x2,...,xn` are the variables needed in the argument list of function `fname`.

APPENDIX A.iv /
Two-Dimensional Plotting

The plotting capabilities of MATLAB are very extensive and can be used to create a huge variety of graphs. Of course, to create a myriad of different graph types requires knowledge of a myriad of new MATLAB commands. However, the commands to produce just a basic simple plot could not be easier.

The MATLAB Plotting Function `plot(x,y)`

The first task in plotting a function is to produce a table of (x, y) values. That is, if $f(x)$ is the function in question, and you are interested in the behavior of the function over the range, $x_0 < x < x_1$, you could use a `for` loop to fill in the table, or more directly,

```
dx = (x1-x0)/50;
x = [x0:dx:x1]';
y = f(x);
```

where the function `f(x)` must be coded as an M-file and written to expect a column vector as input.

Once the table of values is completed, to plot the results, simply enter

```
plot(x,y)
```

and MATLAB will handle the rest. The ranges of both x and y values are determined, and the axes are scaled automatically. The input table to `plot` must consist of vectors, x_i and y_i, of the same length. The vectors must be both column vectors, or both row vectors. In our applications, we will always assume that the input consists of column vectors.

The plot will appear automatically on the terminal screen. If you are running a windows version of MATLAB, a separate graph window will be opened.

Several plots can be graphed at once using `plot`. Once again, you must first create tables of values to be plotted. Suppose that we have computed 4 sets of x, y values and that they are stored in $x_i^{(\alpha)}$ and $y_i^{(\alpha)}$, for $\alpha = 1, \ldots, 4$ and $i = 1, \ldots, n$. Or, in the notation of MATLAB, they are stored by *columns* in the $n \times 4$ matrices x and y. In this case, `plot(x,y)` will graph, in order, the columns of y versus the columns of x. In situations where you wish to plot two or more sets of data simultaneously, but the columns are not of the same length, the process is just as easy. Say the column vectors x_i and y_i are of the same *or different* length than u_i and v_i; then `plot(x,y,u,v)` will first plot y versus x, then v versus u on the same graph. Any number of pairs of column vectors (up to the memory capacity of MATLAB) can be added to the argument list of `plot`.

If running in Version 3.5 in a non-Windows mode, depressing any key will cause the graph to disappear, and you will be returned to MATLAB. By entering `shg`, the graph will be displayed once again.

ADDING TEXT TO A GRAPH

Next, we could add an overall title to the graph by using the statement `title('Title of my Plot')`, where the graph title is enclosed in single quotes. Similarly, the labels for the x and y axes are added using `xlabel('x axis labels')` and `ylabel('y axis labels')`. The y-axis labels will be displayed in a vertical column.

Additionally, text may be added to the graph area itself. The statement

$$\text{text}(\text{x}, \text{y}, \textit{text-string})$$

will position the character string contained in *text-string* at the point specified in the values for (x, y).

CHANGING THE SCALING

Because MATLAB automatically scales the graph, you may wish to change the limits on the x or y axes. Entering `axis` will return a row vector containing the limits of the plotting area, $[x_{\min}, x_{\max}, y_{\min}, y_{\max}]$. These values can be changed by simply retyping `axis` with the new values included in the argument list *as a 4-element row vector*. Thus,

```
axis([0 10 -5 5])
```

will manually rescale the graph to $0 \leq x \leq 10$ and $-5 \leq y \leq 5$. Entering `axis('auto')` once will restore the automatic scaling. If you wish, grid lines can be added to the graph by entering `grid on`. To remove the grid lines, use `grid off`. The axis lines themselves can be removed with `axis('off')`. Another common option is `axis('equal')`, which forces the scaling along the x and y axes to be the same. For more options associated with `axis`, see the MATLAB reference manual or the help file.

The basic MATLAB plotting capabilities are summarized in Table A.iv.1.

plot(x,y)	If x and y are column vectors, y versus x will be graphed with the axes scaled automatically.
	If x and y are matrices of the same size, the columns of y will be graphed versus the columns of x as separate curves on the same graph. If your terminal supports multicolor, each of the curves will be a different color.
	If x is a column vector, and y is a matrix of column vectors of the same length, the columns of y are each plotted versus x as separate curves on the same graph.
plot(x,y,u,v)	If (x,y) and (u,v) are pairs of equal-length column vectors, y versus x and v versus u will each be plotted as separate curves on the same graph.
plot(x,y,'+')	A plus sign will be displayed at each of the points of (x,y). There are two types of graph options available: *line* types and *point* types, and you may choose a different one for each of the sets of (x,y) in the argument list. Also, they may be combined with a *color* option to change the color of the graphs. For example, 'xg' will produce a green plot with the data points marked with x's. The available options are listed as follows:

Line types		Point types		Colors	
solid	–	point	.	red	r
dashed	– –	plus	+	green	g
dotted	:	star	*	blue	b
dash-dot	–.	oh's	o	white	w
		x's	x	invisible	i

title('*text*')	Will display a *text* as a title for the graph at the top.
xlabel('*text*')	Will display the label *text* on the horizontal axis. ylablel has the same function for the vertical axis.
text(x,y,'*text*')	Positions the character string contained in *text* at the point (x, y) on the graph.
grid on	Adds grid lines to a graph; grid off removes them.
axis	Returns a 1×4 row vector containing the x, y limits of the plotting area; i.e., x_{min}, x_{max}, y_{min}, y_{max}. The automatic scaling feature of plot can be overridden by entering axis(r), where r is a row vector containing the new limits. Also, the argument of axis may be one of the strings, 'off', 'on', 'auto', 'normal', 'equal', 'square', 'image', 'xy', or 'ij'. See the help file for an explanation of each.

Table A.iv.1   *Summary of elementary plotting capabilities*

Printing a Graph

The instructions to have the graph printed depend on the version of MATLAB you are using.

Printing Graphs in the Student Edition of MATLAB Version 3.5 If you have a printer connected to your computer, you must use the *screen dump* features of MATLAB. This is accomplished by typing prtsc while the graph is on the screen, or by depressing the *SHIFT* and *Print Screen* keys simultaneously.

Printing Graphs in the Professional Edition of MATLAB and in Versions 4.0 and later The MATLAB command

print filename

will create a *PostScript* file named *filename* that contains the current contents of the plot window. The name of the file should end with the suffix .ps identifying it as a PostScript file. This form of the print command ordinarily *does not print* the file. You must then enter system-dependent commands to have the PostScript file sent to a printer. The MATLAB statement

[PCmmd, Dev, Pname] = printopt

will return the various commands and destination codes used for your particular installation. The character string returned in PCmmd will be the form of the print command on your system that is used to send the file to a printer. The string contained in Dev is optional, and it identifies the default device option associated with a print command. The string contained in Pname identifies a specific printer.

To directly print a graph, simply enter print, and the current contents of the print window will be sent to the default printer.

EXAMPLE A.iv.1
Plotting Ballistic Trajectories

If a bullet is fired from a gun with a muzzle velocity v_0 at an angle of θ with respect to the horizontal, the trajectory of the bullet is given parametrically as function of time as $x(t) = v_0 t \cos\theta$, $y(t) = v_0 t \sin\theta - \frac{1}{2}gt^2$, where $g = 9.8\text{m/s}^2$ is the acceleration due to gravity. The bullet will hit the ground $[y(t = t_{\text{hit}}) = 0]$ at $t_{\text{hit}} = 2v_0 \sin\theta/g$. Thus, for a range of t values, $0 < t < t_{\text{hit}}$, we would compute values for x and y.

```
v0 = 10:                        % muzzle speed in m/s
g = 9.8;                        % gravitational acceleration
theta = 50;                     % angle in degrees
theta = theta*pi/180;           % angle in radians
thit = 2*v0*sin(theta)/g;
dt = thit/50;
t = [0:dt:thit]';               % times are a column vector
x = t*v0*cos(theta);
y = t*v0*sin(theta) - .5*g*t.^2;
```

The graph of the trajectory is obtained by adding

```
plot(x,y)
title('Trajectory for an angle of 50 degrees')
xlabel('x (m)')
ylabel('height')
grid on
```

This graph is shown in Figure A.iv.1.

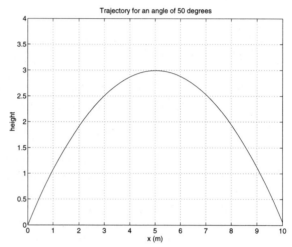

Figure A.iv.1   *A ballistic trajectory for a projectile fired at 50°.*

Next, suppose that we wish to investigate the trajectory as a function of the firing angle, θ, say for $\theta = 30°$, $35°$, $40°$, $45°$, $50°$, and $55°$. Thus, the statements defining **theta** would be replaced by

```
theta = [30:5:55]          % A row vector
theta = theta*pi/180;
thit = max(2*v0*sin(theta)/g); % The longest of the
                               % flight times
```

Recall that the values for **t** were stored in a column vector. Thus, a statement like

```
x = v0*t*cos(theta);
```

is a *matrix* product of a column vector times a row vector, and it will produce an $n \times 6$ matrix, where n is the number of t values. The matrix x will then store each of the trajectories for the different angles by columns. Arranging the equation for y in a similar way,

```
y = v0*t*sin(theta) - .5*g*t.^2*ones(1,length(theta));
```

we now have the trajectories $x(t, \theta)$, $y(t, \theta)$ as function of both t and θ stored by columns for value of θ in the matrices x and y. `plot(x,y)` produces the graph depicted in Figure A.iv.2.

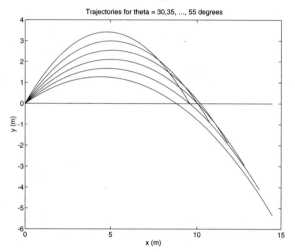

Figure A.iv.2 *Several trajectories as a function of angle.*

However, this is not quite what we want. Some of the trajectories have continued on for negative y values. It would be preferable to replace all negative y values by zero. This can be accomplished by using the `find` function described earlier. (See page 549.)

index = `find(y<0)` returns the *location*; i.e., the index, of each of those elements of the matrix y that is less than zero, and stores them in the vector **index**. The statements that replace the negative y values by zero are

```
index = find(y<0);
y(index) = zeros(1,length(index));
```

Finally, as the graphs have been terminated when $y < 0$, the original coordinate scaling is no longer appropriate. From the graph we can estimate $x_{\min} = y_{\min} = 0$, $x_{\max} \approx 15$, and $y_{\max} \approx 3.5$. Thus, the final version of the graph is obtained with

```
axis([0 15 0 3.5])
plot(x,y)
title('Ballistic trajectories for theta = 30,35, ..., 55 degrees')
xlabel('Range (m)')
ylabel('Height')
```

The final version of the graph is depicted in Figure A.iv.3.

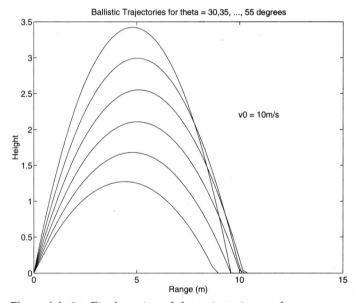

Figure A.iv.3   *Final version of the trajectories graph.*

Problems

REINFORCEMENT EXERCISES

PA.iv.1 MATLAB Plotting Exercises

a. For x = [.5:.1:2.5]' use **polyval** to evaluate the polynomial $f(x) = 25x^4 - 135x^3 + 256x^2 - 256x + 231 - 121$.

b. Use **plot** to graph this polynomial over this range and include a horizontal line for the x axis.

c. Add a **grid** and estimate the value of the multiple root.

d. Use the function **ginput(1)** and move the cross hairs to the position of the multiple root and read off the more accurate estimation of the root.

e. Rescale the plot to $2 \leq x \leq 2.4$ and $-5 \leq y \leq +5$, and repeat the estimation of the multiple root.

PA.iv.2 Plotting Multiple Graphs. Produce a set of ellipses on a single plot. That is, for $\theta = 0$ to 2π, use a **for** loop to step through **a = 0.5:0.5:4.5**, and for each value of a, compute $x = a\cos\theta$, $y = \sqrt{5^2 - a^2}\sin\theta$, and plot x versus y. Use **hold on/hold off** to save or release the plot from the screen. Also, after the plots are drawn, use **axis('image')** to get a *true* image of the graph.

PA.iv.3 Plotting Trajectories Rewrite the MATLAB code for trajectories on page 561 to incorporate "bounces." That is, plot the trajectory from $t = t_{\text{hit}}$, then reduce $v_0 \rightarrow \frac{3}{4}v_0$, and start the trajectory over from that point, $(x_{\text{hit}}, y = 0)$. Continue this for three bounces.

Summary

MATLAB PLOTTING TECHNIQUES

Two-dimensional graphs

The MATLAB function **plot** is designed to provide an easy and interactive capability for the programmer to visualize the behavior of one or more functions of a single variable. The basic function is **plot(x,y)**, which will graph the elements of two vectors x and y of equal size, y versus x, automatically scaling the axes. This graph can then be further customized by adding a title and labels for the two axes (functions **title**, **xlabel**, and **ylabel**), by varying the symbols used for the individual points, by altering the type of the lines connecting the points, or even by changing the colors used for the display. Several graphs can be added to the same set of axes, (**plot(x1,y1,x2,y2)**, etc.), the scales are easily adjusted by using the function **axis**, and grid lines can be added or deleted with the function **grid**.

MATLAB Functions Used

plot

The basic MATLAB program for plotting two-dimensional graphs of **x** and **y** that are two vectors of the same size; **plot(x,y)** will graph the elements of **y** (vertical) versus **x** (horizontal). Multiple plots are possible by including more vector pairs in the argument list. Also, the color and type of line used can be varied by including a third item along with each vector pair.

title, xlabel, ylabel

Used to append a title and axis labels to a plot.

axis

The scale of a graph can be adjusted by the function **axis**. The use is **axis(r)**, where, for a two-dimensional plot, the vector $r = [x_{\min}, x_{\max}, y_{\min}, y_{\max}]$. **axis** alone will return the current value of the axis scales. There are numerous other possibilities, such as forcing the axes to be scaled equally, **axis('equal')**, or so that the graphing pixels are square rather than rectangular, **axis('image')**. See the MATLAB reference manual.

grid

Adds grid lines to a two- or three-dimensional graph. **grid on** adds the grid lines; **grid off** removes them.

ginput

The x, y position on a graph can be read directly from the screen using the function `ginput(n)`. The use is `[x,y] = ginput(n)` *followed by a return.* The integer n is the number of data points to be read. Next, the cursor is positioned at the first desired point, and either the mouse is clicked or any key is depressed. This process is continued for the remaining $n - 1$ points, and their coordinates will be returned in the vectors, x and y.

print

The basic operation of this function is to create a *PostScript* file image of the graph window and to write it to a designated file. The simplest use is then `print` *filename*`.ps`. This file, which must end with the suffix `.ps`, may then be printed on a variety of printers.

shg

If running MATLAB in a non-Windows mode, this command will redisplay the graph screen. Entering any key will then return to the MATLAB workspace.

prtsc

In the student edition, Version 3.5, the only way to obtain printed results is to "print the screen" with either the Print Screen key or, equivalently, by using this MATLAB command.

APPENDIX A.v /
Constructing and Plotting
Multivariable Functions

PREVIEW   Previously, when plotting a function, $y = f(x)$, using the MATLAB function `plot`, the only requirement was to have a *table* of values (x_i, y_i), $i = 1, \ldots, n$. The preliminary steps in producing the graph were to (a) determine the range of x values of interest and select a sampling of values in that range, x_i, $i = 1, \ldots, n$, and (b) for each x_i compute $y_i = f(x_i)$, thereby producing a table consisting of two *vectors* of numbers, (x_i, y_i). After these preliminary steps, MATLAB took over and produced a variety of graphs using these two vectors.

    The plotting of functions of two variables, $z = f(x, y)$, will require three dimensions, x, y, and z, and the procedure is similar to the two-dimensional case. The first task is to establish a rectangular "grid" in the xy plane containing points at which the function $f(x, y)$ will be evaluated. Then, for each grid point, (x_i, y_j), the function is evaluated as $z_{ij} = f(x_i, y_j)$, thereby producing a *matrix* of values, one for each grid point. The two requirements for the graph are then (a) a *matrix* or *grid* of x and y values, and (b) a *matrix* containing the values of the function on the grid. Once these are obtained, the easy-to-use MATLAB commands produce a variety of three-dimensional graphs.

    As with one-dimensional functions, it pays to vectorize. All M-file functions to evaluate functions of two variables, $f(x, y)$, should be written so as to expect a *matrix* of values for x and a separate matrix of values for y, and then to return a *matrix* of values for $z = f(x, y)$. The arithmetic contained in the definition of the function will largely consist of element-by-element operations on the matrices x and y.

    The first task is to understand how to construct the matrices corresponding to x and y. Once the matrix of values of $f(x, y)$ is obtained, we describe the various MATLAB alternatives for graphically displaying the results.

Establishing a Grid of x, y Values

Suppose you wish to evaluate a function $f(x, y)$ for x and y values in the range $-2 \le x \le 0$ and $0 \le y \le 5$, and the points at which the function is to be evaluated are defined by

$$\mathbf{x} = -2 : 0.5 : 1; \quad x = \begin{pmatrix} -2 & -\frac{3}{2} & -1 & -\frac{1}{2} & 0 & \frac{1}{2} & 1 \end{pmatrix}$$

$$\mathbf{y} = [5 : -1 : 0]'; \quad y = \begin{pmatrix} 5 \\ 4 \\ 3 \\ 2 \\ 1 \\ 0 \end{pmatrix}$$

Notice that the y values have been assigned in descending order to correspond to a *grid* in the xy plane of the form

and the function, $f(x, y)$, will return a matrix of values, one for each of the grid points. The values of x and y at each of the grid points are then represented by the matrices

$$x = \begin{pmatrix} -2 & -3/2 & -1 & -1/2 & 0 & 1/2 & 1 \\ -2 & -3/2 & -1 & -1/2 & 0 & 1/2 & 1 \\ -2 & -3/2 & -1 & -1/2 & 0 & 1/2 & 1 \\ -2 & -3/2 & -1 & -1/2 & 0 & 1/2 & 1 \\ -2 & -3/2 & -1 & -1/2 & 0 & 1/2 & 1 \\ -2 & -3/2 & -1 & -1/2 & 0 & 1/2 & 1 \end{pmatrix} \quad y = \begin{pmatrix} 5 & 5 & 5 & 5 & 5 & 5 & 5 \\ 4 & 4 & 4 & 4 & 4 & 4 & 4 \\ 3 & 3 & 3 & 3 & 3 & 3 & 3 \\ 2 & 2 & 2 & 2 & 2 & 2 & 2 \\ 1 & 1 & 1 & 1 & 1 & 1 & 1 \\ 0 & 0 & 0 & 0 & 0 & 0 & 0 \end{pmatrix}$$

So, if the vectors representing x and y are first expanded into *matrices* of the same size as the grid, then *element-by-element* arithmetic operators can be used to evaluate the function, $f(x, y)$, for the complete set of grid points. Thus

$$\mathbf{x} = \mathbf{ones(y) * x} \quad x = \begin{pmatrix} 1 \\ 1 \\ 1 \\ 1 \\ 1 \\ 1 \end{pmatrix} \cdot \begin{pmatrix} -2 & -\frac{3}{2} & -1 & -\frac{2}{2} & 0 & \frac{1}{2} & 1 \end{pmatrix}$$

$$\mathbf{y} = \mathbf{y * ones(x)} \quad y = \begin{pmatrix} 5 \\ 4 \\ 3 \\ 2 \\ 1 \\ 0 \end{pmatrix} \cdot \begin{pmatrix} 1 & 1 & 1 & 1 & 1 & 1 & 1 \end{pmatrix}$$

and an "m"-function to represent, say $f(x,y) = e^{-xy}/\sqrt{x^2 + y^2}$, is simply

```
function z = f(x,y)
f = exp(-x.*y)./sqrt(x.^2 + y.^2)
```

THE MATLAB FUNCTION meshgrid(x,y)

The MATLAB toolbox function to construct the *matrices* for the x and y values at each point of the grid from the vectors representing x and y is called meshgrid. The use of the function is

```
[X,Y] = meshgrid(x,y)
```

where x and y are vectors containing the grid (or *mesh*) points along the x and y axes (in the normal order). Thus, the call to meshgrid for the preceding values might be

```
[X,Y] = meshgrid(-2:.5:1,0:5),
```

or

```
[X,Y] = meshgrid([-2:.5:1],[0:5]),
```

or

```
x = [-2:.5:1];
y = [0:5]';
[X,Y] = meshgrid(x,y).
```

The function then returns two *matrices*, X and Y, that contain the values of x and y at each of the grid points. The values of x will increment left to right, and those of y will increment from top to bottom. Using these matrices, it is then an easy matter to construct *element by element* arithmetic expressions to evaluate a function at all points on the grid.

For example, the function $j_1(r) = (\sin r - r \cos r)/r^2$ is called a spherical Bessel function of order 1, and r is the length of the radius vector, which, in two dimensions, is $r = \sqrt{x^2 + y^2}$. The evaluation of this function over the range $-8 \le x \le 8$ and $-8 \le y \le 8$ proceeds as follows:

```
x = [-8:.5:8];
y = [-8:.5:8];
[X,Y] = meshgrid(x,y);
r = sqrt(X.^2 + Y.^2);
z = (sin(r)-r.*cos(r))./(r.^2);
```

To get an appreciation for the behavior of this and other functions of two variables, we will now investigate the two basic options in MATLAB for graphing two-variable functions.

Three-Dimensional Perspective Graphs

To obtain a graph of the preceding Bessel function, simply add the statement

```
mesh(z)
```

to the code and sit back and watch the beautiful three-dimensional perspective graph of this function unfold. For many purposes, that's all that you will need to know to produce a three-dimensional graph. If A is any $m \times n$ matrix, then `mesh(A)` will produce a perspective drawing of the elements of A, plotting one point for each element, with the "height" of each point proportional to the value of that element of A.

ADDING x, y, z COORDINATES ON THE PLOT

In the simplest application of the function `mesh(z)`, no information is supplied regarding the range of the x and y values, and the graph of z is simply plotted versus the column and row numbers of x and y. To add the appropriate scaling on the x and y axes, simply include these matrices in the argument list[1] of `mesh` as

```
mesh(x,y,z)
```

This can be done by either using the *vectors* x and y or their corresponding matrices X and Y. Of course, when using the vectors x and y, the length of x must equal the number of columns of z, and the length of y must equal the number of rows of z. The plotted points on the graph correspond to x_i, y_j and $z(x_i, y_j)$. Also, the function `meshz` performs exactly the same as `mesh`, except that vertical "drop offs" are added to the figure at the edge of the xy domain, providing a "wall effect."

The ease with which you are able to produce these graphs in MATLAB makes this function quite addictive, and you might wish to experiment with a variety of functions before continuing. We now describe several examples that you may find produce interesting three-dimensional graphs. In several of the suggested plots, you will first have to code the function as an M-file function, compute the grid of x, y values over the indicated range using `meshgrid`, and then plot the function using `mesh`.

---

[1]A feature of MATLAB 4.0.

1. A plot of the 12×12 identity matrix, z = eye(12), is obtained by simply entering mesh(eye(12)). In Figure A.v.1, the x and y axes correspond to the row/column numbers of the matrix.

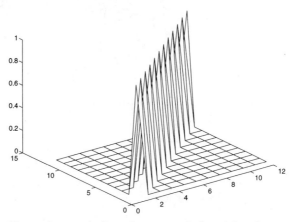

Figure A.v.1 *A three-dimensional plot of the identity matrix.*

2. A plot of the spherical Bessel function $j_0(x, y) = \sin(r)/r$ centered at (x_0, y_0). This function is defined by the equation,

$$j_0(r) = \frac{\sin r}{r}$$

with $r = \sqrt{(x - x_0)^2 + (y - y_0)^2}$. Note that because of the possibility of dividing by zero, the values of computed values of r are displaced by the machine tolerance, eps. The graphical presentation is given in Figure A.v.2.

```
% The spherical Bessel function j0 =============
function z = j0(X,Y,X0,Y0)
%j0 is the zeroth order spherical Bessel function
  r = sqrt((X-X0).^2 + (Y-Y0).^2) + eps;
  z = sin(r)./r;
%==========================================
%   The above is written as an M-file
%   and the code below executed in MATLAB
%==========================================
x = [-8:.5:8]';
y = x;
[X,Y] = meshgrid(x,y);
z = j0(X,Y,1,1);
```

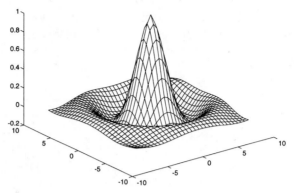

Figure A.v.2   *The spherical Bessel function j_0.*

3. A plot of the spherical Bessel function $j_1(r)$. This function is defined by the equation[2]

$$j_1(x,y) = \frac{\sin r - r \cos r}{r^2}$$

with r defined as it was previously. The plot is shown in Figure A.v.3.

```
% The spherical bessel function j1 =============
function z = j1(X,Y,X0,Y0)
%j1 is the order 1 spherical Bessel function===
  r = sqrt((X-X0).^2 + (Y-Y0).^2) + eps;
  z = (sin(r) - r.*cos(r))./r.^2
%=========================================
%    The above is written as an M-file
%    and the code below executed in MATLAB
%=========================================
  x = [-8:.5:8]';
  y = x;
  [X,Y] = meshgrid(x,y);
  z = j1(X,Y,1,1);
  mesh(X,Y,z)
```

---

[2]Spherical Bessel functions are an important class of functions that arise naturally from the solution of Laplace's equation, $\nabla^2 F = 0$, in spherical coordinates. Spherical Bessel functions of arbitrary order, ℓ, may be computed using the function Sbes(ell,x) included on the accompanying program disk.

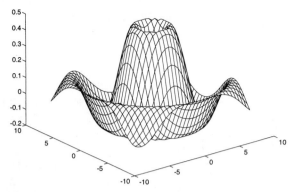

Figure A.v.3 *The spherical Bessel function j_1.*

4. A three-dimensional plot of a combination of spherical Bessel functions. The function defined by

$$f(r) = \frac{j_1(r)}{\sqrt{j_0^2(r) + j_1^2(r)}}$$

will produce a rather exotic plot. See Figure A.v.4.

```
% A Combination of Bessel functions ======
%=========================================
 x = [-8:.5:8]';
 y = x;
 [X,Y] = meshgrid(x,y);
 z0 = j0(X,Y,1,1);
 z1 = j1(X,Y,1,1);
 z3 = sqrt(z0.^2 + z1.^2);
 z = z1./z3;
```

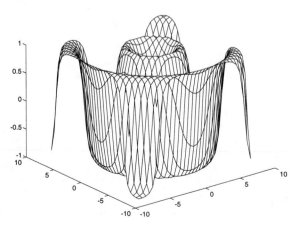

Figure A.v.4 *A combination of spherical Bessel functions.*

5. Graph of an ellipsoid. An ellipsoid is defined by the equation

$$\frac{x^2}{a^2} + \frac{y^2}{b^2} + \frac{z^2}{c^2} = 1$$

We give the code to plot the upper half of the ellipsoid, with $a = b = 8$ and $c = 7$. Note that for some values of x, y on the grid, there is *no* real solution for z. These locations correspond to the points where the value computed for z^2, using the preceding equation, is negative. We can use the function find to locate these points, and set $z = 0$ in this region. An alternative is to plot the graph in polar coordinates, ρ and ϕ. Procedures to do this are described later in this chapter. Finally, to better illustrate the ellipsoid, the values of z on the grid corresponding to both x and y negative have been set equal to zero, resulting in a missing slice from the ellipsoid. The plot is shown in Figure A.v.5.

```
% The Upper Half of an Ellipsoid  =====
% with a Piece Missing            =====
%=======================================
  x = [-8:.5:8]';   y = x;
  [X,Y] = meshgrid(x,y);
  z = (64 - X.^2 - Y.^2)/64;
  i = find(X<0 & Y<0 | z <0);
  z(i) = zeros(size(i));
  z    = 7*sqrt(z);
```

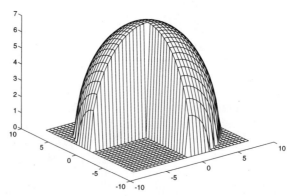

Figure A.v.5   *An ellipsoid missing a quadrant.*

6. The graph of a function near a saddle point. The function defined by $z = x^2 - y^2$ has a *saddle point* at the origin. At this point, the tangents to the function are horizontal, but the function is curving "up" in the x direction, and "down" in the y direction. For added effect, the function has been set to zero for points $|x| > 6$ and $|y| > 6$. See Figure A.v.6.

```
% A Saddle Point at the Origin =====
x = [-8:.5:8]'; y = x
[X,Y] = meshgrid(x,y);
z = X.^2 - Y.^2;
i = find(abs(X)>6 | abs(Y)>6 );
z(i) = zeros(size(i));
```

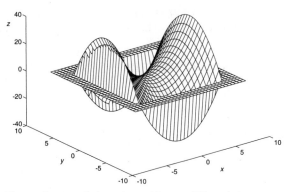

Figure A.v.6 *A function with a saddle point.*

COLOR-SHADING THE GRAPH

If you are using MATLAB Version 4.0 or later, the output from the function **mesh** is colored according to height. The relationship between color and height is controlled by a fourth parameter in the argument list, C, and it is unlikely that you will have occasion to change the factory setting, so it is best to omit this parameter for most uses.[3] However, there is an additional feature that you will very likely find useful for improving the appearance of your graphs. This is the option of using either *interpolated* or *flat* shading on the graph. That is, the "skeleton" of colored lines produced by **mesh** can be replaced by a solid object with the regions bordered by intersecting mesh lines being colored in. There are two choices for coloring each "patch" of the graph. They are implemented by the following MATLAB commands:

shading interp   This statement causes the color between mesh lines to vary smoothly. That is, the color is *interpolated* between the corners of each region bounded by intersecting mesh lines.

---

[3]If you are interested in experimenting with the color-shading, see the help files under the names **COLOR, COLORMAP** (after graphing a function, try **colormap(map)** where **map** is assigned the name of the 10 types of colormaps listed in **help color**). If your printer does not support color output, you can see the plot in an enhanced contrast black-and-white version by using the sequence of statements **cmap = contrast(z);** **colormap(cmap)**.

shading faceted   Faceted (or *flat*) shading determines the color of each patch of the graph by the color of the corner of the patch with the smallest values of the indices. Black grid lines then replace the original colored lines produced by mesh. (The command shading flat will produce a similar graph without the black grid lines.)

A three-dimensional perspective graph can also be shaded directly by using the function surf(x,y,z). This function operates in precisely the same way as mesh; except that the patches between grid lines are now color shaded.

CHANGING THE PERSPECTIVE WHEN USING mesh

The default perspective angle used by mesh when plotting an array, $z(x, y)$, is obtained by using a viewpoint that is rotated about the z axis by an angle (the *azimuthal* angle), az = $-37\frac{1}{2}^{\circ}$. Note that this corresponds to rotating the *object* in an opposite sense. This is shown in Figure A.v.7.

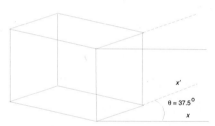

Figure A.v.7   *Default perspective angle obtained by* mesh *when plotting an array,* $z(x, y)$.

The default elevation angle is such that the perspective is one of looking down at the figure by an angle of $el = 30°$. (el = 90° is directly overhead.) Both az and el can easily be changed by defining a two-element vector, M = [az el], where az and el are the new values of the rotation angle and the elevation angle (in *degrees*) to be used by mesh. Remember, a positive value for az will rotate the object about the z axis in a *clockwise* direction.

To alter these values in MATLAB 4.0, use the function view(az,el). Thus, [az,el] = view will return the current values of the rotation and viewing angle, whereas view(new_az,new_el) will replace them with the values contained in new_az and new_el.[4]

You may wish to follow a 90° rotation of one of the preceding figures as follows. After a call to mesh, enter the following:

---

[4]In MATLAB 3.5, this is done by defining a vector M = [az el] and including the vector in the argument list of mesh, as mesh(z,M).

```
for az = 0:15:90
view(az,30);
pause
end
```

CHANGING THE SCALE WHEN USING mesh

The function mesh proceeds by finding the minimum and maximum of the arrays x, y, and z, and then adjusts the scale to fill out the graph region. The limits on the three axes can be determined or altered by using the function axis. After the graph has been drawn with mesh, a statement like s = axis will return a 6-element vector, s, containing the limits on the x, y, and z axes. That is, $s = [x_{min}, x_{max}, y_{min}, y_{max}, z_{min}, z_{max}]$. These values may be changed by entering axis(new_s), where new_s contains the replacement values for the axis limits. Thus, by adjusting the limits on the z axis you can exaggerate or diminish the height of the graph. However, because mesh will graph *all* of the elements of the matrix z, regardless of what you set the x and y limits to be, this is *not* an effective way of altering the xy domain of a plot. To plot the function for a different xy range, you should start afresh and recalculate z for the new range before trying to plot the function.

Similar to plot, you can use gtext to place text on the figures. However, the function ginput will simply return the coordinates of the *cross hairs* on the screen, not on the figure, and therefore, it is useless in connection with mesh.

Perspective plots are extremely useful in providing a qualitative picture of what a function of two variables is doing over a given region. However, it is very difficult to obtain numerical values for the position of a minimum or a maximum from these graphs. For this purpose, a different representation of the function, a *contour* plot, is used.

Contour Graphs

You have probably seen contour maps that give the elevation of the land above sea level as closed curves drawn on a normal map. If x represents the east-west position on the map and y represents the north-south position, the elevation of any point of land can be written as a function of x and y: elevation $= E(x, y)$. If all those points with an elevation of, say 100 m, are connected, we would get a curve of constant (100 m) elevation. Elevations of 200 m, 300 m, and so on, are similarly connected, resulting in a set of closed curves for a selection of elevation values, as illustrated in Figure A.v.8.

The MATLAB function contour will produce a colored contour plot. The use is as follows:

contour(x,y,z)   Creates a contour plot with x and y axes specified by the vectors x and y.

contour(x,y,z,n)   The same as above, except that the user specifies how many contour lines (n) are to be drawn.

contour(x,y,z,V)	The vector V contains values at which to draw contours. One contour line will be drawn for each value contained in V.
contour(z)	Contour lines are drawn with x and y axes specified by the column/row numbers of the matrix z, not by the values of x and y.
contour(z,n)	The same as above, except that the user specifies how many contour lines (n) are to be drawn. These last two contour options are the only ones available in MATLAB 3.5.

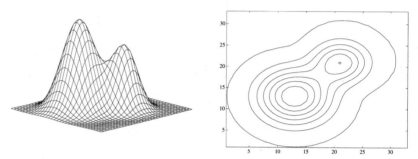

Figure A.v.8   *A three-dimensional perspective graph of a function with two peaks, and the same function represented as a contour graph.*

Also, the function **meshc** performs exactly like **mesh**, except that a contour map is drawn in the xy plane below the figure.

The function **ginput** can be used with a contour map to "pick off" the value of the coordinates of a point on the graph, and it is useful for locating the maximum and minimum of a function of two variables.

Non-Cartesian Coordinates

CYLINDRICAL COORDINATES

Frequently, the function you wish to graph is most naturally expressed in terms of cylindrical, or polar, coordinates, ρ, ϕ, as $z = f(\rho, \phi)$. The relationship between these coordinates and ordinary rectangular, or Cartesian, coordinates x, y, and z is given by the relations,

$$\begin{aligned} x &= \rho \cos \phi \\ y &= \rho \sin \phi \\ z &= z \end{aligned} \tag{A.v.1}$$

where $0 \le \phi \le 2\pi$, and $x^2 + y^2 = \rho^2$. This relationship is illustrated in Figure A.v.9.

Of course, the function could be rewritten in terms of x and y and graphed over a range of x and y values, as was done in the previous sections. A more

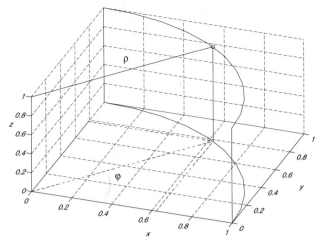

Figure A.v.9 *Relationship between cylindrical and Cartesian coordinates.*

satisfactory approach is to construct a grid of ρ and ϕ points and to directly evaluate the function on this grid. This is best illustrated by an example.

ILLUSTRATIVE PROBLEM A.v.1

Perspective Plot of a Normal Mode of a Drum Head

When a circular drum of radius R is struck, the resulting surface vibrations can be described in terms of what are called the *normal modes* of oscillation of the surface. The set of normal modes is distinguished by two integer parameters, ℓ and m, and they are given by the equation,

$$Z_{\ell m}(\rho, \phi) = J_\ell \left(\Gamma_{\ell m} \frac{\rho}{R} \right) \cos(m\phi) \qquad (\text{A.v.2})$$

where J_ℓ is an ordinary Bessel function of the first kind of order ℓ, and $\Gamma_{\ell m}$ is the mth root of J_ℓ.[5]

The first root of $J_1(x)$ is $\Gamma_{11} = 3.8317\ldots$. Thus, one of the normal modes of a drumhead of radius 3 is

$$Z(\rho, \phi) = J_1(1.2772\rho) \cos\phi$$

To produce a graph of this function, we first construct a column vector containing the ρ values and a row vector containing the values of ϕ:

```
rho = [0:0.2:3]';
phi = 0:pi/20:2*pi;
```

The products `rho*ones(1,length(rho))`,`ones(length(phi),1)*phi`, will be *matrices* corresponding to the grid of values over which the function $Z(\rho, \phi)$ is to be computed. Thus,

---

[5]The ordinary Bessel functions arise from the solution of Laplace's equation, $\nabla^2 F = 0$, in cylindrical coordinates. These functions are an important part of the MATLAB library. The function $J_\ell(x)$ may be computed using the function, `bessel(ell,x)`. A table of the roots of J_ℓ can be obtained from the function, `GammaNul(ell,m)`, which is supplied on the accompanying program disk.

```
arg = 1.2772*rho;
J1  = bessel(1,arg);
Z   = J1*cos(phi);
```

Finally, to plot this function using **mesh**, we must specify the Cartesian, (x, y), coordinates:

```
X = rho*cos(phi);
Y = rho*sin(phi);
```

and the graph is displayed by

```
mesh(X,Y,Z)
```

The resulting graph is given in Figure A.v.10.

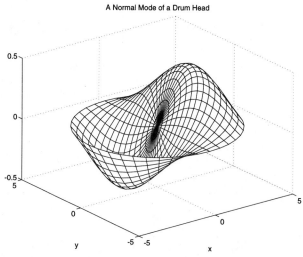

Figure A.v.10   *A normal mode of a circular drum head.*   ■

WHAT IF?   You should duplicate this calculation and plot the result for higher modes of the drumhead oscillation. (See also Problem PA.v.5.)

As a second example, suppose you wish to draw a flared cylinder with a radius of $\rho = 8$ at the top, and a radius of $\rho = 4$ at the bottom, where the bottom is at $z = 0$, and the top is at $z = 10$. There is a direct relationship between z and ρ, namely, $z = 2.5\rho - 10$, so we want to produce a *grid* of all ρ and θ values and then compute the values of z from this equation. Thus, the MATLAB code to graph the flared cylinder is

```
n = 30;   a = 0.4;    b = 0.8;
theta = (0:n)/n*(2*pi);
rho   = a + (0:n)/n*(b-a);
[Theta,Rho] = meshgrid(theta,rho);
X = Rho.*cos(Theta);
Y = Rho.*sin(Theta);
Z = 2.5*Rho - 10;
mesh(X,Y,Z)
```

Variations on this code are also used in Problem PA.v.4.

SPHERICAL COORDINATES

A point in three-dimensional space is located by the coordinates (x, y, z) relative to the ordinary Cartesian axes. The same point can also be located relative to *spherical* coordinates, (r, θ, ϕ), as shown in Figure A.v.11.

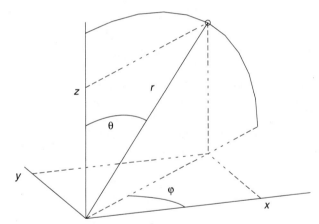

Figure A.v.11   *Relation between spherical and Cartesian coordinates.*

The relationship between the two coordinate systems is given by the following equations:

$$
\begin{aligned}
x &= r \cos \phi \sin \theta \\
y &= r \sin \phi \sin \theta \\
z &= r \cos \theta
\end{aligned}
\tag{A.v.3}
$$

where the ranges of the angles are $0 \le \theta \le \pi$, $0 \le \phi \le 2\pi$, and $0 \le r < \infty$. A sphere in these coordinates is then characterized by the single equation $r = R_0$, for all angles.

To plot the sphere in MATLAB, we first assemble matrices for the complete grid of θ and ϕ values

```
RO = 2;
n = 30;
theta = (0:n)/n*(pi);
phi   = (0:n)/n*(2*pi);
[Theta,Phi] = meshgrid(theta,phi);
```

and then use these matrices to compute *three* matrices containing the values of x, y, and z:

```
X = RO*sin(Theta).*cos(Phi);
Y = RO*sin(Theta).*sin(Phi);
Z = RO*cos(Theta);
```

The perspective graph of the sphere is then obtained by

```
mesh(X,Y,Z);
axis('image')
```

where the last line is needed to give a realistic representation of the sphere. (See script file *ORB.m* on the accompanying program disk.)

Plotting Three-Dimensional Trajectories

A function $z = f(x, y)$ defines a surface in three-dimensional space. The variables x and y are independent variables, and z is the dependent variable. That is, for each value of x and y, there is a value of z defined by the function f. If, however, there are *three* independent functions for the variables x, y, and z, each dependent on a single variable t, as

$$x = x(t)$$
$$y = y(t)$$
$$z = z(t)$$

t is the single independent variable. Each value of t will determine all three values of x, y, and z. This characterization of the variables x, y, and z is called a parametric representation. As t varies continuously, these equations will then trace a *trajectory* or curve in three-dimensional space. The MATLAB function plot3 will trace this trajectory superimposed on a three-dimensional coordinate axis system. The use of the function is as follows:

plot3(x,y,z)	If x, y, z are *vectors* of the same length, this function will plot a continuous curve through the points (x_i, y_i, z_i) for $i = 1, \ldots, n$, the common length of the vectors.
plot3(X,Y,Z)	If X, Y, Z are *matrices* of the same size and shape, the resulting plot will contain a separate trajectory for each of the columns. That is, the first set of columns of X, Y, and Z are used for the first curve, the second set of columns for the second curve, and so on.

plot3(x,y,z,s)   The same as above, except that the variable s contains a 1, 2, or 3 character string specifying the color and/or the symbol to be used at each point. For example, s = 'b+', will produce a blue curve with "plus" signs at each point. See the discussion of the plot function for other options.

plot(x1,y1,z1,s1, x2,y2,z2,s2, x3,y3,z3,s3,...)

Permits plotting several trajectories with different symbols and colors simultaneously.

For example, the plot produced by the statements

```
t = 0:pi/25:10*pi;
z = t;
r = 5 + abs(z-5*pi);
x = r.*cos(2*t);    y = r.*sin(2*t);
plot3(x,y,z)
```

will produce a "bed-spring"-like trajectory. It may take some imagination on your part to interpret the trajectory in a three-dimensional sense. To help in this regard, you can add grid lines to the figure with the command, grid; reducing the elevation angle and the azimuthal angle may also help. Figure A.v.12 was drawn with az = el = 15°.

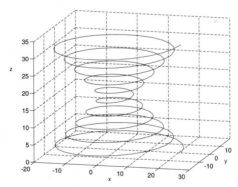

Figure A.v.12   *A "bed-spring" trajectory produced by* plot3.

Four-Dimensional Plots

A four-dimensional plot would correspond to plotting the value of a function $v = f(x, y, z)$, for a range of values of the three independent variables $x, y,$ and z. Clearly, this is impossible to accomplish in a three-dimensional figure. Well, MATLAB takes a stab at doing just this by assigning the fourth variable to be the *color* of the figure. That is, each point in the three-dimensional domain is assigned a color that depends on the value of v. Even so, if the figure is then plotted in a two-dimensional perspective graph, similar to mesh, the interior colors will be hidden by those on the outside. To overcome this limitation, the function designed by MATLAB to produce 4-dimensional plots will display the colors representing v on a selection of planar slices of the x, y, and z domain. This is accomplished by the function slice. Indeed, the graphs produced by this function are quite beautiful, but not surprisingly, they are very difficult to interpret, and so we forgo the use of this function. You may, however, wish to experiment with a few examples. Try a few variations of the examples listed in the help file for slice.

Problems

REINFORCEMENT EXERCISES

PA.v.1 Plotting Problems. For each of the following functions, code a separate M-file function and plot the function using mesh.

a. $z = (1 - x^2 - y^2)^{-1/2}$ for $|x|$ and $|y| \leq 0.7$

b. The atomic electron densities in the $n = 3$ shell are functions of three space variables, x, y, and z. A slice of these densities along the $z = 0$ plane yields the following functions:

$$\rho_{3s} = 243Ce^{-(2/3)r}\left(1 - \frac{2}{3}r + \frac{2}{27}r^2\right)^2 \qquad \text{Angular momentum } \ell = 0$$

$$\rho_{3d} = \frac{1}{5}Cr^4e^{-(2/3)r} \qquad \text{Angular momentum } \ell = 2$$

with $C = 5/3^8 4\pi$, and $x^2 + y^2 = r^2$. Plot these functions over the range $|x|$ and $|y| \leq 8$ and "cap" the ρ_{3s} density if it is greater than 10^{-3}. That is, if $\rho_{3s} > 10^{-3}$, then $\rho_{3s} \to 10^{-3}$.

c. A slice of the same electron densities along the $y = 0$ plane yields

$$\rho_{3s} = 243Ce^{-(2/3)r}\left(1 - \frac{2}{3}r + \frac{2}{27}r^2\right)^2 \left(\frac{2z^2 - x^2}{r^2}\right)^2$$

$$\rho_{3d} = \frac{1}{5}Ce^{-(2/3)r}x^4$$

with $C = 5/3^8 4\pi$, and $x^2 + z^2 = r^2$. Use the range $-8 \leq x \leq 8$, and $-8 \leq z \leq 8$, and "cap" the ρ_{3s} density if it is greater than 10^{-3}. That is, if $\rho_{3s} > 10^{-3}$, then $\rho_{3s} \to 10^{-3}$.

d. Van der Waals' Imperfect Gas. If the pressure (P), volume (V), and temperature (T) of a gas are expressed in terms of scaled variables, $p = P/P_c, v = V/V_c$, and $t = T/T_c$, where the quantities P_c, V_c, and T_c are the values at the critical point of the gas, then a fairly accurate description of the gas is given by the van der Waals' equation of state,

$$p = \frac{8}{3}t\left(v - \frac{1}{3}\right)^{-1} - \frac{3}{v^2}$$

Plot this function over the range $0.9 \leq t \leq 1.1$ and $\frac{1}{2} \leq v \leq 1$.

e. Dipole antenna radiation pattern

$$P(x, y) = P_0 \frac{\cos^2\left(\frac{\pi}{2}\cos\theta\right)}{\sin^2\theta}$$

with $|z| \leq 8$, $0 \leq x \leq 8$, $\theta = \tan^{-1}(x/z)$, and $P_0 = 10$.

f. Electric Potential Near a Dipole. Two charges of equal magnitude but opposite sign, separated by a distance of $2a$, constitute an electric dipole. The expression for the electric potential due to the dipole is then,

$$V(x, y) = V_0 \left[\frac{1}{r_+} - \frac{1}{r_-} \right]$$

where $r_\pm = \sqrt{(x \pm a)^2 + y^2 + z^2}$. Use the values $V_0 = 5$, $a = 0.2$, and plot the potential over the range $|x|, |y| \leq 3$, and $z = 0$. In addition, for all points x and y such that $r_\pm \leq 0.05$, replace with $r_\pm \to 0.05$.

PA.v.2 Contour Graphs. For several of the functions in Problem PA.v.1, draw separate contour graphs.

PA.v.3 The Functions view and shading. For one of the graphs in Problem PA.v.1, follow the rotation of the figure through $90°$ by using the code on page 576. Also, experiment by altering the shading in a variety of ways.

PA.v.4 Cylindrical Coordinates. The MATLAB code

```
function f = PIPE(a,b,L)
 th = [0:pi/10:2*pi]';
 p = [0 L L 0]; r = [b b a a];
 z = ones(size(th))*p;
 mesh(x,y,z)
```

will draw a "pipe" of inner radius a, outer radius b, and height L, oriented along the z axis.

The following equations represent rotations of the coordinate axes:

$$\begin{aligned} x' &= x\cos\theta_z - y\sin\theta_z \quad &\text{Rotation about } z \text{ by } \theta_z \\ y' &= y\cos\theta_z + x\sin\theta_z \end{aligned}$$

$$\begin{aligned} y' &= y\cos\theta_x - z\sin\theta_x \quad &\text{Rotation about } x \text{ by } \theta_x \\ z' &= z\cos\theta_x + y\sin\theta_x \end{aligned}$$

$$\begin{aligned} z' &= z\cos\theta_y - x\sin\theta_y \quad &\text{Rotation about } y \text{ by } \theta_y \\ x' &= x\cos\theta_y + z\sin\theta_y \end{aligned}$$

Write MATLAB code to

a. Draw a pipe of inner and outer radii 5 and 6, respectively, and length 20 that is rotated about the y axis by $45°$. [Use **axis('square')**.]

b. Draw the pipe initially rotated by $45°$ about the y axis, followed by a rotation by $-45°$ about the x axis.

EXPLORATION PROBLEMS

PA.v.5 Normal Modes of a Drumhead. The displacement of a circular drumhead of radius R is given in terms of a linear combination of the normal modes $Z_{\ell m}(\rho, \phi)$ of Equation A.v.2, each multiplied by an oscillatory function of time, t, of the form $\cos(\Gamma_{\ell m} \frac{ct}{R})$, where c is the speed of the sound waves on the drum surface. Use the values $R = 0.25$ m and $c = 500$ m/s to accomplish the following:

    a. Graph each of the two normal modes, $Z_{01}(\rho, \phi)$ and $Z_{11}(\rho, \phi)$.

    b. Assume the actual displacement, z, at time, $t = 0$, is:

$$z(\rho, \phi) = Z_{01}(\rho, \phi) + 0.3 Z_{11}(\rho, \phi)$$

    and graph $z(\rho, \phi)$ at some value of t in the interval, $0 \le t \le 0.001$ s.

    c. Investigate the "movie-making" capabilities of MATLAB (start with `help movie`), and record 10 frames of the movie, with each frame being a graph of $z(\rho, \phi)$ at successive values of t over the interval $0 \le t \le 0.001$ s.

PA.v.6 Deformed Spheres. Ignoring gravity, a drop of water or an atomic nucleus can be described in terms of the vibrations and rotations of a nearly spherical liquid drop. In this case, the surface of the liquid drop is given by Equation A.v.3, but in this instance, the radius, r, is itself a function of θ and ϕ. If the deformation is small, a good approximation is

$$r(\theta, \phi) = R_0 [1 + S_{\beta, \gamma}(\theta, \phi)]$$

where $S_{\beta, \gamma}(\theta, \phi)$, the deviation from sphericity, is a function that depends on θ and ϕ, as well as on two parameters, β and γ, that characterize the size and the shape of the deformation, respectively. The explicit expression for this function is

$$S_{\beta, \gamma}(\theta, \phi) = \sqrt{\frac{5}{16\pi}} \, \beta \left[\cos \gamma (3\cos^2 \theta - 1) + \sqrt{3} \sin \gamma \sin^2 \theta (\cos^2 \phi - \sin^2 \phi) \right]$$

Draw perspective graphs for the following sets of parameters, β and γ:

β	γ	Description	Symmetry axis
0.8	0	prolate, "cigar-shaped"	along z axis
−0.8	0	prolate, "cigar-shaped"	along z axis
0.8	$\frac{2}{3}\pi$	prolate, "cigar-shaped"	along x axis
0.8	π	oblate, "pill box-shaped"	along z axis
0.8	$\frac{1}{3}\pi$	oblate, "pill box-shaped"	along x axis
0.8	$\ne \frac{n\pi}{3}$	ellipsoid	none

PA.v.7 The Earth and the Moon. The MATLAB function `[x,y,z] = sphere(n)` will generate three $(n+1) \times (n+1)$ matrices containing the coordinates of the surface of a unit sphere. Use this function and the function `mesh` to graph a sphere of radius 6.4×10^6 m, representing the earth. The sphere is located at the origin. Next, add a second sphere of radius 1.7×10^6 m to the graph, and locate the second sphere at $X_m = Y_m = 2.7 \times 10^7$ m (this is one-tenth the actual earth-moon distance). You will have to manually adjust the z-axis scale to have spheres appear as spheres. Try rotating the coordinate system.

PA.v.8 Satellite Trajectories. A satellite's orbit about the earth is an ellipse with the earth at one of the foci. That is,

$$x(t) = a \cos t - f$$
$$y(t) = b \sin t$$
$$z(t) = 0$$

where a and b are the semimajor, semiminor axes of the ellipse, and $f = \sqrt{a^2 - b^2}$ is the distance from the center to either focus. Next, if the trajectory is inclined by an angle θ out of the xy plane, the equations become

$$x(t) = a \cos t - f$$
$$y(t) = b \cos \theta \sin t$$
$$z(t) = b \sin \theta \sin t$$

Use $a = 12R_0$ and $b = 9R_0$ with $R_0 = 1$, incline the satellite trajectory by an angle $\theta = 12\frac{1}{2}°$, and plot the satellite trajectory. (The period of the orbit is $T = 2\pi$.) Next, add a sphere of radius R_0 representing the earth to the graph. As a final exercise, assume that the orbit *degrades* with time by multiplying each of the expressions for $x(t), y(t)$, and $z(t)$ by a factor $e^{-t/\tau}$, where $\tau = 20$. Plot the trajectory and follow the satellite as it ultimately impacts the earth.

Summary

CONSTRUCTING GRIDS

meshgrid

A function of two variables, $z = f(x, y)$, is defined over a rectangular region in the xy plane defined by the range of values in the vectors x and y. If X and Y are *matrices* containing the values of x and y at each point of this grid, these matrices can be computed automatically using the function meshgrid. The values of the function $f(x, y)$ can be computed by coding the function as an M-file function that accepts *matrices* as inputs for x and y and computes z using element-by-element operators.

THREE-DIMENSIONAL PERSPECTIVE GRAPHS

mesh, surf

First, matrices are constructed that contain the values of the independent variables x, y on a rectangular grid. Next, the values of a function of two variables, $z = f(x, y)$, are evaluated at each point on the grid, and finally, the function is automatically graphed using either mesh(x,y,z) or surf(x,y,z).

CONTOUR GRAPHS

contour, meshc,
surfc

In place of a perspective graph, a contour map of the function can be obtained using contour(x,y,z).

THREE-DIMENSIONAL TRAJECTORIES

plot3

A three-dimensional graph of the trajectory of a set of three parametric equations $x(t)$, $y(t)$, and $z(t)$ can be obtained using plot3(x,y,z). This function can also be used to draw simple lines on a three-dimensional figure.

MATLAB Functions Used

meshgrid	If x and y are two vectors containing a range of points for the evaluation of a function, [X,Y] = meshgrid(x,y) returns two rectangular matrices containing the x and y values at each point of a two-dimensional grid. The contents of X will be identical rows, each containing the elements of x, and the contents of Y will be identical columns, each containing the elements of y.
mesh(X,Y,z)	If X and Y are rectangular arrays containing the values of the x and y coordinates at each point of a rectangular gird, and if z is the value of a function evaluated at each of these points, mesh(X,Y,z) will produce a three-dimensional perspective graph of the points. The same results can be obtained with mesh(x,y,z), where x and y are the vectors used earlier to establish the two-dimensional grid.
meshc, meshz	If the xy grid is rectangular, these two functions are merely variations of the basic plotting program mesh, and they operate in an identical fashion. meshc will produce a corresponding contour plot drawn on the xy plane below the three-dimensional figure, and meshz will add a vertical wall to the outside features of the figures drawn by mesh.
surf	This function will produce a three-dimensional perspective drawing. Its use is usually to draw surfaces, as opposed to plotting functions, although the actual tasks are quite similar. The output of surf will be a *shaded* figure. The most common use is to plot parametric functions. For example, if row vectors of length n are defined by $x = r \cos\theta$ and $y = r \sin\theta$, with $0 \le \theta \le 2\pi$, they correspond to a circle of radius r. If $\vec{r}$ is a *column* vector equal to r = [0 1 2]';, then z = r*ones(size(x)) will be a rectangular, $3 \times n$, array of 0's, 1's, and 2's, and surf(x,y,z) will produce a shaded surface bounded by three circles; i.e., a cone.
surfc	This function is related to surf in the same way that meshc is related to mesh.
colormap	Used to change the default coloring of a figure. See the MATLAB reference manual or the help file.
shading	Controls the type of color shading used in drawing figures. See the MATLAB reference manual or the help file.
view	view(az,el) controls the perspective view of a three-dimensional plot. The view of the figure is from an angle "el" above the xy plane with the coordinate axes (and the figure) rotated by an angle "az" in a clockwise direction about the z axis. Both angles are in degrees. The default values are $\mathtt{az} = 37\frac{1}{2}^\circ$ and $\mathtt{el} = 30^\circ$.
axis	Is used to determine or change the scaling of a plot. If the coordinate axis limits of a two-dimensional or three-dimensional graph are contained in the row vector $r = [x_{\min}, x_{\max}, y_{\min}, y_{\max}, z_{\min}, z_{\max}]$, axis will return the values in this vector, and axis(r) can be used to alter them. The coordinate axes can be turned *on* and *off* with axis('on') and axis('off'). Some other possible string constant inputs to axis and their effects are listed as follows:

axis('equal')	x and y scaling are forced to be the same.
axis('square')	The box formed by the axes is square.
axis('auto')	Restores the scaling to default settings.
axis('normal')	Restores the scaling to full size, removing any effects of *square* or *equal* settings.
axis('image')	Alters the aspect ratio and the scaling so the screen pixels are square shaped rather than rectangular.

contour

The same matrices that are used to produce a three-dimensional perspective plot can be used to create a two-dimensional *contour* plot. The use is contour(x,y,z). A default value of $N = 10$ contour lines will be drawn. An optional fourth argument can be used to control the number of contour lines that are drawn. contour(x,y,z,N), if N is a positive integer, will draw N contour lines, and contour(x,y,z,V), if V is a vector containing values in the range of z values, will draw contour lines at each value of $z = V$.

plot3

Used to plot lines or curves in three dimensions. The use is similar to plot. If x,y, and z are vectors of equal length, plot3(x,y,z) will draw, on a three-dimensional coordinate axis system, the lines connecting the points. An optional fourth argument, representing the color and symbols to be used at each point, can be added in exactly the same manner as with plot.

grid

grid on adds grid lines to a two-dimensional or three-dimensional graph; grid off removes them.

slice

Used to draw "slices" of a volume at a particular location within the volume. See the MATLAB reference manual or the help file for more information concerning slice.

APPENDIX A.vi / Elementary Input/Output in MATLAB

PREVIEW MATLAB can be used with considerable ease to process a prodigious amount of data. However, it is up to you to arrange for the input of the data. The easiest way is to add a few lines to an ordinary ASCII data file to make it an M-file assigning values to the elements of a matrix. The file should consist of a number of lines of data with the same number of values on each line and each value separated from the next by at least one blank space. We will also describe other mechanisms for having a MATLAB procedure directly read data either from the keyboard or from a file.

Although some of MATLAB's major strengths are the graphical display and analysis of data, its procedures for the input/output of numbers are rather primitive. This is intentional, because the principal aim is to provide an interactive environment for computing rather than to produce lists of computational results. Nonetheless, you may have occasion to input or output data to or from a file, so we briefly summarize these procedures here.

In this chapter we describe the very limited set of commands in MATLAB for reading data from an external file into a MATLAB matrix, or for the writing of numbers computed in MATLAB into such an external file. Although it may surprise you, the input/output capabilities of MATLAB are much less versatile than those of, say an older language like FORTRAN. But there is a good reason for this. The goal of MATLAB is to provide an environment whereby your computed results can easily be viewed in terms of a variety of graphs or labeled by the characterizing statistics described in Part IV.

If you need to produce a neat table of results with a variety of textual headings, the modern preference, even in FORTRAN, is to have the computer program produce a *raw* table of numbers written to a file. Then, afterwards, the table is tidied up and text headings are added using a normal editor program. This procedure has the advantage of bypassing all the tedious editing commands necessary to get the numbers positioned correctly on the page.

We will also outline the rudiments of creating a separate file to contain numbers, and then demonstrate how to both read numbers from such a file and write numbers into the file.

Connecting a Data File to MATLAB: The `fopen` Statement

If you wish to have MATLAB read or write a separate data file consisting of numerical values, you have to *connect* the file to your executing MATLAB program. This is done with the MATLAB statement `fopen` in the form

$$\texttt{fid} \ = \ \texttt{fopen}('Filename', permission\ code);$$

where, if `fopen` is successful, `fid` will be returned as a positive integer greater than 2. (The numbers 0,1, and 2 are reserved, respectively, for the standard input, the keyboard; the standard output, the terminal screen; and the standard error file.) If the attempt to open the file is unsuccessful a value of -1 is returned. The *Filename* is a string enclosed in single quotes and, if necessary, contains the *path* locating the file. The *permission code* is also a string enclosed in single quotes and can be a variety of flags that specify whether or not the file can be written to, read from, appended to, or a combination of these. The most common codes are:

Code	Meaning
`'r'`	read only
`'w'`	write only (will create the file if necessary)
`'r+'`	read and write (file must already exist)
`'a+'`	read and append (will create the file if necessary)

Thus, the statement

```
InData = fopen('BankBal.dat','r')
```

will associate an integer, labeled by `InData`, with the existing data file named `BankBal.dat`. The file is identified as "read-only" and the pointer is positioned at the top of the file.

The reverse statement

```
[filename,permission] = fopen(fid)
```

will return the name and permission code of the file labeled by the integer `fid`. There are numerous other possibilities. See the `help fopen` file in MATLAB.

The `fopen` statement positions the file at the beginning; so after reading or writing to the file, you will have to reissue the statement if you wish to reuse the file from the top.

The statement `fclose(fid)` closes, or disconnects the MATLAB program from the file with identification number `fid`. `fclose('all')` will close all files except the standards, `fid = 0, 1, 2`.

Reading Data from an Existing File: The `fscanf` Statement

The MATLAB statement for reading ordinary numerical data from a standard ASCII data file is `fscanf`. The usage is

$$\texttt{A} = \texttt{fscanf}(\texttt{fid}, FORMAT, SIZE)$$

where *FORMAT* specifies the types of numbers (integers, reals with or without exponent, character strings) and their arrangement in the data file, and optional *SIZE* determines how many quantities are to be read and how they are to be arranged into the matrix **A**. If *SIZE* is omitted, the entire file is read.

For example, suppose your data file bears the name `mydata.dat` and consists of three columns of numbers. The first column contains integers, and the remaining two contain *floating-point* numbers; i.e., numbers with a decimal point, with or without exponents. It might appear as

1	1.230	0.243
2	2.412	0.147
3	4.564	0.078
4	7.731	2.34e−2
5	9.923	4.55e−3
6	11.312	3.04e−3
$\vdots$	$\vdots$	
65	35.285	1.41e−7

The file is opened with `fopen` as a *read-only* file and positioned at the beginning with `fid = fopen('mydata.dat','r')`. The values contained in the file can be read into a matrix **A** by using the MATLAB statement `fscanf` as follows:

$$A \ = \ \texttt{fscanf(fid,'\%3i \%7e \%7e\textbackslash n',[3,65])}$$

The three input fields to `fscanf` are separated by commas and correspond in the present example to

File ID	`fid`	The identification number of the data file to be read.
FORMAT	`'%3i %7e %7e\n'`	A series of C-language format specifications for each of the numbers on a line of data. Each specification begins with % followed by an integer giving the *width* of the *widest* number in that column. The letter corresponds to the *type* of data to be read. The most common are

     i or d   integer
         f   floating point, no exponent
         e   floating point with or
            without an exponent
         s   character string of
            arbitrary length

The final `\n` designates the end of a line of data. The entire format specification is a string that must be enclosed in quotes.

SIZE	[3,65]	The size of the array to be filled. [*Note:* The first index refers to the number of *columns* and the second the number of *rows in the data file*, the reverse of the normal notation. Upon execution of the statement, the matrix **A** will turn out to be the transpose of the data file!]

The *SIZE* field is optional, and if omitted, the entire data file is read, line by line, and stored in a single column of **A**. If you don't know how many lines of data there are, `[3,inf]` will read 3 numbers per line until the end of the file is reached.

Also, because all the numbers are read into a single array, and because in MATLAB all elements of an array must be of the same type, the numbers read as integers in this example are actually converted to reals when they are stored in the matrix A. If all the numbers were of the same type, there will be no conversion. One consequence of this is that character strings and numerical values cannot be read with the same `fscanf` statement.

The result of the `fscanf` statement is thus:

```
A = fscanf(fid,'%3i %7e %7e \n',[3,65])4
A =
```

1.0000	2.0000	3.0000	4.0000	5.0000	6.0000	$\cdots$	65.0000
1.2300	2.4120	4.5640	7.7310	9.9230	11.3120	$\cdots$	35.2850
0.2430	0.1470	0.0780	0.0234	0.0046	0.0030	$\cdots$	0.0000

To reproduce the structure of the example data file, as is, with the values stored by columns, the transpose of the result of the `fscanf` operation could be used; i.e.,

$$A \; = \; \texttt{fscanf}(\texttt{fid},'\,\%3i \; \%7e \; \%7e \backslash n', [3,65])'$$

See the `help fscanf` file for more details.

Output from MATLAB Using `fprintf`

To write the contents of a matrix directly to a data file, the procedures are similar. First, the output data file must be opened with a `'w'` permission flag. The MATLAB statement to write the values to the file is then `fprintf`. Its usage is

$$\texttt{fprintf}(\texttt{fid}, FORMAT, \texttt{A});$$

The format specifications that appear in the *FORMAT* field have the same meaning as described previously. Without the semicolon, `fprintf` will display the total number of bytes of information written. If desired, additional matrices may be added to the argument list and printed.

However, once again, this statement performs a transpose of information in the transfer to the data file. This means that the format specifications refer to the elements of the *rows* of **A**. Thus, after reading our data file, the matrix **A** is a 3×65 array. If this matrix is written to `newdata.dat` by the following statements:

```
fid = fopen('newdata.dat','w');
fprintf(fid,'%3i %7e %7e\n',A);
```

the data file `newdata.dat` will contain 65 rows of 3 numbers each. The first number in each row will be an integer (3-wide), followed by two floating-point numbers (7 digits wide). That is, except for the width of the numbers, it duplicates the original data file.

Thus, the `fprintf` statement will print the elements of **A**, in this case three numbers per line, by cycling through the *columns* of **A**. If **A** is a $3 \times n$ matrix, the result will be the transpose of **A** or an $n \times 3$ matrix. However, if **A** is instead represented by columns in the data file; i.e., an $n \times 3$, the output will still be $n \times 3$, *but* will be a rearrangement of the elements, printing all the elements of the first column of **A** first, then the elements of the second column, etc. As with the `fscanf` statement, the values can be stored in the data file as they appear in the matrix **A** by using the transpose of the matrix in the `fprintf` statement.

Problems

REINFORCEMENT EXERCISES

PA.vi.1 Writing Tables. Produce a table of $\sin\theta$, $\cos\theta$, for $\theta = 0°$ to $45°$ in steps of $5°$. (Remember, the MATLAB functions `sin, cos` expect arguments in radian measure.) Write the elements of this table to a file called `trigtable`. Next, temporarily leave MATLAB and call up your editor program using the !-operator; read the newly created data file into the editor program and clean up the table by adding headings and perhaps vertical and horizontal lines.

PA.vi.2 Creating M-files for Input of Data. In MATLAB, produce a table called `logtable` of $x, \ln x$ for $x = 1$ to $x = 3$ in steps of 0.1. Write this table to the data file so that the x and the $\ln x$ values are in columns in the data file. Next, exit MATLAB and read the data file into your editor program. Add semicolons at the end of each data line but the last. Append a right bracket (]) to the end of the last line. Insert the following two lines at the beginning:

```
function A = dataname
A = [ ...
```

and save the file as an M-file. This file can then be used in later MATLAB programs to assign the data elements to a matrix.

EXPLORATION PROBLEM

PA.vi.3 Adding Semicolons Automatically to the Data File. If a is a MATLAB character string containing a single character, say `a = ';'`, then the statement

```
fprintf(1,'%5f\n',a)
```

will cause the ASCII value of the symbol **a** to be printed to the screen (fid = 1). On my computer, the ASCII value of the semicolon is 59. If we next express this value in base 8, or the octal system, we obtain $59 = 7 \cdot 8 + 3 = (73)_8$ as the octal signature of the symbol ';'. This can then be added to the *format* field of an **fprintf** statement as

```
fprintf(fid,'%6f %6f \073\n',A)
```

and the result will be that each line of the output will end with a semicolon. Use this method to reproduce the data file of Problem PA.vi.2.

MATLAB Functions Used

fopen

Used to connect an existing file to MATLAB or to create a new file from MATLAB. The use is

$$\text{fid} \ = \ \text{fopen}('Filename', permission \ code);$$

where, if **fopen** is successful, **fid** will be returned as a positive integer greater than 2. If the attempt to open the file is unsuccessful, a value of -1 is returned. Both the file name and the permission code are string constants enclosed in single quotes. The permission code can be a variety of flags that specify whether or not the file can be written to, read from, appended to, or a combination of these. The most common codes are:

Code	Meaning
'r'	read only
'w'	write only
'r+'	read and write
'a+'	read and append

The **fopen** statement positions the file at the beginning.

fclose

Will disconnect a file from the operating MATLAB program. The use is **fclose(fid)**, where **fid** is the *file identification number* of the file returned by **fopen**. **fclose('all')** will close all files.

fscanf

Used to read opened files. The use is

$$\text{A} \ = \ \text{fscanf}(\text{fid}, FORMAT, SIZE)$$

where *FORMAT* specifies the types of numbers (integers, reals with or without exponent, character strings) and their arrangement in the data file, and optional *SIZE* determines how many quantities are to be read and how they are to be arranged into the matrix **A**. If *SIZE* is omitted, the entire file is read. The *FORMAT* field is a string (enclosed in single quotes) specifying the form of the numbers in the file. The *type* of each number is characterized by a percent sign (%), followed by a letter (**i** or **d** for integers, **e** or **f** for floating-point numbers with or without exponents). Between the percent sign and the type code, you can insert an integer specifying the maximum width of the field.

fprintf

Used to write to files previously opened. The use is

$$\text{fprintf}(\text{fid}, FORMAT, \text{A})$$

where **fid** and *FORMAT* have the same meaning as for **fscanf**, with the exception that for output formats the string must end with **\n**, designating the end of a line of output.

Answers

CHAPTER 1

P1.1 a. $(\frac{1}{2}, \frac{1}{4})$ c. $(2, 3, 3, 2)$ e. $(-c, c-1, 1-c)$, for any c

P1.5 a. False   b. True   c. True   d. True   e. False   f. False

P1.7 a. $\begin{pmatrix} 1 & 0 \\ 0 & -1 \end{pmatrix}$

P1.13 a. $\vec{a}^T\vec{b} = 4$, $\vec{a}^T\vec{c} = -2$, $\vec{b}^T\vec{c} = 3$, $|\vec{a}| = \sqrt{14}$, $|\vec{b}| = \sqrt{29}$, $|\vec{c}| = 1$
b. $\theta_{ab} = 78.5°$, $\theta_{ac} = 122.3°$, $\theta_{bc} = 56.1°$

P1.15 a. If $\vec{a}' = \mathbf{U}\vec{a}$, then $\vec{a}' \cdot \vec{b}' = \vec{a}'^T\vec{b}' = \vec{a}^T\mathbf{U}^T\mathbf{U}\vec{b} = \vec{a}^T\vec{b} = \vec{a} \cdot \vec{b}$

P1.19 a. $\hat{e}\hat{e}^T\vec{a} = \frac{4}{3}[1 \quad -1 \quad 2]^T$, which is proportional to $\hat{e}$.
b. $a_\perp = (\mathbf{I} - \hat{e}\vec{e}^T)\vec{a} = \frac{1}{3}[8 \;\; \text{-2} \;\; \text{-5}]^T$

P1.25 a. $P = 230.9$ b. $T = 4942.7$

P1.29 a. $\vec{k} \cdot \vec{\delta} = \vec{k}^T\vec{\delta} = \vec{m}^T\mathbf{B}^T\mathbf{A}\vec{n} = \vec{m}^T\vec{n} = p$, where p is a positive integer. And, because $e^{2\pi i p} = +1$, $e^{2\pi i \vec{k} \cdot (\vec{r} + \vec{\delta})} = e^{2\pi i \vec{k} \cdot \vec{r}} e^{2\pi i \vec{k} \cdot \vec{\delta}} = e^{2\pi i \vec{k} \cdot \vec{r}}$.

CHAPTER 2

P2.1 a. $\mathbf{I}_2$ b. $\mathbf{I}_3$ c. Both MATLAB and Gauss-Jordan elimination yield $\begin{pmatrix} 1 & 1 & 1 \\ 0 & 0 & 0 \\ 0 & 0 & 0 \end{pmatrix}$

d. $\begin{pmatrix} 1 & 0 & 2 \\ 0 & 1 & -3 \\ 0 & 0 & 0 \end{pmatrix}$ e. $\mathbf{I}_3$ f. $\mathbf{I}_4$

P2.3 a. If $\alpha = 0$, an infinite number of solutions are possible, none otherwise.   b. If $\alpha = 1$, an infinite number of solutions are possible, none otherwise.

P2.5 a. $\begin{pmatrix} 2 & -1 \\ -3 & 2 \end{pmatrix}$ c. $\frac{1}{3}\begin{pmatrix} 2 & 1 & 0 & 0 \\ 1 & 2 & 0 & 0 \\ 0 & 0 & 2 & 1 \\ 0 & 0 & 1 & 2 \end{pmatrix}$ e. $\frac{1}{2}\begin{pmatrix} 0 & 0 & 0 & 0 & 1 \\ 0 & 1 & 0 & 0 & 0 \\ 0 & 0 & 0 & 1 & 0 \\ 1 & 0 & 0 & 0 & 0 \\ 0 & 0 & 1 & 0 & 0 \end{pmatrix}$

P2.7 `expm(A)*expm(-A) = eye(3)`

P2.11 a. $f_1 = \dfrac{|x_1^T\mathbf{A}^T\mathbf{A}x_1|}{x_1^Tx_1} = 25 + 2\epsilon + \epsilon^2$

$f_2 = \dfrac{|x_2^T(\mathbf{A}^{-1})^T\mathbf{A}^{-1}x_2|}{x_2^Tx_2} = (25 + 2\epsilon + \epsilon^2)/(16\epsilon^2)$

$\text{cond}(\mathbf{A}) \approx \sqrt{f_1 f_2}$

b.

ϵ	$\sqrt{f_1 f_2}$	cond(A)
$\frac{1}{2}$	13.025	13.048
$\frac{1}{4}$	25.487	25.523
$\frac{1}{8}$	50.469	50.512
$\frac{1}{16}$	100.459	100.506
$\frac{1}{32}$	200.455	200.503
$\frac{1}{64}$	400.452	400.501

P2.17 a. $f(z) = (26z^2 + 14z + 5)/(1 + z^2)$ has a maximum at $z_{\max} = 3.3708$, given by $f(z_{\max}) = 28.1186$. $\|A\| = \sqrt{28.1186} = 5.3028$, which is the value returned by `norm(A)`.

CHAPTER 3

P3.1 a. $\lambda^2 = \hat{e}^T \mathbf{R}\mathbf{R}^T \hat{e}/\hat{e}^T \hat{e} = 1$,   (ii) $1 = |\mathbf{R}^T \mathbf{R}| = |\mathbf{R}^T||\mathbf{R}| = |\mathbf{R}|^2$

P3.3 a. $\hat{e}_0 = \frac{1}{\sqrt{6}} \begin{pmatrix} 1 \\ 1 \\ -2 \end{pmatrix}$ b. $\mathbf{P}_0 = \frac{1}{6} \begin{pmatrix} 1 & 1 & -2 \\ 1 & 1 & -2 \\ -2 & -2 & 4 \end{pmatrix}$, yes.

P3.5 a. $\lambda_i = \qquad 0 \qquad\qquad 1 \qquad\qquad 2$

$$\hat{e}_i = \frac{1}{\sqrt{2}} \begin{pmatrix} 0 \\ 1 \\ -1 \end{pmatrix} \quad \begin{pmatrix} 1 \\ 0 \\ 0 \end{pmatrix} \quad \frac{1}{\sqrt{2}} \begin{pmatrix} 0 \\ 1 \\ 1 \end{pmatrix}, \quad \text{orthogonal.}$$

c. $\lambda_i = \qquad 1 \qquad\qquad 2 \qquad\qquad 3$

$$\hat{e}_i = \begin{pmatrix} 1 \\ 0 \\ 0 \end{pmatrix} \quad \frac{1}{\sqrt{2}} \begin{pmatrix} 1 \\ 1 \\ 0 \end{pmatrix} \quad \frac{1}{\sqrt{29}} \begin{pmatrix} 3 \\ 4 \\ 2 \end{pmatrix}, \quad \text{not orthogonal.}$$

P3.9 a. $x_\parallel = (\hat{e}\hat{e}^T)\vec{x}$ and $x_\perp = (\mathbf{I} - \hat{e}\hat{e}^T)\vec{x}$ b. $\vec{x}_{\text{ref}} = \vec{x}_\parallel - \vec{x}_\perp = (2\hat{e}\hat{e}^T - \mathbf{I})\vec{x} = \mathbf{R}\vec{x}$, $\mathbf{R}^2 = 4\hat{e}\hat{e}^T \hat{e}\hat{e}^T - 2\hat{e}\hat{e}^T + \mathbf{I} \neq \mathbf{R}$ c. $\mathbf{R}\vec{x}_\parallel = \vec{x}_\parallel$ and $\mathbf{R}\vec{x}_\perp = -\vec{x}_\perp$.

P3.13 The (x, y) motions are *decoupled*. Because $\sum \mathbf{P}_i = \begin{pmatrix} 1 & 0 \\ 0 & 2 \end{pmatrix}$, the differential equation becomes

$$\begin{pmatrix} \ddot{x} \\ \ddot{y} \end{pmatrix} = -\frac{k}{m} \begin{pmatrix} 1 & 0 \\ 0 & 2 \end{pmatrix} \begin{pmatrix} x \\ y \end{pmatrix}$$

and the normal mode solutions appropriate to the given initial conditions are

$$x(t) = 0.1 \cos\left(\sqrt{\frac{k}{m}}t\right); \quad y(t) = 0$$

$$y(t) = 0.2 \cos\left(\sqrt{2\frac{k}{m}}t\right); \quad x(t) = 0$$

CHAPTER 4

P4.1 a. $f(x) = 32(x - \frac{1}{2})^5$, $f_{\text{series}}(0) = -1$

c. $f(x) \approx \cos 1[1 - 2x^2 + \frac{2}{3}x^4 + \cdots] - 2\sin 1[x - \frac{2}{3}x^3 + \frac{2}{15}x^5 + \cdots]$

e. $f(x) \approx \sqrt{2}[1 - (x - \frac{\pi}{4}) + \frac{3}{2}(x - \frac{\pi}{4})^2 - \frac{11}{6}(x - \frac{\pi}{4})^3 + \frac{57}{24}(x - \frac{\pi}{4})^4 + \cdots]$

P4.4

i	$s_i^{(0)}$	$s_i^{(1)}$	$s_i^{(2)}$	$s_i^{(3)}$	$s_i^{(4)}$	$s_i^{(5)}$
1	1.0					
2	6.1					
3	19.1	−2.3				
4	41.2	−12.5				
5	69.4	−61.3	0.4			
6	98.2	−1368.2	−10.6			
7	122.6	261.1	−643.0	0.6		
8	140.4	188.2	191.3	−283.3		
9	151.8	171.7	166.9	167.6	−109.1	
10	158.2	166.6	164.3	164.0	164.0	
11	161.5	164.9	164.0	164.0	164.0	164.0

$e^{5.1} = 164.02\ldots.$

P4.8

	Chic.	Los A.	Mont.	Lond.	RioJ.	Melb.	Vlad.	J-brg
Chicago	0	1697	745	3951	5288	6481	2274	6031
Los Ang.		0	2421	5369	6317	5211	815	7232
Montreal			0	3246	5090	7171	2901	5705
London				0	5765	10411	5468	5606
Rio de J					0	5923	7133	993
Melbrn						0	5319	6528
Vldvstk							0	8041

CHAPTER 5

P5.1 a. $f(x) = xg(x) = x(x^3 + 6x^2 + 3x - 10)$ and $g(x)$ has one positive real root and either zero or two negative real roots. Also, $g(-1) = 8$, $g(0) = -10$, $g(1) = 0$. So, by long division, we obtain $f(x) = x(x - 1)(x^2 + 7x + 10) = x(x - 1)(x + 2)(x + 5)$

c. $h(x)$ has either zero, two, or four real positive roots. Because $h(1) = h'(1) = 0$, we can write $h(x) = (x - 1)^2 g(x)$. Then, by long division, we obtain $h(x) = (x - 1)^2(x + 1)^2$

P5.3 $f(x) = x^3 - 2x - 5, 1 < x_r < 3$

i	x_{lft}	x_{mid}	x_{rt}	$f(x_{\text{lft}})$	$f(x_{\text{mid}})$	$f(x_{\text{rt}})$
1	1	2	3	-5	-1	16
2	2	2.5	3	-1	5.625	16
3	2	2.25	2.5	-1	$1.8906\ldots$	5.625
4	2	2.125	2.25	-1	$0.3457\ldots$	1.8906
5	2	2.0625	2.125	-1	$-0.3513\ldots$	$0.3457\ldots$
6	2.0625	$2.0938\ldots$	2.125	$-0.3513\ldots$	$-0.0089\ldots$	$0.3457\ldots$

P5.5 The improvement is $\Delta x = -m(f/f')$, and the initial guess is $x_0 = 2$; first with $m = 2$, and then with $m = 4$.

	$m = 2$	
i	x	Δx
1	2.0	$0.57\ldots$
2	$2.57\ldots$	$0.22\ldots$
3	$2.79\ldots$	$0.104\ldots$
4	$2.90\ldots$	$0.050\ldots$
5	$2.950\ldots$	$0.0247\ldots$
6	$2.975\ldots$	$0.0122\ldots$
7	$2.987\ldots$	$0.0061\ldots$
8	$2.9939\ldots$	$0.0030\ldots$
9	$2.9969\ldots$	$0.0015\ldots$
10	$2.9984\ldots$	$0.00076\ldots$

	$m = 4$	
i	x	Δx
1	2	$1.14\ldots$
2	$3.14\ldots$	$-0.14\ldots$
3	$3.0021\ldots$	$-0.0021\ldots$
4	$3.00000045\ldots$	$-0.000000014\ldots$
5	$3.00000044\ldots$	$-0.000000012\ldots$

P5.9

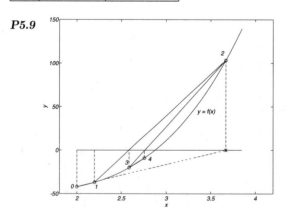

Four steps of the secant method applied to $x^4 - 12x - 34$.

CHAPTER 6

P6.1 a. $e^{-1} = 0.367879\ldots$ c. $0.340737\ldots$

P6.3 The derivative of the polynomial has roots at

$$r_i = \begin{bmatrix} 3.451\ldots \\ -1.4512\ldots \\ 1 \\ 0.84193\ldots \\ 1.1581\ldots \end{bmatrix}$$

The second derivative is positive at the first three values and negative at the last two.

P6.5

```
function r = Fminmin(fnc,xa,xb,ya,yb,tol)
%Fminmin finds a minimum of the function fnc by
%  successively finding the minima in perpendicular
%  directions.  A minimum must be known to be within
%  the rectangular region (xa,xb,ya,yb).  Each of
%  these quantities must be a scalar.  The convergence
%  tolerance is tol, and the coordinates, (x,y),
%  of the minimum are returned in the two element
%  vector r.  The maximum number of steps is 1000.
%****************************************************
 global tx ty fname
 fname = fnc
 x0 = .5*(xa+xb); y0 = .5*(ya+yb);
 Dx = abs(yb-ya)+abs(xb-xa);
                    count = 0;
 while Dx > tol
    tx = x0; ty = y0;
    ty = fmin('Fy',ya,yb);
    tx = fmin('Fx',xa,xb);
    Dx = abs(x0-tx)+abs(y0-ty);
    x0 = tx; y0 = ty
    count = count + 1;
 end
 if count>1000; error('did not converge');end
 r = [x0 y0]';
```

CHAPTER 7

P7.1

```
function T = TRAP(a,b,k,fnc)
%TRAP will evaluate the integral of fnc(x) over a<x<b
%   using the trapezoidal rule.  The number of panels is
%   2^k and the entire set of trapezoidal rule approx-
%   imations for m = 1:k+1 is returned in the vector T.
%=========================================================
 Nmax = 2^k;
 x = zeros(Nmax,1);       T  = zeros(k+1,1);   x = [a b]';
 f = sum(feval(fnc,x)); dx = (b-a);
```

```
T(1) = 0.5*dx*f;
for m = 1:k
    dx = dx/2;
    n  = 2^m;
    x  = a + dx*[1:2:n-1]';
    T(m+1) = 0.5*T(m) + dx*sum(feval(fnc,x));
end
```

P7.4

```
function r = ROMB(T)
%ROMB will produce a triangular table of values for the
%    integral of a function assuming that the trapezoidal
%    approximations for the integral are contained in the
%    vector T.  The values in T are for intervals succes-
%    sively reduced by half.  The values returned in r
%    are the diagonal elements of this table.
%=========================================================
 n = length(T);    R = zeros(n,n);
 R(:,1) = T;
 for k = 2:n
    i = k:n;
    R(i,k) = R(i,k-1)+(R(i,k-1)-R(i-1,k-1))/(4^(k-1)-1);
 end
 r = diag(R);
```

The fractional error in the computed values for integral No. 1 are then

k	Trapezoidal	Romberg
0	$0.30\ldots$	$0.30\ldots$
1	$0.078\ldots$	$0.0028\ldots$
2	$0.020\ldots$	$0.0000066\ldots$
3	$0.0049\ldots$	$0.0000000035\ldots$

P7.7 Let $t = -\ln x$. If the t integral is partitioned into four intervals, $[0 \leq t \leq \gamma]$, $[\gamma \leq t \leq 3\gamma]$, $[3\gamma \leq t \leq 9\gamma]$, $[9\gamma \leq t \leq \infty]$, and if $I_1 \gg I_2 \gg I_3$, we can assume that I_4 is not significant. With $\gamma = 4$, using a $k =$ seventh-order Romberg integration, we obtain $I_1 = -0.908421806\ldots$, $I_2 = -0.091498320\ldots$, and $I_3 = -0.000079875\ldots$, so that $I = -.999999999\ldots$.

P7.8

```
function g = GAUSS10(a,b,fnc)
%GAUSS10(a,b,fnc) will return the value of the integral
%   of the function, fnc, over the range a<x<b using a
%   10 point Gauss quadrature algorithm.
%=========================================================
```

```
x = [0.1488743390;0.4333953941;0.6974095683;...
     0.8650633667;0.9739065285];
w = [0.2955242247;0.2692667193;0.2190863625;...
     0.1494513492;0.0666713443];
t = .5*(b+a)+.5*(b-a)*[-flipud(x);x];
W = [flipud(w);w];
g = sum(W.*feval(fnc,t))*(b-a)/2;
```

Using this function on the three integrals yields

No.	Gauss10	Fractional error
1	0.88611...	0.00013
2	−3.42133...	0.00023
3	2.29884...	0.268

CHAPTER 8

P8.3   a. $\hat{n}(0,0) = -\hat{k}$ c. $\hat{n}(1,1) = \frac{1}{2}(\hat{\imath} + \hat{\jmath} - \sqrt{2}\hat{k})$

P8.5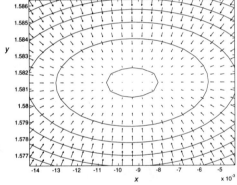

Contour lines and gradient vectors near a maximum of **peaks**.

P8.8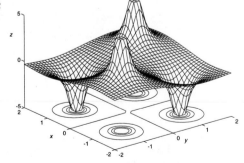

Electric potential from a quadrupole distribution.

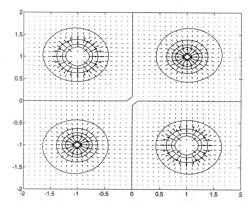

Contour lines and gradient vectors for the quadrupole distribution.

CHAPTER 9

P9.1 a. $x^2 + y^2 = n\pi$ b. $x + y = (n + \frac{1}{2})\pi$ c. $x = -y$ or $x = -1$

P9.3 a.$(\frac{\pi}{4}, \frac{\pi}{4})$ b. $(\pm 1.9861, \pm 1.4353)$.

P9.5 Ground state. $(x, y) = (2.3923\ldots, 2.5570\ldots)$, $E = -29$ MeV
First excited state. $(x, y) = (1.113\ldots, 3.315\ldots)$, $E = -17.4$ MeV

CHAPTER 10

P10.1 a. $(x_{\min}, y_{\min}) = (\frac{1}{4}, \frac{5}{4})$ b. $(x_{\min}, y_{\min}) = (\frac{5}{6}, 0)$

P10.2

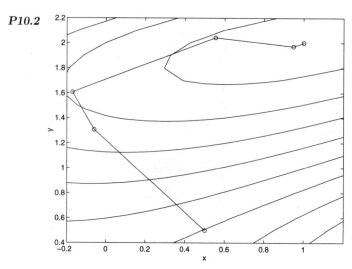

A contour plot of $z = 3x^2 - 2xy^2 + y^4 + 2x - 6y^2$ with the path followed by function **QNEWTN**.

P10.5 The derivative of $|f|$ is discontinuous at the very point of the minimum, so procedures based on gradients will likely fail. However, simplex methods, such as those used by `fmins`, will usually find the root successfully.

P10.7 For the function $f(x, y, z) = (x - 1)^2 + (y - 2)^2 + (z - 3)^2$, `fmins` requires about 4700 flops, whereas if one of the exponents is changed to 4, 6300 flops are needed. `QNEWTN` needed 1400 and 10,300 flops, respectively.

P10.10 $(Z, N) = (27.4, 33.8)$

P10.11 The optimum parameters are $\alpha = 0.2660\ldots$, $\beta = 0.3872\ldots$, which give a value of $\lambda_{\min} > 1.814$, and an estimate of the corresponding eigenvector of

$$\hat{e}_{\min} \approx \begin{pmatrix} 0.700\ldots \\ 0.619\ldots \\ -.097\ldots \\ -.342\ldots \end{pmatrix}$$

The actual minimum eigenvalue is 1.586, which has a corresponding eigenvector

$$\hat{e}_{\min} = \begin{pmatrix} 0.271\ldots \\ 0.653\ldots \\ 0.653\ldots \\ 0.271\ldots \end{pmatrix}$$

Thus, although the estimate of the eigenvalue is reasonably accurate, the estimate of the eigenvector is quite poor.

CHAPTER 11

P11.1 a. The MATLAB code for this integral is

```
I = quad8('G',1,2);
%=====================================
function f = G(x)
  x = x(:);      n = length(x);
  f = zeros(size(x));
  for i = 1:n
     f(i)= quad8('Arg',1,x(i)^2,[ ],[ ],x(i));
  end
%=====================================
function f = Arg(y,tx)
    f = y.^2 + tx^2;
  end
```

and the value of the integral is $9.58095\ldots$.

b. $(x_c, y_c) = \left(\frac{32}{\sqrt{55}}, 3\sqrt{\frac{39}{55}} \right)$, and the common area is $A = 53.82779\ldots$.

c. $\int_1^2 dy \int_y^{y^2} \ln(x/y)$. The integral has a value of $0.237281\ldots$.

P11.4 a. $128/3$ b. $\frac{1}{2}\pi^2 r^4$

P11.5 The MATLAB code to evaluate the triple integral of Problem 11.4(a) is:

```
I = quad('H',0,8);
%***************************************
function f = H(z)
% H(z) is the integral over x and y, of Fyz
%::::::::::::::::::::::::::::::::::::::::::
  z = z(:);   n = length(z);
  f = zeros(size(z));
  for i = 1:n
     YLO = 0;   YHI = z(i)/2;
     f(i) = quad('Fyz',YLO,YHI),[~],[~],z(i));
  end
%***************************************
function f = Fyz(y,tz)
% Fyz is the integral of the actual, arg,
%     integrand over x.
%::::::::::::::::::::::::::::::::::::::::::
  y = y(:);   n = length(y);
  f = zeros(size(y));
  for i = 1:n
     XLO=0;   XHI = tz - 2*Y(i);
     f(i) = quad('arg',XLO,XHI,[~],[~],y(i),tz);
  end
%***************************************
function f = arg(x,ty,tz)
%  arg is the integrand expressed as a function
%  of vector x, and scalars ty and tz.
%::::::::::::::::::::::::::::::::::::::::::
  x=x(:);
  f = ones(size(x));
%***************************************
```

b. The mass is 0.6911; the integral required approximately one-half million flops.

c. $I = \displaystyle\int_0^8 dz \int_0^{z/2} \mathrm{erf}(2z - 4y) e^{-(y^2+z^2)/4} dy.$

CHAPTER 12

P12.1 The average of several runs yields:

n	σ_n	$\sigma_n/\sqrt{n}$
100	$2.99\ldots$	$0.299\ldots$
1000	$9.62\ldots$	$0.304\ldots$
10,000	$30.82\ldots$	$0.308\ldots$

P12.3 For the xy plot, the points will be evenly distributed over the ellipse. However, for the $\rho - \theta$ plot, the points will concentrate near $\rho \approx 0$. This behavior illustrates that, for a Monte Carlo integration in the coordinates ρ and θ, the integrand must be multiplied by the Jacobian; in this case, ρ. See also file `prblzp3` on the applications disk.

P12.5 `V = length(find(sum(rand(3,N).^2)<1))/N;`. The exact answer is $V_0 = 2 - \sqrt{2}$. With $N = 10^3$, this Monte Carlo integration yields $\approx 0.52\ldots$.

CHAPTER 13

P13.1 Chebyshev's theorem states that at least $\left(1 - \frac{1}{k^2}\right) N$ of the points will be within k standard deviations of the mean. Typical results for the Gaussian distribution are:

k	Fraction within $k\sigma$	$\left(1 - \frac{1}{k^2}\right)$
1.5	0.857	0.556
2.0	0.950	0.750
2.5	0.983	0.840
3.0	0.998	0.889

Because the standard deviation is larger for the flat distribution, the theorem is easily satisfied.

P13.3 Both quantities will fluctuate between zero and one for large N.

P13.5 The distribution has narrow peaks at $x = 6, 12$.

P13.7 b. $\sigma^2 = \frac{1}{100}\sigma_i^2$, where σ_i is the standard deviation of 900 random numbers, $\sigma_i = \frac{1}{2\sqrt{3}}\sqrt{1 + \frac{2}{900}}$. Thus, $\sigma \approx 0.0289\ldots$.

P13.8 c.

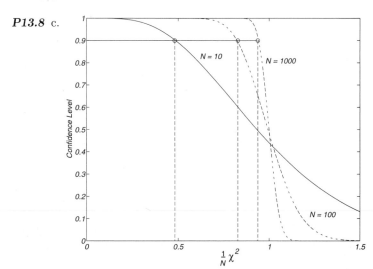

The confidence level, Q, plotted versus χ^2/N for $N = 10$, 100, 1000 points.

P13.9 b. 1 - gammainc(6/2,2/2) = 0.05, $N = (\chi^2 - 2)/33/\delta p^2 = 3812$. The reduction of the number of degrees of freedom from 38 to 2 causes a large increase in the confidence function.

P13.11 b.

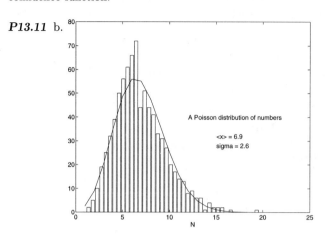

An approximate Poisson distribution of random numbers.

$\overline{x} \approx 7$, $\sigma \approx 2.6$, skewness ≈ 0.8

CHAPTER 14

P14.1

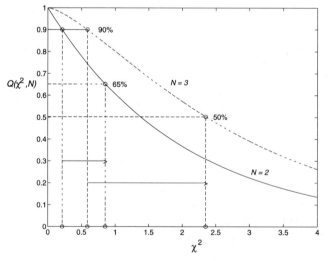

The confidence level versus χ^2 for a linear ($N = 2$) and for a quadratic ($N = 3$) fit.

a. For the linear fit, $\chi^2(90\%) \approx 0.213$. If the errors, Δy, are decreased by half, χ^2 will increase by a factor of 4. And from the figure, $4\chi^2 = 0.85$ corresponds to a new confidence level of only 65%. b. The confidence in the quadratic fit would drop to 50%.

P14.3

n	c
5	23.3
6	20.0
7	16.5
8	12.8
9	9.05
10	5.15
11	1.15
12	−2.95

P14.8 The results are

ν	χ^2	Q
2	25.97	0.00
3	5.17	0.16
4	4.71	0.32
5	3.59	0.61
6	2.58	0.86
7	2.47	0.93

indicating that $\nu = 4$ or $\nu = 5$ are satisfactory fits.

CHAPTER 15

P15.1 Using the function $y(x) = ae^{-x^2/b^2}$:
a. $a = 3.053\ldots, b = 1.990\ldots$
b. $a = 2.998\ldots, b = 2.013\ldots$
c. $a = 2.998\ldots, b = 2.013\ldots$
These numbers will vary in each run because the scatter in the data is random.

P15.2 The MATLAB code to compare the full five-dimensional minimization with the partial linearization is:

```
%Prb15p2 Script File
%Prb15p2 is a partial linearization of Example 15-2.
%  The original fitting function contains five parameters,
%  a(1), ..., a(5).  Using a=[100 1 30 2 .5]', the function
%  is computed over the interval 0<Xdata<4, and random noise
%  is added to simulate noise in the data.  An estimate for
%  the parameters is contained in a0.
%  The first calculation is a full minimization of Chi^2 to
%  find the best fit parameters which are displayed along
%  with the number of kilo-flops.
%  However, the function is linear in a(1) and a(3).  Using
%  estimated values for the remaining a's, these two elements
%  are computed using back-slash division.  The 5-parameter
%  Chi^2 function (HddnPeak) is then rewritten as a function
%  (HddnPkB) of only the three non-linear parameters,
%  (d=[a(2) a(4) a(5)]'), and a minimum is found.  These five
```

```
%  parameters are then displayed along with the number of
%  flops/1.e3.  This code calls HddnPkB.
%================================================================
% Compute the pseudo-data
%
 global Xdata Ydata dy c
 dy = 1.35;
 a = [100 1.0 30 2.0 .5 ]';              %actual values
 disp(a')
 Xdata = [0:.1:4]';   n = length(Xdata);
 Ydata = a(1)*exp(-a(2)*Xdata) + ...
         a(3)*((Xdata-a(4)).^2 + a(5)).^(-1);
 Ydata = Ydata + 5*(rand(n,1)-.5);
 a0 = [107 .63 20 1.9 .75]';             %estimated values
 disp(a0')
%--------------------------------------------------------------
%  The full 5-d minimization
%
 flops(0);    q = fmins('HddnPeak',a0);
 disp([q' flops/1.e3])
%--------------------------------------------------------------
%  Partial Linearization [linear in a(1), a(3)]
%
 flops(0)
 c  = [a0(1) a0(3)]';  d = [a0(2) a0(4) a0(5)]';
 d  = fmins('HddnPkB',d);
 f1 = exp(-d(1)*Xdata);
 f2 =((Xdata-d(2)).^2+d(3)).^(-1);
 c = [f1 f2]\Ydata;
 a(1) = c(1); a(3) = c(2);
 a(2) = d(1); a(4) = d(2); a(5) = d(3);
 disp([a' flops/1.e3])
```

```
function E = HddnPkB(d)
%HddnPkB is used in Problem 15.2, and it is called by the
% code Prb15p2.  It is merely the Hidden Peak function
% rewritten as a function of only the "non-linear'' para-
% meters.  The vector d contains the "non-linear" para-
% meters of HddnPeak(a).  That is d(1) = a(2), d(2) = a(4),
% d(3) = a(5).  The two linear parameters are c(1) = a(1)
% and c(2) = a(2).
%--------------------------------------------------------------
  global Xdata Ydata c
  dy = 1.35;
  x=Xdata;y=Ydata;
  a(1)=c(1);    a(3)=c(2);
  a(2)=d(1);    a(4)=d(2); a(5)=d(3);
  Y=a(1)*exp(-a(2)*x)+a(3)*((x-a(4)).^(2)+a(5)).^(-1);
  E=sum(((y-Y).^2)./dy.^2);
```

The results of the program are:

	a_1	a_2	a_3	a_4	a_5	flops
Model values	100.0	1.000	30.00	2.000	0.500	
Initial guess	107.0	0.630	20.00	1.900	0.750	
Full minimization	101.2	0.977	28.15	2.005	0.468	279k
Partial linearization	103.0	0.901	20.59	2.032	0.340	68k

P15.3 b. **H** would be singular.

c. Test `log(rcond(H))/10/eps < 1`.

CHAPTER 16

P16.1

n	$|M(n)|$
5	$-1.13e-06$
6	$-1.13e-09$
7	$2.73e-13$
8	$1.56e-17$
9	$-2.10e-22$
10	$-6.66e-28$

P16.3 a. With $x_i = x_1 + (i-1)\Delta x$ and $x \equiv \kappa \Delta x + x_1$, we obtain

$$
\begin{aligned}
P_{i,i+1} &= y_i + [\kappa - (i-1)](y_{i+1} - y_i) \\
P_{i,i+1,i+2} &= P_{i,i+1} + \frac{1}{2}[\kappa - (i-1)](P_{i+1,i+2} - P_{i,i+1}) \\
&\vdots \\
P_{i,\dots,i+m} &= P_{i,\dots,i+m-1} + \frac{1}{m}[\kappa - (i-1)](P_{i+1,i+m} - P_{i,i+m-1})
\end{aligned}
$$

b. The Neville interpolation of the function at the point $x = \frac{3}{4}$ yields the following:

x_i	y_i	$P_{i,i+1}$	$P_{i,\ldots,i+2}$	$P_{i,\ldots,i+3}$	$P_{i,\ldots,i+4}$
0	1				
		$\frac{49}{16}$			
$\frac{1}{2}$	$\frac{19}{8}$		$\frac{125}{32}$		
		$\frac{67}{16}$		$\frac{247}{64}$	
1	6		$\frac{61}{32}$		$\frac{247}{64}$
		$\frac{43}{16}$		$\frac{247}{64}$	
$\frac{3}{2}$	$\frac{101}{8}$		$\frac{131}{32}$		
		$-\frac{47}{16}$			
2	23				

c. Three interpolation steps involving any four points will exactly duplicate a cubic polynomial.

P16.5 The values obtained by fitting Padé polynomials are:

x_i	y_i	$F_1(0)$	$F_2(0)$	$F_3(0)$
11	−5.2174	0.3204	0.3660	1.0000
9	−4.2105	0.3368	0.3977	1.0000
7	−3.2000	0.3636	0.4560	1.0000
5	−2.1818	0.4156	0.6154	
3	−1.1429	0.5714		
1	0			

whereas the Bulirsch-Stoer algorithm gives:

N	1	2	3	4
$y(0)$	1.1429	0.6154	0.6465	1.0000

P16.7

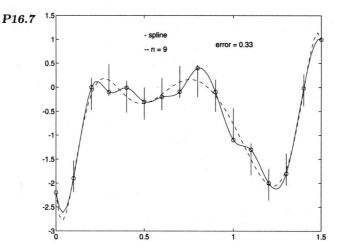

A cubic spline is interpolated through the circle points, and a 9th-degree polynomial is also fitted through the points. The maximum absolute deviation is taken as the error.

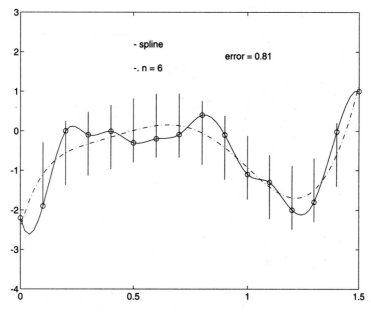

The same calculation using a 6th-degree polynomial.

CHAPTER 17

P17.1 The MATLAB code is:

```
function S = FourierX(fnc,L,n)
%FourierX(fnc,L,n) computes the first n coefficients of
%   the Fourier expansion of the function 'fnc' and plots
```

```
%    the function and the series approximation for
%    -L < x < L
%==========================================================
dx = L/20;   x = [-L:dx:L]'; tol = 1.e-3; TRACE=0;
y = feval(fnc,x);
plot(x,y); hold on
a0 = mean(y);
for k = 1:n
    A(k) = quad8('CosIntg',-L,L,tol,TRACE,fnc,k)/L;
    B(k) = quad8('SineIntg'  ,-L,L,tol,TRACE,fnc,k)/L;
end
A=A(:); B=B(:);
S = a0 + (cos(x*[1:n]*pi/L)*A + sin(x*[1:n]*pi/L)*B);
plot(x,S,'-.')
title('Fourier Expansion')
xlabel('x');ylabel('y')
hold off
%****************************************************
function f = SineIntg(x,fnc,n)
  f = sin(n*pi*x).*feval(fnc,x);
%****************************************************
function f = CosIntg(x,fnc,n)
  f = cos(n*pi*x).*feval(fnc,x);
```

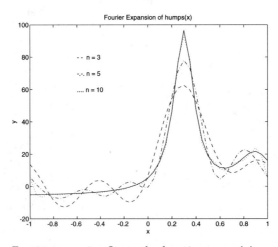

Fourier expansion fits to the function humps(x).

P17.7 $\mathcal{F}(\delta(x-a)) = \dfrac{1}{\sqrt{2\pi}}e^{-i\omega a}$

P17.11

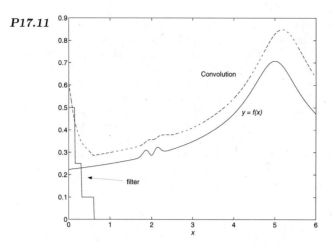

Convolution of a function with two small features near $x = 2$ that are "washed out" by the filter.

CHAPTER 18

P18.1 a. $y = \sqrt{1 + cx}$ c. $y = \dfrac{x}{\ln x + C}$ e. $y = \sqrt{\dfrac{\pi}{2}} e^{x^2} [\operatorname{erf}(x) + C]$

P18.3 $f(x, y) = 1 - y, \qquad y(0) = 0$

$$y_0 = 0$$

$$y_1 = 0 + \int_0^x (1)dx \qquad\qquad = x$$

$$y_2 = 0 + \int_0^x (1 - x)dx \qquad\qquad = x - \frac{x^2}{2}$$

$$y_3 = 0 + \int_0^x \left(1 - x + \frac{x^2}{2}\right) dx = x - \frac{x^2}{2} + \frac{x^3}{3!}$$

$$\vdots \qquad\qquad \vdots \qquad\qquad\qquad \vdots$$

$$y_n = x - \frac{x}{2!} + \frac{x^3}{3!} + \cdots + (-1)^{n+1}\frac{x^n}{n!}$$

$$y = x - e^{-x}$$

CHAPTER 19

P19.1 The error is found to be $\Delta y = y_{\text{exact}} - y_{\text{calc}} \approx 0.95\Delta x$.

P19.2

```
function [x,y] = EulrAdpt(a,b,y0,err,fnc)
%EulrAdpt solves the first order differential equation
%   y' = fnc(x,y) by a simple stepping procedure using the
%   first two terms of a Taylor expansion of the function
```

```
%   y(x).  The function contained in the string variable
%   fnc must expect SCALAR input for both x and y and re-
%   turns a SCALAR.  The step size is adapted to maintain
%   a percentage error in each step of err.  The program
%   is not suitable for realistic solutions of a differen-
%   tial equation. Other input values are the range of x's
%   [a<=x<=b] and the initial value of y, [y0].  Row vec-
%   tors of computed values are returned in x,y.  The use is
%        [x,y] = EulrAdpt(a,b,y0,err,fnc).
%=========================================================
  X = a; dx = 10*(b-a)*err;
  x(1) = a;  y(1)=y0;   i = 1;
  while x < b
    X = x(i) + dx;      R = 0.001;
    while R <= 1
        DY = feval(fnc,X,y(i))*dx;
        Y1 = y(i)+DY;
        Y1a= y(i)+DY/2;
        Y1b= Y1a + feval(fnc,X+dx/2,Y1a)*dx/2;
        dY = abs(Y1-Y1b);
        R = sqrt(err*(1+abs(Y1b/dY)));
        dx = 0.90*R*dx;
    end
    i = i + 1; x(i) = X;   y(i) = Y1b;
  end
  x = x(:); y = y(:);
```

To maintain a fractional accuracy of 10^{-4} using Euler's method requires on the order of 24,000 flops. Using an adaptive step size will require approximately 6000 flops to maintain the same accuracy.

P19.3 The exact solution is $y(x) = \frac{1}{3}(x^2 - 1/x)$, so that $y(2) = 7/6$. A comparison of the accuracy of the two methods for a variety of step sizes is then:

N	Euler $y(2)$	Error	Modified midpoint $y(2)$	Error
10	1.098...	0.0693...	1.1675	0.000833...
100	1.1560...	0.00669...	1.166675	0.00000833...
1000	1.1660...	0.0006669...	1.16666675	0.0000000833...

P19.6

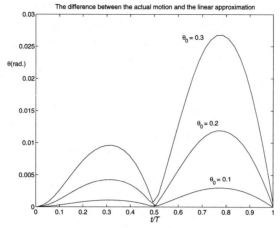

The absolute difference between the solution of the pendulum equation and the linear approximation divided by θ_0 for three initial displacements θ_0.

CHAPTER 20

P20.1 a. Nonunique solution, $y(t) = A\sin 2t$ for any A c. Nonunique solution, $y = A\sin 2t - \cos 2t + 1$,   e. Unique, $y(t) = \frac{\sqrt{3}}{2}\sin 2t + \frac{1}{4}$

P20.3 $(v_{x0}, v_{y0}) = (201.3567\ldots, 197.6620\ldots)$

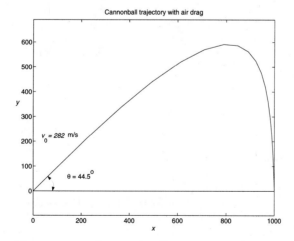

The trajectory of a cannonball including the effects of air drag. The trajectory is such that the cannonball hits the target at $x = 1000$ m in 25 seconds.

P20.7

$$
\begin{aligned}
y_1 &= y & y_1' &= y_2 & y_1(0) &= 0 \\
y_2 &= y' & y_2' &= 5\sin\frac{x}{2} - \frac{y_1}{9} & y_2(0) &= \beta
\end{aligned}
$$

Next, a function, $f(\beta)$, is constructed that integrates these equations from $x = 0$ to $x = \pi$ and returns the value of $f(\beta) = y(\pi)$. The root of this equation corresponds to the missing condition on y' at $x = \pi$. The exact value is $\beta = 4\sqrt{3}$.

P20.9   a. The code

```
r = Secant(1,.05,'Prb20p9a',1.e-4)
function f = Prb20p9a(z)
  y0 = [0 z]';
  [t,y] = ode45('Pb20p9aD',0,pi/2,y0);
  n = length(t);
  f = y(n,1);
function f = Pb20p9aD(t,y)
  f(1) = y(2);
  f(2) = 2*exp(-t) - y(1);
  f = f(:);
```

returns the value $y'(0) = -1.2078796\ldots$. The exact result is $-(1 + e^{-\pi/2})$.

b. The finite difference method applied to this equation yields

$$(\mathbf{A} + h^2\mathbf{I})\vec{y} = 2h^2 e^{-\vec{n}h}$$

with $\mathbf{A}$ as defined in Problem 20.6.

c. The finite element method applied to this equation yields

$$\left(\mathbf{A} + \frac{h^2}{6}\mathbf{B}\right)\vec{y} = 2h^2 e^{-\vec{n}h}$$

with $\mathbf{A}$ and $\mathbf{B}$ as defined in Problem 20.6.

d. For this equation, the finite difference method is approximately six times more accurate than the finite element method across the entire range of x.

P20.11 The missing initial condition is $y'(0) = -2.3831\ldots$.

P20.13 The code for this problem is:

```
function dy = P20p13
n = 1000;
m = n-1; I = ones(n-2,1); h = pi/n;
A = sparse(eye(m)*(-2+h^2/4)+diag(I,1)+diag(I,-1));
b = sparse( h^4*[1:m]'.^2);
y = A\b; x = [h:h:pi-h]';
Y = 32*(sin(x/2)+cos(x/2)-1)+4*(x.^2-pi^2*sin(x/2));
y = full(y);
dy = mean(abs((Y-y)./Y));
```

The mean fractional error is computed to be 3.1421×10^{-7}.

CHAPTER 21

P21.2 $y_{\text{exact}} = 0.1x^3 - 0.4245x^2 + \dfrac{0.1467}{x} + \dfrac{8}{45}\sqrt{x}$

P21.3   a. $y_{\text{exact}}(x) = Ax + Be^x - x^2 - x - 1$

b. $y_{\text{exact}}(x) = Ax^2 + xx^3 - \dfrac{1}{2}x(2x-1)\ln x - x^2 + \dfrac{3}{4}x$

c. $y_{\text{exact}}(x) = A\sin(\ln(x)) + B\cos(\ln(x)) + \dfrac{x^2}{5}$

CHAPTER 22

P22.1 The matrix form of Equation 22.5, as applied to this problem, is

$$\vec{T}_{j+1} = \begin{pmatrix} 1-2\gamma & \gamma & 0 & 0 & \cdots & 0 \\ \gamma & 1-2\gamma & \gamma & 0 & \cdots & 0 \\ 0 & \gamma & 1-2\gamma & \gamma & \cdots & 0 \\ \vdots & \vdots & \vdots & \vdots & \ddots & \vdots \\ 0 & 0 & 0 & 0 & \cdots & 1-2\gamma \end{pmatrix} \vec{T}_j + \begin{pmatrix} 100\gamma \\ 0 \\ 0 \\ \vdots \\ 0 \end{pmatrix}$$

The calculation begins to deteriorate after approximately 100 iterations for the first time step; after 200 iterations it is clear that the calculation is diverging.

P22.2 A comparison of the exact solution with the solution of Example 22.1 is shown in the figure for the first two time steps.

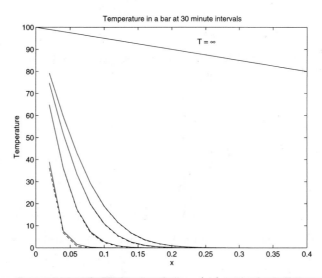

Comparison of the exact solution (−.) with the solution obtained in Example 22.1.

P22.9 A possible ordering of the grid points is illustrated in the figure.

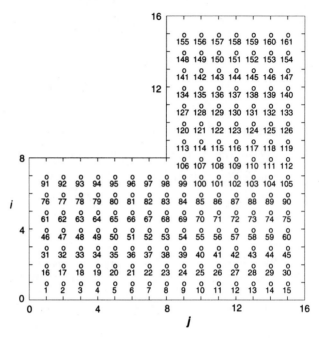

A possible linear sequencing of the grid points on the irregularly shaped region.

Writing Equation 22.5 for each of the 161 points of the grid we arrive at a matrix equation of the form

$$
\vec{V}^{\text{new}} =
\begin{pmatrix}
\mathbf{A} & \mathbf{I} & 0 & \cdots & 0 & 0 & 0 & 0 & \cdots & 0 & 0 \\
\mathbf{I} & \mathbf{A} & \mathbf{I} & \cdots & 0 & 0 & 0 & 0 & \cdots & 0 & 0 \\
\vdots & \vdots & \vdots & \ddots & \vdots & \vdots & \vdots & \vdots & \ddots & \vdots & \vdots \\
0 & 0 & 0 & \ddots & \mathbf{I} & \mathbf{A} & \mathbf{i} & 0 & \cdots & 0 & 0 \\
0 & 0 & 0 & \ddots & 0 & \mathbf{i} & \mathbf{B} & \mathbf{i} & \cdots & 0 & 0 \\
\vdots & \vdots & \vdots & \ddots & \vdots & \vdots & \vdots & \vdots & \ddots & \vdots & \vdots \\
0 & 0 & 0 & \cdots & 0 & 0 & 0 & 0 & \cdots & \mathbf{i} & \mathbf{B}
\end{pmatrix}
\vec{V}^{\text{old}} + \vec{V}_0
$$

where $\mathbf{A}$ and $\mathbf{B}$ are 15×15 and 7×7 banded matrices of the form

$$
\begin{pmatrix}
0 & 1 & 0 & \cdots & 0 & 0 \\
1 & 0 & 1 & \cdots & 0 & 0 \\
0 & 1 & 0 & \cdots & 0 & 0 \\
\vdots & \vdots & \vdots & \ddots & \vdots & \vdots \\
0 & 0 & 0 & \cdots & 1 & 0
\end{pmatrix}
$$

and $\mathbf{I}$ and $\mathbf{i}$ are identity matrices also of size 15×15 and 7×7 respectively. The most difficult part is in determining the form of the inhomogeneous boundary value vector,

$\vec{V}_0$. It is found to be

$$\vec{V}_0 = \begin{pmatrix} \vec{a}+\vec{b}+\vec{f_1} \\ \vec{b}+\vec{f_1} \\ \vdots \\ \vec{b}+\vec{f_1} \\ \vec{c}+\vec{b}+\vec{f_1} \\ \vec{d}+\vec{f_2} \\ \vec{d}+\vec{f_2} \\ \vdots \\ \vec{d}+\vec{f_2} \\ \vec{e}+\vec{d}+\vec{f_2} \end{pmatrix}$$

where $\vec{a}$, $\vec{b}$, $\vec{c}$, and $\vec{f_1}$ are 15-element vectors of the form

$$\vec{a} = V_A \begin{pmatrix} 1 \\ 1 \\ 1 \\ \vdots \\ 1 \end{pmatrix} \qquad \vec{b} = V_B \begin{pmatrix} 1 \\ 0 \\ 0 \\ \vdots \\ 0 \end{pmatrix} \qquad \vec{c} = V_C \begin{pmatrix} 0 \\ 0 \\ \vdots \\ 1 \\ 1 \end{pmatrix} \qquad \vec{f_1} = V_F \begin{pmatrix} 0 \\ 0 \\ 0 \\ \vdots \\ 1 \end{pmatrix}$$

and $\vec{e}$, $\vec{d}$, and $\vec{f_2}$ are 7-element column vectors of the form

$$\vec{e} = V_A \begin{pmatrix} 1 \\ 1 \\ 1 \\ \vdots \\ 1 \end{pmatrix} \qquad \vec{d} = V_B \begin{pmatrix} 1 \\ 0 \\ 0 \\ \vdots \\ 0 \end{pmatrix} \qquad \vec{f_2} = V_F \begin{pmatrix} 0 \\ 0 \\ 0 \\ \vdots \\ 1 \end{pmatrix}$$

It is not difficult to construct these matrices in MATLAB and to compute the solution on the grid. Of course, the 161 computed points must finally be converted back into (i, j) coordinates to plot the result.

Answers to Appendix Problems

APPENDIX A.i

PA.i.1 a. $\begin{bmatrix} 1 & 2 & 3 & 4 \\ 4 & 3 & 2 & 1 \\ 1 & 2 & 3 & 4 \\ 4 & 3 & 2 & 1 \end{bmatrix}$

c. $\begin{bmatrix} 0 & 0 & 1.0000 & 0 \\ 0.1000 & 0.0998 & 0.0995 & 0.1003 \\ 0.2000 & 0.1987 & 0.9801 & 0.2027 \\ 0.3000 & 0.2955 & 0.9553 & 0.3093 \\ \vdots & \vdots & \vdots & \vdots \\ 3.1000 & 0.0416 & -.9991 & -.0416 \end{bmatrix}$

e. $\begin{bmatrix} 16 & 2 & 3 & 13 & 0 \\ 5 & 11 & 10 & 8 & 0 \\ 9 & 7 & 6 & 12 & 0 \\ 4 & 14 & 15 & 1 & 0 \\ 0 & 0 & 0 & 0 & 1 \end{bmatrix}$

PA.i.3 The products in a through d each equal **5*eye(4)**; e. Only **Q** and **W** commute.

PA.i.5 a.   expm([1 0;0 -1]) $= 0.5403 \begin{bmatrix} 1 & 0 \\ 0 & 1 \end{bmatrix} + 0.8415i \begin{bmatrix} 1 & 0 \\ 0 & -1 \end{bmatrix}$

$$= \mathbf{I}_2 \cos(1) + i\mathbf{P}\sin(1)$$

b.   expm(A) $= \mathbf{I}_4 + \mathbf{A} + \frac{1}{2}\mathbf{A}^2 + \frac{1}{6}\mathbf{A}^3 + 0$

APPENDIX A.ii

PA.ii.1 The ratio of times is approximately $140 : 75 : 1$.

PA.ii.3 a.

```
b = zeros(m,1)
for i = 1:m
   b(i) =
   for k = 1:m
      b(i) = b(i)+A(i,k)*x(k);
   end
end
```

c.

```
D = zeros(n1,n2,n3,n4)
for i1 = 1:n1
for i4 = 1:n4
  for i2 = 1:n2
  for i3 = 1:n3
    D(i1,i4) = D(i1,i4) + ...
       A(zeros(i1,i2)*B(i2,i3)*C(i3,i4);
   end
   end
  end
  end
```

PA.ii.5 a. $x^2 - 6x + 1$ b. $x(x+1)^2$

PA.ii.7 b. $x_j k_i$ c(i) sum(K.*(X.^K))) c(ii) sum((X.^(K.^(-1)))')

PA.ii.9 a. cumprod(k) b. sum(cumprod(k).^(-1))
c. sum((2*ones(1,length(k))).^k./cumprod(k))

APPENDIX A.iii

PA.iii.1 a. a > b & b > c c. (c<d & a >= b) | (a<b & c>=d) e. all(B == inv(A))
g. (abs(a-b)<=c) & (c<=a+b)

PA.iii.3 Using $x = \left[\ln\left(\dfrac{3}{x}\right)\right]^{1/3}$, a root is found at $x = 1.032691\ldots$. Using $x = 3e^{-x^2}$, the method diverges.

PA.iii.5

```
function c = STEP(fnc,a,b,dx)
   fa = feval(fnc,a);
   x  = a + dx; f = feval(fnc,x);
   while (sign(f*fa)>0)
   x = x + dx; f = feval(fnc,x);
     if x > b;
         error('No crossing');
         return;
     end
   end
   c = [x-dx x]';
```

APPENDIX A.iv

PA.iv.1 The polynomial is equal to $25(x-1)(x-2.2)^2(x^2+1)$ and has a multiple root at $x = 2.2$. This can be found graphically or by using the function `roots`.

PA.iv.3 The MATLAB code

```
function [X,Y] = Bounce(th,V0,f)
 g=9.8; theta = th*pi/180;
 X=0; Y=0; x0 = 0; t0 = 0;
 for i = 1:4
    Vx = V0*cos(theta); Vy = V0*sin(theta);
    tf = 2*Vy/g;   dt = tf/50;
    t = [dt:dt:tf]';
    X = [X;x0 + Vx*t];
    Y = [Y;Vy*t - .5*g*t.^2];
    T = [T;t+t0];
    t0 = tf; V0 = f*V0; x0 = x0 +Vx*tf;
 end
 xf = x0;
 plot(X,Y)
 axis('equal');hold on
 xlabel('x'),ylabel('y')
 title('A bouncing trajectory')
 axis([0 xf 0 xf])
 axis('square')
 hold off
```

produces the following graph:

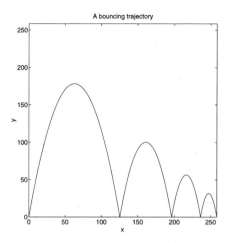

The trajectory of a bouncing ball. Initial velocity, $v_0 = 60$m/s, initial angle, $\theta = 80°$, coefficient of bounce, $f = 0.75$.

APPENDIX A.v

PA.v.1

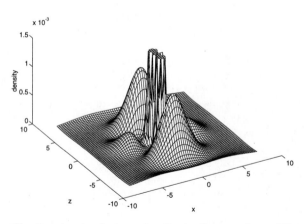

The 3s atomic electron density in the xz plane. The peaks near the origin have been "capped."

PA.v.2

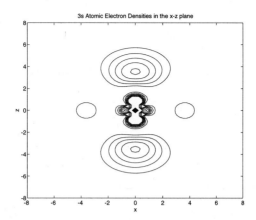

Contours of the 3s atomic electron density in the xz plane.

PA.v.4

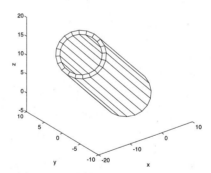

A pipe rotated about the y axis by 45°.

PA.v.6 The MATLAB code,

```
function [X,Y,Z] = DfrmdDrp(b,g)
 R0 = 1;  C = sqrt(5/16/pi);
  th=[0:pi/40:pi]';ph=[0:pi/20:2*pi];
  It = ones(length(th),1);
  Ip = ones(1,length(ph));
  S1= (3*cos(th).^2-It)*Ip;
  S2= (sin(th).^2)*(cos(ph).^2-sin(ph).^2);
  S = cos(g)*S1 + sqrt(3)*sin(g)*S2;
  S = C*b*S;
  R = R0*(S+ones(size(S)));
```

```
X = R.*(sin(th)*cos(ph));
Y = R.*(sin(th)*sin(ph));
Z = R.*(cos(th)*Ip);
mesh(X,Y,Z)
axis('square')
```

was used to produce the following figure:

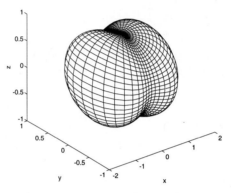

A deformed liquid drop $[\beta = 0.8$ and $\gamma = \frac{2\pi}{3}]$.

PA.v.7 The MATLAB code is

```
function f = PA5p6
 Re = 6.4e6;  Rm = 1.7e6; Dm = 3.85e7;
 [x,y,z]=sphere(20);
 xe=Re*x; ye = Re*y; ze=Re*z;
 xm=Rm*x + Dm/sqrt(2); ym=Rm*y + Dm/sqrt(2);
 zm=Rm*z;
 mesh(xe,ye,ze);hold on
 mesh(xm,ym,zm)
 axis([-Re Dm -Re Dm -Re Dm/2])
 hold off
```

PA.v.8 The code to use **plot3** to plot the satellite trajectory is

```
function f = SatOrbit
  f=0; R0 = 1;
  T = 10*pi; dt = pi/20;
  t = [0:dt:T]';
```

```
a = 12*R0; b=9*R0;
f = sqrt(a^2-b^2);
th = 12.5*pi/180;
E = exp(-t/20);
x = E.*(a*cos(t)-f);
y = E.*(b*cos(th)*sin(t));
z = E.*(b*sin(th)*sin(t));
plot3(x,y,z)
axis([-12*R0 12*R0 -12*R0 12*R0 -6*R0 6*R0])
hold on
[X,Y,Z] = sphere(20);
X=R0*X;Y=R0*Y;Z=R0*Z;
mesh(X,Y,Z)
hold off
```

Index